Devonian Change: Case Studies in Palaeogeography and Palaeoecology

IUGS/GSL publishing agreement

This volume is published under an agreement between the International Union of Geological Sciences and the Geological Society of London and arises from IGCP 499 project on "Devonian land–sea interaction: evolution of ecosystems and climate".

GSL is the publisher of choice for books related to IUGS activities, and the IUGS receives a royalty for all books published under this agreement.

Books published under this agreement are subject to the Society's standard rigorous proposal and manuscript review procedures.

It is recommended that reference to all or part of this book should be made in one of the following ways:

Königshof, P. (ed.) 2009. *Devonian Change: Case Studies in Palaeogeography and Palaeoecology*. Geological Society, London, Special Publications, **314**.

George, A. D., Trinajstic, K. M. & Chow, N. 2009. Frasnian reef evolution and palaeogeography, SE Lennard Shelf, Canning Basin, Australia. *In*: Königshof, P. (ed.) *Devonian Change: Case Studies in Palaeogeography and Palaeoecology*. Geological Society, London, Special Publications, **314**, 73–107.

GEOLOGICAL SOCIETY SPECIAL PUBLICATION NO. 314

Devonian Change: Case Studies in Palaeogeography and Palaeoecology

EDITED BY

P. KÖNIGSHOF
Forschungsinstitut und Naturmuseum Senckenberg, Germany

2009
Published by
The Geological Society
London

THE GEOLOGICAL SOCIETY

The Geological Society of London (GSL) was founded in 1807. It is the oldest national geological society in the world and the largest in Europe. It was incorporated under Royal Charter in 1825 and is Registered Charity 210161.

The Society is the UK national learned and professional society for geology with a worldwide Fellowship (FGS) of over 9000. The Society has the power to confer Chartered status on suitably qualified Fellows, and about 2000 of the Fellowship carry the title (CGeol). Chartered Geologists may also obtain the equivalent European title, European Geologist (EurGeol). One fifth of the Society's fellowship resides outside the UK. To find out more about the Society, log on to www.geolsoc.org.uk.

The Geological Society Publishing House (Bath, UK) produces the Society's international journals and books, and acts as European distributor for selected publications of the American Association of Petroleum Geologists (AAPG), the Indonesian Petroleum Association (IPA), the Geological Society of America (GSA), the Society for Sedimentary Geology (SEPM) and the Geologists' Association (GA). Joint marketing agreements ensure that GSL Fellows may purchase these societies' publications at a discount. The Society's online bookshop (accessible from www.geolsoc.org.uk) offers secure book purchasing with your credit or debit card.

To find out about joining the Society and benefiting from substantial discounts on publications of GSL and other societies worldwide, consult www.geolsoc.org.uk, or contact the Fellowship Department at: The Geological Society, Burlington House, Piccadilly, London W1J 0BG: Tel. +44 (0)20 7434 9944; Fax +44 (0)20 7439 8975; E-mail: enquiries@geolsoc.org.uk.

For information about the Society's meetings, consult *Events* on www.geolsoc.org.uk. To find out more about the Society's Corporate Affiliates Scheme, write to enquiries@geolsoc.org.uk

Published by The Geological Society from:
The Geological Society Publishing House, Unit 7, Brassmill Enterprise Centre, Brassmill Lane, Bath BA1 3JN, UK

(*Orders*: Tel. +44 (0)1225 445046, Fax +44 (0)1225 442836)
Online bookshop: www.geolsoc.org.uk/bookshop

British Library Cataloguing in Publication Data

A catalogue record for this book is available from the British Library.
ISBN 978-1-86239-273-1

Typeset by Techset Composition Ltd., Salisbury, UK
Printed by MPG Books Ltd, Bodmin, UK

Distributors

North America
For trade and institutional orders:
The Geological Society, c/o AIDC, 82 Winter Sport Lane, Williston, VT 05495, USA
Orders: Tel +1 800-972-9892
Fax +1 802-864-7626
Email gsl.orders@aidcvt.com

For individual and corporate orders:
AAPG Bookstore, PO Box 979, Tulsa, OK 74101-0979, USA
Orders: Tel +1 918-584-2555
Fax +1 918-560-2652
Email bookstore@aapg.org
Website http://bookstore.aapg.org

India
Affiliated East-West Press Private Ltd, Marketing Division, G-1/16 Ansari Road, Darya Ganj, New Delhi 110 002, India
Orders: Tel. +91 11 2327-9113/2326-4180
Fax +91 11 2326-0538
E-mail affiliat@vsnl.com

Contents

Acknowledgements

On behalf of the leaders of the IGCP 499 I would like to express my thanks to the various institutions and numerous colleagues who helped to organize many meetings. For all participants the workshops and accompanying field trips offered a unique opportunity to get an insight into complex Devonian sequences at many places around the world. Cordial thanks are expressed to the local specialists for the organization and guidance of the field trips. We are grateful to these and many other individuals who helped make the workshops both successful and highly enjoyable.

I would like to express my thanks to the following companies and institutions for sponsoring: UNESCO/IUGS, German Science Foundation (DFG), German Federal Foreign Office.

I am particularly grateful to the following colleagues (and of course several anonymous reviewers) who have assisted by providing referee reports on manuscripts submitted for this Special Publication of the Geological Society: T.R. Becker, C. Brett, A. Embry, E. Gischler, R. Haude, U. Jansen, U. Linnemann, J. Marshall, B. Mistiaen, K. Oekentorp, J. Over, S. Peralta, R. Prokop, M. Roux, S. Schöder, G. Sevastopulo, P. Steemanns, M. Streel, J. Talent, K.-W. Tietze, R. Wicander and M. Zhu.

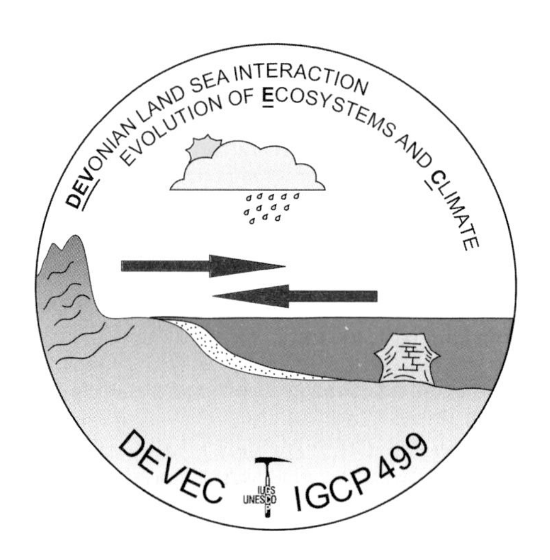

Devonian change: case studies in palaeogeography and palaeoecology – an introduction

P. KÖNIGSHOF*

Forschungsinstitut und Naturmuseum Senckenberg, Senckenberganlage 25, 60325 Frankfurt am Main, Germany

**Corresponding author (e-mail: Peter.Koenigshof@senckenberg.de)*

Gaining knowledge about behaviour and development of ecosystems and climate in the geological past is one of the most interesting tasks for future geological/palaeontological research because our knowledge of past ecosystems and climate is at best rudimentary: the further we look back, the greater the uncertainties. Such knowledge is essential for the understanding of Earth's history and palaeobiodiversity. The Devonian period is one of the most interesting systems in Earth history because it was a time of global greenhouse climates, lacking any significant ice shields, and was characterized by extensive shallow marine and continental lowland areas. Therefore, a wide range of different habitats is preserved in the sedimentary record.

Following the work of numerous individuals and regional groups that have been established in the International Geological Correlation Programme Project 421 (IGCP 421), led by John Talent and Raimund Feist, a group of scientists proposed a new successor project (IGCP 499) in order to develop a better understanding of marine and terrestrial ecosystems, climate change and the influence on sedimentation not only in terrestrial but also marine realms (Königshof *et al.* 2004). One aspect that is still important concerned the neritic–pelagic correlation; although many new results have been published in recent years this is still a focus of research. Another focus of the project concerned the interrelated evolution of terrestrial and marine palaeoecosystems with respect to biotic and abiotic factors requiring high-resolution stratigraphic control and detailed analysis of various facies. Biotic and abiotic factors of palaeoecosystems are controlled by both earthbound and extraterrestrial triggers, causing cyclicity and/or distinct events; this is a special feature in the Late Devonian but it is becoming increasingly evident that the Middle Devonian was also characterized by local stability as well as episodes of abrupt changes (Brett & Baird 1995; Brett *et al.* 2008). The work on the rapid evolution of early life on land and its interaction with sedimentary processes, climate and palaeogeography, both on land and in marine settings, was scheduled through the duration of the project (2004–2008). For that reason studies included individual palaeoecosystems and their components as well as their palaeobiogeographic distribution (Figs 1–3) which should provide a better understanding of the Devonian system with respect to the evolution of palaeoecosystems and to palaeogeographical and palaeoclimatic changes. The project has been developed in close collaboration with the Subcommission on Devonian Stratigraphy (SDS) and many workshops have been organized as a joint IGCP 499/SDS meeting.

IGCP 499 meetings and workshops

The inaugural meeting took place in Rabat followed by a field workshop in the Dra Valley of the Anti Atlas Mountains (Morocco) which was organized by the SDS in Rabat (El Hassani 2004). During the technical part of the meeting the project was introduced for the first time. In October 2004 the first business meeting of the project was held at the University of Göttingen, Germany, and was attended by about 40 colleagues from nine countries. Presentations and discussions focused on the activities during the initial phase of the project and how forthcoming activities would be co-ordinated. In oder to channel the work within IGCP 499, which covers a wide range of scientific disciplines, we started to establish different working groups/regional co-ordinators which provided a better co-ordination of forthcoming field meetings and workshops. In the following years numerous workshops and conferences took place in different countries (Fig. 4). In the tradition of successful joint meetings and field trips of Devonian IGCP projects and the SDS, a meeting was held in 2005 at the Institute of Petroleum Geology, United Institute of Geology and Mineralogy of the Russian Academy of Sciences, Siberian Branch, in Novosibirsk on 'Devonian terrestrial and marine environments: From continent to shelf'. About 75 scientists presented 35 oral lectures and four posters. In the same year a workshop was organized in Istanbul on 'Depositional environments of the Gondwanan and Laurussian Devonian'. The Devonian is of special interest because it comprises

From: KÖNIGSHOF, P. (ed.) *Devonian Change: Case Studies in Palaeogeography and Palaeoecology*.
The Geological Society, London, Special Publications, **314**, 1–6.
DOI: 10.1144/SP314.1 0305-8719/09/$15.00

Fig. 1. (**a**) The field trip group in the Ghadamis Basin, Libya.

Laurussian and Gondwanan components on different tectonic blocks and there might be some faunal affinities with other areas, such as Bulgaria and even Libya. The official part of the workshop and field trip was followed by an extended field workshop in the framework of a bilateral Turkish–German co-operation project (2005–2008). In the meantime additional small workshops and field trips took place.

Some other meetings have been organized in conjunction with International Congresses, such as the 2nd International Palaeontological Congress (IPC) in China in 2006, and the International Geological Congress (IGC) which was organized recently in Oslo. Others have been incorporated in smaller congress events. Embedded in the meeting of the Commission International de Microflore du Paleozoique (CIMP) that took place in Praha 2006, a symposium was held on 'Palaeozoic palynology', as some of the palynological topics of the meeting appeared to be very important for recognition of marine–non-marine sequences and for correlation between different shallow water areas. Several other field meetings and workshops followed and in 2007 two meetings were organized. A workshop was held between 14 and 22 May in San Juan, Argentina. The Devonian of the San Juan and Mendoza Precordillera and the San Rafael Block are of special interest for the project regarding the palaeogeographic and biostratigraphic importance in the Devonian Malvinokaffric Realm, but also in the tectonic-sedimentary evolution of the Precordillera terrane. In September a joint meting of SDS and IGCP 499 was held in Eureka, Nevada (USA), with a special focus on global change in the Devonian, land–sea interactions, sequence stratigraphy and events in the Middle Devonian. This workshop was followed by a meeting in Libya which was devoted to the stratigraphic evolution of Devonian sequences in the Awaynat Wanin area, Southern Ghadamis Basin, with a focus on sequence stratigraphy, sedimentology and facies, and palaeoecology (Königshof *et al.* 2008). The Devonian of Libya is of special interest for the project because it contains the excellent sequences mainly representing very shallow water environments as well as fluvial sequences. Finally, a joint SDS/IGCP 499 meeting took place in Uzbekistan between 24 August and 3 September 2008. Subjects emphasized in the conference and the field trip are cyclicity and sedimentary markers for intra- and interbasinal correlations, neritic–pelagic associations and their interrelations as well as global

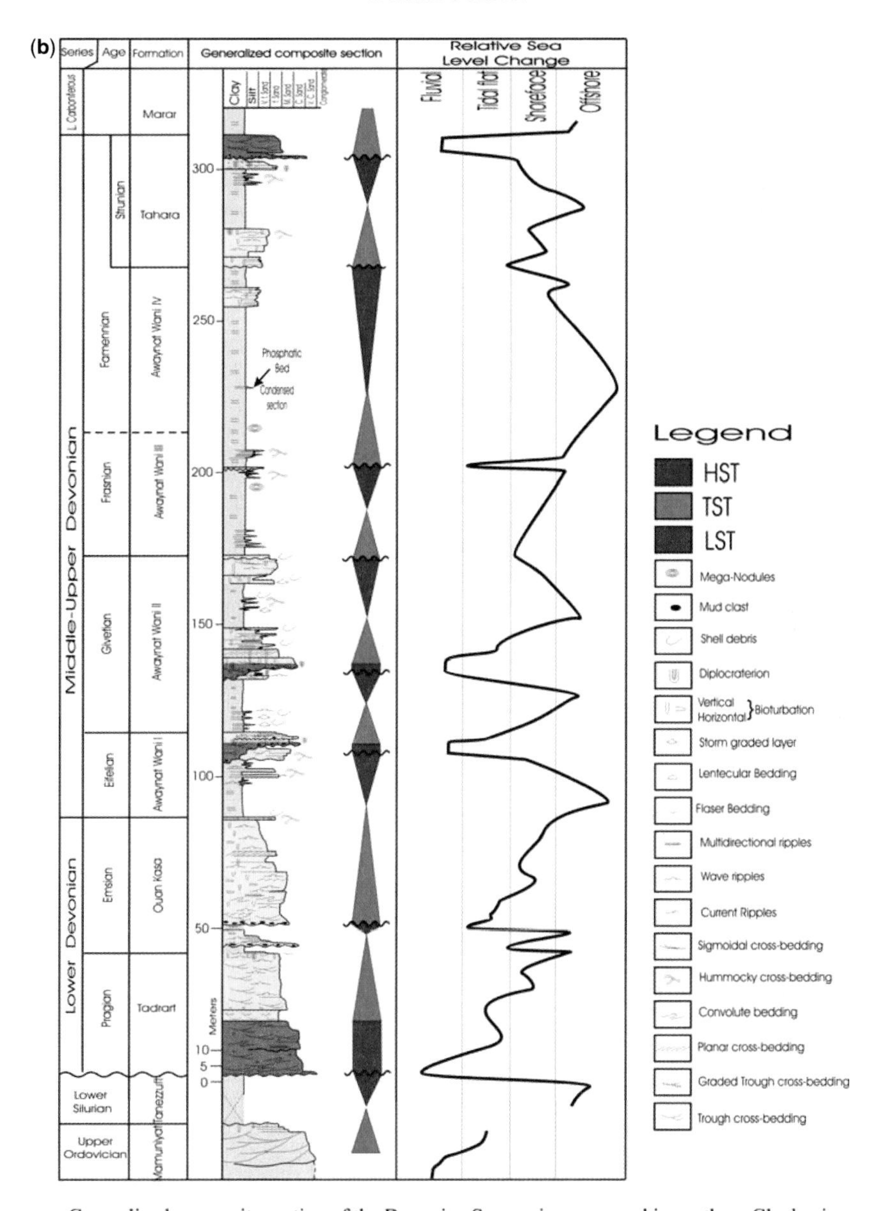

Generalized composite section of the Devonian Successions exposed in southern Ghadamis Basin, Awaynat Wanin area. The formation ages and boundaries are based on: Belhaj, F. (1996); Gundobin, V. M. (1985); Massa, D. (1988).

Fig. 1. (**b**) Generalized composite section of the Devonian successions exposed in the Southern Ghadamis Basin, Awaynat Wanin area (Ben Rahuma *et al.* 2008). Formation names and ages after Belhaj (1996), Massa (1988) and Gundobin (1985).

sedimentary and biotic events. The final joint IGCP 497 on the Rheic ocean and IGCP 499 was held in October 2008 in Frankfurt, Germany.

Relationships of regional geological features and open questions with respect to biostratigraphy, facies interpretation and depositional environment, especially land–sea transitional settings, generated vivid discussions in all of these meetings and workshops, and helped identify topics requiring future research. The workshops acted as a catalyst for

Fig. 2. Field work in the Taurid Mountains, Turkey (DEVEC–TR project).

future collaborative research between groups all over the world, as well as offering a network for collaboration between researchers involved in IGCP 499.

Additionally, there have been many activities in the regional and local working groups in different countries. More than 450 papers in international journals have been published to date, and many others are to be expected, including special volumes on different objectives. All in all, even if there are many open questions left which especially concern the land–sea transitional settings or sequence stratigraphy versus global bioevents, it should be stressed that this project has been a highly successful IGCP.

Case studies in palaeogeography and palaeoecology

The papers presented in this volume are mainly the result of case studies in palaeogeography and palaeoecology from the various meetings mentioned above and generally follow this theme, although individual papers deal with more than one of the topics mentioned above. It was a pity that not all submitted contributions could be included in the volume for various reasons and they will be published elsewhere. Below I introduce the different papers, in the order that they are assembled in this Special Publication on 'Devonian Change: Case Studies in Palaeogeography and Palaeoecology'.

Brett *et al.* provide an update and discussion of ecological–evolutionary subunits (EESUs) or faunas of the latest Silurian to mid-Late Devonian interval in the Appalachian Basin. New data compilations for successive formation scale intervals of approximately third-order sequences permit a statistical characterization of the faunas. A total of nine EESUs are recognized in the Early to mid-Devonian interval of the Appalachian Basin. Most of the EESU boundaries occur within early transgressions, to highstand intervals rather than at sequence boundaries, and the majority are coincident with global bioevents.

Schäfer & Stets present a paper on land–sea transitional settings at the northern margin of the Rhenohercynian basin. In this case study they describe different subenvironments which belong to a huge delta complex of the Lower Devonian south of the Old Red Continent. Steady sediment supply in combination with accumulation and subsidence have been the main factors controlling the long-term sedimentary processes.

George *et al.* provide a comprehensive paper on Frasnian reef evolution and palaeogeography from the Canning Basin, Australia. Based on a sequence stratigraphic approach, underpinned by facies analysis and well documented biostratigraphic data, they established an interesting history of reef complex evolution, in particular the palaeogeography and role of tectonism in controlling relative sea-level changes. They identified seven phases of early-middle Frasnian platform growth which are bounded by third-order flooding surfaces associated with backstepping of platform margins.

Copper & Edinger present a preliminary report from a remote Devonian reef area of western Arctic Canada. The huge Frasnian reef complex covers several thousand square kilometres and occurs in a distal megadelta setting. The reef builders are mainly stromatoporoids and corals showing a high abundance but low diversity, which is similar to other early and middle Frasnian reefs worldwide.

Zapalski sheds new light on the development of commensalism/parasitism in tabulate corals even if some interpretations remain speculative. This topic is very interesting with respect to symbiont/parasite equivalents in cnidarians and should be a focus for further research.

Fig. 3. Devonian deltaic sediments of the Rio Seco de los Castaños Formation, San Rafael Block, Mendoza Province, Argentina.

Fig. 4. Locations of IGCP 499 field meetings and workshops (2004–2008).

Webster & Becker document Emsian to Frasnian crinoids of the Dra Valley, Western Anti Atlas Mountains, Morocco, providing useful new palaeobiogeographic data of this area. This is of prime importance for the correlation of Devonian nearshore and offshore facies and faunas, complemented by description of several new taxa.

In another paper dedicated to Famennian echinoderm diversity, **Waters & Webster** reinterpret patterns of Late Devonian extinction and rebound. According to the authors, Famennian echinoderm communities are less diverse than Middle Devonian or Early Mississippian ones, but the decline is not as dramatic as previously believed. They also recognize that Famennian species are endemic and that Famennian genera can be divided into two palaeogeographic settings, with few genera occurring in both realms.

Streel compares the stratigraphical occurrence of 38 Upper Devonian miospore taxa with some miospore and conodont zones in 28 intercalibrated levels with a special focus on correlation using a Correlation Quality Index.

Amenabar presents a paper on Middle Devonian microfloras from the Precordillera, NW Argentina. The data presented are important because they improve our knowledge of an area for which there is sparse information.

Di Pasquo *et al*. provide new data on Middle Devonian floras from western Argentina and southern Bolivia. The assemblages are discussed in their palaeogeographical and palaeoecological context.

The contribution by **Manassero *et al*.** provides a case study of siliciclastic sediments in the southern hemisphere. Based on sedimentological, geochemical and isotope data they describe a marine platform–deltaic realm of Silurian–Devonian age from the Argentine Precordillera, complemented by a discussion on provenance analysis and tectonic position.

Ma *et al.* give an overview of Devonian strata and biostratigraphy as far as known from diverse terranes within present-day China. Moreover, they provide a relative sea-level curve for the entire Devonian based on the best known sections in China. The paper provides a good basis of data, mainly in Chinese literature that may be obscure to many colleagues and it is therefore a good reference source.

Webster & Waters present a comprehensive paper on Late Devonian echinoderms from northwestern China which is an important contribution to the evolutionary and biogeographically significant echinoderm faunas. New collections and re-evaluation of earlier collections result in description of several new taxa. Additionally, palaeogeographic affinities are discussed.

The meetings of IGCP 499 and some of the papers selected for this Special Publication, clearly show that there is a huge potential for special topics requiring future research and perhaps a successor project of IGCP 499. Furthermore, the last five years have shown that successful research is based on multidisciplinary co-operation. On the other hand, there are many open questions and there are some disciplines where we should concentrate research in the Devonian. Based on a profound knowledge of organisms, sequence stratigraphic correlations should be brought into agreement with available biostratigraphic data. Another focal point could be a better correlation between terrestrial events and the marine realms, and in the neritic settings we need a better biostratigraphic and sedimentary record. Palynomorphs may be a useful tool in this context. Another interesting aspect could be the evolution of terrestrial vegetation and the interaction with the arthropods as well as the evolution of early terrestrial ecosystems.

We sincerely thank all the authors and reviewers for their great efforts in ensuring a special issue of high quality and on behalf of the IGCP 499 project leader I would like to express our thanks to all contributors for a long-lasting cooperation. All of them helped to make the IGCP 499 project both successful and highly enjoyable. Finally, I would like to thank UNESCO and IUGS for their financial support during the last years.

References

BELHAJ, F. 1996. Palaeozoic and Mesozoic stratigraphy of Eastern Ghadamis and Western Sirt Basin. *In*: SALEM, M. J., MOUZUGHI, A. J. & HAMMUDA, O. S. (eds) *The Geology of the Sirt Basin*, Elsevier, Amsterdam, I, 57–96.

BEN RAHUMA, M., PROUST, J.-N. & ESCHARD, R. 2008. *The Stratigraphic Evolution of the Devonian Sequences, Awaynat Wanin area, Southern Ghadamis Basin: a Fieldguide Book*. Libyan Petroleum Institute (LPI), Tripoli, Libya, 1–67.

BRETT, C. E. & BAIRD, G. C. 1995. Coordinated stasis and evolutionary ecology of Silurian to Middle Devonian marine biotas in the Appalachian basin. *In*: ERWIN, D. H. & ANSTEY, R. L. (eds) *New Approaches to Speciation in the Fossil Record*. Columbia University Press, New York, 285–315.

BRETT, C. E., DESANTIS, M. K., BARTHOLOMEW, A. J., BAIRD, G. C., SCHINDLER, E. & KÖNIGSHOF, P. 2008. *Middle Devonian eustasy, paleoclimate, and bioevents: Toward an integrated model*. 20th International Senckenberg and 2nd Geinitz Conference, 30 September–3 October 2008. Abstract volume.

EL HASSANI, A. (ed.) 2004. Devonian neritic–pelagic correlation and events in the Dra Valleys (Western Anti Atlas, Morocco). *Documents de l'Instiitut Scientifique*, **19**, 1–100.

GUNDOBIN, V. M. 1985. *Geological Map of Libya, 1 : 250,000: Qararat Al Marar NH 33-13*, Industrial Research Centre, Tripoli.

KÖNIGSHOF, P., LAZAUSKIENE, J., SCHINDLER, E., WILDE, V. & YALCIN, N. 2004. The new IGCP Project 499: 'Devonian land–sea interaction: evolution of ecosystems and climate' (DEVEC). *Facies*, **50**, 347–348.

KÖNIGSHOF, P., BEN RAHUMA, M. & PROUST, J.-N. 2008. IGCP 499: Devonian land–sea transitional settings. *Episodes*, **31**, 438–439.

MASSA, D. 1988. *Paleozoique de Libye occidentale – Stratigraphie et Paleographie*, PhD Thesis, Université Nice.

Devonian ecological–evolutionary subunits in the Appalachian Basin: a revision and a test of persistence and discreteness

C. E. BRETT[1]*, L. C. IVANY[2], A. J. BARTHOLOMEW[3], M. K. DESANTIS[4] & G. C. BAIRD[5]

[1]*Department of Geology, University of Cincinnati, 500 Geology/Physics Building, Cincinnati, OH, USA*

[2]*Department of Geology, Syracuse University, Syracuse, NY, USA*

[3]*Department of Geology, SUNY College at New Paltz, New Paltz, NY, USA*

[4]*Department of Geology, University of Cincinnati, OH, USA*

[5]*Department of Geosciences, SUNY Fredonia, Fredonia, NY, USA*

**Corresponding author (e-mail: Carlton-Brett@uc.edu)*

Abstract: New data compilations for successive formation scale intervals, approximately third-order sequences, permit a statistical characterization of the ecological–evolutionary subunits (EESUs) or faunas of the Latest Silurian to mid-Late Devonian interval in the Appalachian Basin. Cluster analysis using the Jaccard coefficient of similarity show that certain formations cluster tightly together on the basis of faunal composition while in other cases units are sharply set off from superseding units. This result indicates both the coherence of faunal composition within EESUs and the discreteness of their boundaries. The results also require minor revision of EESUs previously delineated, including the addition of three new units. The Esopus Formation is designated as a distinct unit separate from the Schoharie on the basis of brachiopods from shallow water facies of the Skunnemunk outlier in New York; in addition, a short-lived Stony Hollow fauna is recognized in shallow shelf facies of the Union Springs Formation and coeval units (formerly referred to as the lower Marcellus Formation) and the lower transgressive beds of the overlying Oatka Creek Formation. This fauna, consisting of subtropical Old World Realm (OWR) emigrants, is distinctive from both the underlying Onondaga fauna and that of the overlying Hamilton Group. Moreover, the Tully Formation presents another case of a short-lived incursion of tropical OWR taxa followed by the extermination of this fauna and reappearance of a suite of typically Hamilton taxa. This case illustrates that EESUs may persist globally despite their local extermination or emigration from a large region, such as the Appalachian Basin.

Review of the broader context of EESU turnover suggests that these crises are geologically rapid and synchronous. Moreover, most of the Devonian EESU boundaries coincide with recognized global bioevents, within the limits of combined biostratigraphic and sequence stratigraphic resolution. Hence, these crises, although perhaps locally accentuated in the Appalachian Basin, are allied to global causes. They appear mostly to be associated with rapid rises in sea level, periods of widespread climatic change and hypoxic events in basinal areas.

In the past decades, Boucot (1975, 1981, 1983, 1990*a*, *b*) has repeatedly proposed that the Phanerozoic can be subdivided into about 12 'ecological–evolutionary units'. Within each of these units, time series of similar, evolving communities, the so-called 'community groups' display little change at family levels. These ecological–evolutionary units are punctuated by major or minor extinction events and/or adaptive radiations that disrupt community structure rather completely. Boucot (1990*a*, *b*) further contended that level bottom communities re-diversify relatively rapidly following extinction events by a process of quantum evolution that he termed 'metacladogenesis'. During the long intervals of stability of an ecological–evolutionary unit, little substantial change takes place in the diversity or the relative abundance of most organism lineages within community groups. Abundant taxa, comprising typically eurytopic, cosmopolitan genera that make up the bulk of communities in terms of shelly biomass, show little or no change within long time intervals. Boucot's conception of an ecological–evolutionary unit is a relatively long time interval, for example from Early Silurian to mid-Late Devonian, spanning about 70 million years and bounded by major extinction events. (Boucot 1982, 1988, 1990*a*, *b*; Boucot & Lawson 1999).

Recently, we have suggested, as did Boucot (1990*a*), that these ecological–evolutionary units can be subdivided into shorter intervals, a few

From: KÖNIGSHOF, P. (ed.) *Devonian Change: Case Studies in Palaeogeography and Palaeoecology*.
The Geological Society, London, Special Publications, **314**, 7–36.
DOI: 10.1144/SP314.2 0305-8719/09/$15.00

million years in duration, during which a majority of species or at least subgenera (closely related species complexes) persist with effective morphological stasis. These finer subdivisions were termed 'ecological–evolutionary subunits' (Fig. 1; Brett & Baird 1995; Ivany 1996; Brett *et al.* 1996; Miller 1997).

A key issue is the contention of the existence of discrete ecological–evolutionary subunits (EESUs). Do these have a strong degree of internal consistency? Do they show significant changes at their boundaries? In short, are they statistically recognizable entities? A second important issue involves the timing and possible synchroneity of turnover between EESUs in different areas. Similar patterns have been recognized at various scales (temporal, taxonomic, geographic), and an interesting question is whether these comparatively smaller turnovers are in fact also global. Baumiller (1996) tested the probability that blocks of stability with characteristics of EESUs could be generated by random extinction and origination using a bootstrap resampling methodology and argued that this probability was vanishingly small; EESUs appear to be real entities. However, the details of their boundaries and a field test of their discreteness have yet to be presented.

This paper has two major objectives: first, to present a partial test of their internal consistency and abrupt boundaries; and second, to document the faunal and ecological characteristics of Early to Middle Devonian EESUs. We do not address the issue of ecological stasis in this paper, but rather the issue of concurrent species-level composition and change in mid- to outer shelf biotas in the Appalachian Basin. The faunal compositions of successive geological formations, essentially third-order depositional sequences, are compared statistically using similarity coefficients and cluster analyses. The results show that certain groups of formations cluster closely in terms of overall composition, whereas other, adjacent units exhibit very different biotas. The re-study requires minor amendments of previously defined EESUs, and these are discussed in this text.

Several further implications of the EESUs are considered in this paper. In particular, we consider the relationship of the turnovers to phenomena of sea-level, climatic and oceanographic changes, and tectonics. Finally, we return to the question of whether or not the boundaries of the EESUs are synchronous in different basins and therefore whether they reflect global or merely local events.

Materials and methods

The Early to Middle Devonian interval in the New York, Pennsylvania, and Ontario outcrop belts is among the best-studied successions in North America. This interval has been subdivided into a set of local stages, many of which have been eclipsed by the usage of European stage terminology. To a large extent, the present contribution is based on the detailed biostratigraphic work of a host of earlier researchers and stands as reaffirmation of the detailed bioevent zonation of the Appalachian basin region.

<table>
<tr><td colspan="4">Lower Devonian</td><td colspan="5">MiddleDevonian</td><td>U.D.</td><td>System</td></tr>
<tr><td colspan="4">Ulsterian</td><td colspan="5">Erian</td><td>Sen.</td><td>Series (NA)</td></tr>
<tr><td>Helderbergian</td><td>Deerpark</td><td colspan="2">Sawkill</td><td>Southwood</td><td>Caz.</td><td>Tioug.</td><td colspan="2">Tag.</td><td>F.L.</td><td>Stages (NA)</td></tr>
<tr><td>Lochovian</td><td>Pragian</td><td colspan="2">Emsian</td><td colspan="2">Eifelian</td><td colspan="3">Givetian</td><td>Fras.</td><td>Stages (E)</td></tr>
<tr><td>Helderberg</td><td>Oriskany</td><td>Esopus</td><td>Schoharie</td><td>Onondaga</td><td colspan="2">Hamilton</td><td colspan="2">Tully</td><td>Gen.</td><td>Units (NY)</td></tr>
<tr><td>Carbonates</td><td>Sandstone</td><td>Siltstone</td><td>Carbonate</td><td>Limestone</td><td colspan="2">Shale/Ls.</td><td colspan="2">Ls.</td><td>Shale</td><td>Lithologies</td></tr>
<tr><td>Helderberg</td><td>Oriskany</td><td>Esopus</td><td>Schoharie</td><td>Onondaga</td><td colspan="2">Mahantango</td><td colspan="2">Tully</td><td>Burket-Harrell</td><td>Units Central App.</td></tr>
<tr><td>Carbonate/Sh.</td><td>Sandstone</td><td>Shale</td><td>Shale</td><td>Ls./Shale</td><td colspan="2">Silt./Ss./Sh.</td><td colspan="2">Ls./Sh.</td><td>Shale</td><td>Lithologies</td></tr>
<tr><td>**Helderberg**</td><td>**Oriskany**</td><td>**Esopus**</td><td>**Schoharie**</td><td>**Onondaga**</td><td>**S. H.**</td><td>**Hamilton**</td><td>**L. T.**</td><td>**Up. Tu.**</td><td>**Gen.**</td><td>E-E Subunits</td></tr>
<tr><td>2: *Tentaculites*
3: Gypidulinid (*Cyrtina*)
4A: *Megakozlowskiella*
4B: *Hedeina*
5: *Coelospira*
6: *Atlanticocoelia*</td><td>2: *Hipparionyx*
3: *Hipparionyx*
4A: *Plicosplasia*
4B: *Leptocoelia*</td><td>2: ________
3: *Hysteriolites*
4: *Etymothyris-Meganterella*
5: *Atlanticocoelia*</td><td>2: *Chonostrophia*
3: Pentam.-*Aemulo.*
4: *Etymothyris-Amphigenia*
5: *Atlanticocoelia*</td><td>2: ________
3: *Heliophyllum*
4A: Diverse. brach.
4B: *Atrypa*
5: *Atlanticocoelia*</td><td>2:
3:
4A:
4B:
5:</td><td>2: ________
3: *Pentam.- Helio.*
4A: Diverse brach.
4B: *Athyris - Medio.*
5: *Ambo.*- chonet.
6: Leiorhynchid</td><td>2:
3:
4A:
4B:
5:</td><td>2:
3:
4A:
4B:
5:</td><td>not sub-divided</td><td>Typical Community Types</td></tr>
</table>

Fig. 1. Revised chart of Lower and Middle Devonian EESUs in the Appalachian Basin. Modified from Brett & Baird (1995).

This compilation is based on a survey of faunal lists of Early to Middle Devonian formations by a small number of palaeontologists (compiled in Linsley 1994). These include monographs by Grabau (1906) and Goldring (1935, 1943) on the Helderberg, Oriskany and Schoharie Formations; and Cooper (1929, 1930, 1933) and Buehler & Tesmer (1963) for the Onondaga and Hamilton Groups in New York. However, the general observations of EESUs also come from two decades of personal observations of faunas in New York, Pennsylvania and Ontario. For each given time interval we attempted to compile a total fauna of megafossils (excluding ostracodes, conodonts and other microfossils).

We have tried to compare the most nearly similar shallow shelf mixed siliciclastic and carbonate facies. This work will be complemented by a much more thorough study of within-biofacies (or within-community) stability and change for the Middle Devonian interval (Brett *et al.* 2007). In general, these comparisons seem to indicate that percentages of faunal consistency or faunal carry-over based upon the whole faunal assemblages (without consideration of facies) are similar to those based on shallow shelf carbonate assemblages and mixed carbonate and siliciclastic assemblages. Conversely, a somewhat higher proportional number of faunal elements typically carry over within the nearshore siliciclastic facies than in the overall more carbonate-based sample facies. In each instance, we have compiled all available data based upon personal observation and identification of previous authors. For a given time interval, generally no more than two authors' identifications were used for best consistency. The comparisons are made from New York, Pennsylvania, Maryland, southern Ontario and Canada.

The ranges of more than 800 species have been compiled based upon the literature sources as well as collection records of Brett & Baird and those of Cooper (1929), and Cooper & Williams (1935). We have confined the database to calcareous, grey mudstones or argillaceous limestones that contain highly fossiliferous, diverse coral, bryozoan, brachiopod, echinoderm faunas. These assemblages represent offshore, shallow shelf mixed siliciclastic–carbonate mudrock settings below fair weather wave base, but above storm wave base, assignable to benthic assemblage (BA) categories 3–4 (Boucot 1975; Brett *et al.* 1993). Excluded from this study are faunas of nearshore or peritidal facies; in addition we have not included faunas of highly progradational clastic wedges. Macrofaunal richnesses of individual faunas range from about 100 to over 300 species. This dataset is subject to various biases; certain taxa are probably not equally well preserved in all intervals. Echinoderms are notoriously subject to Lagerstätte effects; they may well be under-represented in many collections. Additionally, bryozoans require thin section study and in several of the faunas they have not received recent detailed study. Hence, these taxa were omitted from one set of analyses.

In order to document blocks of relative stability or EESUs, we have compared transgressive to early highstand systems tracts of successive formations from the Lower Devonian (Lochkovian) Helderberg Group through the Middle Devonian Genesee Group. These formations, rather than being strictly lithological units are bounded packages, typically commencing with minor to major disconformities and thus they approximate third-order depositional sequences. Thus, for example, both the two major subformations of the Onondaga Limestone and the formations of the Hamilton Group were defined as being bracketed between thin, widespread carbonate-rich units that are interpretable as transgressive systems tracts (Brett & Baird 1985, 1986, 1994, 1996; Ver Straeten & Brett 1995; Ver Straeten *et al.* 1995). Such packages probably each represent about 1 to 3 million years of deposition.

Our basic approach was to compile data on presence–absence of species from within offshore oxic mudrock facies of each successive formation or subgroup. This type of data recording was dictated, in part, by the way in which data are reported in the literature. In some cases only a single third-order sequence is represented within a subunit. This represents, in part, simplicity of certain intervals. For example, the Lower Devonian (Pragian) 'Oriskany' fauna is based on a single depositional sequence, the Glenerie Formation. However, in other cases, more detailed sampling is required. Thus, the Esopus, and Schoharie formations have not received detailed study and were treated as single third-order cycles. Both the Esopus and Schoharie formations are more complex and each can be subdivided into two or more sequences (Ver Straeten & Brett 2006). Nonetheless, our observations indicate that the faunas, as cited, are found both high and low within the formations in appropriate facies. In addition, only brachiopod data were available for the Esopus Formation, so that this interval was only compared with other units for the brachiopod subset.

The faunas of each successive formation (roughly each depositional sequence) were then compared using Jaccard similarity coefficients. Because of potential biases against most common taxa, we re-ran the analyses for two subsets: (a) only brachiopod data; (b) brachiopods, molluscs (excluding cephalopods).

Results

A dendrogram resulting from Q-mode cluster analysis of Jaccard coefficients shows relationships among the faunas of Early–Middle Devonian formations (Fig. 2). Note that the coefficients range from low values (e.g. Helderberg compared to Oriskany = 0.03) to extremely high values, underscoring the near identity of faunas (in terms of presence–absence of species) from adjacent formations. For example, Hamilton 2 and Hamilton 3 show a very high similarity, near 1.0, indicating nearly identical species lists.

Note that certain distinctive groupings emerge. The major cycles of the (a) Helderberg Group, (b) Onondaga Group, and (c) Hamilton Group cluster tightly together showing their internal similarities. Conversely, in many other cases adjacent faunas show very low similarities to one another. The groupings provide a quantified test of part of the model of co-ordinated stasis. In general, the tightly clustered groupings represent blocks of stability previously designated as EESUs by Brett & Baird (1995; Fig. 1).

However, a few deviations from the original designations of Brett & Baird (1995) are also noted. In particular, the lowest Hamilton sample (from the Stony Hollow Member of the Union Springs Formation), previously grouped with the Onondaga as unit 9A, is clearly set off from both the subjacent Onondaga and the superjacent Hamilton faunas, and warrants designation as a separate EESU, albeit one of relatively limited duration. Samples from the immediately overlying Halihan Hill bed sample are clearly and tightly clustered with the main Hamilton Group, while the underlying, highly discrete, Stony Hollow fauna shows very little in common, despite similar lithofacies.

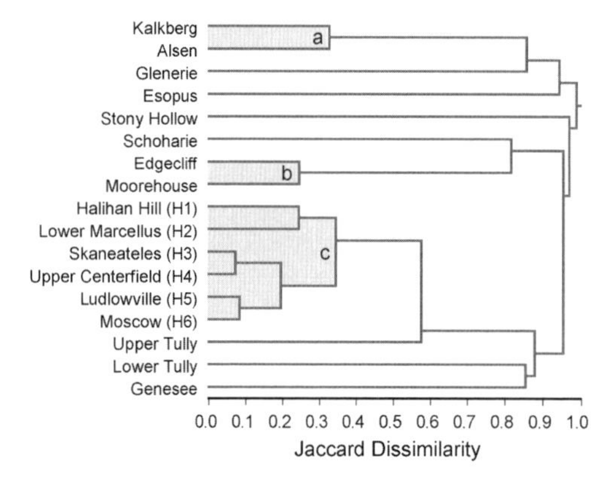

Fig. 2. Q-mode cluster analysis of successive Lower and Middle Devonian formations (*c.* third-order sequences). Note strong grouping of (a) Helderberg, (b) Onondaga and (c) Hamilton formations as in previously recognized EESUs. Upper Tully shows stronger affinities for the Hamilton cluster than does lower Tully.

The Tully faunas present another distinctive situation. The lower Tully fauna is clearly set off from that of the underlying upper Hamilton, as it has a large number of unique brachiopod species. Conversely, the upper Tully fauna shows greater affinity to the upper Hamilton sample than to the lower Tully. However, the Jaccard coefficient is only moderate for the Hamilton–upper Tully comparison. Detailed study of coral beds showed over 80% of species in the West Brook bed of the upper Tully in common with the upper Hamilton Windom Shale coral beds, but with higher diversity in the Hamilton samples (Bonelli *et al.* 2006), Thus, while we had previously united the Hamilton–upper Tully as a single EESUs, the latter is also set off from the Hamilton cluster. This result reflects not so much a basic difference in fauna – although about five upper Tully taxa are genuinely unique, the majority of species are shared with the underlying Hamilton (see Bonelli *et al.* 2006) – but rather a substantially lower diversity of the presently recorded Tully fauna. Either a number of typical Hamilton forms did not survive into the Tully or they have yet to be discovered in the rather limited set of appropriate facies.

EESUs within the Early to Middle Devonian interval of the Northern Appalachian Basin

The Early to Mid-Late Devonian (Lochkovian to Frasnian) interval has been regarded as a single ecological–evolutionary unit by Boucot (1983) and Sheehan (1985). However, this interval can clearly be differentiated into ten distinctive subdivisions or EESUs (Figs 2 and 3).

Keyser fauna (Late Silurian, Pridoli Series to Earliest Devonian)

The Keyser fauna is known from muddy carbonates and skeletal limestones from the West Virginia–Maryland and central Pennsylvania region. Equivalent beds include peritidal, shallow marine carbonates of the Rondout Formation of very Latest Silurian age in New York State. The upper Keyser Formation of Maryland, West Virginia and Pennsylvania is equivalent to a portion of the Helderberg Group and is of earliest Devonian age. Not surprisingly, owing to somewhat restricted conditions in the north, the Keyser faunas of the Pennsylvania–Maryland regions are substantially more diverse than that of New York. Sixty-nine species are known from this area and not from

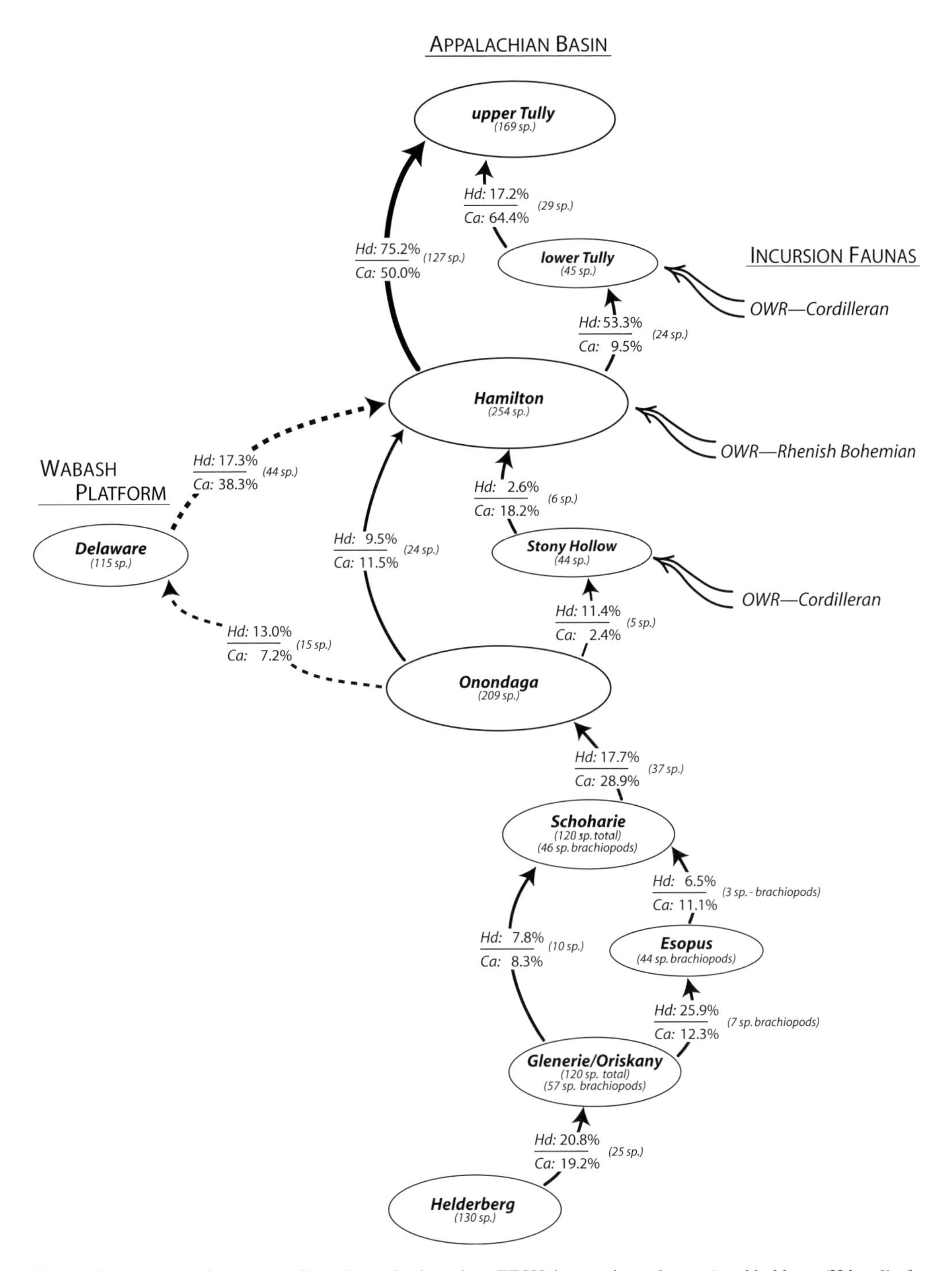

Fig. 3. Comparisons of carryover (Ca = % species in a given EESU that persist to the next) and holdover (Hd = % of species in a given ESSU that are held over from the previous one) of successive Appalachian Basin faunas; in a few instances arrows indicate carryover/holdover from an EESU older than the immediately preceding one to point up similarities. In the early Givetian comparisons are made among the Onondaga EESU, the Stony Hollow EESU and contemporaneous Delaware fauna of the Wabash Platform in Ohio, as well as with the Hamilton EESU. Size of ellipses around the EESU names indicates relative size of the species pool being compared.

New York State, although 54 species are found in both areas. With a total of 123 species of coral, brachiopods, bryozoans, pelmatozoans and others, the Keyser fauna is moderately diverse, a return to conditions more typical of medial Silurian faunas following the restrictive conditions (hypersalinity) that characterized the Late Silurian Salina Group. Nonetheless, it is distinctive: of the widespread 54 species that occur in New York Rondout as well as the typical Keyser of Pennsylvania and Maryland, only seven (13%) carry through. In the Pennsylvania–Maryland area, however, 24 of the 69 local taxa (35%) persist through. Overall, about 19.7% (31 out of 124) species carry over to the overlying Helderberg Group across the Silurian–Devonian boundary. Hence, again a major change is obvious here, despite very similar facies in the Keyser and Helderberg Groups. This coincides with Boucot's Silurian–Devonian boundary bioevent.

Helderberg fauna (Lochkovian to early Pragian)

In New York as well as central Pennsylvania and Maryland, the Silurian–Devonian boundary lies just below the base of the carbonate rocks of the Helderberg Group. The Helderberg formations have been correlated fairly firmly across the New York–Pennsylvania region. In West Virginia, Maryland and central Pennsylvania, the Silurian–Devonian boundary actually lies within the upper portion of the Keyser Formation. Facies on either side of the system boundary are quite similar carbonates. However, as noted, the fauna displays the marked change.

The Helderberg fauna has been relatively well-characterized by a suite of brachiopods, including the large spiriferid *Megakozlowskia macropleura* (Fig. 4). A fairly diverse fauna of 130 species of larger invertebrates was tallied from the faunal lists of Grabau (1906) and Goldring (1935, 1943) from the Helderberg, Oriskany and Schoharie faunas in southeastern New York State. Of these, only about seven species (5.4%) constitute hold-overs from the immediately underlying Rondout fauna. However, in the central Appalachians of Pennsylvania and Maryland, Swartz (1913) lists a total of 132 species of which 31 (23.5%) are hold-over taxa from the underlying lower Keyser beds. In turn, a comparable percentage of 26 species (19.7%) of the Helderberg faunal elements carry over into later Oriskany fauna. In New York the percentage is very similar: 25 of 130 common Helderberg species (19.2%) carry over into the next superjacent Oriskany assemblages (Fig. 3). The Helderberg Group constitutes most or all of the Lochkovian and possibly a part of the Pragian. All told this EESU probably spans some 8 to 10 million years. Nonetheless, there is considerable consistency from the lower Helderberg cycles to the upper. The well-known transgressive mega-cycles from Coeymans (shoal facies) to deeper shelf Kalkberg and below-wave-base New Scotland shaly limestones, have their parallel in the upper Helderberg group with the Becraft, Alsen and Port Ewen Formations (Laporte 1969). These very similar lithofacies provide a degree of control in comparing assemblages. Comparison of the diverse offshore (BA-3 to BA-4) biotas of the Kalkberg and New Scotland with the Alsen and Port Ewen Formations indicates that at least 70% of the diverse fauna is persistent from lower to upper Helderberg. Nonetheless, a distinctive suite of species does appear in the high Helderberg. Thus, this interval is subdivisible into an early

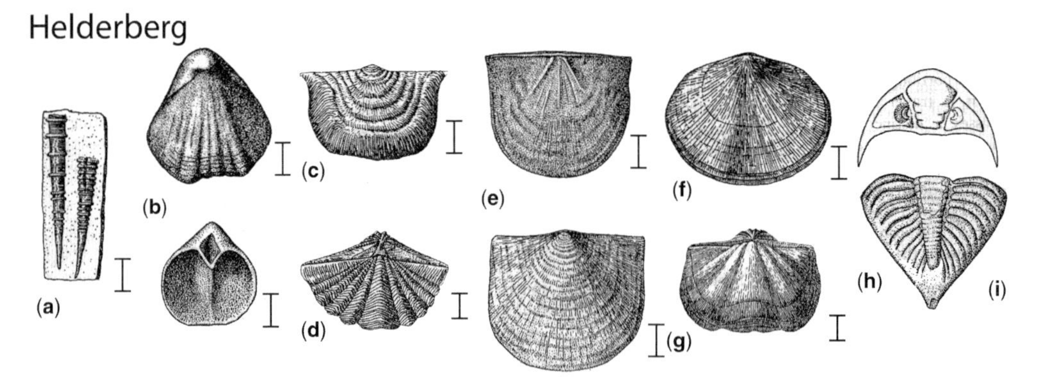

Fig. 4. Typical fossil taxa of the Helderberg fauna (EESU D-1); Hudson Valley region of New York State. (**a**) *Tentaculites gyracanthus*; (**b**) *Gypidula coeymansensis* (external and internal views); (**c**) *Leptaena rhomboidalis*; (**d**) *Howellella perlamellosus*; (**e**) *Leptostrophia*; (**f**) *Rhipidomella*; (**g**) *Megakozlowskia macropleura*; (**h**) trilobite: *Odontocephalus*. Figures from Linsley (1994) after illustrations in Hall (1867).

stage and late stage with a minor subdivision occurring approximately at the New Scotland–Becraft boundary.

Oriskany fauna (Pragian)

In most locations, the Helderberg Group is terminated by the major Wallbridge Unconformity. Sloss (1963) used this unconformity to mark out the boundary between his Tippecanoe and Kaskaskia megasquences. Nonetheless, in portions of the central Appalachian Basin in the lower Hudson Valley and central Pennsylvania, the break between these packages is relatively minor to nearly conformable. In many areas the Oriskany Formation (Pragian age) is represented by a rather unusual facies of shallow shelf quartzose sandstone. This may, in part, account for the distinctive character of its fauna, which Boucot (1975) characterized as the 'Big Shell fauna.' Nonetheless, in the Hudson Valley, Oriskany age sediments are represented by silty to sandy carbonates of the Glenerie Formation, which resemble in most aspects the Kalkberg or Alsen facies of the underlying Helderberg Group. Hence, by combining data from the Glenerie with other calcareous Oriskany fossil assemblages, a somewhat better biofacies spectrum is represented. The Oriskany fauna is dominated by large, robust brachiopod shells, as noted, including *Costispirifer arenosus*, *Acrospirifer*, *Hipparionyx*, *Rennsselaeria* and *Megastrophia* (Fig. 5). Relatively few corals or bryozoans are known from this interval, even from within the Glenerie, which suggests that comparisons with the Helderberg Group may not be completely justified. In Maryland and central Pennsylvania, the Ridgely Formation of Late Oriskany age carries some 63 species, many of which are endemic and cannot be compared with anything outside of the area.

In New York State, the overall fauna consists of about 120 common species. The time of Oriskany deposition during a major transgression, and constituting approximately half of the Pragian Stage, was a time of fairly abrupt change. In New York, about 25 of the 120 Oriskany species are direct carryovers from the Helderberg Group. A higher percentage is evident from the data of Swartz (1913) on the Oriskany of Pennsylvania, where some 26 of 52 taxa are carryovers; however, this does not include the rich and largely endemic Ridgely fauna.

Esopus fauna (early Emsian)

The Oriskany Sandstone is directly overlain by thick, dark, silty mudrock of the Esopus Formation in New York and by dark shales of the Needmore Formation in Pennsylvania. These formations are sparingly fossiliferous, except for the trace fossil *Zoophycos*, which dominates ichnofossil assemblages of these beds and appears for the first time in great abundance in the Appalachian Basin (see Bordeaux *et al.* 1990). Certain brachiopods and other shelly fossils, which do occur rarely in the Esopus, are carried through from the Oriskany fauna, including *Pacificocoelia*, which is probably the dominant Esopus brachiopod (Boucot & Rehmer 1977); however, a number of unique species of brachiopods have been reported from the Skunnemunk Outlier in southeastern New York (Fig. 6; Boucot 1959; Boucot *et al.* 1970, Boucot & Rehmer 1977). The major change to the overlying Schoharie fauna thus appears to occur during or after the time of the Esopus transgression. The duration of the Esopus fauna is somewhat in doubt, largely because of a lack of zonally important conodonts. Ver Straeten & Brett (2006) have recognized at

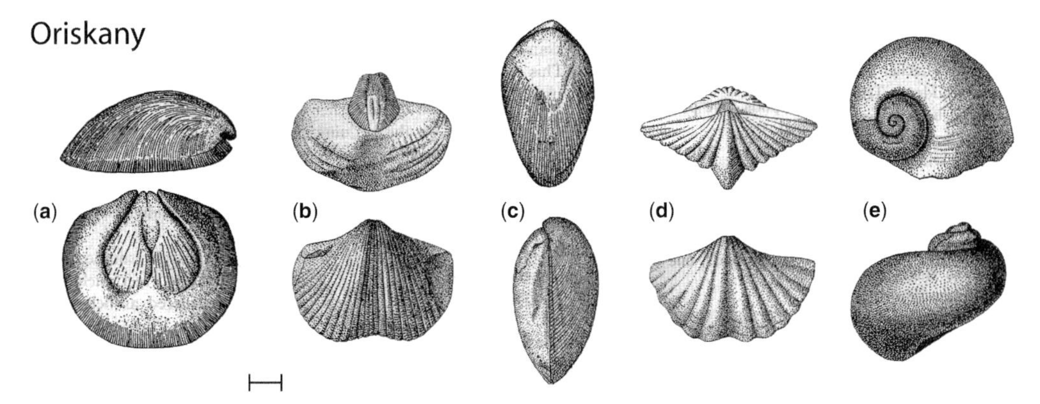

Fig. 5. Typical fossil taxa of the Oriskany fauna (EESU D-2); Hudson Valley region of New York State. (**a**) *Hipparionyx proximus*; (**b**) *Costispirifer arenosus*; (**c**) *Rennsselaria*; (**d**) *Acrospirifer murchisoni*; (**e**) *Naticonema* sp. Figures from Linsley (1994) after illustrations in Hall (1867).

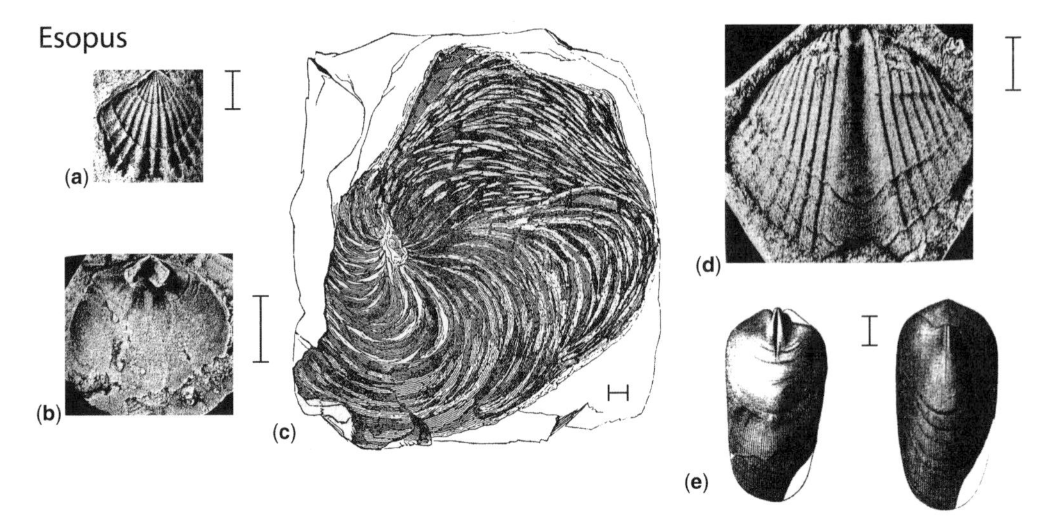

Fig. 6. Typical fossil taxa of the Esopus fauna (EESU D-3); Skunemunk outlier of southern New York State. (**a**) *Atlanticocoelia*; (**b**) *Meganterella*; (**c**) trace fossil *Zoophycos*; (**d**) *Hysterolites*; (**e**) *Etymothyris*. Figures (a), (b) and (d) from Boucot (1959); (c) from Goldring (1935); (e) from Linsley (1994) after illustrations in Hall (1867).

least three sequences within the formation. Moreover, the Emsian is now believed to be quite a long stage, perhaps as much as 17 million years (Tucker *et al.* 1998; Kaufman 2006).

Schoharie fauna (Emsian)

The silty limestone and calcareous mudstones of the Schoharie Formation (or Needmore of Pennsylvania) display a return to conditions very similar to those seen in the deeper portions of the Helderberg Group, such as in the Kalkberg and New Scotland Formations. Like the Helderberg, the Schoharie carries a very rich fauna with an abundance of brachiopods (Fig. 7), some corals, diverse trilobites and particularly a rich fauna of nautiloid cephalopods. This fauna is highly distinctive from that of the comparable muddy carbonate facies of the underlying Glenerie assemblage. Only approximately ten of 30 brachiopods carry through from the Esopus

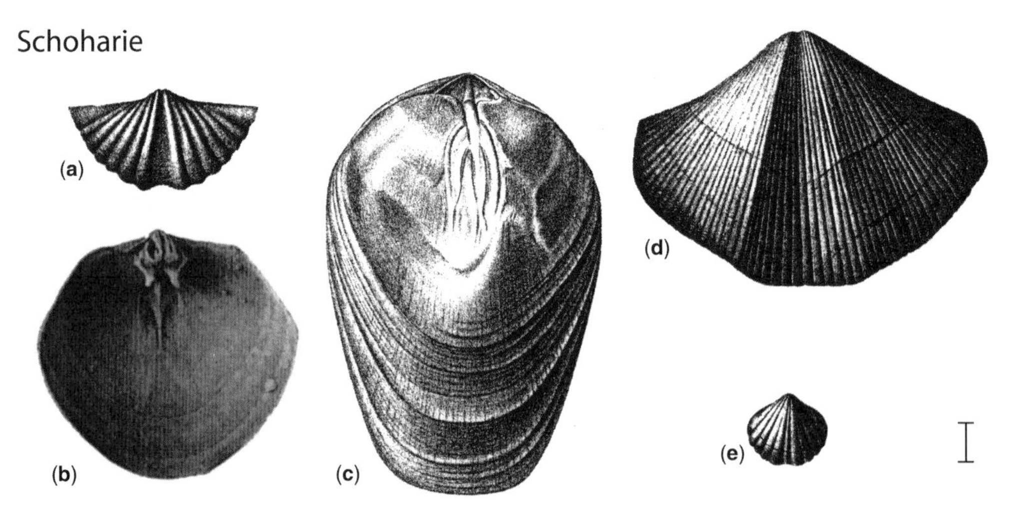

Fig. 7. Typical fossil taxa of the Schoharie fauna (EESU D-4); Hudson Valley region of New York State. (**a**) *Acrospirifer duodenaria*; (**b**) *Cloudithyris postovalis*; (**c**) *Amphigenia elongata*; (**d**) *Fimbrispirifer divaricatus*; (**e**) *Coelospira camilla*. Figures, except (b), from Linsley (1994) after illustrations in Hall (1867); (b) from Boucot (1959).

fauna into the Schoharie; ten of the 120 Glenerie–Oriskany species (8.3%) are represented in the Schoharie, and here they constitute only about 7.8% of the rich 128 species fauna of the Schoharie. The Schoharie Formation also shows the incursion of a large number of new elements, including the brachiopod *Amphigenia*, which persists into the overlying Onondaga fauna, as well as several, such as *Elita fimbriata, Nucleospira concinna* and *Cyrtina hamiltonensis*, which persist all the way through the Givetian Hamilton fauna. Overall about a third of the Schoharie species, 37 of the 128 taxa (Fig. 3), appear to carry over into the overlying Onondaga Formation of early Middle Devonian (Eifelian) age. The carryover value is lowered substantially because of the presence of a large number of unique Schoharie cephalopods and trilobites. If brachiopods alone are compared, the Schoharie and Onondaga faunas appear to have much more in common. Some 21 of 34 brachiopod species (62%) carry over directly to the Onondaga, and these include most of the abundant Onondaga brachiopods. At a generic level, these values appear somewhat higher even for other taxonomic groups: among genera of gastropods, five continue up to the Hamilton; among 18 cephalopods, five continue to the Hamilton; five trilobites of 14 continue up to the Onondaga and three into the Hamilton. Nonetheless, the Schoharie fauna does stand out as being sufficiently differentiated to warrant its own designation. The fauna appears to be relatively consistent throughout the thickness of the Schoharie, but no detailed studies are available to verify this quantitatively. As with the Esopus, the Schoharie lacks zonally important conodonts; it is generally suggested to be upper Emsian, probably *serotinus* to *partitus* zones. The duration of this portion of the Emsian is in question, but may approximate 5 to 8 million years based on recent estimates of 17 million years for the duration of the Emsian (Kaufman 2006).

Onondaga fauna (latest Emsian to late mid-Eifelian)

The Early–Middle Devonian (latest Emsian through much of the Eifelian) interval, constituting approximately 4 million years, is represented by widespread shallow shelf carbonates to deeper shelf, muddy carbonates of the Onondaga Group. Portions of the lower Onondaga (Edgecliff Member) carry small to moderate size bioherms; corals are a prominent element of the Onondaga fauna throughout. Of the over 200 species of Onondaga brachiopods, corals, crinoids, gastropods and other benthic taxa, 37 (about 18%) are holdovers from the Schoharie assemblages (Fig. 3). For brachiopods, this number is considerably higher: 15 of 32 common Onondaga brachiopods, or nearly half (47%), are retained from the underlying Schoharie. However, a majority of Onondaga taxa are unique and are found throughout the formation (Fig. 8). There is considerable internal consistency in comparing the similar lower shallow shelf fauna of the Edgecliff Member to those of the upper Onondaga (Moorehouse Member) fauna. Approximately 78.6% of the faunal elements carry through. However, only 24 Onondaga species continue through into the Hamilton fauna (Fig. 3) where

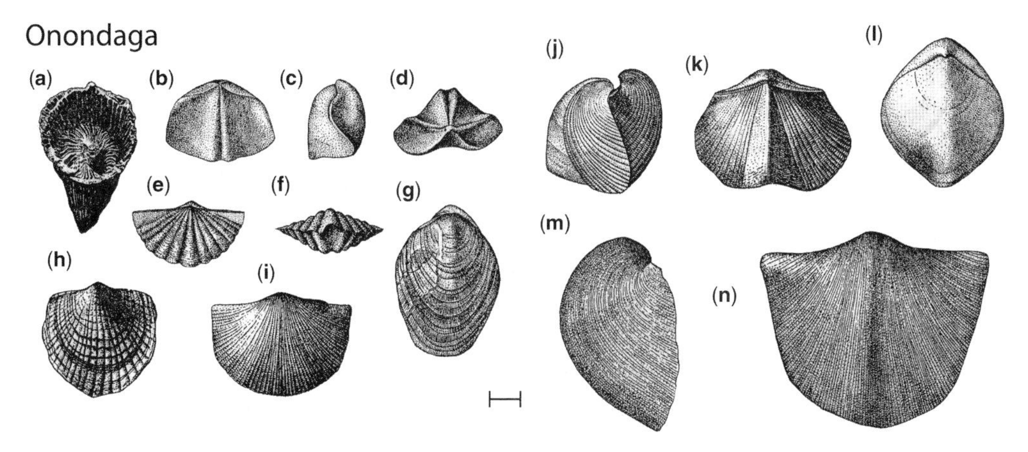

Fig. 8. Typical fossil taxa of the Onondaga fauna (EESU D-5); western and central New York State. (**a**) *Heterophrentis prolifica*; (**b–d**) *Pentagonia unisulcata*; (**e, f**) *Brevispirifer gregarius*; (**g**) *Amphigenia*; (**h**) *Pseudoatrypa* cf. *devoniana*; (**i**) *Strophodonta* sp.; (**j, k**) *Paraspirifer acuminatus*; (**l**) *Meristella*; (**m, n**) *Megastrophia concava*. Figures from Linsley (1994) after illustrations in Hall (1867).

they constitute a relatively small proportion of that highly diverse fossil assemblage. Subdividing this by taxonomic group yields similar results. Of coral species, nine of 47 (19%) carry through into the Hamilton coral beds; amongst brachiopods, nine of 46 (19.5%) carried through. At a generic level, corals appear somewhat stronger with 60% carry-over, whereas brachiopods display about 47% carry-over of genera. Part of the discrepancy here could represent an artifact of comparison of carbonate-dominated facies in the Onondaga with siliciclastics in the Hamilton. However, to facilitate comparisons, only the calcareous western facies of the Hamilton Group have been considered in this faunal tabulation.

Lower Marcellus (or Stony Hollow) fauna

Although technically a part of the Hamilton Group, the lower portion of the Marcellus subgroup, the Union Springs Formation (Ver Straeten & Brett 2006), has a distinctive fauna (Figs 9–11); previously considered simply a subdivision of the Onondaga fauna (Brett & Baird 1995), we now realize that it has little in common with that fauna. Indeed, only five of the 40 or so species of the lower Marcellus Shale are derived from the Onondaga (Fig. 3).

Most of the lower Marcellus fauna in central and western New York State and Pennsylvania is sparse, because the Union Springs Formation is mainly represented by black shale facies. However, in the Hudson Valley region of New York, some shallower water facies occur in the lower Marcellus (Stony Hollow Formation; see Griffing & Ver Straeten 1991). A thin (10–20 cm) limestone, the Chestnut Street bed (Griffing & Ver Straeten 1991; Ver Straeten *et al.* 1994) also carries the unusual Stony Hollow fauna in central to western New York. This bed lies slightly below or is adnate to the more prominent Cherry Valley Limestone and was previously termed the 'Proetid bed' (because of abundant *Dechenella haldemani*) or '*Werneroceras* bed'; although it contains specimens of the goniatite *Cabrieroceras* (formerly *Werneroceras*) these appear to be reworked from an underlying horizon (Brett & Kloc in Anderson *et al.* 1986).

In these beds the brachiopod fauna is almost completely distinctive from typical Onondaga or Hamilton assemblages, including such elements as *Subrennsselandia*, *Carinatrypa*, *Variatrypa arctica*, *Pentamerella* cf. *wintereri* and *Emanuella* spp., as well as the solitary rugose coral *Guerichiphyllum*, the trilobite *Dechenella haldemanni*, and the minute crinoid *Haplocrinites* (Fig. 10). Elements of this fauna occur widely in eastern North America during the *kockelianus* and *eiflius* to lower *ensensis* Zones (Koch 1979); it appears to reflect the incursion of a distinctive warm water biota from the Canadian Arctic Old World Realm. The major changeover occurs at, or slightly above, the Cherry Valley Limestone, a widespread deeper water carbonate facies, which contains a pelagic-dominated assemblage characterized by large *Agoniatites*. The Cherry Valley Member is of late Eifelian age and assigned to the *kockelianus* Zone. This changeover also involves the extinction of several long-ranging lineages, including the *Amphigenia* and *Rennsselaeria* brachiopods. It also may coincide with the termination of the cold water Malvinokaffric realm fauna and the incursion of the widespread *Tropidoleptus* into South America and Africa (Isaacson & Perry 1977; Troth *et al.* 2009). This suggested to Boucot (1990*a*) that an incursion of warmer water into the Appalachian basin region and elsewhere may have taken place at this time, an idea that receives some minor support from oxygen isotopic studies of conodonts (Joachimski *et al.* 2004).

Hamilton fauna (Givetian)

A substantial event marks the termination of the peculiar Stony Hollow fauna and the incursion of the so-called Hamilton fauna. This event occurs at least one major sedimentary cycle into the Hamilton Group, i.e. in the lower part of the Marcellus, Oatka Creek Formation above the Cherry Valley Limestone (Figs 11, 12). The changes of the Onondaga, Stony Hollow and the overlying Hamilton fauna are the subject of ongoing study (Koch & Day 1996; DeSantis *et al.* 2007). As in the case of the Stony Hollow fauna, the appearance of much of the Hamilton fauna in the Appalachian Basin coincides with a major transgressive black shale, the East Berne Member of the Oatka Creek Formation. It represents a major incursion of new taxa into the Appalachian basin, including a large number from the so-called Rhenish–Bohemian region of the Old World Realm (see Boucot 1975; Koch 1979; McIntosh 1983). This event has been recognized worldwide as the Kačák bioevent (Walliser 1990; Walliser *et al.* 1995). Boucot suggested that it represents a relatively rapid decline in global climatic gradient resulting in changes in oceanic circulation patterns as well.

The major changes of the Kačák-*otomari* bioevent occur at or near the Eifelian–Givetian boundary (Walliser 1986, 1990, 1996). This occurs in the Appalachian basin within the siliciclastic-dominated Hamilton Group. Major changes occurring low in the Hamilton Group, near the earliest Givetian Stage, resulted in the setup of the distinctive suite of biofacies, which are described in more detail elsewhere. Brett *et al.* (1990) recognized about 12 distinct biofacies which can be demonstrated through 10 to 12

Fig. 9. Typical fossil taxa of the Stony Hollow fauna (EESU D-6); shared between Appalachian and Michigan basins. (**a**) *Variatrypa arctica*; (**b**) *Carinatrypa*; (**c**) *Subrennsselandia*; (**d, e**) *Dechenella* cf. *haldemanni*. Photos from Ehlers & Kesling (1970).

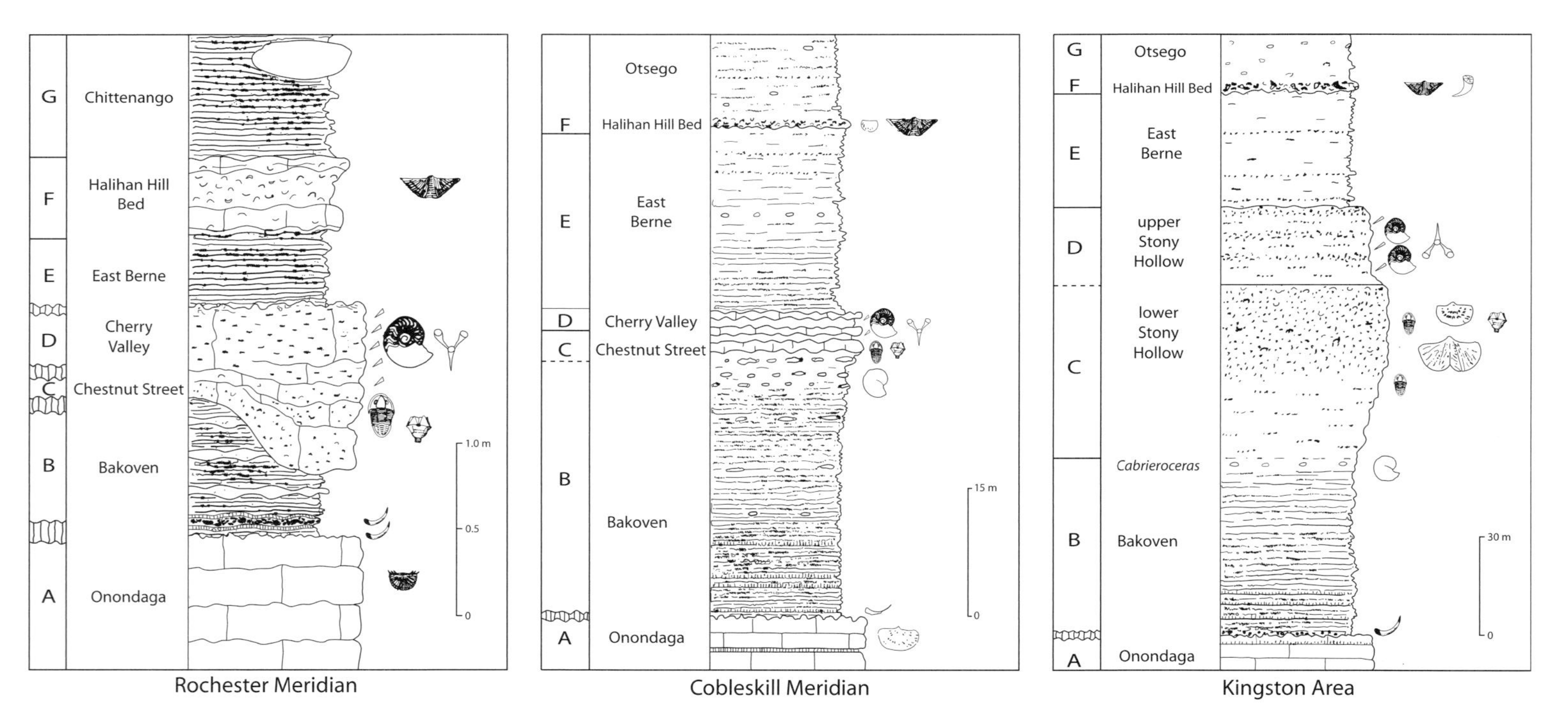

Fig. 10. Late Eifelian to earliest Givetian bioevents in western and eastern New York State. Fossil symbols from lower to upper: brachiopod *Hallinetes lineatus* typical of upper Onondaga fauna; onychodid fish teeth; goniatite: *Cabrieroceras*; trilobite: *Dechenella haldemanni* and crinoid *Haplocrinites clio*, representative of the Stony Hollow fauna. *Styliolina fissurella*, goniatite *Agoniatites vanuxemi* and auloporid corals typical of the Cherry Valley Member; brachiopods *Mediospirifer audaculus* and *Hallinetes* sp., typical of the Halihan Hill, initial Hamilton fauna. Note similarity of pattern across New York State (*c*. 300 km) despite differences in thickness.

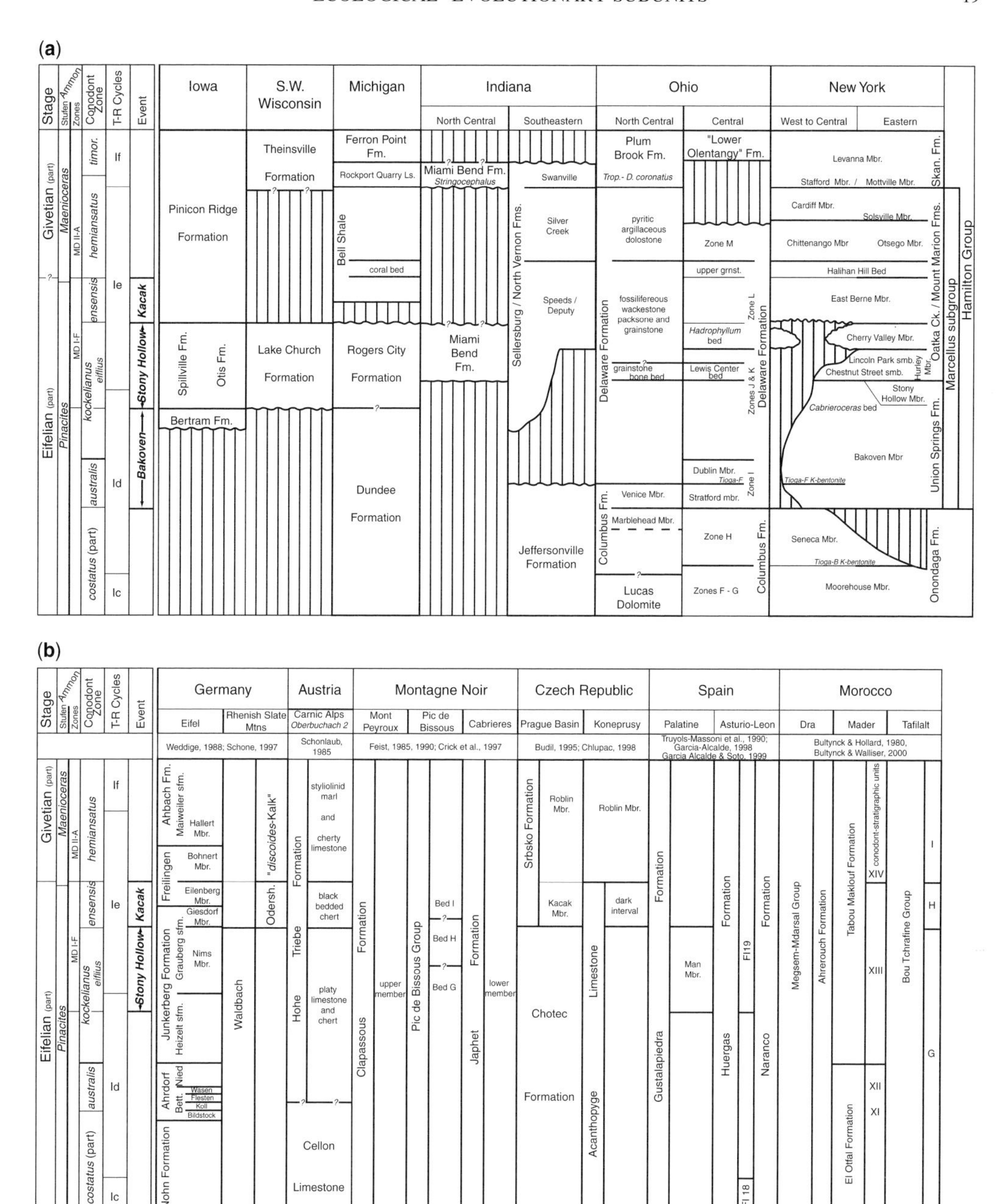

Fig. 11. (**a**) Correlation of Stony Hollow and Kacak bioevents in eastern North America. (**b**) Correlation of Stony Hollow and Kačák bioevents in Europe and North Africa.

major cycles within the Hamilton Group. Within each biofacies, there is a strong similarity between the lowest and the highest fossil assemblages (Brett & Baird 1995; see Figs 2, 12). Over 80% of the species in most taxonomic groups are known to persist throughout this time with little extinction. Fewer than ten species lineages are known to become extinct in the interval and very little introduction of new faunal elements occurs. The total Hamilton fauna has not been tabulated, but from the area of relatively thin calcareous mudstone deposits and thin carbonates of western New York, a list of over 250 species was compiled. Among these, almost none carry over from the underlying Stony Hollow

Hamilton-upper Tully

Fig. 12. Typical fossil taxa of the Hamilton fauna (EESU D-7); western and central New York State. (**a**) *Favosites hamiltoniae*; (**b**) trilobite *Eldredgeops rana*; (**c**) solitary rugose corals *Sterolasma rectum*; (**d**) *Pseudoatrypa devoniana*; (**e**) *Mediospirifer audaculus*; (**f**) *Athyris spiriferoides*; (**g**) *Ambocoelia umbonata*; (**h**) *Eumetabolotoechia multicostum*.

fauna; in contrast, 24 appear to be direct carryovers from the Onondaga fauna (Fig. 3), which evidently were reintroduced into the Appalachian Basin, perhaps from the nearby Wabash Platform (Ohio and Indiana) following a period of outage. The carryover of genera constitutes only 11.5% of the Onondaga fauna and less than 10% of the Hamilton faunas. Many of the distinctive new arrivals, which characterize the Hamilton Group, such as *Mucrospirifer* and *Tropidoleptus* among brachiopods, the *Eldredgeops rana* lineage, as well as *Greenops* and *Dipleura* among trilobites, and numerous pelmatozoans, appear to be direct incursions from the Old World Rhenish–Bohemian Region (Fig. 3). Their appearance in the Appalachian Basin may represent modified patterns of oceanic circulation, as well as climate change, according to Boucot (1990*a*). This event coincides with the local extinction of nearly all of the Stony Hollow fauna and may thus indicate a shift in climatic/oceanic conditions more similar to those of the Onondaga even though most species do not return.

Tully faunas

The Tully Formation, which overlies the Hamilton Group, signals the return to a dominance of carbonate sedimentation in the western New York region, although siliciclastic equivalents have been recognized (Figs 13–15). The lower member of the Tully carries a distinctive fauna, the so-called *Tullpothyridina* fauna, consisting of the brachiopods *Tullypothyridina*, *Rhyssochonetes aurora*, and several other taxa (Baird & Brett 2003). These taxa do not occur in either underlying or overlying units (Fig. 3). Most of the Tully fauna also recurs in distinctive, fine-grained carbonate facies in the lower Tully, which is not preserved in the older Hamilton or even the Onondaga Group. However, this fauna also occurs in dark grey shales and even siltstones in east central New York, Pennsylvania, Kentucky and elsewhere (e.g. Heckel 1973; Brett *et al.* 2004). Regardless, this fauna characterizes the lower Tully (upper *ansatus* Zone) throughout eastern North America. Thus, it appears likely that the onset of this fauna represents a new incursion of warm water taxa into the Appalachian Basin. This lower Tully incursion apparently coincides with the early phase of the tripartite Taghanic Events, and thus with a major faunal restructuring in other parts of the world (Aboussalam 2003; Aboussalam *et al.* 2001, 2004; Fig. 16).

The recurrence of most typical Hamilton species in the upper portion of the Tully, which carries

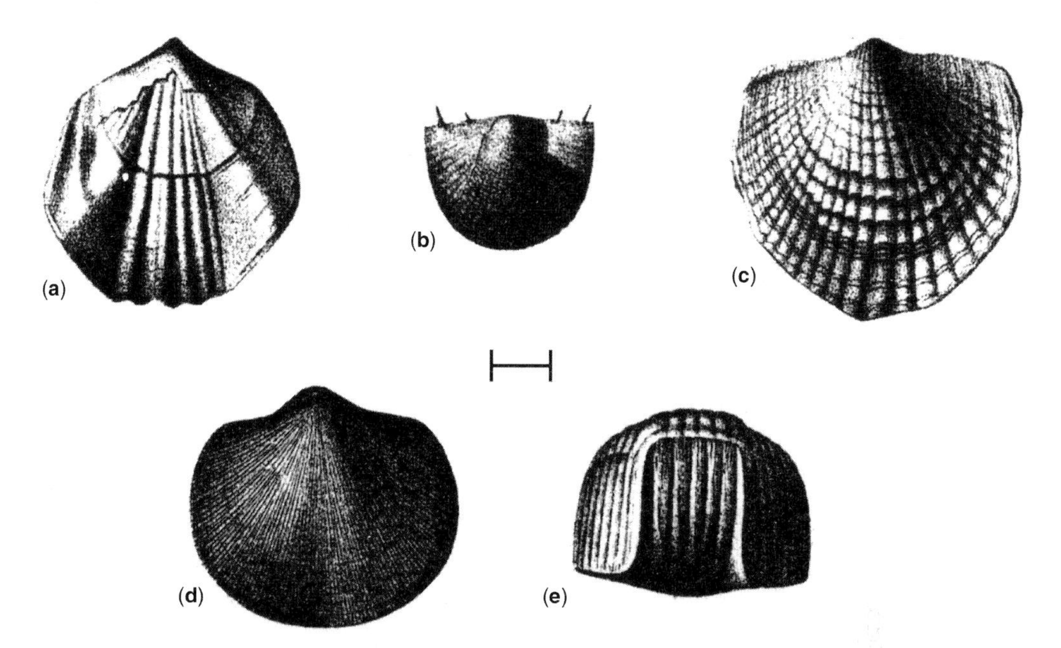

Fig. 13. Typical fossil taxa of the Tully fauna (EESU D-8); Hudson Valley region of New York State. (**a**) *Leiorhynchus mesocoastale*; (**b**) *Rhyssochonetes aurora*; (**c**) *Spinatrypa* cf. *spinosa*; (**d**) *Schizophoria tulliensis*; (**e**) *Tullypothyridina venutula*. Magnification ×2.5. Figures from Linsley (1994) after illustrations in Hall (1867).

facies more nearly similar to some of the muddy carbonates in the Hamilton and Onondaga Groups, suggests that community structure has not been drastically altered in the change between Hamilton and Lower Tully. Many of the offshore coral, brachiopod, bryozoan, trilobite and pelmatozoan species make their last appearance in the beds of the upper Tully. In particular, the Bellona coral bed (Heckel 1973) carries a very similar assemblage of brachiopods and corals to that seen in the underlying Moscow Formation of the Hamilton Group (Cooper & Williams 1935; Bonelli *et al.* 2006). However, the upper Tully (Hamilton) fauna appears to be dramatically decimated with the incursion of black Genesee Formation facies.

At least 35 of the 310 species of the Hamilton–Tully shallow shelf taxa (approximately 11%) reappear in the overlying Genesee Formation. However, this suite of holdover taxa makes up a substantial proportion of at least the common faunal elements of the Genesee and overlying Sonyea Group shallow shelf assemblages (Zambito *et al.* 2007). This is apparently because much of the Hamilton diversity has not been replaced by the origination of new species by *in situ* evolution or immigration.

Genesee–Sonyea fauna

The Genesee–Sonyea fauna appears to represent a relatively modest diversity assemblage, approximately 50 species, dominated by brachiopods and bivalves. The late-Middle Devonian Taghanic event appears to have decimated entire groups of organisms, which flourished up through the Tully subfauna; these include certain rugose and tabulate corals, phacopid trilobites, which barely continue into the Frasnian Ithaca beds but then disappear, as well as many common Hamilton–Tully brachiopods and pelmatozoan taxa. This extinction without complete replacement appears to coincide with the major transgression of the Devonian and the incursion of a large mass of anoxic water into the Appalachian basin. Also, it is associated with major progradational phases of the Catskill deltaic complex, which may have forced out many of the low turbidity carbonate calcareous mudstone biofacies that were prominent in the west during Hamilton–Tully time.

Certain Givetian taxa appear to have persisted in the mid-continent. For example, the Rapid Member of the Cedar Valley Formation in Iowa displays a fauna that carries many of the same genera, if not the same species, present in the older Hamilton–Tully fauna to the east, but even this is terminated by early Frasnian time. Hence, the decimation event may have been somewhat diachronous, but it basically occurs at or near the Givetian–Frasnian boundary. However, certain elements of the Hamilton fauna do recur within the Frasnian Genesee Group; these assemblages occur within the lower portion of

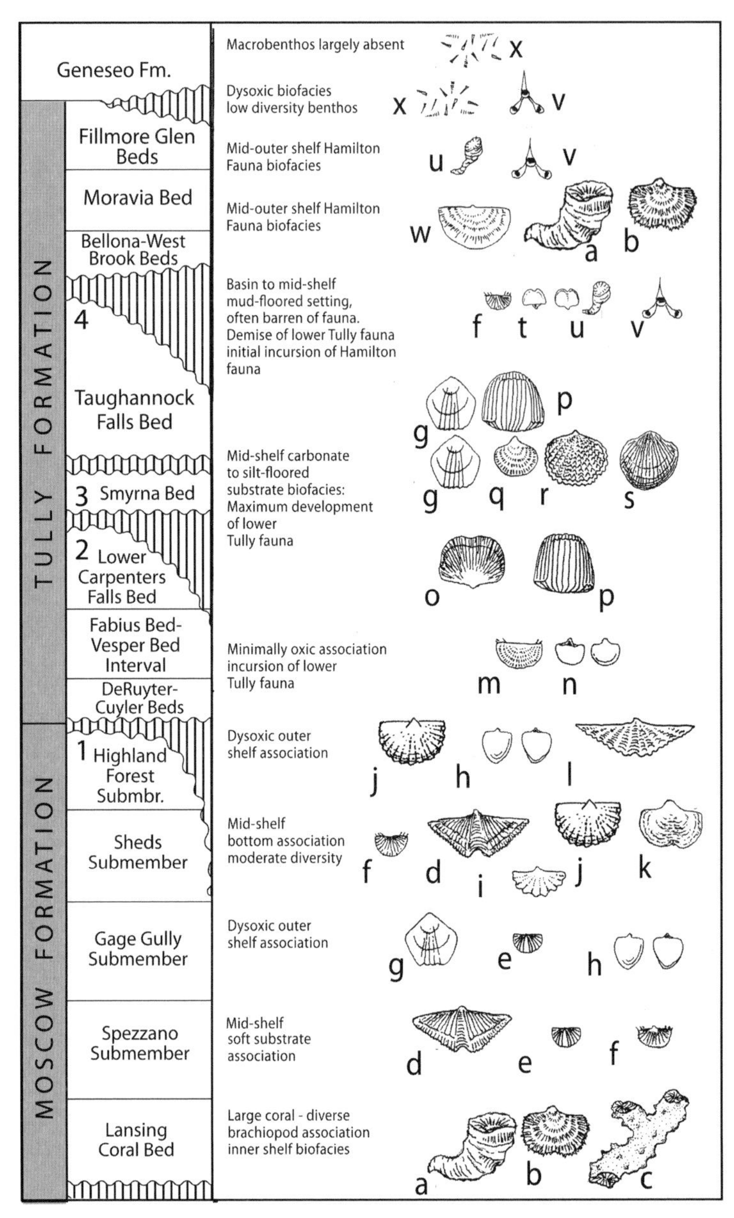

Fig. 14. Generalized faunal succession in the upper Windom Member and Tully Formation succession in east central New York showing the two-phased influx of the Tully fauna ('lower Tully bioevent') at the base of the Tully Formation, re-establishment of the diverse Hamilton fauna association ('second Tully bioevent'), and the decline of faunal diversity in the highest Tully owing to the onset of transgression-related dysoxia ('third Tully bioevent'). Key lowstand unconformities include: the base-Tully (1), base-middle Tully (2), and base-West Brook (4) sequence disconformities. Maximum flooding surface contacts include: the top-Smyrna Bed unconformity (3) and the top-Fillmore Glen corrosional discontinuity (5). Diagnostic and/or common taxa include: (**a**) *Heliophyllum halli*; (**b**) *Spinatrypa spinosa*; (**c**) large bryozoans; (**d**) *Mediospirifer audaculus*; (**e**) *Allanella tullius*; (**f**) *Devonochonetes scitulus*; (**g**) *Camarotoechia* (*'Leiorhynchus'*) *mesocostale* (in the Tully, this taxon is restricted to the TFCCS in eastern New York and Pennsylvania); (**h**) *Emanuella praeumbona*; (**i**) *Pustulatia* (*Vitulina*) *pustulosa*; (**j**) *Tropidoleptus carinatus*; (**k**) *Athyris spiriferoides*; (**l**) *Mucrospirifer spiriferoides*; (**m**) *Rhyssochonetes aurora*; (**n**) *Emanuella subumbona*; (**o**) *Schizophoria tulliensis*; (**p**) *Tullypothyridina venustula*; (**q**) *Echinocoelia ambocoeloides*; (**r**) *Spinatrypa* sp.; (**s**) *Pseudoatrypa devoniana*; (**t**) *Ambocoelia umbonata*; (**u**) small rugosan; (**v**) auloporids; (**w**) *Leptaena rhomboidalis*; (**x**) styliolines (modified from Baird & Brett 2003).

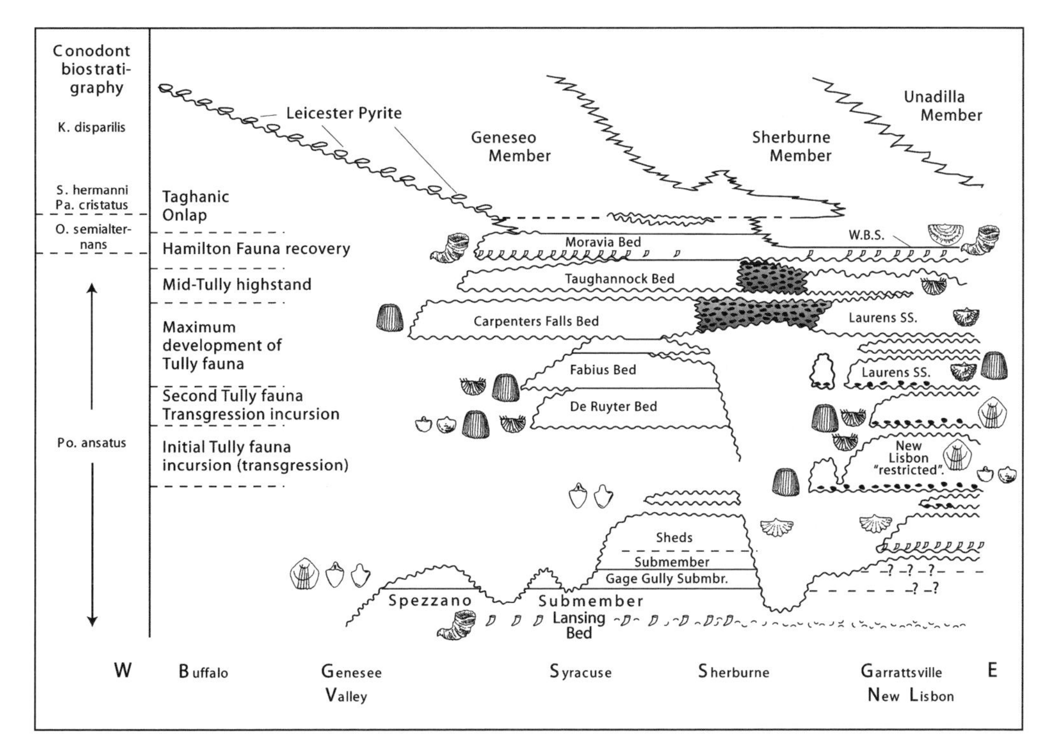

Fig. 15. Wheeler correlation chart of upper Moscow, Tully and Genesee units in New York State (west on left side, east on right), showing complex succession of faunas and physical events, in relation to sequence stratigraphy during the later *ansatus* through *hermanni* zones. Fossil images as in Figure 14.

the Ithaca Formation and in a few beds in higher units (Williams 1913; Zambito *et al.* 2007). These recurrent assemblages, typified by generalized Hamilton species such as *Tropidoleptus carinatus* and *Ambocoelia umbonata*, survived under conditions that were similar to those of the Appalachian Basin in the Late Devonian (Williams 1913; Zambito *et al.* 2007).

The typical Ithaca or Genesee fauna, characterized by *Tylothyris mesocostale*, *Orthospirifer mesastrialis*, '*Pugnoides*' sp. and *Schizophoria impressa*, appears abruptly, associated with a minor transgression in the upper middle Ithaca beds (Zambito *et al.* 2007). The basic community setup of the Genesee (Ithaca) and most of the common species (nearly 85%) persist into the overlying Sonyea Group (e.g. Thayer 1974). This constitutes the bulk of the Frasnian Stage and probably represents at least 5 to 6 million years of geological time.

Late Frasnian faunas

In the later Frasnian, as represented by the West Falls and Java groups, at least three short-lived faunas appear to have replaced one another in succession (Dutro 1981; McGhee 1990). The first of these is the early West Falls or Rhinestreet fauna, which appears to be largely made up of species carrying through from the underlying Genesee–Sonyea fauna. The so-called *Cyrtospirifer* fauna, which characterized the upper West Falls Group, particularly the Gardeau and Nunda sandstone, represents a suite of brachiopod species that appeared relatively briefly in the Appalachian Basin, but may be derived from mid-continent Frasnian faunas. This assemblage is relatively short-lived and most of the species, which appear abruptly in the upper portion of the West Falls Group, disappear as abruptly near the top of that group. The latest Frasnian or Java fauna, with relatively low diversity, at least in terms of brachiopods, appears to be made up largely of genera, which have persisted through much of the Frasnian, or, in some cases, all the way from the Givetian (e.g. *Tropidoleptus carinatus* and several atrypids, including *Spinatrypa* and *Pseudoatrypa*). It is particularly significant that these long-ranging lineages are abruptly terminated at or near the Frasnian boundary. This is just one aspect of the very widespread terminal Frasnian

(a)

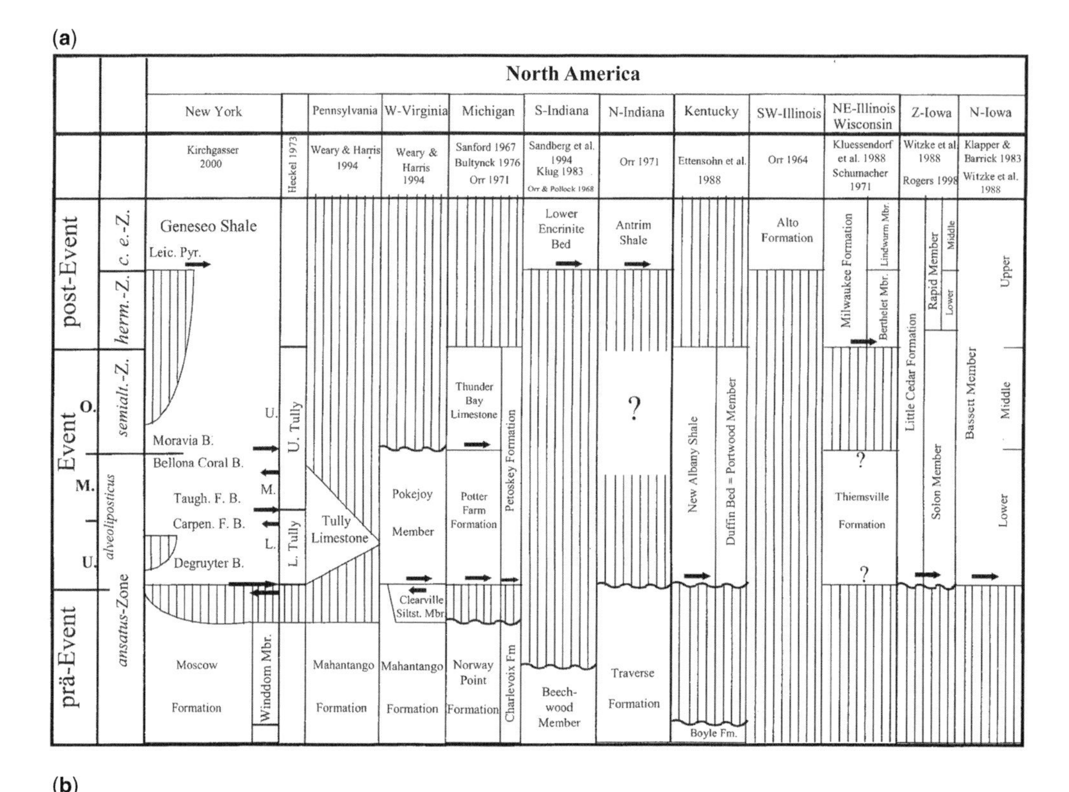

(b)

Germany | N France | Armorica | S France (Montagne Noire) | W Pyrenees | Cantabrian Mountains | Celtiberia | Italy

Nordsauerland | Bergisches Land | Harz (Elbingeroder Riff) | Boulonnais | Rade de Brest | Mont-Peyroux | Pic de Bissos | Cabrières | Asturien | Leon | Karnische Alpen

May & Becker 1996 | Kleinebrinker 1992 | Fuchs 1987 | Brice et al. 1979 | Carls 1988 Morzadec & Weyant 1982 | Feist 1983, 1985 | Feist & Klapper 1985 | Wirth 1967 | Garcia-Lopez et al. 2002 | Garcia-Lopez & Sanz-Lopez 2002 | Carls 1988 | Spaletta & Perri 1998

post-Event
Event
prä-Event
c. e.-Z.
herm.-Z.
semialt.-Z.
ansatus-Zone
O.
M.
U.
Bänder-Schiefer
Tannen-kopf-Schiefer
Flinz
Massenkalk
?
Hornstein-Horizont
Unterer Platten-Kalk
Riff-Kalke
Eisenerz-Schichten
Vulkanite
Blacourt-Formation
Couderouse-Member
Griset-member
Kergarvarn Formation
Lanvoy Formation
Kerbelec Formation
Coumiac Formation
Clapassous Formation
Pic de Bissous Group
Japhet Formation
Tonschiefer
sandige Kalke
Tonschiefer + dünne Kalke
Korallenkalk
massiger kalk
Candas Formation
Mbr. D
Mbr. C
Mbr. B
Portilla Formation
Mbr. C
Mbr. B
Cabezo Agudo Formation
Salobral Formation
Rudite
Tonschiefer
Rudite
Arenite
Rudite
Arenite
Rudite

Fig. 16. (**a**) Correlation of Taghanic bioevents in eastern North America. (**b**) Correlation of Taghanic bioevents in Europe. Adapted from Aboussalam (2003).

extinction event. However, as McGhee (1990) has demonstrated, the later Frasnian was a time of substantial faunal instability and the occurrence of the three distinctive faunas within just 1 to 3 million years suggests substantial and rapid environmental change at this time. The Frasnian–Famennian extinction event is widely understood to be the second or third largest extinction in the history of the Phanerozoic. Its many possible causes have been discussed in detail by McGhee (1988, 1990). This event effectively terminated the entire Ecological–Evolutionary Unit 5 of Sheehan (1985). It provides a convenient upper boundary for the consideration of detailed bioevents in the Appalachian basin.

Summary

Counting the three latest Frasnian events, there are 15 distinctive faunas in the 70-million-year interval that separates the earliest Silurian from the mid-Late Devonian (Brett & Baird 1995), that is, between the two major extinction boundaries of the Late Ordovician and the late Frasnian. In general, these faunas appear to be quite persistent and in most cases, within a particular block of time or interval, 70 to 80% of the species persist throughout. Relatively few new species immigrate into the basin, except at the boundaries of these intervals. Furthermore, very few species go extinct during these times, except perhaps for the rare taxa, as predicted by Boucot (1990). Rare taxa mainly undergo pseudoextinction, which results in the production of new species by phyletic evolution within a block of time. The boundaries of these events are abrupt, following 1 to 7 million years of stability. Periods of time that must represent only a few tens of thousands of years are characterized by very rapid incursion of new taxa, substantial extinctions (in most cases 60 to 70%), and major reorganization of community structure. Within each of the blocks, particularly notably in the case of the Onondaga and Hamilton faunas, a great deal of similarity exists. This can be demonstrated at the level of relative abundance of taxa within a particular community and at the stability of species at the 70 to 85% level overall integrated across all facies for a particular interval of time. It is this pattern of blocks of 'co-ordinated stasis', coinciding roughly with the traditional stages, which requires a substantial evolutionary–ecological explanation.

Discussion: implications for evolutionary palaeoecology

Nature of EESU boundaries

The pattern of faunal change within the Appalachian Basin is staccato, with brief intervals of intensified change alternating with periods of more gradual, little change. A major difference among the various recent studies is the degree to which organisms appear stable within the blocks or to follow biofacies fidelity during the sedimentary cycles. This implies differences in the degree of stability during 'background' intervals that may vary from conservation of a majority of species and even ecological associations for several million years (Brett & Baird 1995), to persistence of a majority of genera (Bennington & Bambach 1996), or to a simple bimodality of rates of species/generic turnover (Patzkowsky & Holland 1997). Thus, the most important underlying aspect of the EESUs is their boundaries: all show some bimodality of rates with considerably accelerated rates of local/global extinction, species evolution, and/or immigration/emigration. This leads to the appearance of a strong biotic turnover. A variety of issues arise if we focus on these boundaries. Are these boundaries an artifact of missing section? Are these boundaries associated with stratigraphically important surfaces such as sequence boundaries or maximum flooding surfaces? Are these boundaries synchronous over several basins? Can some or any be correlated with global bioevents?

Relationship to stratigraphic surfaces

The results of this study indicate both a strong degree of change at certain third-order sequence boundaries or flooding surfaces and very little change at others. Not all biofacies were compared but only diverse, mid-shelf BA 3 to 4 assemblages. Thus, the appropriate biofacies do not, in each case, occur at the base of depositional sequences. In fact, we chose maximum flooding surfaces as the bases of the comparative units rather than the lowest beds because basal transgressive portions of sequences were typically heavily reworked lag beds with poorly preserved and heavily time-averaged fossil assemblages.

Even though gaps in the record may lead to artificial sharpening of boundaries (Holland 1996), it is the differences among such boundaries that are important here. The essential fact is that the offshore shelf facies of some sequences show evidence of marked change with respect to the next subjacent sequence, while those of others do not. Those which do not show marked changes are nonetheless separated from one another by discontinuities. The lack of much change in these successive formations shows that not every gap produced either real or artifactual differences. Conversely, in some cases, major abrupt changes occur despite the absence of a major discontinuity.

For example, much of the Hamilton fauna in the Appalachian Basin first occurs within the Halihan Hill bed, a thin 0.3 to 1.5 m fossil-rich bed. This interval is as sharply defined in thick Hudson Valley sections as in western New York where the underlying black shales are only a few centimetres thick. Hence, the sharpness of this transition appears to be real and not an artifact. Likewise, detailed studies of thick siliciclastic successions in the upper Hamilton through Tully interval (Baird & Brett 2003, 2005) have failed to reduce the sharpness of changes into the lower Tully fauna and its passage into the recurrent Hamilton (upper Tully). Both events remain sharply defined within thick, essentially conformable successions. Such examples indicate that EESU boundaries are biologically real and not simply artifacts of incomplete records.

A second issue is the relationship between sequence stratigraphy and biotic turnovers. The best-resolved series of biotic turnovers are those of the late Eifelian to Givetian (Fig. 12). A major biotic turnover separates the diverse Onondaga–Columbus Limestone fauna from the rather less diverse Stony Hollow fauna (DeSantis *et al.* 2007). Both occur in limestone facies, although the latter is also known from calcareous siltstones of the Stony Hollow Member; the shallower biofacies show an unusual suite of brachiopods, including *Variatrypa arctica*, which appear to represent immigrants from the tropical environments represented in present-day NW Canada (Koch & Day 1996). At present, the precise position of this turnover within the upper Onondaga–Union Springs third-order sequence is unknown. It is clear that it did not occur within the transgressive limestone interval (Seneca Member–basal Union Springs), which still carries a diverse Onondaga fauna. The overlying Bakoven Member generally comprises anoxic black basinal shale facies. Therefore it is not clear exactly when the immigrant taxa characteristic of the Stony Hollow first arrived in the Appalachian Basin. However, this must have occurred no later than in beds above the first appearance of *Cabrieroceras*, in a nearly continuous regressive succession. Interestingly, this fauna appears to span a sequence boundary into lowstand to transgressive limestones of the Hurley and Cherry Valley members of the Oatka Creek Formation (Fig. 11).

The Hamilton fauna appears very abruptly in the overlying highstand deposits of the East Berne Member (Oatka Creek Formation). Again, the precise timing of incursion may be masked by the black, anoxic shale facies that typifies the lower East Berne Member. Studies of this interval in the Hudson Valley region of New York have revealed a diverse assemblage with typical Hamilton taxa, such as *Tropidoleptus* and *Mediospirifer*, at the top of a minor shallowing cycle some 19 metres below the Halihan Hill bed (Bartholomew *et al.* 2007). Again, it is clear that the faunal turnover occurs at or just after a fourth-order maximum flooding event and not at a sequence boundary.

The transition from the Hamilton to the lower Tully fauna (Figs 12–14) now also appears to occur abruptly, associated with widespread dark shale facies (Baird & Brett 2003), although there are some hints of a change in one of the faunal elements (the brachiopod *Emanuella*) associated with an older, highstand dark shale (Sessa *et al.* 2002). The main influx of the unusual lower Tully faunal elements (*Tullypothyridina venustus*, *Rhyssochonetes aurora*, *Camarotoechia mesocostale*, *Spinatrypa* sp., *Nervostrophia*, *Schizophoria tulliensis*), again from more tropical areas of western Laurentia, is now known to occur near the base of a previously little known dark shale (New Lisbon Shale of Cooper & Williams 1935; Baird & Brett 2005). This dark shale overlies the highest calcareous siltstone/limestone (Shedds submember of Windom Shale Member) with a rich, typical Hamilton brachiopod–coral fauna, which represents a fifth-order transgressive shell bed (Figs 14, 15). Thus, again, the faunal change, involving local extinction and immigration, occurs not at a sequence boundary or lowstand, but at or slightly above a maximum flooding event, associated with dark, dysoxic shale facies (Baird *et al.* 2003). The lower Tully fauna itself crosses two major erosive sequence boundaries at the bases of the lower and middle Tully depositional sequence and is common in overlying compact transgressive limestones. The introduction of *Emanuella* and reduction of the typical *Ambocoelia umbonata* in dark shales two cycles earlier, as noted by Sessa *et al.* (2002), is an exception that proves the rule: introduction of new taxa within highstands associated with dysoxic facies.

Finally, the demise of the lower Tully fauna and reintroduction of the typical Hamilton occurs above a transgressive limestone (Smyrna bed; see Fig. 14), which contains a very well developed lower Tully fauna, in the overlying dark shales and nodular micritic limestones of the Taughannock Falls Bed (Heckel 1973; Baird & Brett 2003), clearly a highstand deposit that overlies a maximum flooding surface. Again, the fauna persists across another minor sequence boundary and has its peak development in the limestones of the overlying Bellona–West Brook transgressive beds before being decimated in overlying dark shales of the Geneseo Formation and replaced by the Genesee fauna.

Thus, at the scale of short (fifth order; 10^1 ka), intermediate (fourth order; 10^2 ka) and long (third order; 10^3 ka) timescales the major turnovers of fauna during the late Eifelian–Givetian interval are characterized as occurring above transgressive limestone intervals in highstands associated with dark, dysoxic facies in the Appalachian foreland basin. In no case does the faunal change appear to occur in lowstand to transgressive successions. Indeed, in all cases, stable faunas typical of each EESU cross erosive sequence boundaries and appear to have had their best development in overlying shell-rich beds of the transgressive systems tract.

Relationship to global bioevents

A further critical issue involves the degree to which the boundaries of the EESUs are contemporaneous across a single basin, between basins within a biogeographic province and among different biogeographic provinces. Perhaps above all, do some or all of the Devonian EESU boundaries of the Appalachian Basin coincide with bioevents that have been recognized in other pats of the world (House 1985; Walliser 1990, 1996)? The answer to this question hinges largely on biostratigraphy, which is unfortunately poorly resolved in much of the Appalachian Basin (but see Klapper 1981; House 1981; and recent summaries in Bartholomew *et al.* 2007 and DeSantis *et al.* 2007).

For the Middle Devonian, it is clear that analogous EESUs were developed in different biogeographic subprovinces of the Appalachian and Michigan basins. Even though a majority of dominant species and genera differ between these basins there are clear similarities in the biotas. Bartholomew (2006) identified onshore–offshore gradients in the two basins that show strong parallels (Fig. 17). Major faunal turnovers in the Michigan Basin are equally abrupt and characterized by similar levels of turnover/holdover as those of EESUs in the Appalachian Basin (Fig. 18). Detailed correlations based on biostratigraphy and sequence stratigraphy indicate that these rapid faunal turnovers are approximately synchronous in the two basins, suggesting a common cause. As yet, other Appalachian Basin EESUs have not been documented in detail in different basins. However, new findings from isotope stratigraphy and sequence stratigraphy permit some tentative correlations at this time.

The base of the Helderberg Group lies close to the Silurian–Devonian boundary, although recent work by Kleffner *et al.* (2006) suggests that the system boundary may actually occur at the unconformable base of the Coeymans Formation. The rather subtle changes from the Salina fauna to that of the Helderberg EESU may thus coincide with Silurian–Devonian boundary or Klonk events that have been recognized elsewhere, particularly in the Czech Republic (Walliser 1996). The presence of a sharp positive $\delta^{13}C$ excursion is observed globally at the Silurian–Devonian boundary (Buggisch & Joachimski 2006; van Geldern *et al.* 2006) and isotopic studies are presently underway in eastern North America to test the synchroneity of this boundary event (M. Saltzman, pers. comm.).

The abrupt termination of many of the Helderberg lineages and the appearance of the Oriskany fauna coincides with a major unconformity at least in North America: Sloss's (1963) Wallbridge Unconformity. There seems little doubt that this represents a partially eustatic sea-level drop (Johnson *et al.* 1985) although the Helderberg–Oriskany boundary is nearly conformable in southeastern New York. However, sea-level lowstand near the Lochkovian–Pragian boundary, a negative $\delta^{13}C$ excursion, and biotic changes (Lochkov-Prag event) are noted in the Czech Republic, Siberia,

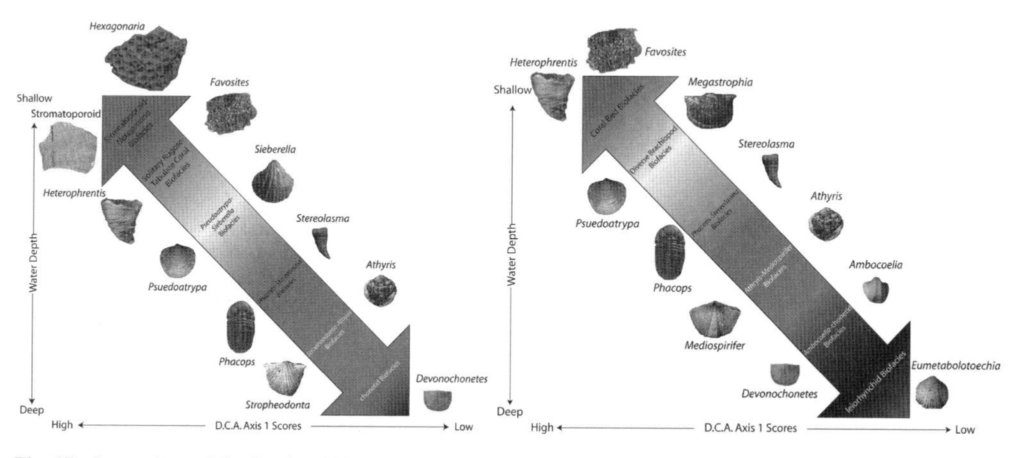

Fig. 17. Comparison of depth-related biotic gradients in the Appalachian and Michigan basins during Givetian time (see Brett *et al.* 2007 for discussion.) Note parallelism, despite differences in dominant genera.

APPALACHIAN BASIN MID-DEVONIAN FAUNAS

Faunas (EE Subunits)	Age Devonian	Hold-over	%	Duration (MA)
C) Genesee-Sonyea	Givet-Frasn.	30:150	20%	3.0
B) Hamilton-Tully	Givetian	3:335*	<1%	5.5
A) Union Springs	L. Eifelian	5:60	8.3%	<1

MICHIGAN BASIN MID-DEVONIAN FAUNAS

Faunas (EE Subunits)	Age Devonian	Hold-over	%	Duration (MA)
C) Squaw Bay-Antrim	Givet-Frasn.	2:15	13.3%	3.0
B) Traverse	Givetian	2:91	2.2%	5.5
A) Rogers City	L. Eifelian	3:31	9.7%	<1

Fig. 18. Comparison of Michigan Basin and Appalachian Basin EESUs during Middle Devonian time.

Morocco and other parts of the world (Walliser 1996; van Geldern *et al.* 2006; Fig. 19). If this bio-event proves to be associated with a widespread low-stand it will provide an exception to the association of most Devonian bioevents with highstands and positive $\delta^{13}C$ excursions.

The widespread deepening and hypoxia associated with the Esopus transgression in New York appears to mark the termination of a majority of Oriskany taxa and incursion of Esopus forms. This may be a local representation of the early Emsian Zlichkov event seen in the Czech Republic, Morocco, Rhenisch areas, and elsewhere (Walliser 1996). Similarly, the turnover to the Schoharie fauna quite possibly records the later Emsian Daleje event near the *inversus/serotinus* conodont zonal boundary; the latter could be associated with a widespread dark shale within the Schoharie, the Aquetuck Member (Ver Straeten & Brett 2006), but the timing is at present too poorly constrained to say more.

The change to the Onondaga Formation occurs in the latest Emsian and does not appear to be directly correlative with a global bioevent, although it is associated with a negative $\delta^{13}C$ excursion. Rather, it seems to precede the widely recognized Choteč event. The Choteč event is recorded in the Appalachian Basin by widespread dark, organic-rich shales in the Nedrow Member of the Onondaga formation (Ver Straeten & Brett 1995, 2006); however, it does not appear to coincide with a major loss in diversity or turnover of benthic taxa. Indeed, as noted above, there is a strong similarity in faunal content of the lower versus upper Onondaga faunas. Thus, this particular bioevent, while leaving a sedimentological signature in the Appalachian Basin, does not appear to be associated with major extinction or incursion of taxa.

The boundaries of the Stony Hollow and Hamilton faunas are very likely associated with global bioevents, including newly recognized Bakoven and Stony Hollow bioevents. The later Middle Devonian bioevents in the Appalachian Basin are the subject of ongoing study (Baird & Brett 2003; Aboussalam 2003; Aboussalam & Becker 2001; Bartholomew *et al.* 2007; DeSantis *et al.* 2007).

Later Middle Devonian EESU boundaries show a strong association with global events (Figs 10, 11). A major change from Onondaga carbonates to black shales of the Union Springs Formation probably reflects a combination of major eustatic rise and tectonically induced subsidence related to the second tectophase of the Acadian Orogeny (Ettensohn 1987, 1998). This series of events is associated in time with the incursion of Old World Realm taxa from the Canadian Arctic area; this incursion is not limited to the Appalachian Basin, as many of these same taxa appear at about the same time in the Illinois and Michigan Basins (Koch & Day 1996; M. K. DeSantis, unpublished data). A second major change in the Appalachian Basin, corresponding to the Kačák Event, involves the near total loss of the Stony Hollow fauna and its replacement by the diverse and relatively stable Hamilton fauna near the

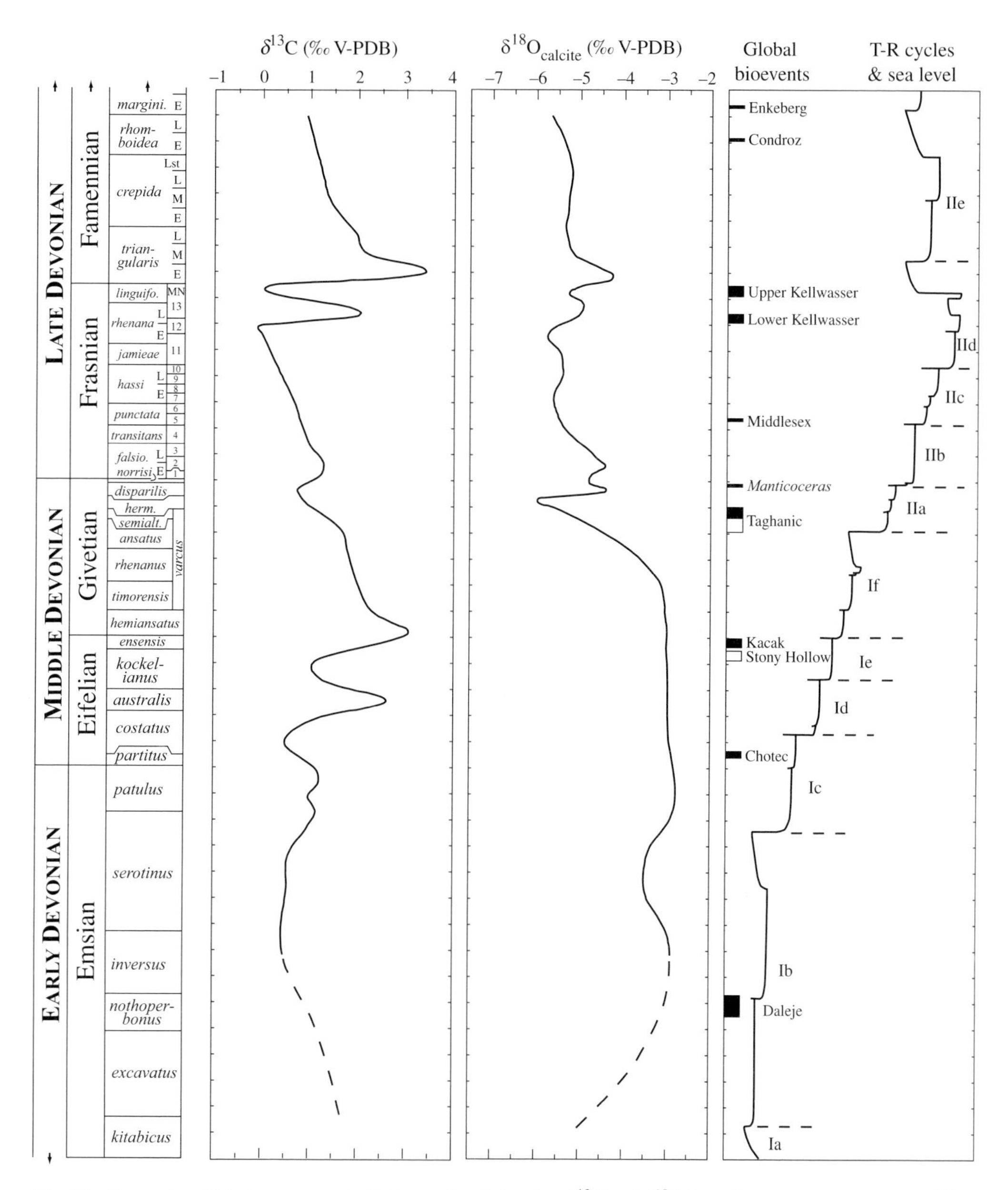

Fig. 19. Composite of biotic events and relative sea level showing δ^{13}C and δ^{18}O isotopic curves. Note positive δ^{13}C excursions associated with both the Stony Hollow and Kacak events and to a minor negative δ^{13}C anomaly at the Taghanic event. Also note negative excursion in δ^{18}O values associated with the Taghanic and suggesting oceanic warming during the late Givetian. Modified from van Geldern *et al.* (2006).

base of the Oatka Creek sequence. The Hamilton fauna, which contains a mixture of endemic Appalachian Basin taxa, mainly holdovers from the Onondaga and newly immigrated species from the Old World Realm Rhenish–Bohemian area (Koch 1979), makes an abrupt appearance low in the Oatka Creek Formation (Halihan Hill Bed; Brett & Baird 1995, 1996).

One signature of the Stony Hollow Event is the widespread appearance of tropical brachiopod taxa in the subtropical to temperate Eastern Americas Realm. Further, there is little evidence for migration going the other way. The only Eastern Americas Realm endemic to reach the Old World Realm Cordilleran Province is *Orthospirifer*, and it is does not occur there until the Late Devonian (J. E. Day pers. comm. 2007). Based on these migration patterns Koch & Boucot (1982), Koch (1983) and Boucot (1990*a*) suggest a marked lowering of the global temperature gradient. Conversely, the end of the Stony Hollow fauna and abrupt incursion of the Hamilton faunas clearly took place in the latest Eifelian time associated with the Kačák or Odershausen events (DeSantis *et al.* 2007).

Oxygen isotope curves (Joachimski *et al.* 2004; van Geldern *et al.* 2006; see Fig. 19) are generally stable around −3‰ throughout most of the Eifelian; however, there are several low amplitude excursions within this interval. Both conodont apatite- and brachiopod calcite-derived curves show a slight (up to 1.5‰) negative excursion in the middle to upper part of the *kockelianus* Zone. Translated into palaeotemperatures, these values suggest an approximate 1 to 2°C warming during the *australis* through upper *kockelianus* Zone interval. This trend is followed by a 2°C cooling, beginning in the upper *kockelianus* Zone and extending into the lower part of the *hemiansatus* Zone.

A strong positive $\delta^{13}C$ excursion (2.1‰) occurs in the latest *costatus* through *australis* Zones (van Geldern *et al.* 2006). This was recorded within material from the Eifel Hills region of Germany and, as such, may be a local phenomenon. However, the excursion is roughly coeval with Bakoven black shales of the Appalachian basin and thus may reflect a global signal. A second, slightly younger, positive $\delta^{13}C$ excursion is associated with the Eifelian–Givetian boundary interval (Kačák Event) in Morocco, as well as in the Barrandian Basin of the Czech Republic (Hladakova *et al.* 1997; Buggisch & Mann 2004) and the Montagne Noire in southern France (Buggisch & Mann 2004).

The late Givetian Taghanic Events (House 1985; Abbousalam 2003) represent a series of relatively significant bioevents that may have had an important impact in decimating long-lasting lineages and indeed may have been as severe as the well studied Frasnian–Famennian event. It is obvious that the crises associated with the end of the Hamilton Group and Tully Formation are not a strictly local effect. Walliser (1986), House (1985) and Aboussalam (2003) have described a major turnover of ammonoids that occurs at approximately this level. They refer to this as the *Pharciceras* or upper Taghanic bioevent and it coincides with the Taghanic transgression in New York State, which ushers in the black shales of the Genesee Formation. This occurs slightly below the Givetian–Frasnian boundary. It is yet another example of a bioevent that coincides with a black shale incursion in the Appalachian Basin, a major transgression and probably altered water mass characteristics. In large measure, it appears to represent decimation of an older Hamilton fauna without complete replacement. Boucot (1990*b*) considers the Taghanic event to represent the disappearance of the Eastern Americas Faunal Realm and development of a lowered global climatic gradient.

Oxygen isotope curves of Joachimski *et al.* (2004) and van Geldern *et al.* (2006) show evidence for a very strong warming trend in the late Givetian (Fig. 19). Brachiopod calcite-derived curves show a strong (up to 4‰) negative excursion, while conodont apatite yields a less pronounced (−1.5 to 2 ‰) but sharp excursion in these values that suggests an approximate 5 to 10°C warming during the *ansatus* through *hermanni* Zone interval, followed by cooling in the latest Givetian. Carbon isotopic studies show a negative trend in $\delta^{13}C$ during this time with a strong negative spike at approximately the *dengleri–disparilis* zone (van Geldern *et al.* 2006), possibly indicative of lowered productivity, despite elevated sea levels and widespread hypoxia at this time. This lowered productivity could be associated with the strong warming trend.

In summary, late Givetian bioevents that brought about the demise of a majority of the Hamilton fauna and ushered in the Ithaca fauna in the Appalachian Basin were associated with rather abrupt climatic changes, together with oceanographic changes of a global nature. Strong warming is associated with the Taghanic event and an abrupt cooling was associated with the subsequent Frasne event.

Summary

A total of nine ecological–evolutionary subunits are recognized in the Early to Mid-Devonian interval of the Appalachian Basin. These are distinctive groupings of taxa that show prolonged concurrent persistence. The boundaries of these EESUs are abrupt and typically occur within a single small-scale (fifth order) cycle. Cluster analysis of successive formations within the Devonian of the Appalachian Basin recognizes discrete groupings based on shared faunal similarities. These results are consistent with the predictions of co-ordinated stasis, that a large number of lineages persist concurrently for a period of 1 to several million years and then show synchronous local extermination, evolution, and/or emigration at times of ecological restructuring.

Most of the EESU boundaries occur within early transgressions to highstand intervals rather than at sequence boundaries. A majority are coincident with global bioevents. Causes of turnovers remain poorly understood. However, common features are an association with sea-level fluctuation, especially rises, widespread anoxia, and evidence for climatic change, especially warming and oceanographic fluctuations as recorded in carbon isotopic anomalies.

Initial studies indicate that EESU boundaries are synchronous in different biogeographic provinces, but much more study is required to document details of these stable blocks in different areas and the nature of local faunal turnovers. Future research is required to compare quantitatively the extent of species-level persistence and the amount of change at EESU boundaries among different basins and biogeographically isolated regions. Ultimately, these studies may lead to a better understanding of the dynamic interactions of the biosphere and other Earth systems and the causes of stability and extinction.

We thank Peter Königshof for organizing this volume and encouraging this paper. The paper benefited by reviews of anonymous referees. Research on this project was supported by NSF grant EAR 9219807 to C. Brett. This paper is a contribution to IGCP 499 'Devonian Land–Sea Interaction: Evolution of Ecosystems and Climate' (DEVEC).

References

ABOUSSALAM, Z. S. 2003. Das Taghanic-Event im höheren Mitteldevon von West-Europa und Marokko. *Münstersche Forschungen zur Geologie und Paläontologie*, **97**.

ABOUSSALAM, Z. S. & BECKER, R. T. 2001. Prospects for an Upper Givetian substage. *Mitteilungen Museum Naturkunde Berlin, Geowissenschaftliche Reihe*, **4**, 83–99.

ABOUSSALAM, Z. S., BECKER, R. T. & SCHULTZ, H. P. 2001. The global Taghanic biocrisis in the Upper Givetian (Middle Devonian). *15th Senckenberg Conference: Mid-Palaeozoic Bio- and Geodynamics, The North Gondwana–Laurussia Interactions*. Frankfurt am Main, Germany, Abstracts, **1**.

ABOUSSALAM, Z. S., BECKER, R. T., BOCKWINKEL, J. & EBBINGHAUSEN, V. 2004. Givetian biostratigraphy and facies development at Oufrane (Tata region, eastern Dra Valley, Morocco). *In*: EL HASSANI, A. (ed.) *Devonian of the Western Anti Atlas: Correlations and Events*. Documents de l'Institute Scientific, Rabat, **19**, 53–59.

ANDERSON, E. J. & BRETT, C. E. *ET AL*. 1986. Upper Silurian to Middle Devonian stratigraphy and depositional controls, east-central New York. *In*: LANDING, E. (ed.) *The Canadian Paleontology and Biostratigraphy Seminar, New York State Museum Bulletin*, **462**, 111–134.

BAIRD, G. C. & BRETT, C. E. 2003. Taghanic Stage shelf and off shelf deposits in New York and Pennsylvania: Faunal incursions, eustasy, and tectonics. *Proceedings of 15th Senckenberg Conference*. Courier Forschungsinstitut Senckenberg, **242**, 141–156.

BAIRD, G. C. & BRETT, C. E. 2005. Timing of onset of Middle Devonian Taghanic bioevents relative to sea level changes, east-central New York and central Pennsylvania. *PaleoBios*, **25**, 17.

BAIRD, G. C., BRETT, C. E. & BARTHOLOMEW, A. J. 2003. Middle Devonian biotic and sedimentation events in east-central New York: Tully Formation clastic correlative succession in the Sherburne-Oneonta area. *In*: JOHNSON, E. L. (ed.) *New York State Geological Association 75th Annual Meeting, Field Trip Guidebook*, Oneonta, NY, 1–54.

BARTHOLOMEW, A. J. 2006. *Middle Devonian faunas of the Michigan and Appalachian basins: Comparing patterns of biotic stability and turnover between two paleobiogeographic subprovinces*. PhD Dissertation, University of Cincinnati.

BARTHOLOMEW, A. J., SCHRAMM, T. & SCHEMM-GREGORY, M. 2007. Faunal turnover between two EESU: Investigating the timing of large-scale faunal turnover in the latest Eifelian of eastern North America. *Geological Society of America, Abstracts with Programs*, **39**, 565.

BAUMILLER, T. K. 1996. Exploring the pattern of coordinated stasis: Simulations and extinction scenarios. *Paleogeography, Palaeoclimatology, Palaeoecology*, **127**, 135–146.

BENNINGTON, J. B. & BAMBACH, R. K. 1996. Statistical testing for paleocommunity recurrence: Are similar fossil assemblages ever the same? *Palaeogeography, Palaeoclimatology, Palaeoecology*, **127**, 107–133.

BONELLI, J. R. JR., BRETT, C. E., MILLER, A. I. & BENNINGTON, J. B. 2006. Testing for faunal stability across a regional biotic transition: Quantifying stasis and variation among recurring coral biofacies in the Middle Devonian Appalachian Basin. *Paleobiology*, **32**, 20–37.

BORDEAUX, Y. L., THAYER, C. W. & BRETT, C. E. 1990. Escalating bioturbation in the Devonian: Global precursor to the Mesozoic? *Geological Society of America Abstracts with Programs*, **22**(7), A357.

BOUCOT, A. J. 1959. Brachiopods of the Lower Devonian Rocks at Highland Mill, New York. *Journal of Paleontology*, **33**(5), 727–769.

BOUCOT, A. J. 1975. *Evolution and Extinction Rate Controls*. Elsevier, Amsterdam.

BOUCOT, A. J. 1981. *Principles of Benthic Marine Paleoecology*. Academic Press, New York.

BOUCOT, A. J. 1982. Ecostratigraphic framework for the Lower Devonian of the North American Appohimchi Subprovince. *Neues Jahrbuch für Geologie und Paläontologie, Abhandlungen*, **163**, 81–121.

BOUCOT, A. J. 1983. Does evolution take place in an ecological vacuum? II. *Journal of Paleontology*, **57**, 1–30.

BOUCOT, A. J. 1988. Devonian biogeography, an update. *In*: MACMILLAN, N. J., EMBRY, A. F. & GLASS, D. J. (eds) *Devonian of the World; Proceedings of the Second International Symposuium on the Devonian System; Volume III: Paleontology, Paleoecology and Biostratigraphy*. Canadian Society of Petroleum Geologists, Memoir, **14**, 211–227.

BOUCOT, A. J. 1990*a*. Silurian and pre-Upper Devonian bioevents. *In*: KAUFFMAN, E. G. & WALLISER, O. H. (eds) *Extinction Events in Earth History*. Springer, Berlin, Lecture Notes in Earth Sciences, **30**, 125–132.

BOUCOT, A. J. 1990*b*. Phanerozoic extinctions: How similar are they to each other? *In*: KAUFFMAN, E. G. & WALLISER, O. H. (eds) *Extinction Events in Earth History*. Springer, Berlin, Lecture Notes in Earth Sciences, **30**, 5–30.

Boucot, A. J. & Lawson, J. D. 1999. *Paleocommunities: A Case Study from the Silurian and Lower Devonian*. Cambridge University Press, Cambridge.

Boucot, A. J. & Rehmer, J. 1977. *Pacificocoelia acutiplicata* (Conrad, 1841) (Brachiopoda) from the Esopus Shale (Lower Devonian) of eastern New York. *Journal of Paleontology*, **51**, 1123–1132.

Boucot, A. J., Gauri, K. L. & Southard, J. 1970. Silurian and Lower Devonian brachiopods of the Green Pond outlier in southeastern New York. *Palaeontographica*, **135**, 1–59.

Brett, C. E. & Baird, G. C. 1985. Carbonate shale cycles in the Middle Devonian of New York: An evaluation of models for the origin of limestones in terrigenous shelf sequences. *Geology*, **13**, 324–327.

Brett, C. E. & Baird, G. C. 1986. Symmetrical and upward shallowing cycles in the Middle Devonian of New York: Implications for the punctuated aggradational cycle hypothesis. *Paleoceanography*, **1**, 1–16.

Brett, C. E. & Baird, G. C. 1994. Depositional sequences, cycles and foreland basin dynamics in the late Middle Devonian (Givetian) of the Genesee Valley and western Finger Lakes Region. *In*: Brett, C. E. & Scatterday, J. (eds) *Field Trip Guidebook – New York State Geological Association, 66th Annual Meeting*. Rochester, New York, 505–586.

Brett, C. E. & Baird, G. C. 1995. Coordinated stasis and evolutionary ecology of Silurian to Middle Devonian marine biotas in the Appalachian basin. *In*: Erwin, D. & Anstey, R. (eds) *New Approaches to Speciation in the Fossil Record*. Columbia University Press, New York, 285–315.

Brett, C. E. & Baird, G. C. 1996. Middle Devonian sedimentary cycles and sequences in the northern Appalachian Basin. *In*: Witzke, B. J., Ludvigson, G. A. & Day, J. (eds) Paleozoic sequence stratigraphy, views from the North American Continent. *Geological Society of America Special Paper*, **306**, 213–241.

Brett, C. E., Baird, G. C. & Miller, K. B. 1990. A temporal hierarchy of paleoecological processes within a Middle Devonian epeiric sea. *Special Publication (Paleontological Society), University of Tennessee*, **5**, 178–209.

Brett, C. E., Boucot, A. J. & Jones, B. 1993. Absolute depths of Silurian benthic assemblages. *Lethaia*, **26**, 25–40.

Brett, C. E., Ivany, L. C. & Schopf, K. M. 1996. Coordinated stasis: An overview. *Palaeogeography, Palaeoclimatology, Palaeoecology*, **127**, 1–20.

Brett, C. E., Baird, G. C. & Bartholomew, A. J. 2004. Sequence stratigraphy of highly variable Middle Devonian strata in central Kentucky: Implications for regional correlations and depositional environments. *In*: Schieber, J. (ed.) *Devonian Black Shales of the Eastern U.S.: New Insights into Sedimentology and Stratigraphy from the Subsurface and Outcrops in the Illinois and Appalachian Basins*. Field guide for the 2004 Annual Field Conference of the Great Lakes Section of SEPM, Indiana Geological Survey Open-File Report, **04–05**: 35–60.

Brett, C. E., Bartholomew, A. J. & Baird, G. C. 2007. Biofacies recurrence in the Middle Devonian of New York State: An example with implications for habitat tracking. *Palaios*, **22**, 306–324.

Brice, D., Bultynck, P., Deunff, J., Loboziak, S. & Streel, M. 1978. Donnees biostratigraphiques nouvelles sur le Givetien et le Frasnien de Ferques (Boulonnais, France). *Annales—Societe Geologique du Nord*, **98**, 325–344.

Budil, P. 1995. Demonstrations of the Kacak Event (Middle Devonian, uppermost Eifelian) at some Barrandian localities. *Věstnik Českého Geologického ústavu*, **70**, 1–24.

Buehler, E. J. & Tesmer, I. H. 1963. Geology of Erie County. *Buffalo Society of Natural Sciences Bulletin*, **21**(3).

Buggisch, W. & Joachimski, M. M. 2006. Carbon isotope stratigraphy of the Devonian of central and southern Europe. *Palaeogeography, Palaeoclimatology, Palaeoecology*, **240**, 68–88.

Buggisch, W. & Mann, U. 2004. Carbon isotope stratigraphy of Lochkovian to Eifelian limestones from the Devonian of central and southern Europe. *International Journal of Earth Sciences*, **93**, 521–541.

Bultynck, P. L. 1976. Comparative study of middle Devonian conodonts from North Michigan (U.S.A.) and the Ardennes (Belgium–France). *Geological Association of Canada, Special Paper*, **15**, 119–141.

Bultynck, P. & Hollard, H. 1980. Distribution comparée de conodonts et goniatites dévoniennes des plaines du Dra, du Ma'der et du Tafilalt (Maroc). *Aardkundige Mededelingen*, **1**, 1–73.

Bultynck, P. & Walliser, O. H. 2000. Devonian boundaries in the Moroccan Anti-Atlas. *Courier Forschungsinstitut Senckenberg*, **225**, 211–226.

Carls, P. 1988. The Devonian of Celtiberia (Spain) and Devonian paleogeography of SW Europe. *In*: McMillan, N. J., Embry, A. F. & Glass, D. J. (eds) *Devonian of the World; Proceedings of the Second International Symposium on the Devonian System; Volume I: Regional Syntheses*. Canadian Society of Petroleum Geologists, Memoir, **14**, 421–466.

Chlupáč, I. 1998. Prague Basin; Devonian. *In*: Chlupáč, I., Havlíček, V., Kříž, J., Kukal, Z. & Štorch, P. (eds) *Palaeozoic of the Barrandian (Cambrian to Devonian)*, Czech Geological Survey, Prague, 101–133.

Cooper, G. A. 1929. *Stratigraphy of the Hamilton Group*. PhD Dissertation, Yale University.

Cooper, G. A. 1930. Stratigraphy of the Hamilton Group of New York: *American Journal of Science*, **19**, 116–134, 214–236.

Cooper, G. A. 1933. Stratigraphy of the Hamilton Group of Eastern New York. *American Journal of Science*, **26**, 537–551; **27**, 1–12.

Cooper, G. A. & Williams, H. S. 1935. Tully Formation of New York. *Geological Society of America, Bulletin*, **46**, 781–868.

Crick, R. E., Ellwood, B. B., El Hassani, A., Feist, R. & Hladil, J. 1997. MagnetoSusceptibility Event and Cyclostratigraphy (MSEC) of the Eifelian–Givetian GSSP and associated boundary sequences in North Africa and Europe. *Episodes*, **20**, 167–175.

DESANTIS, M. K., BRETT, C. E. & VER STRAETEN, C. A. 2007. Persistent depositional sequences and bioevents in the Eifelian (early Middle Devonian) of eastern Laurentia: North America. Evidence of the Kacak Events? *In*: BECKER, R. T. & KIRCHGASSER, W. T. (eds) *Devonian Events and Correlations*. Geological Society, London, Special Publications, **278**, 83–104.

DUTRO, J. T. JR. 1981. Devonian brachiopod biostratigraphy of New York State. *In*: OLIVER, W. A. JR. & KLAPPER, G. (eds) *Devonian Biostratigraphy of New York. Part 1 Text*. International Union of Geological Sciences, Subcommission on Devonian Stratigraphy, Washington, DC, 67–82.

EHLERS, G. M. & KESLING, R. V. 1970. *Devonian Strata of Alpena and Presque Isle Counties, Michigan*. Guidebook for the Northeastern Sectional Meeting of the Geological Society of America, Michigan State University.

ETTENSOHN, F. R. 1987. Rates of relative plate motion during the Acadian Orogeny based on spatial distribution of black shales. *Journal of Geology*, **95**, 572–582.

ETTENSOHN, F. R. 1998. Compressional tectonic controls on epicontinental black shale deposition: Devonian-Mississippian examples from North America. *In*: SCHIEBER, J., ZIMMERLE, W. & SETHI, P. S. (eds) *Shales and Mudstones, vol. 1 (Basin Studies, Sedimentology and Paleontology)*. E. Schweizerbart' sche Verlagsbuchhandlung (Nagele u. Obermiler), Stuttgart, 109–128.

ETTENSOHN, F. R., BARRON, L. S., DILLMAN, S. B., ELAM, T. D., GELLER, K. L., MARKOWITZ, G., MILLER, M. L., SWAGER, D. R. & WOOCK, R. D. 1988. Characterization and implications of the Devonian–Mississippian black-shale sequence, eastern and central Kentucky, USA; pycnoclines, transgression, regression and tectonism. *In*: MCMILLAN, N. J., EMBRY, A. F. & GLASS, D. J. (eds) *Devonian of the World; Proceedings of the Second International Symposium on the Devonian System; Volume II: Sedimentation*. Canadian Society of Petroleum Geologists, Memoir, **14**, 323–345.

FEIST, R. 1983. *The Devonian of the Eastern Montagne Noire (France)*. International Union of Geological Sciences, Subcommission on Devonian Stratigraphy Guidebook, Field Meeting Montagne Noire.

FEIST, R. 1985. Devonian stratigraphy of the southeastern Montagne Noire (France). *Courier Forschungsinstitut Senckenberg*, **75**, 331–352.

FEIST, R. 1990. *The Frasnian–Famennian Boundary and Adjacent Strata of the Eastern Montagne Noire, France*. Field Meeting Guidebook, International Subcommission on Devonian Stratigraphy, Montpellier.

FEIST, R. & KLAPPER, G. 1985. Stratigraphy and conodonts in pelagic sequences across the Middle-Upper Devonian boundary, Montagne Noire, France. *Palaeontographica. Abteilung A: Palaeozoologie-Stratigraphie*, **188**, 1–18.

FUCHS, A. 1987. Conodont biostratigraphy of the Elbingerode reef complex, Harz Mountains. *Acta Geologica Polonica*, **37**, 33–50.

GARCÍA-ALCALDE, J. L. 1998. Devonian events in northern Spain. *Newsletters in Stratigraphy*, **36**, 157–175.

GARCÍA-ALCALDE, J. L. & SOTO, F. 1999. El límite Eifeliense/Givetiense (Devónico Medio) en la Cordillera Cantábrica (N de España). *Revista Española de Paleontología*, (unnumbered) Homenaje al Prof. J. Truyols, 43–56.

GARCÍA-LÓPEZ, S., SANZ-LÓPEZ, J. & SARMIENTO, G. N. 2002. The Paleozoic succession and conodont biostratigraphy of the section between Cape Penas and Cape Torres (Cantabrian coast, NW Spain). *Cuadernos del Museo Geominero*, **1**, 125–162.

GARCÍA-LÓPEZ, S. & SANZ-LÓPEZ, J. 2002. Devonian and Lower Carboniferous conodont biostratigraphy of the Bernesga Valley section (Cantabrian Zone, NW Spain). *Cuadernos del Museo Geominero*, **1**, 163–206.

GOLDRING, W. 1935. Geology of the Berne Quadrangle. *New York State Museum Bulletin*, **303**, 238.

GOLDRING, W. 1943. Geology of the Coxackie Quadrangle, NY. *New York State Museum Bulletin*, **332**, 374.

GRABAU, A. W. 1906. Geology and paleontology of the Schoharie Valley. *New York State Museum Bulletin*, **92**, 386.

GRIFFING, D. H. & VER STRAETEN, C. A. 1991. Stratigraphy and depositional environments of the Marcellus Formation (Middle Devonian) in eastern New York. *In*: EBERT, J. R. (ed.) *Fieldtrip Guidebook, New York State Geological Association 63rd Annual Meeting*, 205–249.

HALL, J. 1867. *Descriptions and Figures of the Fossil Brachiopoda of the Upper Helderberg, Hamilton, Portage, and Chemung Groups, Natural History of New York Part VI: Paleontology, Vol. 4 Parts I and II*, Van Benthuysen and Sons, Albany.

HECKEL, P. H. 1973. *Nature, Origin, and Significance of the Tully Limestone*. Geological Society of America, Special Paper, **139**.

HLADIKOVA, J., HLADIL, J. & KIBEK, B. 1997. Carbon and oxygen isotope record across Pridoli to Givetian stage boundaries in the Barrandian basin (Czech Republic). *Palaeogeography, Palaeoclimatology, Palaeoecology*, **132**, 225–241.

HOLLAND, S. M. 1996. Recognizing artifactually generated coordinated stasis: Implications of numerical models and strategies for field tests. *Palaeogeography, Palaeoclimatology, Palaeoecology*, **127**, 147–156.

HOUSE, M. 1981. Lower and Middle Devonian goniatite biostratigraphy. *In*: OLIVER, W. A. JR. & KLAPPER, G. (eds) *Devonian Biostratigraphy of New York. Part 1 Text*. International Union of Geological Sciences, Subcommission on Devonian Stratigraphy, Washington, DC, 3–37.

HOUSE, M. R. 1985. Correlation of Mid-Paleozoic ammonoid evolutionary events with global sedimentary perturbations. *Nature*, **313**, 17–22.

ISAACSON, P. E. & PERRY, D. G. 1977. Biogeography and morphological conservatism of *Tropidoleptus* (Brachiopoda, Orthida) during the Devonian. *Journal of Paleontology*, **51**, 1108–1122.

IVANY, L. C. 1996. Coordinated stasis or coordinated turnover? Exploring intrinsic vs. extrinsic controls on pattern: *Palaeogeography, Palaeoclimatology, Palaeoecology*, **127**, 239–256.

JOACHIMSKI, M. M., VAN GELDERN, R., BREISIG, S., BUGGISCH, W. & DAY, J. 2004. Oxygen isotope evolution of biogenic calcite and apatite during the Middle and Late Devonian. *International Journal of Earth Sciences (Geol. Rundschau)*, **93**, 542–553.

JOHNSON, J. G., KLAPPER, G. & SANDBERG, C. A. 1985. Devonian eustatic fluctuations in Euroamerica. *Geological Society of America Bulletin*, **96**, 567–587.

KAUFMAN, B. 2006. Calibrating the Devonian timescale: A synthesis of U–Pb ID–TIMS ages and conodont stratigraphy. *Earth-Science Reviews*, **76**, 175–190.

KIRCHGASSER, W. T. 2000. Correlation of stage boundaries in the Appalachian Devonian, Eastern United States. *Courier Forschungsinstitut Senckenberg*, **225**, 271–284.

KLAPPER, G. 1981. Review of New York Devonian conodont biostratigraphy. *In*: OLIVER, W. A. JR. & KLAPPER, G. (eds) *Devonian Biostratigraphy of New York. Part 1 Text*. International Union of Geological Sciences, Subcommission on Devonian Stratgraphy, Washington, DC, 55–66.

KLAPPER, G. & BARRICK, J. E. 1983. Middle Devonian (Eifelian) conodonts from the Spillville Formation in northern Iowa and southern Minnesota. *Journal of Paleontology*, **57**, 1212–1243.

KLEFFNER, M. A., BARRICK, J. E., EBERT, J. R. & MATTESON, D. K. 2006. Conodont biostratigraphy, $\delta^{13}C$ chemostratigraphy, and recognition of Silurian/Devonian boundary in the Appalachian basin at Cherry Valley, New York. *Geological Society of America, Abstracts with Programs*, **38**, 146.

KLEINEBRINKER, G. 1992. Conodonten–Stratigraphie, Mikrofacies und Inkohlung im Mittel- und Ober-Devon des Bergischen Landes. *Geologisches Institut der Universität Köln, Sonderveröffentlichung*, **85**, 99 S.

KLUESSENDORF, J. J., CARMAN, M. R. & MIKULIC, D. G. 1988. Distribution and depositional environments of the westernmost Devonian rocks in the Michigan Basin. *In*: MCMILLAN, N. J., EMBRY, A. F. & GLASS, D. J. (eds) *Devonian of the World; Proceedings of the Second International Symposium on the Devonian System; Volume I: Regional Syntheses*. Canadian Society of Petroleum Geologists, Memoir, **14**, 251–264.

KLUG, C. R. 1983. Conodonts and biostratigraphy of the Muscatatuck Group (Middle Devonian), south-central Indiana and north-central Kentucky. *Transactions of the Wisconsin Academy of Sciences, Arts and Letters*, **71**, 79–112.

KOCH, W. F. II. 1979. *Brachiopod paleoecology, paleobiogeography and biostratigraphy in the upper Middle Devonian of eastern North America: An ecofacies model for the Appalachian, Michigan, and Illinois basins*. PhD Dissertation, Oregon State University.

KOCH, W. F. II. 1983. Late Eifelian paleobiogeographic boundary fluctuations in North America. *Geological Society of America Abstracts with Programs*, **15**, 171.

KOCH, W. F. II & BOUCOT, A. J. 1982. Temperature fluctuations in the Devonian Eastern Americas Realm. *Journal of Paleontology*, **56**, 240–243.

KOCH, W. F. & DAY, J. 1996. Late Eifelian–Early Givetian (Middle Devonian) brachiopod paleobiogeography of eastern and central North America. *In*: COPPER, P. (ed.) *Proceedings of the Third International Brachiopod Congress*, 135–143.

LAPORTE, L. 1969. Recognition of a transgressive carbonate sequence within an epeiric sea: Helderberg Group (Lower Devonian) of New York State. *In*: FREEMAN, G. M. (ed.) *Depositional Environments in Carbonate Rocks: a Symposium*. Society of Economic Paleontologists and Mineralogists, Special Publications, **14**, 98–119.

LINSLEY, D. M. 1994. *Devonian Paleontology of New York*. Paleontological Research Institution, Special Volume, **21**.

MCGHEE, G. 1988. Evolutionary dynamics of the Frasnian–Famennian extinction event. *In*: MACMILLAN, N. J., EMBRY, A. F. & GLASS, D. J. (eds) *Devonian of the World: Proceedings of the Second International Symposuium on the Devonian System*. Canadian Society of Petroleum Geologists, Memoir, **14**, 23–28.

MCGHEE, G. 1990. The Frasnian–Famennian mass extinction record in eastern United States. *In*: KAUFFMAN, E. G. & WALLISER, O. H. (eds) *Extinction Events in Earth History*. Springer, Berlin, Lecture Notes in Earth Sciences, **30**, 161–168.

MCINTOSH, G. C. 1983. Crinoid and blastoid biogeography in the Middle Devonian (Givetian) of eastern North America. *Geological Society of America Abstracts with Programs*, **15**, 171.

MAY, A. & BECKER, R. T. 1996. Ein Korallen-Horizont im Unteren Bänderschiefer (höchstes Mittel-Devon) von Hohenlimburg-Elsey im Nordsauerland (Rheinisches Schiefergebirge). *Berliner Geowissenschaftliche Abhandlungen*, **E18**, 209–241.

MILLER, A. I. 1997. Coordinated stasis or coincident relative stability? *Paleobiology*, **23**(2), 155–164.

MORZADEC, C. & WEYANT, M. 1982. Lithologie et Conodontes, de l'Emsien au Famennien, dans la rade de Brest (Massif Armoricain): Lithology and conodonts of the Emsian to Famennian of Brest Roads, Armorican Massif. *Geologica et Palaeontologica*, **15**, 27–46.

ORR, R. W. 1964. Conodonts from the Devonian Lingle and Alto formations of southern Illinois. *Illinois State Geological Survey, Circular*, **361**, 1–28.

ORR, R. W. 1971. Conodonts from Middle Devonian Strata of the Michigan Basin. *Indiana Geological Survey, Bulletin*, **45**.

ORR, R. W. & POLLOCK, C. A. 1968. Reference sections and Correlation of Beechwood Member (North Vernon Limstone, Middle Devonian) of Southern Indiana and Northern Kentucky. *American Association of Petroleum Geologists, Bulletin*, **52**, 2257–2262.

PATZKOWSKY, M. & HOLLAND, S. M. 1997. Patterns of turnover in Middle and Upper Ordovician brachiopods of the eastern United States: A test of coordinated stasis. *Paleobiology*, **23**, 420–443.

ROGERS, F. S. 1998. Conodont biostratigraphy of the Little Cedar and lower Coralville formations of the Cedar Valley Group (Middle Devonian) of Iowa and significance of a new species of *Polygnathus*. *Journal of Paleontology*, **72**, 726–737.

SANDBERG, C. A., HASENMUELLER, N. R. & REXROAD, C. B. 1994. Conodont biochronology, biostratigraphy, and biofacies of the Upper Devonian Part of New Albany Shale, Indiana. *Courier Forschungsinstitut Senckenberg*, **168**, 227–253.

SANFORD, B. V. 1968. Devonian of Ontario and Michigan. *International Symposium on the Devonian System, Calgary, 1967 [Proc.]*, **1**, 973–999.

SCHÖNE, B. R. 1997. Der *otomari*-Event und seine Auswirkungen auf die Fazies des Rhenherzynischen Schelfs (Devon Rheinisches Schieferegeberge), *Göttinger Arbeiten zur Geologie und Paläontologie*, **70**, 1–140.

SCHONLAUB, H. P. 1985. Devonian conodonts from section Oberbuchach II in the Carnic Alps (Austria). *Courier Forschungsinstitut Senckenberg*, **75**, 353–374.

SCHUMACHER, D. 1971. Conodonts from the Middle Devonian Lake Church and Milwaukee Formations. *In*: Conodonts and Biostratigraphy of the Wisconsin Paleozoic. *Wisconsin Geological and Natural History Survey, Information Circular*, **19**, 55–67, 90–99.

SESSA, J., BRETT, C. E., MILLER, A. I. & BAIRD, G. C. 2002. The dynamics of rapid, asynchronous biotic turnover in the Middle Devonian Appalachian Basin of New York. *Geological Society of America, Abstracts with Programs*, **34**(2), A117.

SHEEHAN, P. M. 1985. Reefs are not so different – They follow the evolutionary pattern of level bottom communities. *Geology*, **13**, 46–49.

SLOSS, L. L. 1963. Sequences in the cratonic interior of North America. *Geological Society of America Bulletin*, **74**, 93–114.

SPALETTA, C. & PERRI, M. C. 1998. Stop 2.1A—Givetian conodonts from the Poccis section (Carnic Alps, Italy). *Giornale di Geologia, serie 3a*, **60**, 184–188.

SWARTZ, C. K. 1913. Correlation of the Lower Devonian. *In*: *Devonian, Lower*. Maryland Geological Survey Publication, Johns Hopkins Press, Baltimore.

THAYER, C. W. 1974. Marine paleoecology in the Upper Devonian of New York. *Lethaia*, **7**, 121–155.

TROTH, I., MARSHALL, J. E. A., RACEY, A. & BECKER, R. T. 2009. Mid Devonian sea-level change at high palaeolatitude: Testing the global sea-level curve. *Palaeogeography, Palaeoclimatology, Palaeoecology*, in press.

TRUYOLS-MASSONI, M., GARCÍA-ALCALDE, J. L., LEYVA, F. & MONTESINOS, R. 1990. The Kacak-*otomari* Event and its characterization in the Palentine Domain (Cantabrian Zone, NW Spain). *In*: KAUFFMAN, E. G. & WALLISER, O. H. (eds) *Extinction Events in Earth History*, Lecture Notes in Earth Sciences, Springer, Berlin, **30**, 133–144.

TUCKER, R. D., BRADLEY, D. C., VER STRAETEN, C. A., HARRIS, A. G., EBERT, J. & MCCUTCHEON, S. R. 1998. New U-Pb zircon ages and the duration and division of Devonian time. *Earth and Planetary Science Letters*, **158**, 175–186.

VAN GELDERN, R. & JOACHIMSKI, M. M. ET AL. 2006. Carbon, oxygen and strontium isotope records of Devonian brachiopod shell calcite. *Palaeogeography, Palaeoclimatology, Palaeoecology*, **240**, 47–67.

VER STRAETEN, C. A. & BRETT, C. E. 1995. Lower and Middle Devonian foreland basin fill in the Catskill Front: Stratigraphic synthesis, sequence stratigraphy, and the Acadian Orogeny. *In*: GARVER, J. I. & SMITH, J. A. (eds) *Fieldtrip Guidebook, New York State Geological Association 67th Annual Meeting*, 313–356.

VER STRAETEN, C. A. & BRETT, C. E. 2006. Pragian to Eifelian (middle Lower to lower Middle Devonian), northern Appalachian Basin—Stratigraphic nomenclatural changes. *Northeastern Geology and Environmental Sciences*, **28**, 80–95.

VER STRAETEN, C. A., GRIFFING, D. H. & BRETT, C. E. 1994. The lower part of the Middle Devonian Marcellus 'shale,' central to western New York State: Stratigraphy and depositional history. *In*: BRETT, C. E. & SCATTERDAY, J. (eds) *Fieldtrip Guidebook, New York State Geological Association 66th Annual Meeting*, 271–306.

VER STRAETEN, C. A., BRETT, C. E. & ALBRIGHT, S. S. 1995. Stratigraphy and paleontolologic overview of the upper Lower and Middle Devonian, New Jersey and adjacent areas. *In*: BAKER, J. E. B. (ed.) *Contributions to the Paleontology of New Jersey, Geological Society of New Jersey Bulletin*, **12**, 229–239.

WALLISER, O. H. 1986. Towards a more critical approach to bio-events. *In*: WALLISER, O. H. (ed.) *Global Bio-events*. Springer, Berlin, Lecture Notes in Earth Sciences, **8**, 5–16.

WALLISER, O. H. 1990. How to define 'Global bio-events'. *In*: KAUFFMAN, E. G. & WALLISER, O. H. (eds) *Extinction Events in Earth History*. Springer, Berlin, Lecture Note in Earth Sciences, **30**, 1–3.

WALLISER, O. H. 1996. Global events in the Devonian and Carboniferous. *In*: WALLISER, O. H. (ed.) *Global Events and Event Stratigraphy*. Springer, Berlin, 225–250.

WALLISER, O. H., BULTNYCK, P., WEDDIGE, K., BECKER, S. M. & HOUSE, M. R. 1995. Definition of the Eifelian–Givetian Stage boundary. *Episodes*, **18**(3), 107–115.

WEARY, D. J. & HARRIS, A. G. 1994. Early Frasnian (Late Devonian) Conodonts from the Harrel Shale, Western Fold- and Thrust Belt, West Virginia, Maryland, and Pennsylvania Appalachians, U.S.A. *Courier Forschungsinstitut Senckenberg*, **168**, 195–225

WEDDIGE, K. 1988. Field Trip A. Eifel Hills. *Courier Forschungsinstitut Senckenberg*, **102**, 103–110, 132–133.

WILLIAMS, H. S. 1913. *Recurrent* Tropidoleptus *Fauna of the Upper Devonian in New York*. United States Geological Survey, Professional Paper, **79**.

WIRTH, M. 1967. Zur Gliederung des hoeheren Palaeozoikums (Givet-Namur) im Gebiet des Quinto Real (Westpyrenaeen) mit Hilfe von Conodonten: The subdivision of the upper Paleozoic (Givetian-Namurian) in the Quinto Real area, western Pyrenees, with the aid of conodonts. *Neues Jahrbuch für Geologie und Paläontologie, Abhandlungen*, **127**, 179–240.

WITZKE, B. J., BUNKER, B. J. & ROGERS, F. S. 1988. Eifelian through lower Frasnian stratigraphy and deposition in the Iowa area, Midcontinent, U.S.A. *In*: McMillan, N. J., Embry, A. F. & Glass, D. J. (eds) *Devonian of the World; Proceedings of the Second International Symposium on the Devonian System; Volume I: Regional Syntheses.* Canadian Society of Petroleum Geologists, Memoir, **14**, 221–250.

Zambito, J. J., Baird, G. C., Brett, C. E. & Bartholomew, A. J. 2007. Examination of the type Ithaca Formation: Correlations with sections in western New York. *In*: McRoberts, C. (ed.) *Field Trip Guidebook, New York State Geological Association, 79th Annual Meeting*, Cortland, NY.

The Siegenian delta: land–sea transitions at the northern margin of the Rhenohercynian Basin

J. STETS & A. SCHÄFER*

Steinmann-Institut für Geologie, Mineralogie und Paläontologie, University of Bonn, Nussallee 8, D-53115 Bonn, Germany

**Corresponding author (e-mail: schaefer@uni-bonn.de)*

Abstract: Analysis, interpretation and reconstruction of different depositional palaeoenvironments in the land–sea transition zone of the huge Lower Devonian siliciclastic delta system at the northern margin of the Rhenohercynian Basin (Central European Variscides, Germany) are the main subjects of this paper. Seven representative columnar sections have been taken at larger outcrops in the Mid-Rhine region. They furnished the main criteria for this analysis. The understanding and interpretation of the different depositional palaeoenvironments have been supported by regional geological lithological and biostratigraphical data.

The sedimentary processes of distal meandering fluvial systems that gradually passed into fluvial-dominated deltaic systems of the elongate bird-foot type have been recognized. Lower delta slope to delta-foot plain conditions have been detected in a subtidal environment probably down to 100 m of former water depth. Thus, the Siegenian delta must have had a low inclined but wide slope down to the basin floor.

Palaeoclimatic conditions at a latitude of about 10 to 30° south, in the basinal as well as in the source area of the Old Red Continent in the north, controlled the input of siliciclastic detritus as well as the depositional processes in the basin. Moreover, steady subsidence and an intermittent sea level played a prominent role during sedimentation. They held this system in balance for more than seven million years.

The area that is the subject of our paper is situated in the northwestern part of the Rheinische Schiefergebirge, which belongs to the Rhenohercynian fold- and-thrust belt within the Central European Variscides (Kossmat 1927; Franke 1992, 2000). This part of the orogen originated from a narrow but elongate mobile belt, nowadays called the Rhenohercynian Basin (Fig. 1). Sedimentation started here in the Gedinnian, but there are very few data available to reconstruct the sedimentary history, palaeoenvironment and palaeogeography of the basin at that time. From the Siegenian to the end of the Lower Carboniferous, i.e. a period of about 85.5 Ma, the basin underwent various stages of development (Franke *et al.* 1978; Meyer & Stets 1980) that can be studied in detail in the different parts of the Schiefergebirge.

One of the features studied is the huge Siegenian siliciclastic delta palaeoenvironment exposed in outcrops near to, and along the Mid-Rhine valley between Bonn and Andernach as well as parallel to the tributaries Sieg and Ahr. Here, a sedimentary succession of the Siegenian up to 3000–5000 m thick is well exposed in various outcrops (Meyer & Stets 1996). They provide a rather complete and representative record from the Early to the Late Siegenian. Thus, the depositional conditions and history of this delta at the northern margin of the Rhenohercynian Basin can be well observed, and distinguished using sedimentological, lithological and palaeontological criteria. We will describe the different siliciclastic palaeoenvironments in this land–sea transition zone, and put it into context with the model of the whole basin that was established by Stets & Schäfer (2002).

Whereas stratigraphic, structural and palaeontological observations and interpretations were the main interest in several thousands of papers published in the last 150 years on the geology of the Rheinische Schiefergebirge, sedimentological research did not start until the 1940s when Bausch van Bertsbergh (1940) used the stimulating ideas of Hans Cloos, Bonn. Not until more than 40 years later, was the idea taken up again, and Lower and Middle Devonian deltaic sedimentation was identified in the northeastern part of the Rheinische Schiefergebirge (Langenstrassen *et al.* 1979; Langenstrassen 1983; Walliser & Michels 1983). Similar conditions have been established for the Early Devonian by Stets & Schäfer (1990), Breil-Schollmayer (1990), Muntzos (1990), Schäfer & Stets (1995), Schindler *et al.* (2001, 2004) and Wehrmann *et al.* (2005). Stets & Schäfer (2002) tied together all the information available, and prepared a model for the whole basin along a NW–SE orientated transect situated in the Mid-Rhine valley from Bingen to Bonn.

From: KÖNIGSHOF, P. (ed.) *Devonian Change: Case Studies in Palaeogeography and Palaeoecology*.
The Geological Society, London, Special Publications, **314**, 37–72.
DOI: 10.1144/SP314.3 0305-8719/09/$15.00

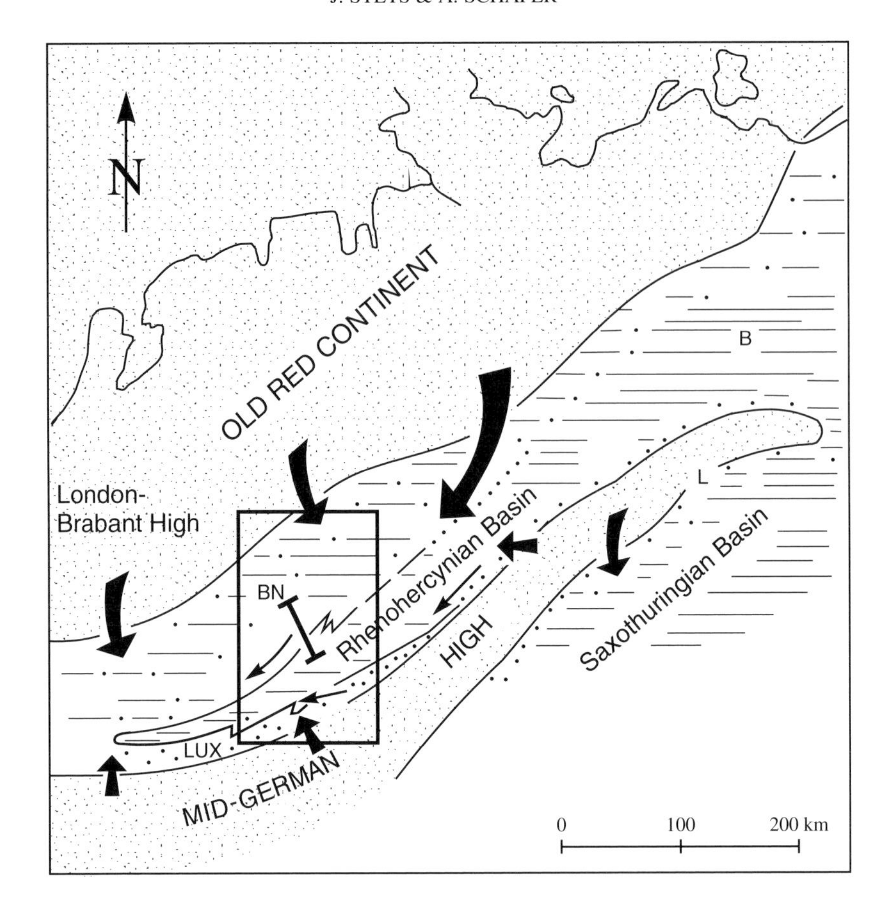

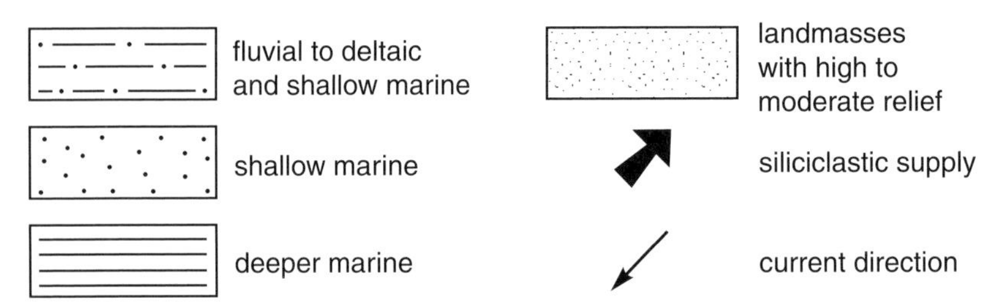

Fig. 1. Palaeogeographic setting of the Rhenohercynian Basin during the Siegenian without palinspastic adjustment (after Ziegler 1990, modified; Stets & Schäfer 2002). The bold line marks the transect under research; the rectangular frame shows the area in Figures 2–3. Abbreviations: B, Berlin; BN, Bonn; L, Leipzig; LUX, Luxembourg.

Studies on the Lower Devonian palaeogeography of the Rhenohercynian Basin were started by Nöring (1939). Later on, concepts on its evolution were published by Kegel (1950), Schmidt (1952), Meyer & Stets (1980), Schievenbusch (1992), Winterfeldt (1994), Winterfeldt *et al.* (1994*a*, *b*), Stroetmann-Heinen (1999) and Oncken *et al.* (2000).

The area of interest underwent relatively low and negligible deformation during the Variscan (Hercynian) orogeny. Therefore, sedimentological features can be well recognized, and thoroughly studied in the field. Our statements on the Siegenian delta focus on the understanding of the different sedimentary palaeoenvironments using lithology, estimated grain size, grain-size distribution, distinctive primary depositional structures and biogenic remains gained in the field as well as from the literature. As the whole sedimentary package exceeds

5000 m, especially in the southern part of the delta, we were forced to select, investigate, and interpret the depositional environments from typical columnar sections taken from larger outcrops. Although these individual sections show only very small segments of the whole sedimentary pile, they represent the sedimentary processes and palaeoenvironments with adequate precision, whenever typical sedimentological units (cycles) within the individual columnar section repeat.

Geological setting

The northwestern part of the Rheinische Schiefergebirge is built up of Lower Devonian sedimentary rocks mostly of Siegenian and Emsian age. These sediments underwent folding, thrusting and shortening during the Variscan (Hercynian) orogeny starting in the late Early Carboniferous. During this process, a huge anticlinal structure originated in the northern central Schiefergebirge called Osteifel Main Anticline (Osteifeler Hauptsattel). It consists of two minor anticlinal structures: the Ahrtal Anticline (Ahrtal-Sattel) in the north, and the Hönningen-Seifen Anticline (Hönningen-Seifen-Sattel) in the south. They are modified by minor folds and thrust faults (Meyer 1958, 1994). All these structures are orientated in a NE–SW direction at about N 50–65°E forming the general strike direction (Fig. 2).

In the core of the Ahrtal and Hönningen-Seifen anticlines, Lower Siegenian sediments are exposed. The thickness of this sedimentary package is more than 570–1000 m (Meyer 1994; LGB 2005). It is overlain by 1300 to more than 2400 m of Middle Siegenian sediments. The large differences in thickness are caused by a huge sedimentary wedge enlarging from the NW towards the centre of the basin in the SE. The Upper Siegenian sediments reach a thickness of about 2000 m at the southeastern flank of the Osteifel Main Anticline. The axis of this huge anticlinal structure plunges towards the SW. Thus, Lower Emsian beds surround this structure from the north towards the west and the south (Fig. 2).

Towards the SE, the Osteifel Main Anticline is cut by a major thrust fault system called the Siegen Main Thrust (Siegener Hauptüberschiebung, Fig. 2). Here, sediments of the Gedinnian and Lower Siegenian in the Hunsrückschiefer facies have been thrust upon Middle and Upper Siegenian sediments towards the NW. The sediments south of the thrust fault belong to the northwestern flank of the huge Mosel Syncline (Mosel-Mulde) having its core near the town of Boppard/Rhine. This synclinal structure is built up of several imbricate structures. It is filled with mostly Lower and Upper Emsian sediments in its core.

General facies distribution and palaeogeography

The Rhenohercynian Basin formed as an extensional belt south of the Old Red Continent. All the molasse-type sediments at the northern margin of this basin have been derived from this source area (Franke *et al.* 1978; Franke 1992; Holl 1995; Wierich 1999; Stets & Schäfer 2002, 2004). They may be assigned to three depositional facies belts that are separated from each other by the predecessors of the Siegen Main Thrust and the Taunuskamm Thrust in the Mid-Rhine valley (Fig. 3). Nowadays, both thrust faults dip steeply at about 80° toward the SE. They are regarded as inverted synrift normal faults incorporated in the syngenetic extension process. During the Early Devonian, these normal faults are thought to have caused areas of different palaeobathymetry close to each other. Thus, three areas of different facies originated (Meyer & Stets 1975, 1980, 1996; Stets & Schäfer 2002) which are from the NW toward the SE (Fig. 3): the Northern Facies Belt, the Central Facies Belt and the Southern Facies Belt. During the Gedinnian and Siegenian, the Rhenohercynian Basin was confined by the Mid-German High in the south which subsided during the Early Emsian. From this time onward, the Southern Facies Belt gradually lost its importance.

The Siegenian delta complex comprises the whole Northern Facies Belt, and northern parts of the Central Facies Belt south of the Siegen Main Thrust. The Northern Facies Belt is built up of the Siegenian Normal Facies (Sandige Siegener Normalfazies: Meyer & Stets 1980, 1996; Meyer 1994). Its sediments are predominantly composed of sand-, silt- and mudstones alternating with each other. They were formed from the siliciclastic detritus directly delivered to the basin from the north or NE (Haverkamp 1991; Holl 1995; Wierich 1999). It was deposited under the changing environmental conditions within the huge delta complex that incorporated fluvial, deltaic to tidal, and fully open marine palaeoenvironments. The influence of marine conditions normally increased towards the south, but transgressions and regressions produced alternating conditions even far within the delta complex. They are proved by the prograding and retreat of typical lithofacies as well as by different faunas and floras developed in the land–sea transition zone.

Meyer & Stets (1975) formulated the idea of a submarine slope just along today's Siegen Main Thrust that caused striking differences in lithofacies between the Siegenian Normal Facies in the north and the adjacent Hunsrückschiefer facies of the Central Facies Belt in the south. Along strike toward the NE and SW, this prominent structural element was overwhelmed towards the south by

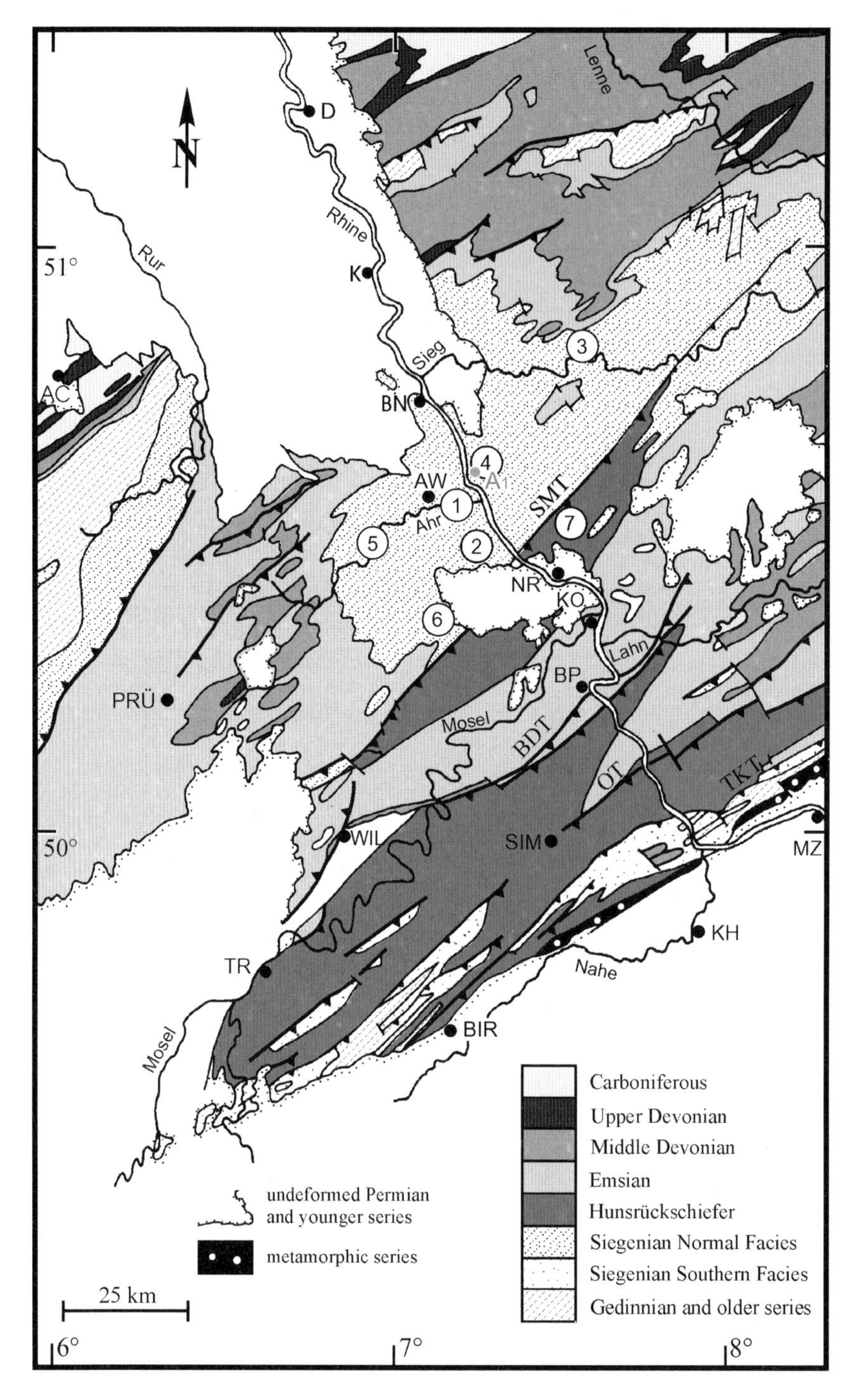

Fig. 2. Simplified geological map of the eastern Ardennes and the western and central Rheinische Schiefergebirge (from Meyer & Stets 1980, modified). Abbreviations: AC, Aachen; AW, Ahrweiler; BDT, Boppard-Dausenau Thrust; BIR, Birkenfeld; BN, Bonn; BP, Boppard; D, Düsseldorf; K, Köln; KH, Kreuznach; KO, Koblenz; MZ, Mainz; NR, Neuwied; OT, Oberwesel Thrust; PRÜ, Prüm/Eifel; SIM, Simmern/Hunsrück; SMT, Siegen Main Thrust; TKT, Taunuskamm Thrust; TR, Trier; WIL, Wittlich/Eifel. Localities of columnar sections: 1, Landskrone; 2, Rheineck; 3, Schladern; 4, Unkel; 5, Altenahr; 6, Bürresheim; 7, Augustenthal; A1, outcrop Kaskade near Unkel.

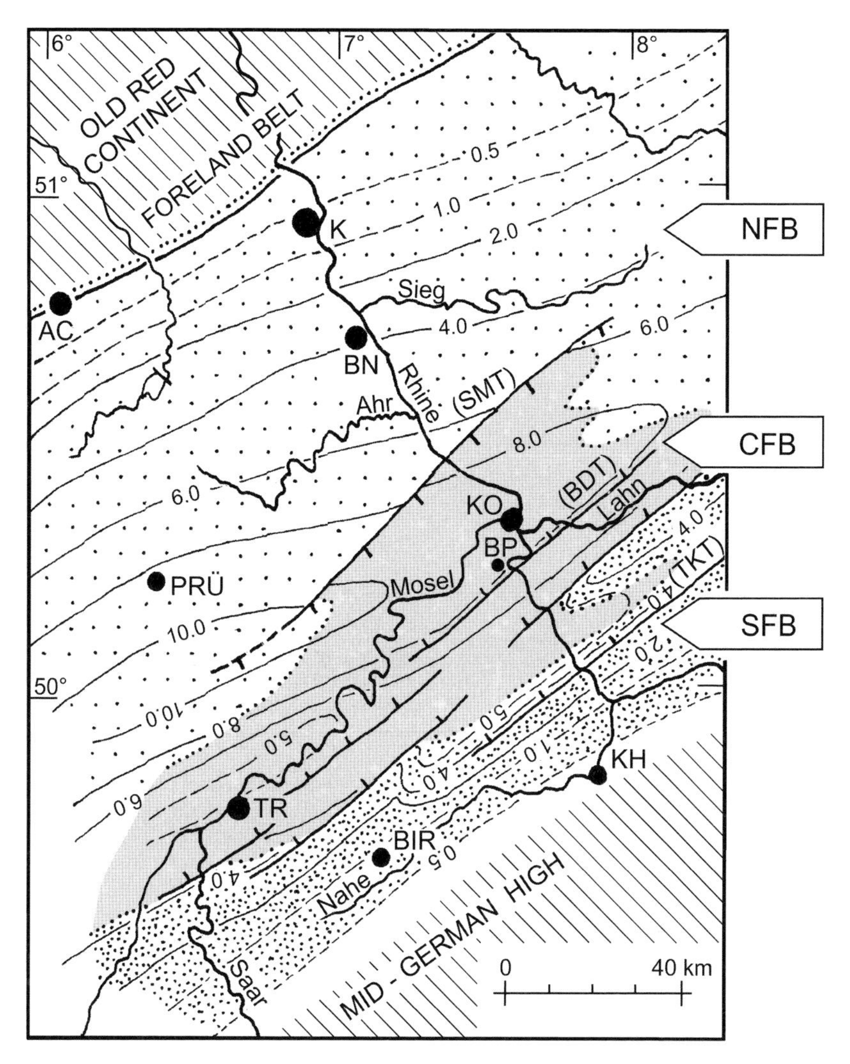

Fig. 3. Simplified palaeogeographical map of the eastern Ardennes and the western and central Rheinische Schiefergebirge showing the three main facies belts during the Early Devonian: the Northern Facies Belt (NFB), Central Facies Belt (CFB) and Southern Facies Belt (SFB). Maximum thickness is given in thousands of metres; hatched areas illustrate ancient landmasses. The map has no palinspastic adjustment (from Stets & Schäfer 2002, modified). Abbreviations: (BDT), later Boppard-Dausenau Thrust; (OT), later Oberwesel Thrust; (SMT), later Siegen Main Thrust; (TKT), later Taunuskamm Thrust; see Figure 2 caption for other abbreviations.

prograding sediments of the Siegenian Normal Facies interfingering with the Hunsrückschiefer in the south (Fig. 3).

The Hunsrückschiefer facies within the Central Facies Belt is built up of predominantly fine-grained silty and clay-rich sediments ranging in time from the Gedinnian (Gad 2005) to the Lower Emsian (Meyer 1965; Meyer & Stets 1980, 1996). They originated under open marine conditions. Close to the Siegen Main Thrust in the north, the sedimentation process was controlled by sediment input that came directly down the delta slope. Pro-delta conditions were installed here during the Siegenian.

During the orogeny, the depositional palaeoenvironments of the two facies belts were tightly stacked together. Palinspastic restoration of this part of the Schiefergebirge (Wunderlich 1964; Schievenbusch 1992; Winterfeld *et al.* 1994*a*, *b*) revealed a shortening of the Northern Facies Belt by about 40% of its former extent. Transitions between the Northern and the Central Facies belts are mostly lacking as they were tectonically suppressed along the Siegen Main

Thrust. Therefore, parts of the lower delta slope can no longer be found today.

Attempts have been made to give rough estimates for the thickness of the sedimentary packages in the facies belts. In the Northern Facies Belt, approximate thicknesses of 3000–5000 m have been estimated for the sediments of the Siegenian Normal Facies within the scope of more than 4000 to about 8000 m for the Lower Devonian as a whole (Meyer & Stets 1980, 1994, 1996; Meyer 1994; Winterfeld *et al.* 1994*a*, *b*; Stets & Schäfer 2002). As the sedimentary conditions remained similar during this time interval of about 7 Ma (DSK 2002), subsidence and detrital input at and near the northern margin of the Rhenohercynian Basin were nearly in balance during all the time under discussion.

Stratigraphic framework

Much work has been done in the past century with respect to the biostratigraphy of the Lower Devonian of the Rheinische Schiefergebirge, for instance by Dahmer (1932, 1934*a*, *b*, 1936, 1937, 1940), Solle (1950, 1951, 1956, 1972), Mittmeyer (1974, 1980, 1982), Jahnke (1971), Carls *et al.* (1982) and Fuchs (1971, 1974, 1982). Nevertheless, good index fossils are rare in the thick Siegenian sedimentary pile except some brachiopods and trilobites. In places, a partial or total absence of fossils can be stated in the Siegenian Normal Facies as well as in the Hunsrückschiefer facies belt. Therefore, a detailed biostratigraphic control has been lacking until today, and age relationships are still open to debate. Nevertheless, the biostratigraphy of the Siegenian sediments of the Normal Facies and of the Hunsrückschiefer facies have been sufficiently defined. Characteristic rock types that may be reliable marker horizons are lacking in both facies belts. Their absence poses difficulties with respect to lithostratigraphy. But using the sand/silt ratio a lithostratigraphical subdivision is possible (Meyer 1958, 1965; Elkholy & Gad 2006).

Siegenian Normal Facies

Within the Siegenian Normal Facies, three subunits in the order of groups (Salvador 1994) were established by the geologists of the former Siegerland iron-ore mining district. They distinguished the Lower Siegen Group (Untere Siegen-Schichten, Tonschiefer-Gruppe), the Middle Siegen Group (Mittlere Siegen-Schichten, Rauhflaser-Gruppe), and the Upper Siegen Group (Obere Siegen-Schichten, Herdorf-Gruppe; Fig. 4). This subdivision into three units, each of them several hundreds of metres thick, can be handled using flora and fauna.

Close to the Rhine river, the Lower Siegen Group consists of sandstones near to the base followed by black shales containing several sandstone layers. Within the black shales, horizons of pure black mudstones occur that contain coal seams and layers rich in relicts of plants ('Pflanzenschiefer', Plant Shale Formation), for instance *Taeniocrada decheniana*, *Zosterophyllum rhenanum*, *Drepanophycus spinaeformis* and *Protolepidodendron wahnbachense*. Others are *Prototaxites* cf. *logani* and *Pachytheca* sp. (Schweitzer 1983, 1987, 2003). Faunal relicts of open marine taxa such as corals, trilobites, crinoids and spirifers are totally lacking. Instead, pteraspids, gigantostrakes, ostracods and bivalves occur. In particular the clam *Modiolopsis* sp., lingulids, and the brachiopod *Crassirensselaeria crassicosta* have often been found. Thus, the dominantly shaly Lower Siegen sediments can best be defined by their lithology. Agnathan 'fishes' occur in places. They range in age up to the Emsian, and *Crassirensselaeria crassicosta* has been found throughout the Siegenian.

The Middle Siegen Group consists of several subunits of sandstones alternating with sandy and silty shales. This unit is also called 'Rauhflaser-Schichten' meaning that within the shales the sand content is confined to small lenses and flasers. In contrast to the Lower Siegen beds, black shales rich in plant relicts and small coal seams ('Pflanzenschiefer') are lacking here. Instead, marine horizons containing the special 'Seifen-Fauna' (Dahmer 1934*a*) occur indicating that open marine conditions prevailed in places. This fauna is distinguished by taxa with thick and large shells such as the Siegenian index fossil *Acrospirifer primaevus* or '*Stropheodonta*', bivalves and corals, such as '*Zaphrentis*' sp. and *Pleurodictyum* sp. Relicts of crinoids have also often been found in the Middle Siegen Group. According to Mittmeyer (1982) and Carls *et al.* (1982) *Boucostrophia herculea*, *Fascistropheodonta sedgwicki*, *Multispirifer solitarius* and *Mauispirifer pterinus* can be used for biostratigraphic purposes in the Middle Siegen Group. Concerning plant remains, the Middle Siegenian is characterized by *Taeniocrada decheniana*, *T. dubia*, *T. longisporangiata*, *Zosterophyllum rhenanum*, *Sciadophyton laxum*, *Drepanophycus spinaeformis* and *Gosslingia cordiformis*. They occur in places where a lithofacies equivalent to that of the Lower Siegenian occurs (Schweitzer 1987, 2003).

The Upper Siegen Group shows two different lithofacies.

(1) North of the Ahrtal Anticline the Wahnbach Formation occurs. It also consists of alternating sand-, silt- and mudstones. They originated under fluvial-deltaic to brackish conditions similar to those of the Lower Siegen Group. Thus, the impoverished fauna is of the *Modiolopsis* type

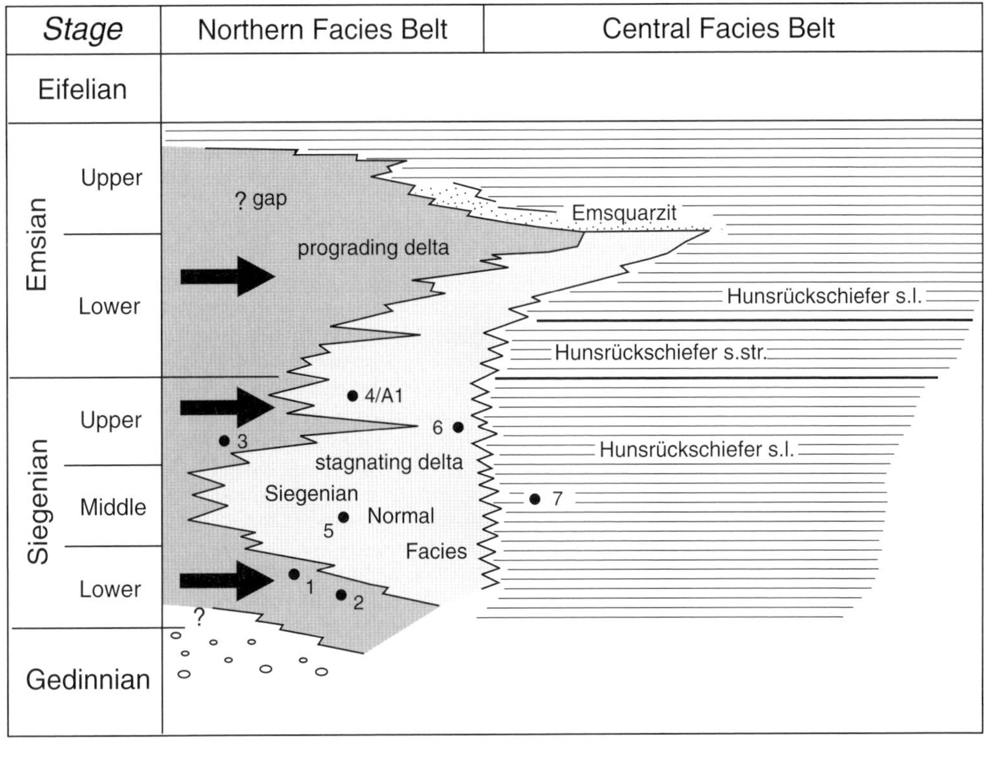

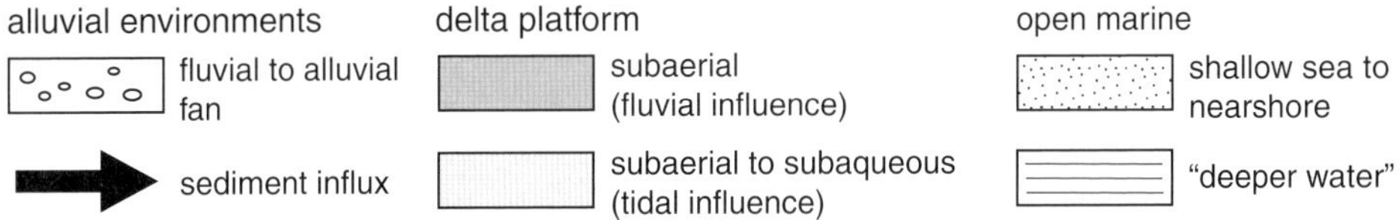

Fig. 4. Diagram showing time and facies relationships of the Lower Devonian northern and central siliciclastic facies belts along the Mid-Rhine valley transect. The stratigraphic position of the columnar sections 1–7 is indicated (from Stets & Schäfer 2002, modified).

(Dahmer 1936). Rich floral findings are described from the famous outcrops in the Wahnbach and Jabach valleys north of Bonn by Steinmann & Elberskirch (1929). Schweitzer (1987) listed the following taxa: *Prototaxites* cf. *logani*, *P. psygmophylloides*, *Pachytheca* sp., *Wahnbachella bostrychioides*, *Platyphyllum fissipartitum*, *Sporogonites exuberans*, *Sciadophyton laxum*, *Hicklingia* sp., *Taeniocrada decheniana*, *T. longisporangiata*, *T. dubia*, *Stockmannsella* (*Taeniocrada*) *langii*, *Zosterophyllum rhenanum*, *Sawdonia ornata*, *Drepanophycus spinaeformis*, *Sartilmania jabachensis*, *Psilophyton burnotense* and *Protolepidodendron jabachense*. This Upper Siegenian floral palaeocommunity is much more diverse than that of the Lower Siegen 'Pflanzenschiefer' Formation although the living conditions were quite similar (Braun & Mörs 2001).

(2) At the southeastern flank of the Osteifel Main Anticline, the Upper Siegen Group was subdivided by Simpson (1940) starting with the Lower Dark Formation (Untere Dunkle Schichten), followed by the Kürrenberg Sandstone Formation (Kürrenberg-Sandstein), the Upper Dark Formation (Obere Dunkle Schichten), the Monreal Quartzite Formation (Monreal-Quarzit) and the Saxler Formation (Saxler-Schichten; Meyer 1994; LGB 2005). This sediment pile as a whole consists mainly of grey silty and sandy shales as well as sandstones with the Kürrenberg Sandstone Formation in the middle part, also containing several suites of sandstones. These five formations contain mostly marine faunas.

Hunsrückschiefer. The Hunsrückschiefer south of the Siegen Main Thrust consists of clay-rich sediments up to 5000 m in thickness. The dominant pelitic sediments are slates. Most of the distinctive primary sedimentary features are delicate, and are mostly masked by the slaty cleavage. Quartzites and quartzitic sandstones occur in certain formations thinning out towards the SW. Sedimentary features in these rocks can still be read easily. Roofing slates also occur in different levels. Large parts of these Hunsrückschiefer are of Siegenian age. They have been classified as Hunsrückschiefer *sensu lato* (Mittmeyer 1980). According to their sand/silt ratio, they can be subdivided into five formations (Meyer 1965): Mayen Formation (thickness exceeds 1200 m); Leutesdorf Formation (thickness about 800 m); Augustenthal Formation (thickness up to 900 m; rich faunas have been found near the locality Augustenthal containing middle Siegenian faunal elements; Dahmer 1932; Carls *et al.* 1982); Rüscheid Formation (thickness up to 2000 m); and Isenburg Formation (thickness up to 2000 m).

Here, a short remark seems necessary concerning the term 'Hunsrückschiefer' (Hunsrück Slate; U. Jansen pers. comm. 2007) as most geologists connect it with the famous 'Fossillagerstätten' at Bundenbach/Hunsrück only (Bartels *et al.* 1998). In 1881, Koch installed the 'Hunsrückschiefer' (formerly Wisperschiefer) in the Taunus region, i.e. the southeastern part of the Rheinische Schiefergebirge. There, this term encloses all those blue-grey to grey slates including roofing slates as well as thin light grey quartzitic sandstones lying above the Siegenian Taunusquartzite Formation (Taunusquarzit; see also Solle 1951). Within the region under consideration here, the Taunusquartzite Formation is totally lacking. In the northern part of the Central Facies Belt, it is replaced by Hunsrückschiefer-type slates of Siegenian age. The application of the term 'Hunsrückschiefer' to this slaty rock pile has been discussed controversially throughout the twentieth century (for more details see Gad 2006). In our opinion, the term 'Hunsrückschiefer' cannot be used in any stratigraphical context as was done by Koch (1881) because the 'Hunsrückschiefer' comprises Siegenian and Lower Emsian rock piles in certain regions. Therefore, we continue to use it for the whole lithofacies unit *sensu* Mittmeyer (1980) here comprising Siegenian (Hunsrückschiefer *sensu lato*) as well as Lower Emsian (Ulmen substage: Hunsrückschiefer *sensu stricto*) Hunsrückschiefer-type sediments. The newly installed Wied Group (Wied-Gruppe; Elkholy & Gad 2006: 66) contains the entire Hunsrückschiefer up to a first pyroclastic layer ('Porphyroid'; Kirnbauer 1991). The controversial debate concerning the 'Hunsrückschiefer problem' has to be continued as in large areas this boundary criterion to the Hunsrückschiefer Group (Hunsrückschiefer-Gruppe; Elkholy & Gad 2006: 66) above is totally lacking. There are almost no 'porphyroids' found in the central and western Hunsrück mountains. In our opinion, the famous 'Fossillagerstätten' at Bundenbach and Gemünden/Hunsrück are special palaeoenvironments and palaeobiotopes within the large 'Hunsrückschiefer' basinal facies belt that can be best understood from the lithological and sedimentological point of view under the scope of the Hunsrückschiefer on the whole (Stets & Schäfer 2002).

Stratigraphic criteria. Within the open marine environments, only *Boucostrophia herculea* is restricted to the Middle Siegenian. *Burmeisteria (Digonus) ornata disornata*, *Burmeisteria rudersdorfensis*, *Acrospirifer primaevus*, *Hysterolites hystericus*, *Athyris avirostris* and *Rhenorensselaeria strigiceps* range from the Middle Siegenian up to the Upper Siegenian as index fossils. *Proschizophoria personata*, *Alatiformia affinis*, *Vandercammenina rhenana*, *Tropidoleptus* sp., *Chonetes unkelensis*, *Cryptonella minor*, *'Uncinulus' frontecostatus* and *Crassirensselaeria crassicosta* constitute the suite for this epoch (Mittmeyer 1982).

Concerning the international timescale, most of the sediments mentioned above belong to the Pragian. As the global Pragian–Emsian boundary (*pireneae–kitabicus* boundary) cannot be correlated with the beginning of the Emsian in the Schiefergebirge, we continue to use the traditional Rhenish stages and substages here. That means that the uppermost part of the Siegenian already belongs to the Zlichovian (DSK 2002).

The Gedinnian–Siegenian boundary in the Schiefergebirge is still open to debate due to changes in the evaluation of pteraspid fossils (Schmidt 1958; Blieck & Jahnke 1980; Carls *et al.* 1982). Nevertheless, we imagine that north of the Siegen Main Thrust, i.e. in the core of the Osteifel Main Anticline, Gedinnian sediments do not crop out. In the Hunsrückschiefer belt, i.e. south of the Siegen Main Thrust, the Gedinnian–Siegenian boundary is still under research using palynomorphs (Gad 2005).

In the marine strata, the Siegenian–Lower Emsian boundary is given by the extinction of *Acrospirifer primaevus*, *Hysterolites hystericus*, *Cryptonella minor* and others. The appearance of *Acrospirifer 'fallax'*, arduspirifers of the *arduennensis* group, *Euryspirifer assimilis* and *Burmeisteria (Digonus) rhenana* as well as *B. (D.) armata* are sure indications (Mittmeyer 1982).

Rocks, grain size and sedimentological features

The Lower Devonian sedimentary rocks within the area under research mainly consist of sandstones, quartzitic sandstones and quartzites alternating with pure shales, silty shales and slates containing various amounts of disseminated sand. Coarse-grained sandstones and conglomerates are totally lacking in the outcrops within the Northern and northern Central Facies Belt along the Mid-Rhine valley. Most of the sandstones are litharenites and sublitharenites with arcosic arenites occurring in places. Pure quartzarenites are lacking. The silt- and mudstones form shales in the Northern Facies Belt, and slates in the Central Facies Belt just south of the Siegen Main Thrust. Here, they underwent higher deformation than the time-equivalent rocks north of the thrust. No severe differences in the mineralogical composition have been found within the Siegenian rock pile (Holl 1995; Wierich 1999). All the sandy detritus delivered to the basin from the north corresponds to that of recycled orogens *sensu* Dickinson *et al.* (1983), with the quartzose material being the most frequent.

The grain size of the sandstones has been determined in the field using a grain-size chart. Grain size of the siltstones had to be estimated roughly. All the grain-size statements of the sandstones are in accordance with the German nomenclature (Füchtbauer 1988) using the threefold distinction of fine-, medium- and coarse-grained measures down to 0.063 mm. For the description and interpretation of the depositional environments, compaction and diagenesis will be ignored in the text. The rocks have been addressed as if they were unlithified sediments.

The grainsize of the siliciclastics within the Northern Facies Belt ranges from medium-grained sands to fine-grained silts, whereas clays are relatively rare. Predominantly fine-grained sands to black silts and clays occur in the northern part of the Central Facies Belt. Processes have to be scrutinized that kept back most of the sandy material still found in the north, and grain-size distribution has to be related to individual energy conditions within the different subenvironments.

In the area under research, the following sedimentary structures developed within the whole range of bedforms. (i) Channel fills normally consist of sandy material, but some of the channels are filled with silty material; at the base of the channel fills, mud chips, intraformational clasts, plant remains, shell fragments or shell hash occur as channel lags. (ii) Large-scale trough cross-bedding with sets up to 30 cm thick have been found within the medium- to fine-grained sands. Large-scale lenticular bedding and planar cross-bedding are rare. (iii) High-energy horizontal planar bedding of the upper flow regime is restricted to fine-grained sands and coarse-grained silts. Within this grain-size range, current ripple lamination, small-scale cross-bedding, flaser and lenticular bedding have been found; wave or oscillation ripples also occur as well as wrinkle marks. (iv) Most of the medium-grained silts are devoid of primary bedding structures. In some places they contain plant remains, bioturbation, desiccation cracks, and small-sized nodules of limonite that generated from former framboids of pyrite or siderite nodules. (v) The medium- to fine-grained silts show parallel lamination, and thin fining-up cycles whenever bedforms can be noticed.

Base level

As summed up by Cross & Homewood (1997), Wheeler (1964) considered the stratigraphic base level as an abstract (non-physical), non-horizontal, undulatory continuous surface that rises and falls with respect to the Earth's surface. As base level rises, intersections of the base level surface and the seaward-inclined Earth surface move uphill. This increases the area of the Earth's surface below base level where sediment may accumulate, and increases the sediment storage capacity in continental environments. As base level falls, the opposite occurs. Stratigraphic base level is a descriptor of the interactions between processes that create and remove accommodation space, and surficial processes that bring sediment to or remove sediment from that space. In effect, but not explicitly, Wheeler (1964) defined stratigraphic base level as a potentiometric energy surface that describes the energy required to move the Earth's surface up or down to a position where gradients, sediment supply and accommodation are in equilibrium.

This concept enabled Cross *et al.* (1993), Ramón & Cross (1997), Cross & Lessenger (1998) and Mjos *et al.* (1998) to handle and interpret depositional sections in the field and in drill holes. The rise or fall of the stratigraphic base level generally conforms with the overall grain-size trends in profile sections (Homewood *et al.* 2000; Schäfer 2005). Moreover, it will provide interpretative ideas. Most important is the consideration of the development of accommodation space in a basin related to the supply of sediments into it. Turnaround points from fall to rise trends and vice versa will give direct correlation markers in concordance with sequence stratigraphic surfaces, and a chronostratigraphic timescale fixes ages to the profile. Base level trends explain the physical (energy-related) development of depositional systems on land and in the sea. In the following, we will use these ideas to interpret our field sections.

Sedimentary palaeoenvironments demonstrated by case studies

The fluvial palaeoenvironment

Two different case studies have been compiled from the Lower Siegen Group (Tonschiefer-Schichten, 'Pflanzenschiefer' Formation), the first beneath the castle Landskrone near Heimersheim/Ahr, and the second at Rheineck close to the Rhine river. A third one is situated in the Upper Siegen Wahnbach Formation in the Sieg valley near Schladern.

Description. The **Landskrone columnar section** (Fig. 5b) has been taken from rocks near the railway station Heimersheim on the left bank of the Ahr river (TK 25 sheet 5409 Linz/Rhein, R. 258343, H. 560204). The sediments exposed here belong to the 'Pflanzenschiefer' Formation in the core of the Ahrtal Anticline. Here, grain size ranges from medium-grained sand to fine-grained silt except for one clayey layer. The two columnar sections (Fig. 5a, b) taken at this locality contain four major fining-up cycles (cycles I–IV). Each of them starts with medium-grained sands just above major erosional surfaces. The sands show large-scale trough cross-bedding in the lower part turning to current ripple lamination occasionally. High-energy parallel bedding of the upper flow regime has been found in the finer sands in the upper parts. Each sandy event ends with medium-grained silts, the latter mostly without any recognizable bedding structures. Above the sands and below each cycle, an irregular alternation of thin fine-grained sands and mostly medium-grained silts occurs. Most of the thinner sandy layers also start above erosional surfaces. They show small-scale cross-bedding. The silts sometimes contain current ripple lamination as well as desiccation cracks, here and there. Shell hash, intraformational pebbles and relicts of plant fossils have also been found, especially in the black muds. Thus, each cycle starts at a base level lowstand with lateral accretion followed by vertical accretion during base level highstands.

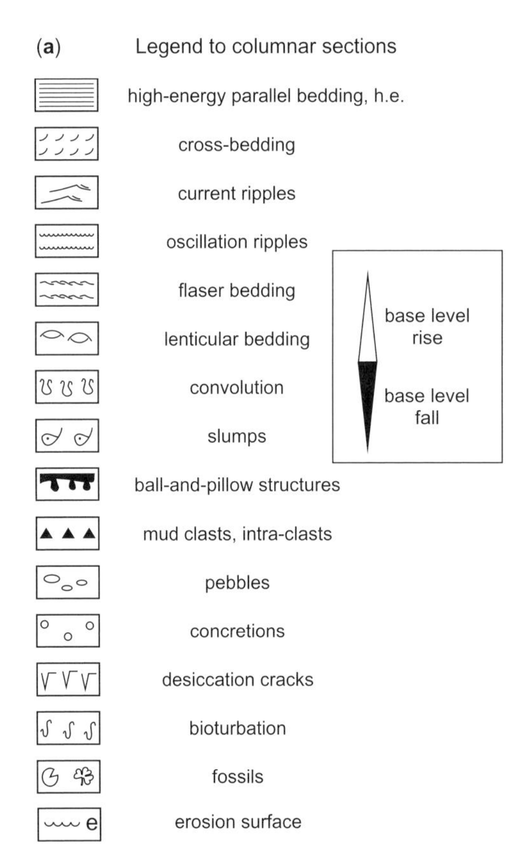

Fig. 5. (**a**) Legend to columnar sections 1–7 (Figs 5B–11, 13).

The **Rheineck columnar section** (Fig. 6) has been taken from a large abandoned quarry situated in the Vinxtbach valley just opposite Rheineck castle (TK 25 sheet 5509 Burgbrohl, R. 259286, H. 559625). This outcrop also belongs to the 'Pflanzenschiefer' Formation but it is situated further south in the core of the Hönningen-Seifen Anticline. It is at nearly the same stratigraphical level as the Landskrone outcrop (W. Meyer, pers. comm. 2006). This section encloses more than 200 m in thickness incorporating seven major cycles (cycles I–VII). They are organized in the same manner as in the Landskrone columnar section (Fig. 5). Grain size ranges again from medium-grained sand to fine-grained silt except for another clayey layer. The bulk sand content amounts to about 39% with a medium-grained sand content of about 13%. There have been three major units of medium-grained sands that are mostly excavated today. Each of them starts above a major erosional surface cutting down into the underlying fine-grained sediments. These sands show large-scale trough cross-bedding. They belong to fining-up cycles as has already been shown in the Landskrone columnar section. In between, smaller fining-up cycles occur in places, ranging from fine-grained sands to medium-grained silts that also show erosional surfaces at their bases. Some of these cycles end with fine-grained silts or even dark clay rich in plant remains ('Pflanzenschiefer' *sensu stricto*). In various places outside the Rheineck section, small coal seams have been reported (Meyer 1958, 1994). The silts and muds show parallel lamination at a millimetre scale. Within larger packages, laminae of fine-grained sands alternate with medium-grained silts. Flasers of fine sands to coarse silts occur. Thicker layers of fine-grained sands or coarse-grained silts show small-scale cross-bedding, and/or current ripple lamination in places. High-energy parallel bedding has been found in the upper parts of cycles V and VII. Within the muds, small limonitic nodules occur.

At a whole, two types of depositional environments are documented in this section: the high-energy sand packages contrast with the low-energy

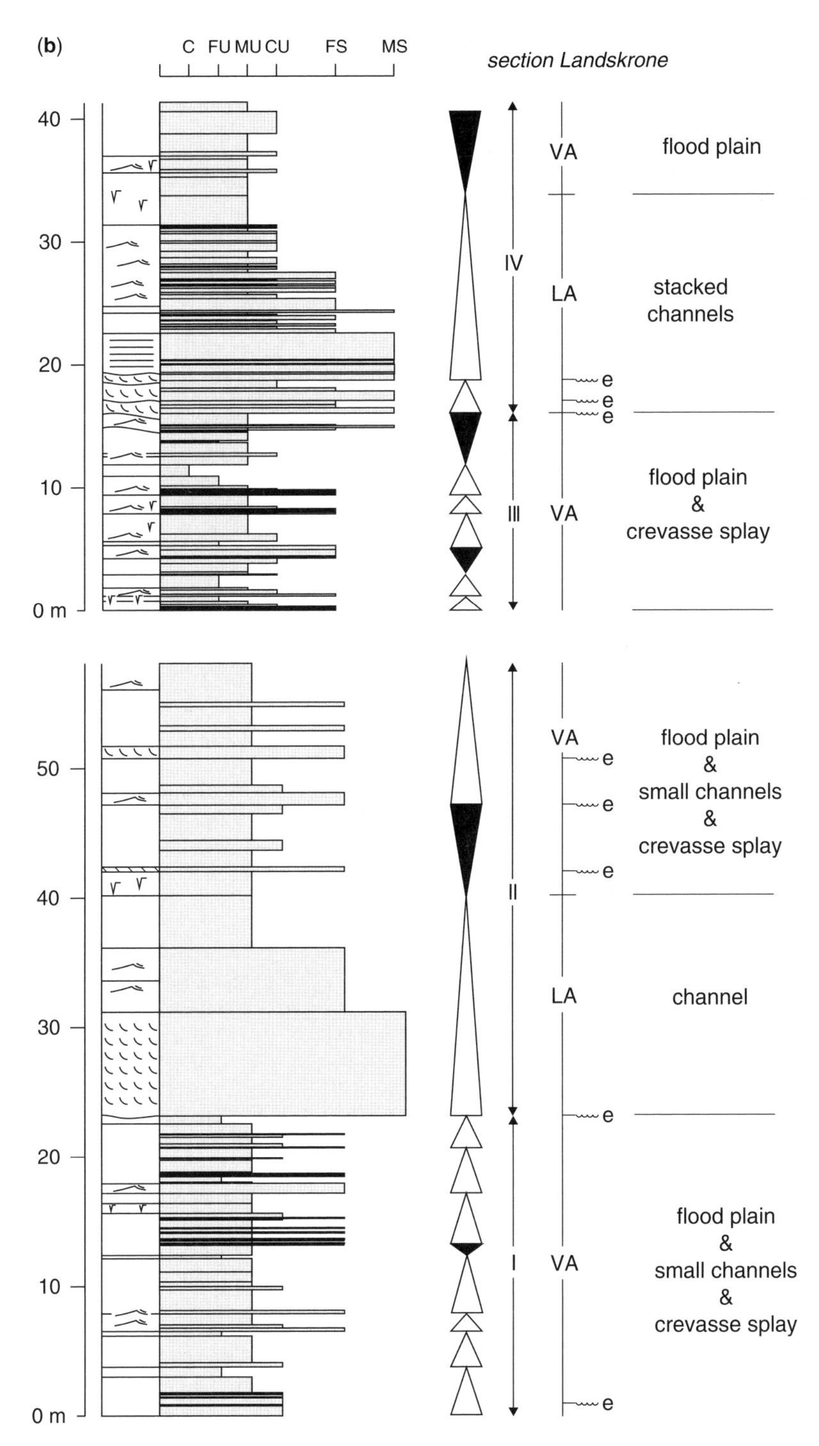

Fig. 5. (*Continued*) (**b**) Landskrone columnar section (locality no. 1; 'Pflanzenschiefer' Formation, Lower Siegen Group); (a, b) measured columnar sections at Landskrone outcrop; I–IV, sedimentary cycles. Abbreviations: C, clay; FU, MU, CU, fine-, medium- and coarse-grained silts; FS, MS, fine- and medium-grained sands; e, major erosional surfaces; LA, laterally accreting channel fill sediments; VA, vertically accreting floodplain sediments.

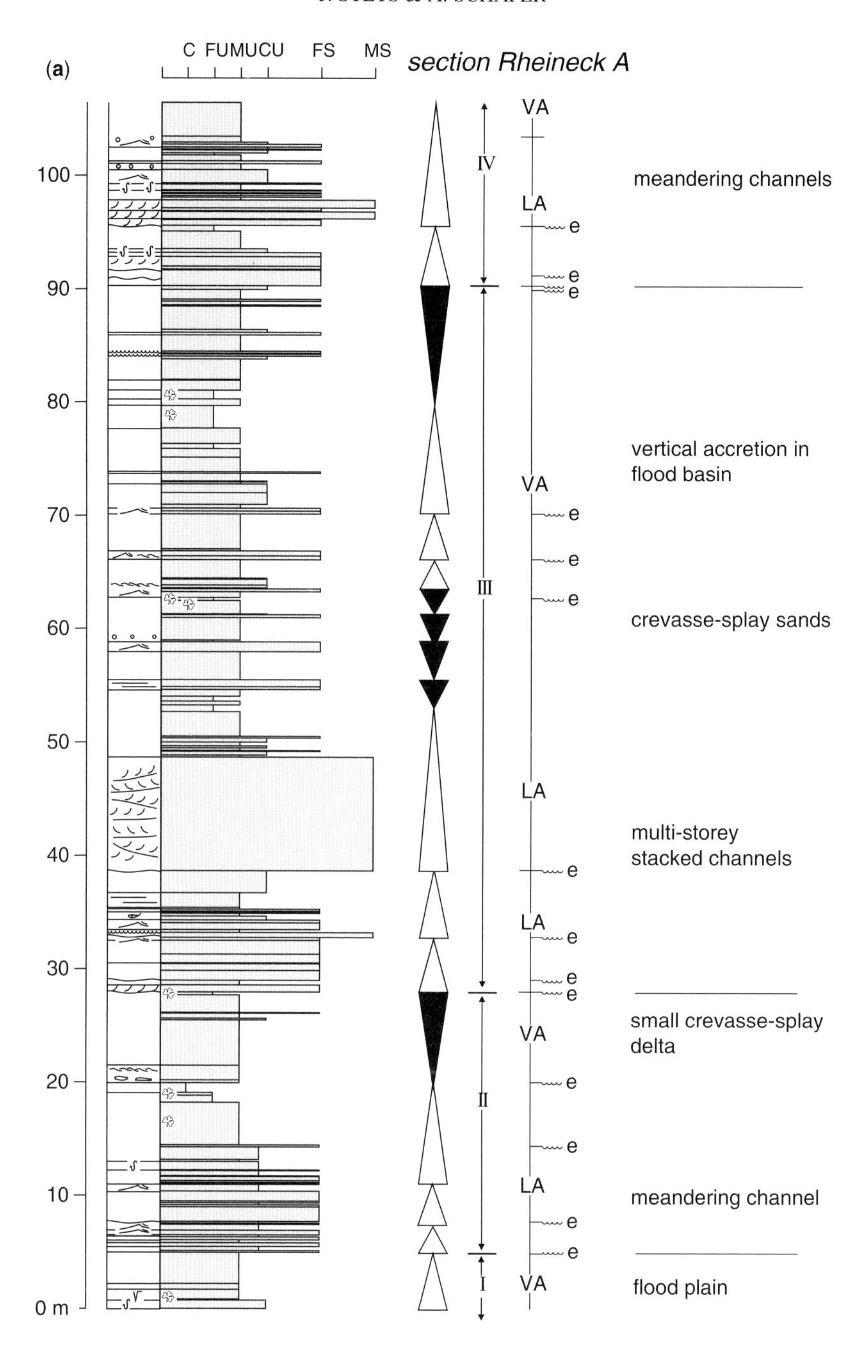

Fig. 6. (**a**, **b**) Rheineck columnar section (locality no. 2; 'Pflanzenschiefer' Formation, Lower Siegen Group). I–VII, Sedimentary cycles. For interpretation see text; for legend and abbreviations see Figure 5.

silts and muds. The latter constitute the majority of the sediment column with a bulk content ranging up to 60%. Each of the seven main cycles starts with lateral accretion of the high-energy type, and grades up to mostly vertically accreted low-energy type sediments. This is also the case even in the smaller cycles.

The **Schladern columnar section** (Fig. 7) has been taken from a road cut at the B 62/256 in the Sieg valley near Schladern (TK 25 sheet 5111

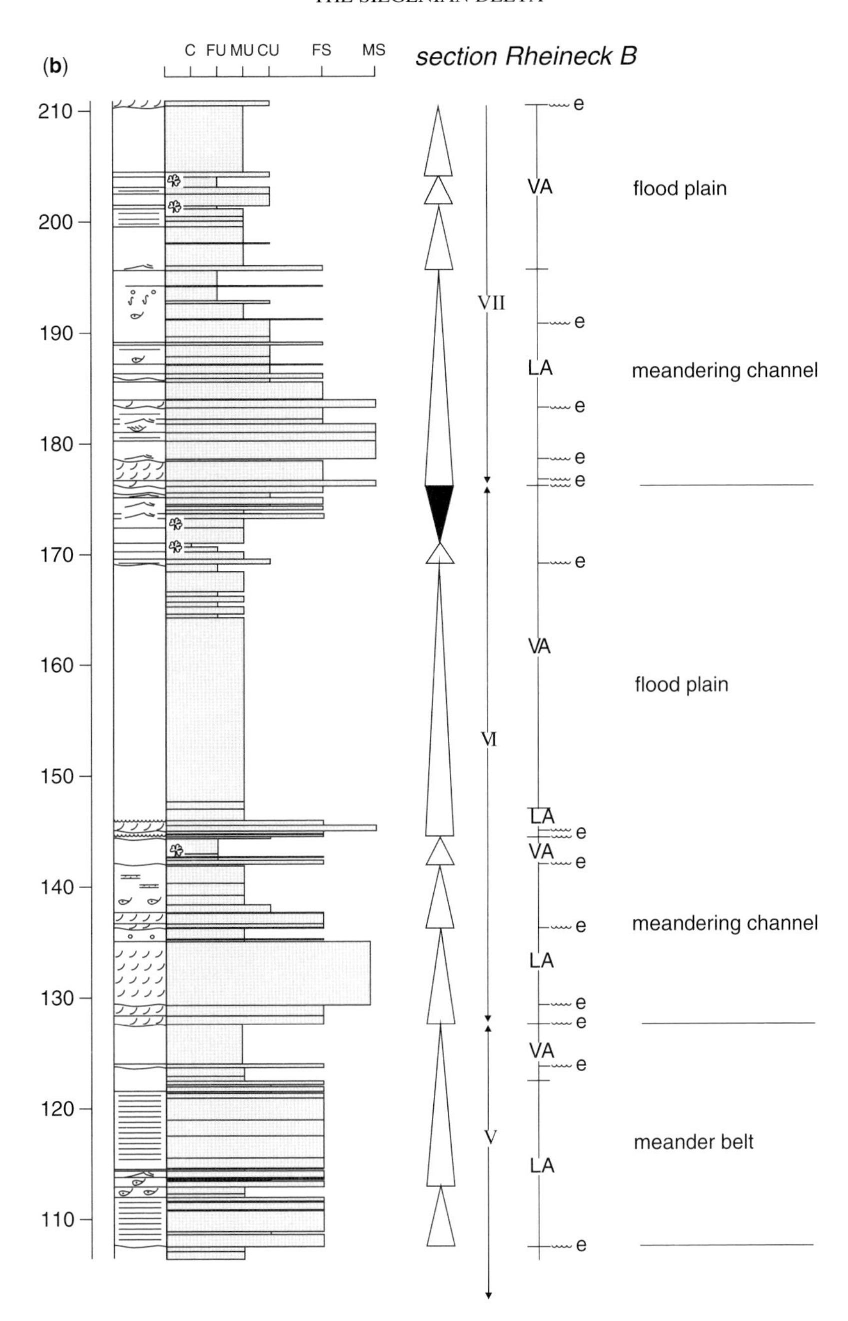

Fig. 6. (*Continued*)

Waldbröl, R. 340125, H. 563130; this outcrop is hidden behind a recently built concrete wall along the road today). Here, the sediments belong to the upper part of the Wahnbach Formation of Late Siegenian age (Grabert 1979). A sediment package up to 100 m thick was exposed here. Grain size ranges from medium-grained sands to fine-grained silts with a bulk sand content of up to 48%. Within this section, six major cycles can be found that are piled up closely. They follow the same principles of stacking as at the localities of Heimersheim and Rheineck, but some of them, e.g. cycle III, are not fully developed with the vertically accreting suite on top. Each of the cycles starts with

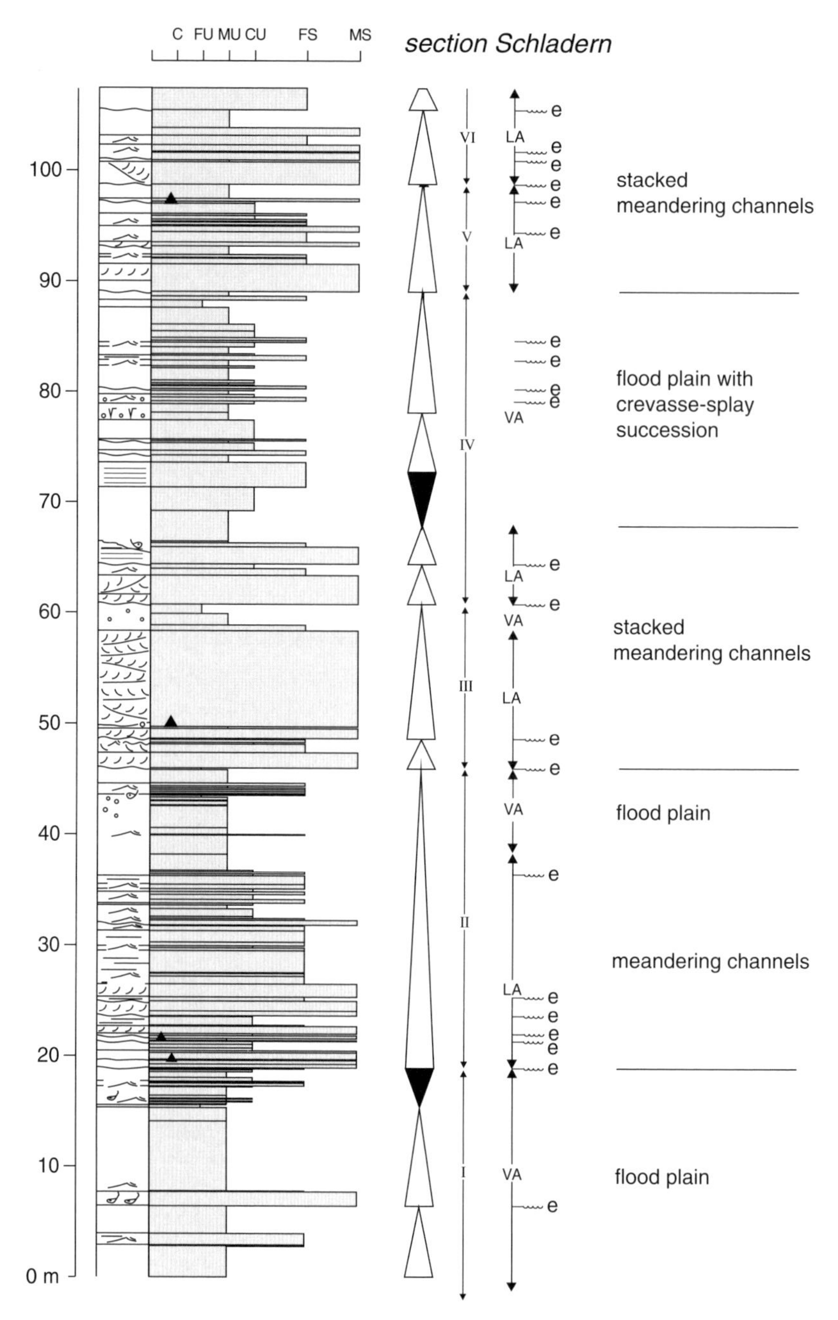

Fig. 7. Schladern columnar section (locality no. 3; Wahnbach Formation, Upper Siegen Group). I–VI, Sedimentary cycles. For interpretation see text; for legend and abbreviations see Figure 5.

medium-grained sands above erosional surfaces. The medium-grained sands normally show large-scale trough cross-bedding again. High-energy horizontal parallel bedding and current ripple lamination are restricted to the fine-grained sands as well as to the coarse-grained silts above. Fine-grained silts and clays are mostly lacking. Calcretes have been found in places together with mud cracks indicating drying up of the sediment surface from time to time. Although this section is younger in age, the sedimentary pattern is similar to those of the Landskrone and Rheineck columnar sections. The only exception is a higher sand content.

Fauna and flora. Faunal and floral content is most important to diagnose the palaeoenvironment of the above sedimentary columns. Obviously, a high sedimentation rate and drying out of the sediment surfaces may have prevented benthic biota from settling on the muddy substrate. Nevertheless, faunal and floral relicts have been found in places (Schweitzer 1983, 1994, 2003; Meyer & Raufi 1991; Schindler *et al.* 2001, 2004). Within this palaeoenvironment, the '*Modiolopsis*' facies occurs. Herein, faunal assemblages contain relicts of 'fishes' (e.g. *Rhinopteraspis dunensis*, *Drepanaspis* sp., *Cephalaspis* sp.), as well as remnants of Eurypterida (e.g. *Pterygotus* sp.) and other arthropods, various species of bivalves (e.g. *Modiolopsis* sp., *Limoptera* sp., *Goniophora* sp., *Myalina* sp.) and gastropods (e.g. *Bellerophon* sp., *Bembexia* sp., *Platyceras* sp.). *Tentaculites* sp. has often been found, but brachiopods are rare except individuals of *Crassirensselaeria crassicosta* indicating a palaeoenvironment located near to the sea.

In addition, there is a rich assemblage of fossil plants already mentioned above, which is best characterized by the term '*Psilophyton*' flora (Kräusel & Weyland 1930; Kräusel 1936; Schweitzer 1983, 2003). The dark Lower Siegenian mudstones rich in plants ('Pflanzenschiefer' *sensu stricto*) proved to be the best indicator for palaeoecological reconstructions. Moreover, there are rhizoids found beneath thin coal seams indicating autochthonous growth of plants in places. The fructifications of *Taeniocrada decheniana* rose above the water table demonstrating very shallow water at its place of growth. *Zosterophyllum rhenanum* even survived dry periods, and *Drepanophycus spinaeformis* may even have been a land plant at that time.

Growth conditions for plants colonizing the land–sea transition zone were also rather good, above are in the sedimentation area of the Upper Siegen Wahnbach Formation. In the Wahnbach valley area nearby, extensive meadow-like assemblages of *Stockmansella (Taeniocrada) langii* have been found as the predominant plant among the '*Psilophyton*' flora. *Drepanophycus spinaeformis* also occurred, indicating land conditions. Schweitzer (1983, 1987, 2003) named further species as there are *Taeniocrada dubia* living submerged in very shallow water, whereas *Zosterophyllum rhenanum* grew on river banks or along lakesides. Moreover, *'Psilophyton' burnotense, Sawdonia ornata, Sartilmania (Dawsonites) jabachensis* and also *Protolepidodendron wahnbachense* existed here. A few individuals of *Prototaxites psygmophylloides* and *Pachytheca* sp., as well as several more-or-less intact exemplars of the extremely delicate *Wahnbachella bostrychioides*, have been found. In the Jabach valley district nearby, Schweitzer (1983, 1987, 2003) noticed individuals of *Stockmannsella (Taeniocrada) langii*, and T. *longisporangiata* which he imagined to live submerged under shallow water conditions. They often occured together with *Prototaxites psygmophylloides*. In the Jabach valley district, plants prefering dryer habitats are mostly lacking.

Interpretation. In the Landskrone, Rheineck and Schladern columnar sections, fluvial sedimentation is obvious (Fig. 8). The predominance of mudstones indicates that muds were deposited on vast mud flats, in lakes and pools or on floodplains of large fluvial systems. These aquatic environments were vegetated in various places by *Taeniocrada* and others that lived under very shallow water conditions. The muds contained former siderite nodules that are mostly oxidized to limonite today indicating anaerobic conditions below the sediment surface. From time to time, these mud flats fell dry as is shown by desiccation cracks. There must have been isles nearby in that vast lowland environment as indicated by *Zosterophyllum*, *Sawdonia* and *Drepanophycus*.

In contrast, laterally accreted channel-fills up to 10 m or more in thickness had been cut into these mud flats. There are three large channels, several tens of metres wide, documented in the Rheineck columnar section. The channel-fills are mostly built up of medium-grained sands with channel lag deposits at their bases in places. These sands are grading up to fine sands. Most of the detritus transported within these channels may have also been fine-grained. Transport was towards the SW as already demonstrated by Bausch van Bertsbergh (1940).

Alongside the channels, thinner layers of fine-grained sands fining up to coarse-grained silts formed thin sand sheets. They are interpreted as crevasse-splay sands that had been spread upon the mud flats nearby. Current ripple lamination has often been found in these sediments. Water level marks and rounded crests of ripples indicate waning floods and exposure.

This can also be shown in the Schladern columnar section. There, another system of multiple channels is documented (Fig. 8). Thin blankets of fine-grained sands covering the muds have also been interpreted as crevasse-splay sediments. The plant remains in the outcrops nearby in the Sieg area indicate closely vegetated perennial lakes, and floodplains with changing conditions. They ranged from shallow water to subaerial conditions.

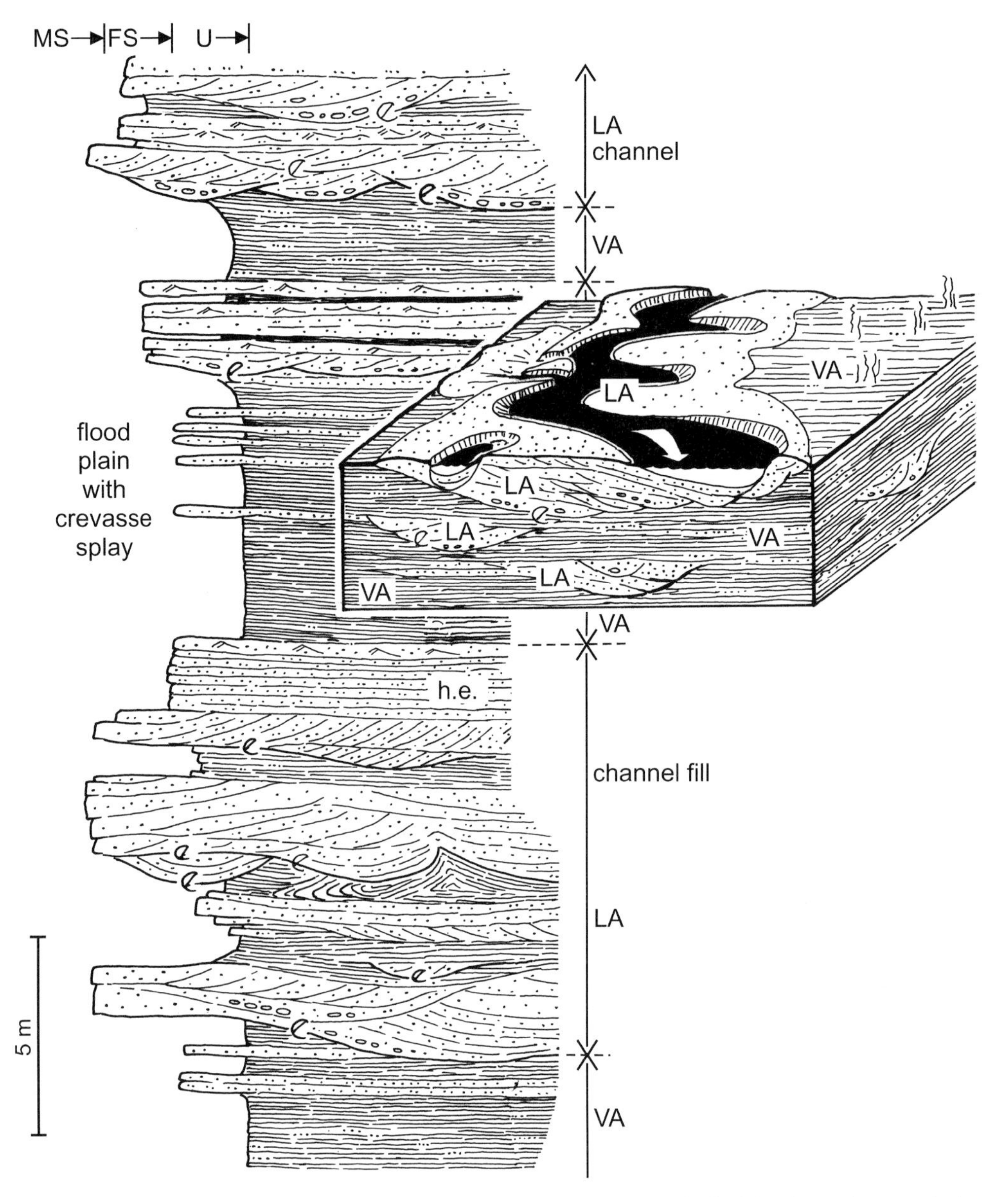

Fig. 8. Generalized columnar section and interpretative block diagram showing the fluvial palaeoenvironment in the Lower Siegen 'Pflanzenschiefer' Formation, and in the Upper Siegen Wahnbach Formation (drawn in accordance with a field sketch of P. Wurster 1985). Abbreviations: h.e., high-energy parallel bedding; U silts; see Figure 5 for other abbreviations.

The three columnar sections belong to a vast lowland palaeoenvironment at the top of the huge Siegenian siliciclastic delta complex of the Northern Facies Belt (Stets & Schäfer 2002, 2004). The rivers that nourished this lowland palaeoenvironment are located still further to the north in the region of Aachen and towards the NE where medium- to fine-grained cross-bedded sand deposits of the Siegenian occur. They have been interpreted as point-bar deposits of a braided fluvial system. We imagine that they belong to a wide belt of braided river systems called the Old Red Continent Foreland Belt (see Fig. 18). It was situated immediately south of the Old Red Continent. These more proximal river systems fed the distal reaches of the Wahnbach Formation in the Northern Facies Belt. Proceeding on the assumption of a very large thickness of the Siegenian delta complex, these rivers must have been major ones. The grain size of the sands left in the Foreland Belt in the Schiefergebirge area is not coarser than medium-grained sands. These findings imply a rather long travel distance from the source area through the Foreland Belt (Breil-Schollmayer 1990).

At a whole, the fluvial sedimentation in parts of the Northern Facies Belt developed in a vast low-relief environment just between the proximal fluvial systems of the Old Red Continent Foreland Belt in the north and the delta front in the south. It was active there throughout the whole Siegenian. High-sinuosity marshland channels cut down into the mud plain of the lowlands (Fig. 8). In this palaeoenvironment, perennial lakes and/or pools allowed living conditions even for 'fishes'. On the other hand, these floodplains were vegetated by a rich '*Psilophyton*' flora living submerged as well as close to the channel courses or on islands. In places, the vicinity of the sea in the south must have controlled the living conditions as a special fauna preferring more brackish environments has been encountered even in the Wahnbach Formation (Schweitzer 1987, 2003; Schindler *et al.* 2001, 2004). A variable sea level due to storms, spring tides, changing subsidence, and/or a changing supply of detritus coming down from the source area in the north may have caused short-term ingressions and regressions. Thus, brackish or even marine faunal elements had a chance to survive in the normal fluvial environment for short intervals.

Deltaic palaeoenvironments

At the southern margin of the river-dominated subaerial delta platform, different deltaic subenvironments can be studied. As the sea prograded intermittently towards the north during the Middle Siegenian, and started to retreat again in parts of the Northern Facies Belt during the uppermost Siegenian to Lower Emsian (Stets & Schäfer 2002), different palaeoenvironments were stored upon each other. Three such environments are described below.

(1) The tidal environment

Changing land–sea conditions can be well studied at the natural steep escarpment near the small town of Unkel/Rhine. A columnar section taken at the gently SE-dipping flank of the famous Unkel anticline (Jankowsky 1955; Meyer & Stets 1996; TK 25 sheet 5409 Linz a. Rhein, R. 258688, H. 560744) includes about 85 m. In nearby localities, rich open-marine faunas have been found in sandstones indicating a Late Siegenian age (Dahmer 1936).

Description. In the **Unkel columnar section** (Fig. 9) two complete cycles are exposed.

The first cycle (Ia) starts just above an erosional surface at 2.2 m with medium-grained sands alternating with fine-grained sands. The medium-grained sands show large-scale trough cross-bedding. Planar horizontal bedding and current ripples have been found in the fine-grained sands above. There are several erosional surfaces within this sandy package (Ia), which is up to 7 m thick. On top of the sands, medium-grained silts occur with some fine-grained sands intercalated showing mostly parallel horizontal bedding (Ib). These silts contain limonite nodules, and show desiccation cracks.

At 33.0 m the second cycle starts above a further erosional surface (IIa). At the bottom of the medium-grained sandy package, mud chips indicate reworking of the underlying silty muds. In contrast to the three columnar sections in fluvial environments mentioned above, the sands can be traced over several hundreds of metres in the field. The medium-grained sands are intensely large-scale trough cross-bedded. The single cross-bedded sets are only centimetres thick and show tidal bundling. Transport direction is toward the SW. This sandy 'event' ends in a cycle 2.5 m thick, grading up from fine-grained sands to medium-grained silts. On top, another well-bedded alternation of medium- and coarse-grained silts (IIb about 35 m thick) occurs containing several layers of fine-grained sands (up to 0.2 m thick). Small coarsening-up cycles are indicated. Within the muds, current ripple lamination has been found as well as bioturbation, and desiccation cracks especially in this cycle. Jankowsky (1955) described several V-shaped sand features, 'Grauwackengänge', in the muds measuring 2–10 cm in width, and 0.3–0.5 m in depth. They are supposed to be deep-reaching mud cracks.

At 74.0 m, the next cyle (IIIa) starts with cross-bedded medium-grained sandstones again.

Fauna and flora. An example of concentration enrichment *sensu* Seilacher (1970) of a rich open-marine fauna has been cited by Dahmer (1936) from a locality named 'Kaskade' near Unkel (Fig. 2, location A1). The fauna has been found here within a sandy sequence corresponding with the uppermost medium-grained sands of the Unkel columnar section (IIIa). Here, *Chonetes unkelensis*, *Hysterolites hystericus*, *Tropidoleptus rhenanus*, *Proschizophoria personata*, *Platyorthis circularis* and several other brachiopods and bivalves have been found enriched within one layer of the sands. Above, another enrichment has been found at the

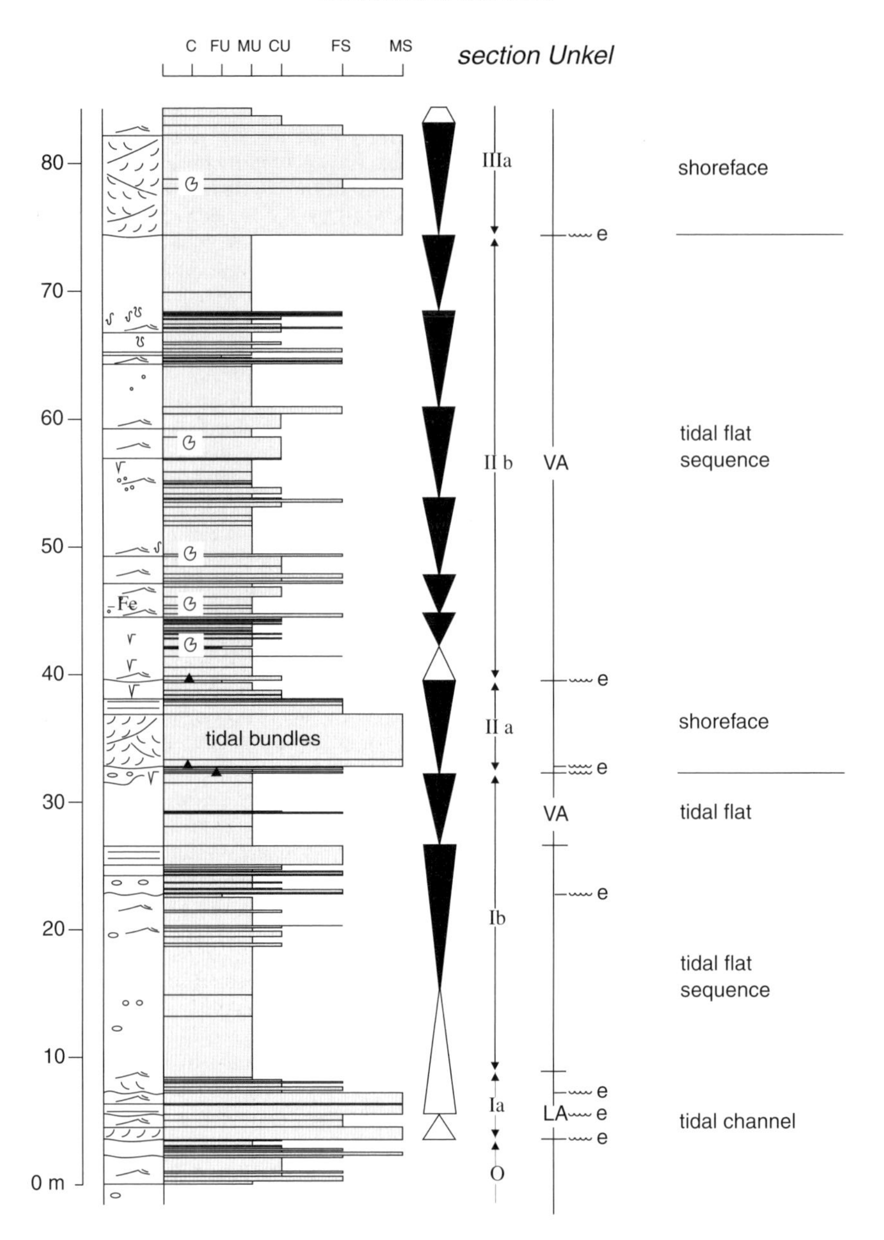

Fig. 9. Unkel columnar section and 'Kaskade' outcrop (localities no. 4; A1, Upper Siegen Group). 0, I–III sedimentary cycles. For interpretation see text; for legend see Figure 5A. Abbreviations: LA, laterally accreting sediments of tidal channels and shore face; VA, vertically accreting sediments of tidal flat; see Figure 5 for other abbreviations.

top of the sands containing about 80 different taxa of a mostly marine fauna.

At the Kaskade locality, the uppermost sands are abruptly overlain by a silty unit that was called 'Bröckelschiefer' by Dahmer (1936). It was impoverished of marine biota. Instead, *Modiolopsis* sp., *Ctenodonta* sp., *Nuculites* sp. and plant remains were found there, as mentioned above relating to the fluvial palaeoenvironment. They indicate that the *'Modiolopsis'* facies appeared again. Within the alternating silts and fine-grained sands, only very few individuals of *Crassirensselaeria crassicosta*, *Tropidoleptus rhenanus*, *Chonetes* sp. and some delicate bivalves have been found, indicating that the muds were located nearer to the sea than had been the case in the localities of Rheineck, Heimersheim and Schladern.

Interpretation. The interpretation of the Unkel columnar section and neighbouring outcrops involves two subenvironments.

(1) Medium- to fine-grained detritus had been deposited in a muddy foreshore subenvironment. These muds were settled by endobenthic biota as is shown by bioturbation in places. They were exposed to subaerial conditions as indicated by desiccation cracks and 'sand dykes'. Limonitic nodules indicate former anaerobic conditions within the deeper parts of the muds. Single delicate brachiopods prove marine conditions on a vast mud flat, and bivalves of the *'Modiolopsis'* facies indicate the influence of the nearby fluvial palaeoenvironment of the delta platform. The thin fine-grained sands are ascribed to sheet sands on the mud flat as indicated by current ripple lamination. Tidal influence is also interpreted from the observation of herringbone cross-stratification. Other sandy layers that show an erosive base have been assigned to small asymmetric meandering tidal channels cut into the muds. From the content of plant and animal fossils as well as from the distinctive sedimentological inventory, we imagine that a large part of the silts and fine-grained sands had been laid down in a tidal flat environment well protected from the nearby sea. Some of the sands may be interpreted as beach foreshore sands, others as wash-over sands. These tidal mud flat deposits built up thick sequences. They proved to be the normal sedimentary environment here during the Late Siegenian. Normally, the chance for their preservation is rather limited. Thus, we imagine that continuous subsidence combined with a high accumulation rate favoured their formation and supported their preservation. The persistance of this subenvironment prone to partial or total erosion had thus been ensured.

(2) The large-scale cross-bedded medium- to fine-grained sands are assumed to be part of a large barrier sand system. The sands might have been formed during a relative sea level rise as all these megaripples normally generate below or near to the low-tide level. Within this high-energy marine environment, the chance of a preservation of faunas is rather low. In places, their relicts had been swept together to build up concentration enrichments in the troughs of the megaripples. The ingression of this shoreface environment was due to a stronger rise of the base level than usual. The sandy shoreface environment had been buried and overwhelmed by the prograding muds whenever the rise of the base level slowed down intermittently.

(2) Interdistributary bays

In the Ahr valley, beds of the Middle Siegen Group (Rauhflaser-Schichten) are exposed in various places within the nearly vertical northwestern flank of the Ahrtal Anticline (Meyer 1958, 1994). Here, another columnar section has been taken at the small bridge crossing the Ahr river near Altenahr (TK 25 sheet 5407 Altenahr, R. 257080, H. 559835). This is one of the best localities to study the typical 'Rauhflaser' environment in this region.

Description. In the **Altenahr columnar section** (Fig. 10) grain size generally ranges from fine-grained sands to fine-grained silts. The bulk sand content amounts to 25%. Medium-grained sands are restricted to three thin layers. They contain mud pebbles, and show trough cross-bedding. They had been deposited above erosional surfaces that indicate very shallow channel fills. Flaser and lenticular bedding of the 'Rauhflaser' type are common. Current ripples and oscillation ripples have often been found. Wrinkle marks occur on top of fine-grained sandy layers indicating very thin water films above the sediment. Convolution is evident at various places. Body fossils are rare. There had been only one layer containing relicts of small crinoids in this locality. On the other hand, bioturbation has often been found in the coarse- to medium-grained silts. In several places, the primary bedding structures are totally destroyed by burrowing endobenthonts. Indications of bar sands have been found three times built up of ripple-laminated fine-grained sands.

Fauna and flora. Different rich faunal assemblages have been documented from exposures in the Ahr region by Dahmer (1937). According to his findings, they contain a fauna of the Middle Siegenian Seifen type. It consists of mostly rather robust brachiopods such as *Acrospirifer primaevus*, *Prosocoelus pesanseris*, *'Uncinulus' frontecostatus*, *Crassirensselaeria crassicosta* and others. Bivalves such as *Myalina* sp., *Cypricardella* sp., *Goniophora* sp. and relicts of crinoids have also been listed. They signal an open marine environment.

On the other hand, Schweitzer (1987) mentioned several species of the *'Psilophyton'* flora from the Ahr region as there are *Sciadophyton laxum*, *Taeniocrada decheniana*, *T. longisporangiata*, *T. dubia*, *Gosslingia cordiformis* and *Drepanophycus spinaeformis*. They all are indicators for the land–sea transition zone. But within the Altenahr columnar section itself, no floral relicts have been found at all.

Interpretation. Concerning the interpretation of the Altenahr columnar section, we have to imagine that most of the fine- to medium-grained silts had been

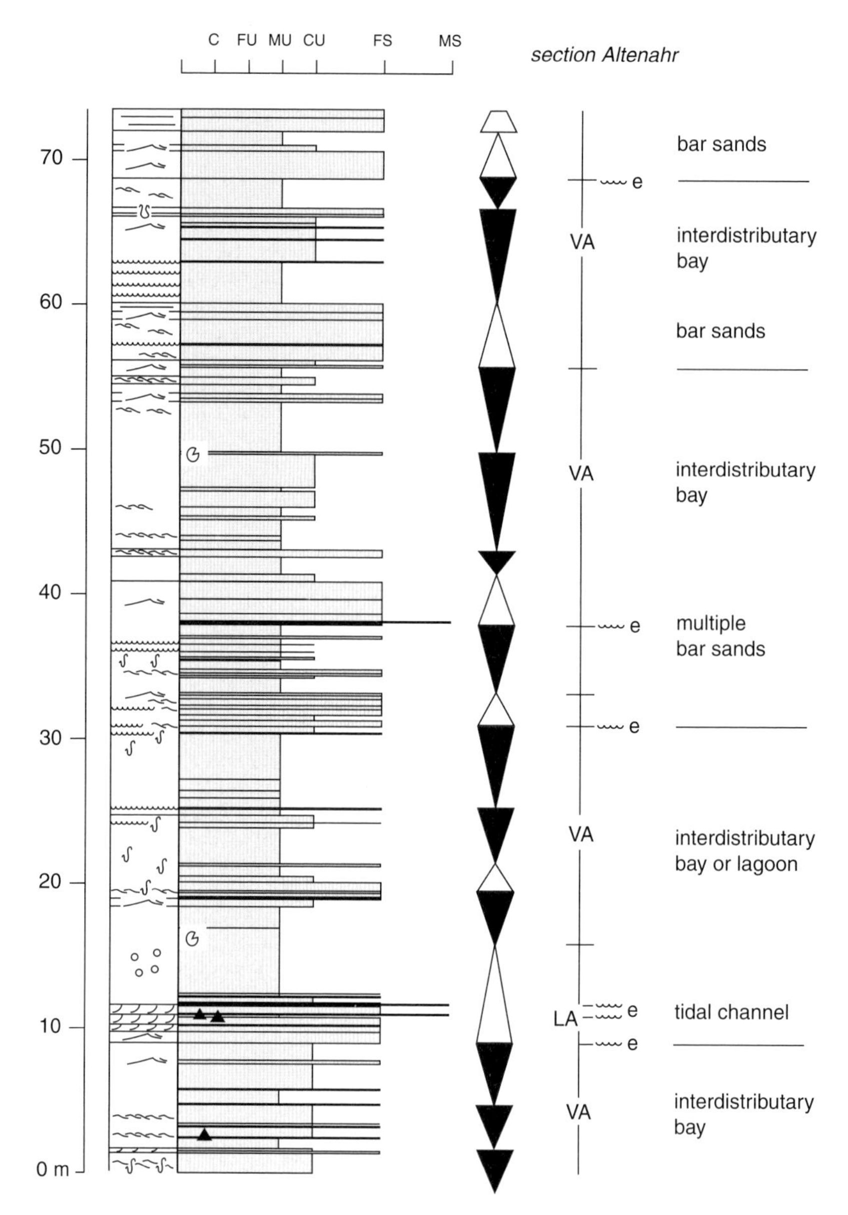

Fig. 10. Altenahr columnar section (locality no. 5; 'Rauhflaser-Schichten', Middle Siegen Group). For interpretation see text; for legend see Figure 5A. Abbreviations: LA, laterally accreting channel fill and bar sands; VA, vertically accreting sediments of lagoons and interdistributary bays; see Figure 5 for other abbreviations.

deposited in a low-relief mud flat just close to the sea. In contrast to the columnar section at Unkel, no desiccation cracks have been observed. Fine-grained sands to fine-grained silts showing wave ripples as well as wrinkle marks indicate only a few centimetres of water depth at times (Reineck & Singh 1980). Ripple- and flaser-bedded muddy sands to sandy muds indicate an intertidal environment. This must have been near to a protected coast or in a wide and rather shallow distributary bay far away from the input of coarser siliciclastic detritus.

Cyclic organization has not been found in this columnar section. Thus, we imagine that this palaeoenvironment had been installed nearshore just between the prograding levees of delta tributaries, maybe in lagoons during an intermittently changing base level. We favour the interpretation as interdistributary bays for the subenvironment because no larger sandy barriers have been noticed here or elsewhere in the Middle Siegenian strata of this region except some smaller bar sands. On the other hand, indications of fluvial activity are suggested by small and shallow channels, splay sands, and small coarsening- and fining-up cycles. They should be attributed to the fluvial activity of the distributary channels of the delta. They contrast with the flaser- and lens-shaped sedimentary 'Rauhflaser' structures.

These observations underline the ambivalence of this environment that existed just in the land–sea transition zone: Splay sands and silts of the fluvial system nourished the interdistributary bay with material that had been reworked by tidal processes forming the 'Rauhflaser'-type sediments. The mixed sand and mud flats had always been covered by the sea, or at least the interstices of the sediments were saturated with water due to a high base level. There had been enough water available either from fluvial or marine activities. Thus, desiccation cracks had no chance to develop.

Delta front to subaqueous delta platform

Just north of the Siegen Main Thrust, the Bürresheim columnar section has been taken to be situated at the southern flank of the Osteifel Main Anticline, just close to the southern border of the Northern Facies Belt. The Bürresheim columnar section belongs to the Upper Siegen Kürrenberg Sandstein Formation (Simpson 1940; Dahmer 1934*b*, 1940; Meyer 1958, 1994). This locality is situated west of the Bürresheim castel near to Mayen/Eifel along the small road uphill to the village of Kirchwald (TK 25 sheet 5609 Mayen, R. 258370, H. 558060).

Description. The **Bürreshein columnar section** (Fig. 11) embraces another 135 m thick segment of an Upper Siegenian sand/silt sequence. Grain-size range is from fine-grained sand to medium-grained silt with only a few layers of medium-grained sand and fine-grained silt. The bulk sand content amounts to about 22% with a medium-grained sand portion less than 1%. This columnar section can be subdivided into several segments that belong to two main cycles.

The section starts with a sand/silt alternation sequence. Parallel bedding and small-scale trough cross-bedding can be noticed as well as current ripple lamination. Shells and shell fragments of small brachiopods have been found in the fine-grained sands. Reworking is indicated by mud clasts.

Cycle I starts at 9.4 m with a rather uniform pile of silts containing shell hash in places. Slumping, ball-and-pillow structures as well as convolute bedding occur in places (Ia).

From 34.5 m up to 62.0 m, an alternation of fine-grained sands and coarse- to medium-grained silts builds up several smaller coarsening-up cycles. The sands show parallel bedding at a centimetre scale, as well as current ripple lamination. Most of the silts are barren with respect to sedimentary structures.

On the top, a third alternation follows (Ib) that is about 10 m thick. The sands were deposited above erosional surfaces. Mud pebbles at their bases indicate reworking of the fine-grained materials below. In certain places, channel-fill sands occur showing trough cross-bedding. Fine-grained sands contain high-energy parallel bedding and current ripples. Additionally, bioturbation has been found there within the silts. We imagine that the entire package, up to 63.5 m thick, forms one major marine cycle (I) shallowing up.

Cycle II starts at 72.0 m (IIa). Most of the sedimentary structures mentioned above are repeated once more; silts are dominant here.

At about 107.0 m to 118.0 m (IIb), sands occur above erosional surfaces that again build smaller fining-up cycles. Trough cross-bedding passes to high-energy parallel bedding indicating rapid sediment transport again.

This is more accentuated above 118.0 m (IIc). Here, medium- to fine-grained, cross-bedded sands built fluvial channel fills as has been shown earlier.

On the top, i.e. at 132.0 m, muds occur again that mark the onset of a third major cycle (IIIa) that has not been measured up.

Fauna and flora. Marine faunas containing *Hysterolites hystericus*, *Rhenorensselaeria strigiceps* and other brachiopods indicate open-marine conditions for most parts of this section, which is of Late Siegenian age (Dahmer 1940). They contrast with impoverished faunal communities containing plant remains in outcrops nearby (Fuchs 1974). To date, no autochthonous growth of plants has been documented by rhizoids. Most of the detritus had been swept in from higher, further landward positions of the delta.

Interpretation. In the Bürresheim columnar section, suspension fall-out, slumping, and transport of fine-grained material down to deeper water seems obvious, indicated by contorted balls and rolls of sand, especially in the lower parts of cycles I and II. This supported the idea of a suspension fallout at or near the delta front. Coarsening-up cycles promote the idea of deltaic progradation conditions. The sands have been interpreted as splay sands just in front of the mouth of distributary channels (Fig. 12).

Twice, the conditions of the delta front (Ia, IIa) passed over to those of a subaqueous delta platform

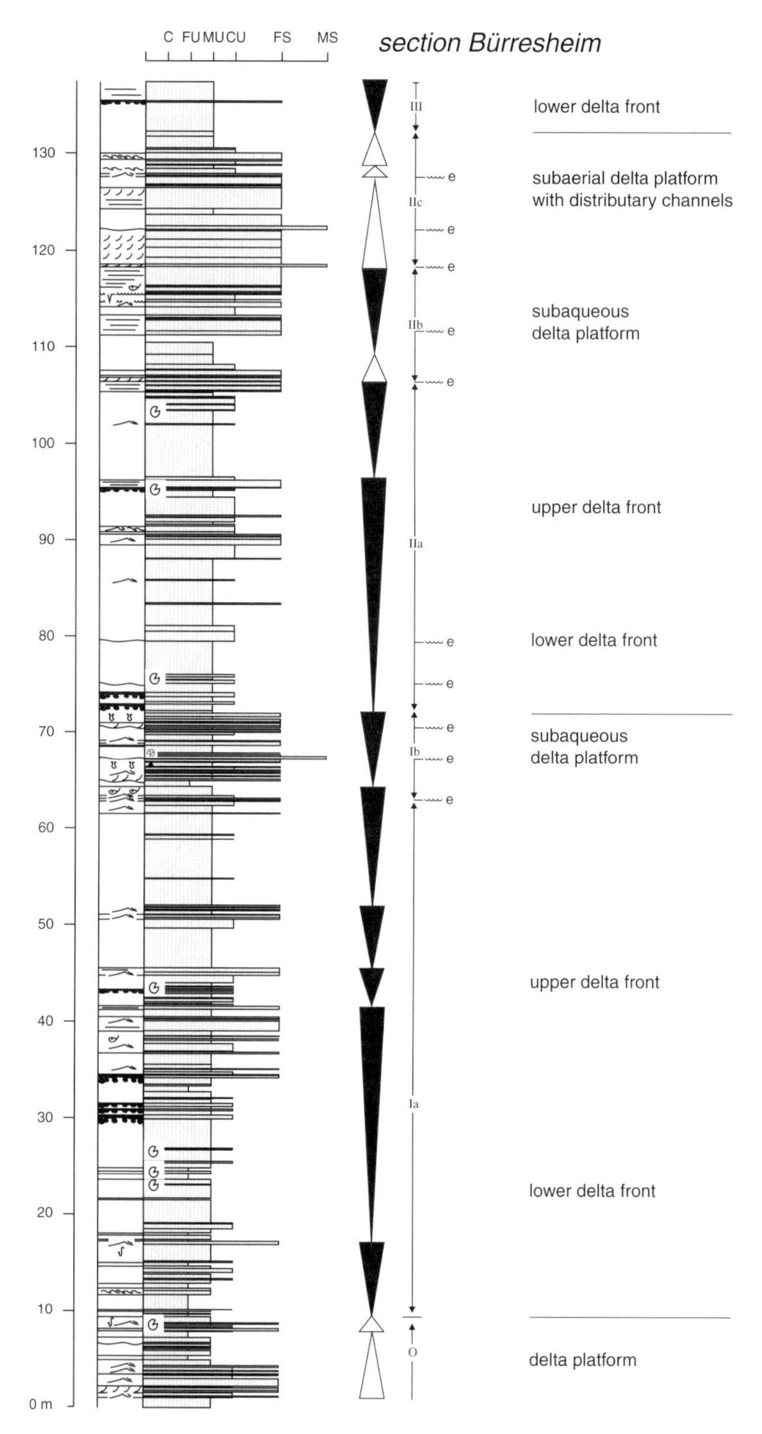

Fig. 11. Bürresheim columnar section (locality no. 6; Kürrenberg-Sandstein Formation, Upper Siegen Group). I–III, major sedimentary cycles. For interpretation see text; for legend and abbreviations see Figure 5.

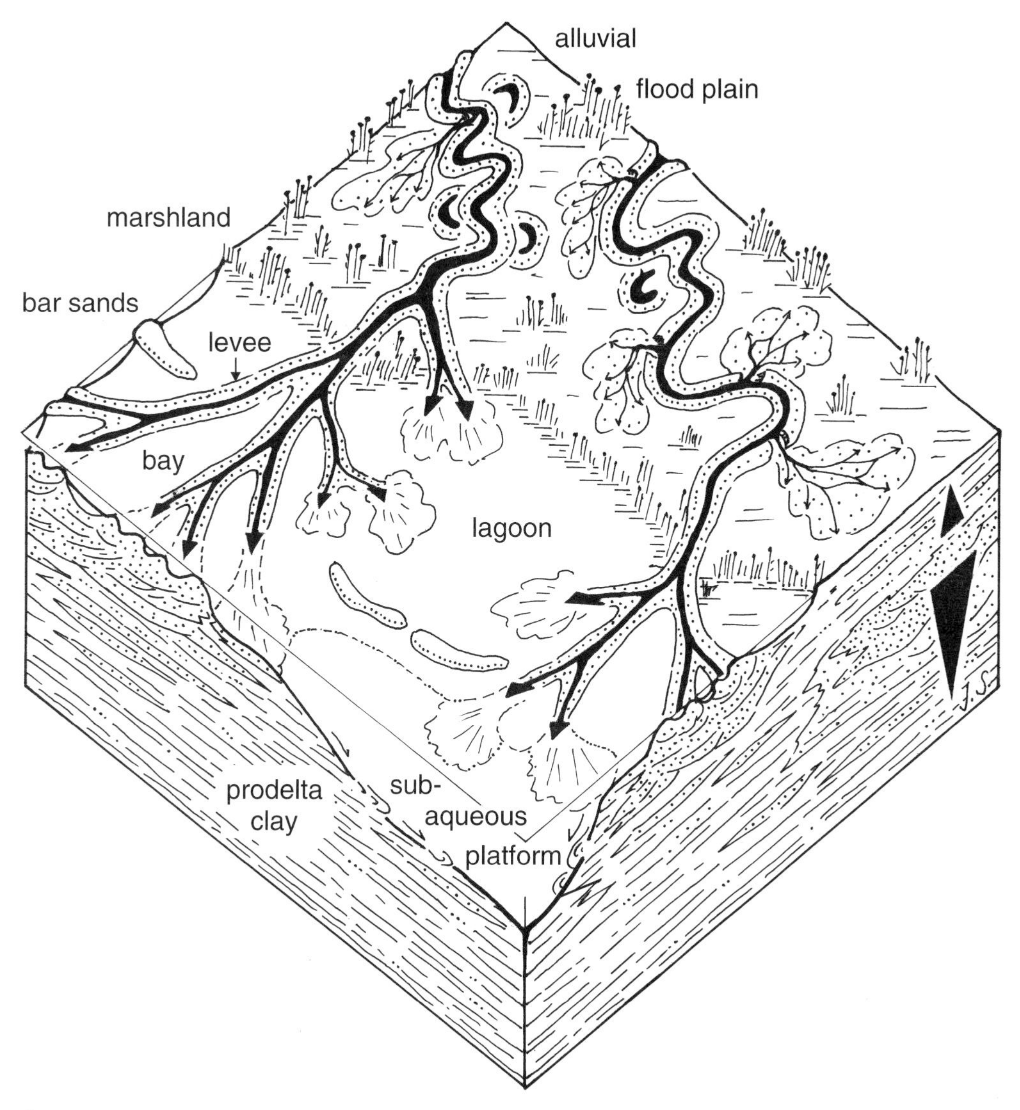

Fig. 12. Block diagramm illustrating the different subenvironments of the Siegenian fluvial-dominated elongate bird-foot delta. Black triangles indicate coarsening- and fining-up trends within the siliciclastic pile of the prograding delta.

(Ib, IIb). The uppermost part of the second cycle (IIc) may even be interpreted in terms of a subaerial delta platform. This is evident from tightly stacked meandering fluvial channel fills of the fluvial type cut into the sediments of a mixed sand/mud flat of the subaerial delta platform.

As the sediment load predominantly consisted of silts and muds here, i.e. more than 75% of the whole section, a rather long passage through the alluvial plains of a subaerial delta platform in the north has to be assumed. As neither beach sands nor larger tidal bar sands have been found in this section or nearby, our idea of a river-dominated delta of the bird-foot-type (Elliot 1986) seems to be corroborated. The delta slope is assumed to have been only gently inclined towards the basin floor in the south. Thus, frontal splay sands could be spread out far down the upper delta front. Slumping indicates that water-saturated material was deposited on an inclined surface, i.e. the delta front.

Due to the high sedimentation rate, life was hindered in this environment. Therefore, the content of biota is rather rare in the thick sedimentary column as a whole. Marine taxa have been found near to the lower and upper delta front in places (Dahmer 1940). Burrows of endobenth organisms are also sparse. Thus, the idea of a delta dominated by strong fluvial input has been supported once more. It was characterized by continuous subsidence and a high sediment supply. An intermittent fall and rise of the base level favoured the accumulation of huge masses of sediments in this part of the delta most of the time.

Prodelta environment

South of the Siegen Main Thrust, the Augustenthal columnar section has been taken in an abandoned quarry in the Wied valley north of Neuwied (TK 25 sheet 5510 Neuwied, R. 260481, H. 559395). This locality is situated in the northern part of the Central Facies Belt. Bedding is mostly vertical at the steep northern flank of the huge Mosel syncline, and slaty cleavage dips toward the NW in the Hunsrückschiefer-type sediments here. According to a rich fauna documented by Dahmer (1932) and Carls *et al.* (1982), the sediments belong to the Middle Siegen Augustenthal Formation (Meyer 1965). This sandy package within the several thousand metres thick Hunsrückschiefer *sensu lato* is time-equivalent to the Middle Siegen Rauhflaser Formation of the Northern Facies Belt (see Altenahr columnar section).

Description. The grain size of the **Augustenthal columnar section** (Fig. 13), which is about 75 m thick, ranges from fine-grained sand to fine-grained silt. Medium-grained sands are totally lacking. The bulk sand content amounts to 22%. Most of the material consists of medium-grained silt (54%). Horizontal planar bedding is most common in the sands, as is horizontal lamination in the silts ('Bänderschiefer' type sediments). There, the thickness of the single layer ranges down to a centimetre- to millimetre scale. In certain segments of this columnar section, for instance in the muddy sectors at 19–25 m or at 66–72 m, small-scale fining-up cycles have been observed that represent distal turbidites building up submarine fan sequences.

The layers of fine-grained sands are quartzitic sandstones today. They mostly show well defined bottom and top surfaces. There are no erosional surfaces noticed beneath the sands as was common in the columnar sections in the Northern Facies Belt. Nevertheless, mud clasts, lens-shaped enrichments of faunal relicts (e.g. the detritus of small crinoids) and undetermined shell hash indicate reworking and transport. This is also evident from current ripple lamination in the sands, and coarse-grained silts. But some ripple-like features on the bedding surfaces of many of the sand- and siltstones are due to later tectonic deformation. Fining- and thinning-up cycles at a scale from 0.25 to 7 m as well as coarsening- and thickening-up cycles of the same scale have been observed. But there are no large-scale cycles of tens of metres as has been found in several columnar sections of the Northern Facies Belt. Slumping has been noticed at 33.0 m. The long axis of the rolls is orientated NE–SW with a transport direction toward the SE. Outside the Augustenthal locality, this kind of slumping often occurs associated with the silts and sandy muds. The front of the slump rolls is rounded. Thus, a gravitational transport down-slope toward the SE is indicated. Slump structures have achieved a diameter in the order of 0.05 to 0.5 m, even up to 1 m.

Fauna and flora. Floral relicts are rare in the Hunsrückschiefer facies belt. They are of very limited distinctive value for environmental analysis because they all are allochthonous. They were washed in from the delta in the north.

Most of the animalia of this Hunsrückschiefer *sensu lato* lived as epibenthics. Anthozoans, crinoids, trilobites, brachiopods and bivalves have been found, as well as worm-shaped taxa (Dahmer 1932). Thus, well-oxygenated conditions must have prevailed down to the bottom of the sea. The faunal palaeocommunity found at Augustenthal belongs to the 'concentration enrichments' *sensu* Seilacher (1970). It is best characterized as a 'Brachiopoda–Crinoidea palaeocommunity'. It is in strong contrast to the fauna found at the classic Hunsrückschiefer *sensu stricto* localities near Bundenbach and Gemünden in the Hunsrück, which can best be characterized by the term 'Echinodermata–Arthropoda palaeocommunity' (Mittmeyer 1980; Bartels *et al.* 1998). Two taxa of the Augustenthal fossil assemblage, i.e. *Strophostylus naticoides* and *Leptaenopyxis bouei*, normally occur in the fully marine Hercynian facies belt (Carls *et al.* 1982; Erben 1962, 1994). They signal faunal connections to the remote open ocean.

Interpretation. Former reconstructions that postulated the Hunsrückschiefer environment to be an extended wadden sea by means of the faunal palaeocommunities, and other indications (e.g. Richter 1931; Kutscher 1931; Solle 1951; Engels 1956), are refuted today, as are water depths of about 600 m and more suggested by Seilacher & Hemleben (1966) and others. Following the faunal indications, we suggest that an intrashelf environment below the storm wave base existed here. A water depth of 100 m or even less seems to be appropriate.

For interpretation it seems important that there are no indications of intertidal to shallow subtidal conditions in this locality as had been reported

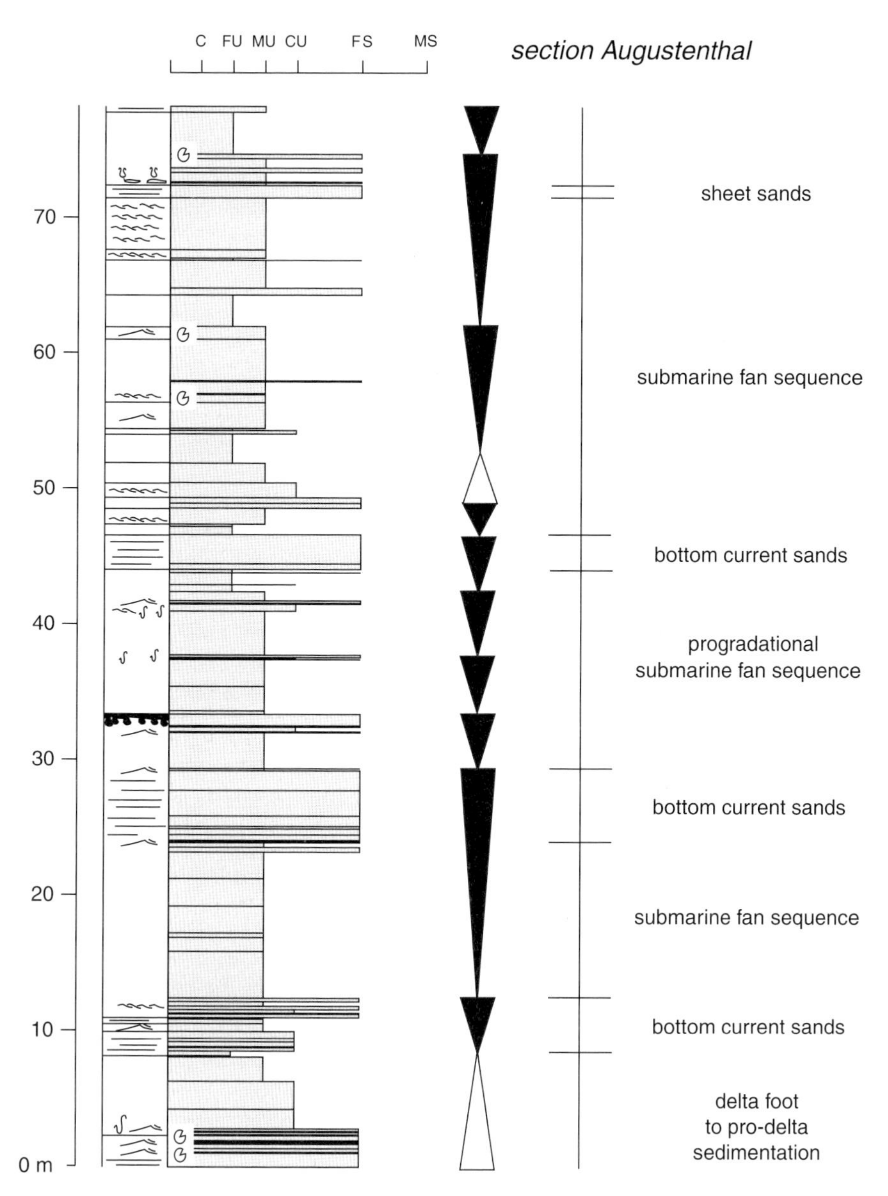

Fig. 13. Augustenthal columnar section (locality No. 7; Augustenthal Formation; Middle Siegenian). For interpretation see text; for legend and abbreviations see Figure 5.

from the Northern Facies Belt. Thus, the deposition of the sands and silts may have taken place below the storm wave base just in front, i.e. south of the huge Siegenian delta described above. We imagine that most of the material was deposited following rapid suspension transport, and fall-out down to the sea floor. This suspension fall-out took place from sediment clouds rich in fine-grained material. This material was rapidly deposited down-slope as indicated by the slump balls and rolls. The fining- and coarsening-up cycles as well as the thinning- and thickening-up cycles reflect a shifting of the distributaries of the delta, and a changing sediment supply down-slope from the delta mouth in the north. In certain sectors of this columnar section, indications of distal turbidites are noticed that gave rise to the idea that sequences of deep water fans originated from the fine-grained material that came down the delta slope in the north.

On the other hand, the horizontal parallel bedding of the sandy strata indicates another transport

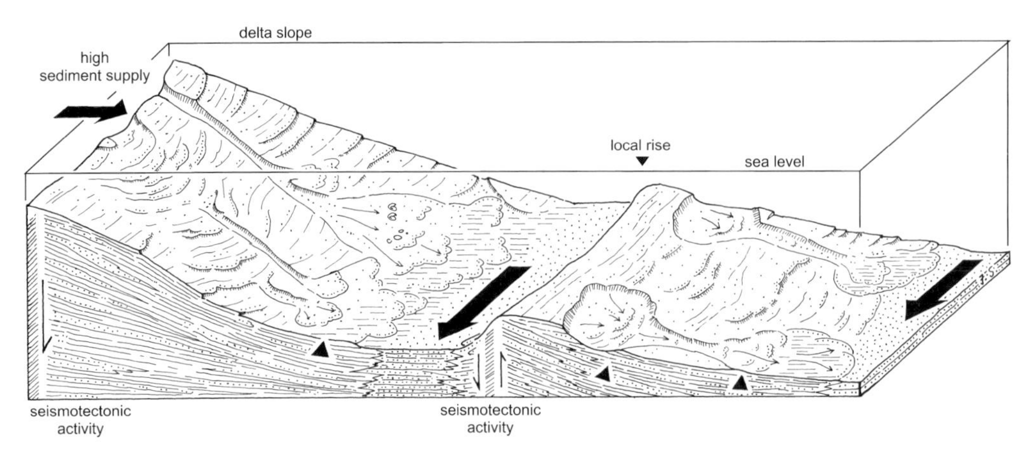

Fig. 14. Block diagram illustrating the Hunsrückschiefer-type environment in the delta foot to pro-delta region. Down-slope transport on the delta slope may have been triggered by seismotectonic shocks along syngenetic normal faults; bottom currents near to the delta foot produce well-sorted fine-grained sandy layers in a high-energy environment. Black triangles illustrate fining-up trends within distal fan sequences.

in an upper flow regime. Reworking resulted in the lens-shaped concentrations of fossil remains. This high-energy palaeoenvironment may have prevailed along bottom currents. The material that came down the delta slope was in competition with the material transported by the bottom currents in front of the delta. Thus, the Augustenthal palaeoenvironment was positioned in a pro-delta environment where submarine fan sequences alternate with bottom current sands and sand sheets spread about the mostly muddy sediments at the bottom of the sea (Fig. 14).

Factors controlling depositional processes

Palaeoclimatic conditions

According to Golonka *et al.* (1994), the Old Red Continent as a part of the huge landmass Laurussia was situated near to the equator in the Early Devonian with the area under research at latitude about 15–20° south. There are differences of about ±10° regarding this value (Witzke 1990; Tait *et al.* 1997). Due to the Caledonian collision, a mountain range had been built up north of the basin running NNE–SSW. It may have favoured perennial rainfall at the eastern and southeastern flanks (Fig. 15).

Concerning the climatic conditions in these latitudes, we have to expect a monsoonal to trade wind coastal climate according to the Köppen classification (Strahler & Strahler 1997). That means that the average temperature each month should have been above 18°C, and that no winter season existed there. Annual rainfall of about 1000 to 2000 mm/a exceeded the annual evaporation. The rainfall may have been produced by moisture-laden maritime tropical air masses, and the precipitation was favoured at the Old Red landmass as the trade winds blew to the SW. The climate within the source area, i.e. the highlands of the Old Red Continent, may have been quite similar to that of the lowlands. Looking for equivalent climatic conditions today, the climate would have been similar to that in Southeast Asia or Indonesia.

Regarding lithic palaeoclimatic indicators, coal seams that have been recorded from various localities within the Siegenian sediment pile of the Northern Facies Belt (Meyer 1994; Fuchs 1974) point to a high water table in the fluvial palaeoenvironment. Moreover, thick sediment sequences of terrigenous origin found in both facies belts indicate humid conditions in the area under research at that time. Tropical rainfall supported a perennial flow towards the depositional sink.

In the Siegenian, plants conquered the land at the northern margin of the Rhenohercynian Basin (Schweitzer 2003). Vegetated surfaces were bound to the lowlands, thus a plant-related hydrologic cycle was restricted at that time, and erosional processes would have been dominant in the source area. Huge amounts of weathered terrigenous material were thus ready for transport to the basin.

Subsidence

The thickness of the basin fill has been compiled from different places within the former Rhenohercynian Basin (Fig. 3; Meyer & Stets 1980, 1996; Stets

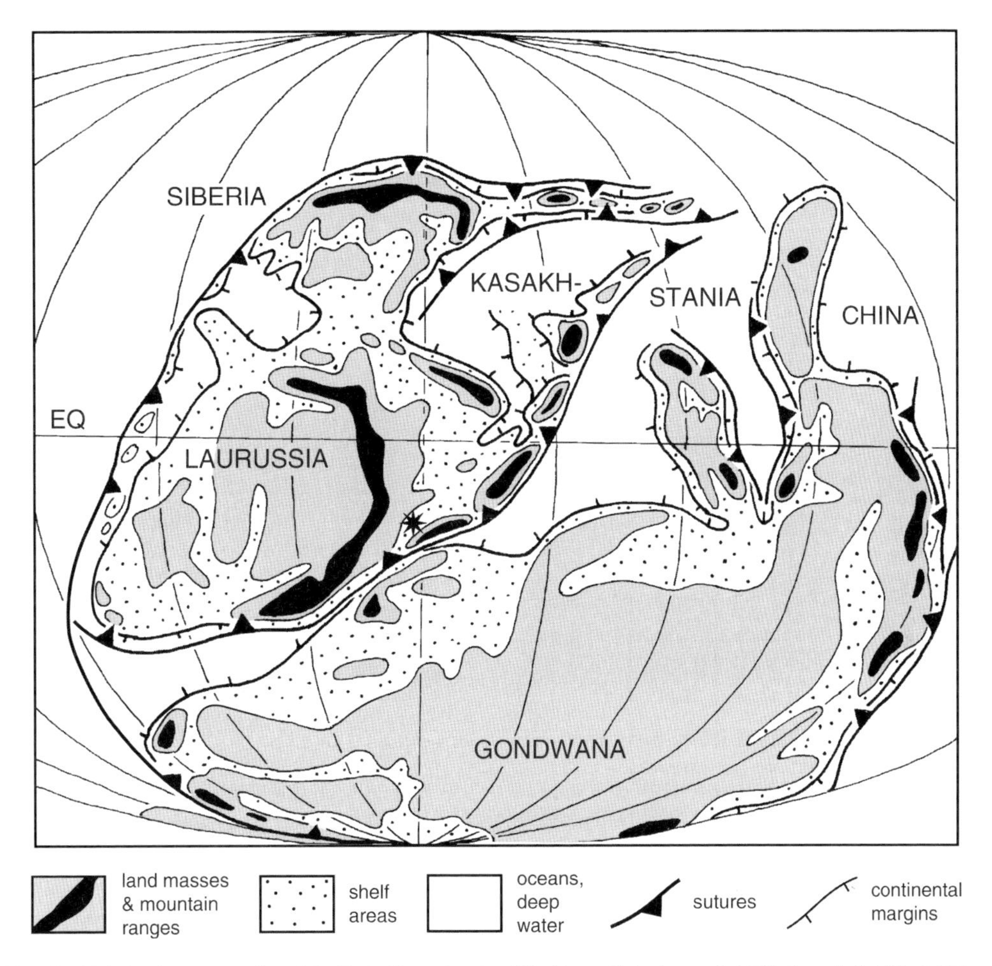

Fig. 15. Global palaeogeography of the Early Devonian (modified from Golonka *et al.* 1994; Stets & Schäfer 2002). The position of the Rhenohercynian Basin is marked by a star; EQ, equator.

& Schäfer 2002). They may give a rough idea of the pre-orogenetic processes that formed the basin. Generally, the subsidence of this basin followed a 'mixed shear' model (Kuznir & Ziegler 1992; Ziegler 1992; Heinen 1996; Einsele 2000) taking into account simple shear and pure shear due to the properties of the lithosphere. Thus, different stages of subsidence have been distinguished during the evolution of the basin (Meyer & Stets 1980; Heinen 1996; Stroetmann-Heinen 1999).

The subsidence during the Siegenian belongs to a first stage that lasted from the Gedinnian to the Early Emsian (inclusive). With thicknesses of up to 3000 to 5000 m for the Siegenian sediments in the Northern and northern Central Facies belts, rather high rates of subsidence occurred during the 7 Ma. The Siegenian interval signalled a period of initial extension and active rifting (Einsele 2000). Subsidence was promoted by syngenetic normal faulting. The predecessor of the Siegen Main Thrust, for example, belonged to this type of fault, producing an area of higher water depth in the northern part of the Central Facies Belt just in front of the delta.

There are almost no signs of syngenetic normal faulting in the Northern Facies Belt. Thus, we imagine that subsidence was lower in the northern part than in the south as indicated by a higher thickness in the south. At a whole, a sedimentary wedge resulted, opening towards the south. Moreover, no severe hiatus has been recorded from the whole sedimentary column of the Siegenian. Thus, subsidence would have taken place nearly continually.

Sea level

A global sea level curve was introduced and discussed by Dennison (1985) and Johnson *et al.* (1985) for the Devonian and it is still valid today

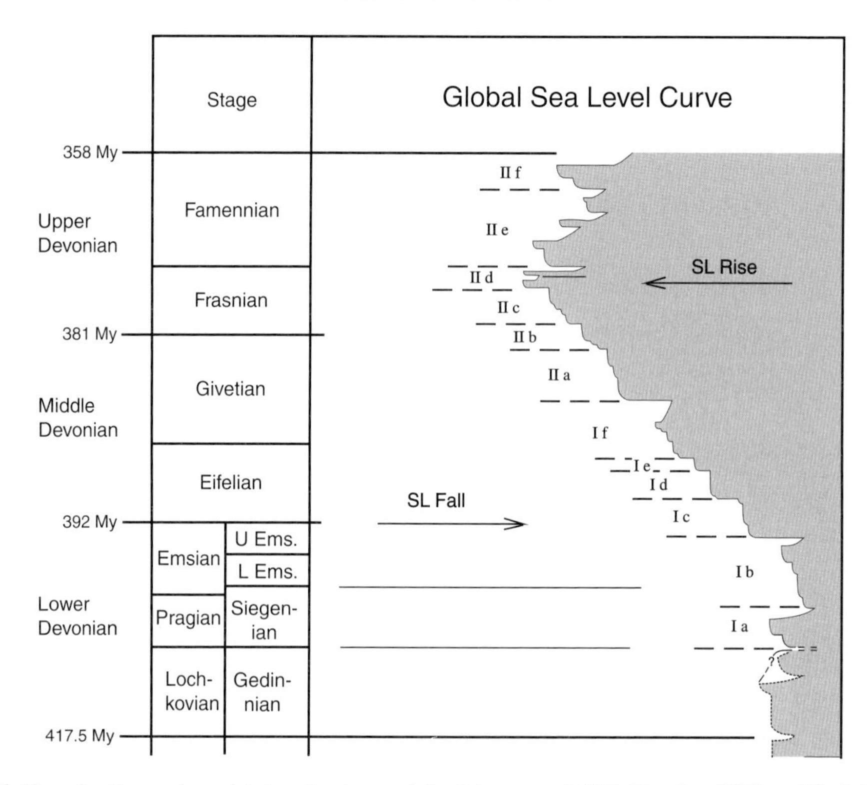

Fig. 16. Devonian Euramerican global sea level curve (after Johnson *et al.* 1985; Dennison 1985, modified). The numerical ages (from DSK 2002) give a rough measure of chronostratigraphy.

(Fig. 16). According to these authors a marked sea level fall happened near to the Silurian–Gedinnian boundary followed by a slightly changing sea level throughout the Gedinnian. Another sea level rise and fall (Ia) took place during the Early and Middle Siegenian. In the Late Siegenian and throughout the Early Emsian (Ib), a consequent sea level rise started with another relapse in the Late Emsian just before the important global sea level rise of the Middle and Late Devonian started.

In the Northern Facies Belt, the course of the relative sea level can be traced solely using the sub-environments of the delta and its fossil assemblages. It does not seem quite similar to the general trend of the global sea level curve. It has to be admitted that coastal onlaps or offlaps are not obvious in the Northern Facies Belt during the Siegenian. Palaeogeographic research indicated a rather stable position of the northern border of the basin in the region from Aachen to Düsseldorf during that time (Hesemann 1975; Meyer & Stets 1980). But the fossil palaeocommunities found within the Northern Facies Belt indicate a rising open-marine influence during the Middle to early Late Siegenian. In the north, it is followed by a relative sea level fall in the Late Siegenian to Early Emsian (Wahnbach Formation, Grey Clastic Lower Emsian series; Fuchs 1974; Meyer 1994; Stets & Schäfer 2002).

As the trend of the relative sea level nearly coincides with the global sea level of the Early to Middle Siegenian, no severe tectonic influence on the sedimentary processes has been supposed in this region at that time. On the other hand, an over-abundant sedimentary supply or a slow-down of subsidence starting in the Late Siegenian to Early Emsian produced an opposing effect with respect to the global sea level curve. Nevertheless, the Siegenian delta system at the northern margin of the Rhenohercynian Basin remained stable. Subsidence, sediment supply as well as sea level changes kept in balance nearly all the time.

Yet, within the Central Facies Belt, persisting subtidal open-marine conditions during the Siegenian and Lower Emsian 'Hunsrückschiefer' sedimentation do not permit recognition of any sea level change.

Siliciclastic input from the north

The siliciclastic supply from the Old Red source area was controlled by the climatic conditions, weathering processes, tectonic uplift, the mineralogical composition of the rocks exposed there and, last but not least, by the fluvial transport down towards the basin. None of the Siegenian sediments in the Northern Facies Belt and in the northern part of the Central Facies Belt of the Rhenohercynian Basin show any marked differences. They are of an uniform and nearly monotonous composition. No severe disruptions can be identified throughout the whole Siegenian, and mineralogical and petrological research did not furnish any severe variations of the detritus delivered. All the results of the detritus research point to the Old Red Continent in the north as the only source area (Haverkamp 1991; Wierich 1999). The material delivered to the basin corresponds to that of recycled orogens *sensu* Dickinson *et al.* (1983). Quartzose material proved to be the most frequent among the sandstones (Holl 1995).

The material delivered to the Northern and northern Central Facies Belts ranges from medium-grained sands to mud. All the coarser material was kept back in the fluvial systems of the Old Red Continent itself and in its Foreland Belt. Thus, a sorting process controlled the input into the fluvial and deltaic systems of the Northern Facies Belt. Within this area, the grain-size distribution pattern depended on the different energy conditions of the supratidal to shallow subtidal palaeoenvironments. Here, another screening effect gave rise to the fact that only fine-grained detritus reached the Central Facies Belt. Thus, no coarse- and medium-grained sands have been found within the sedimentation area of the later Hunsrückschiefer *sensu lato* and *sensu stricto*.

Conclusions

The relationships between the different subenvironments in the land–sea transition zone of a huge delta complex have been demonstrated using field data of selected outcrops. Biostratigraphical data derived from the Siegenian flora and from the mostly benthic faunas allowed linking of the different columnar sections that stem from a 3000 to 5000 m thick sedimentary package of Siegenian age. Unfortunately, it does not contain any stratigraphic gaps, key surfaces, condensed sections or specific marker horizons. Subsidence, steady sediment supply, and accumulation during tropical to subtropical climatic conditions, and a slowly changing relative sea level were the main factors controlling the uniform long-term sedimentary processes. Thus, subsidence and detrital input were in balance with each other all through the Siegenian.

The sequence stratigraphy interpretation model of Mjos *et al.* (1998) may be used to explain the development of facies in the Early Devonian in the northern part of the Rhenohercynian Basin. A maximum flooding stage may have occurred in the Gedinnian (central broken line in Fig. 17). The facies tracts recorded in our field sections altogether show a pronounced small-scale cyclicity. This is coincident with a prominent base level rise trend in the proximal fluvial Old Red Continent Foreland and parts of the Northern Facies belts. In the distal reaches of the Northern Facies Belt, showing fluvial and deltaic depositional environments, a base level fall trend is obvious.

The small-scale cyclicity, derived from our field sections, is tentatively located in the prograding highstand systems tract of the traditional sequence stratigraphy model, above the maximum flooding surface, central to the model of Mjos *et al.* (1998). This model coincides with an intense progradational trend observed in our field sections, providing numerous rise/fall patterns of the small-scale cycles (Fig. 17).

No precise stratigraphical measures are available in this part of the Lower Devonian siliciclastic wedge of the Rhenohercynian Basin. The global sea level curve (Fig. 16) gives an overall second-order trend only, tentatively resolved to give a third-order systems tracts cyclicity. Thus, a sequence stratigraphy model in the Rhenohercynian Basin cannot line up to a fourth-order cyclity. But this will be necessary to show parasequences in a 0.5 to 0.01 Ma range that should be specified in a coastal marine setting. Therefore, base level cycles form as a replacement; they give a depositional trend not only in detail, but also on a larger scale (Schäfer 2005). In combination with the model of Mjos *et al.* (1998), the position of the depositional environments can be located in the progradational highstand systems tract of the third-order hierarchy of the standard sequence stratigraphy model of Baum & Vail (1988). Yet, a maximum flooding surface below and a sequence boundary above, as demanded in the model shown in Figure 17, cannot be specified for the moment.

Summary

According to the lithological record, wide distal fluvial systems developed in a vast lowland landscape near the northern margin of the Rhenohercynian Basin in front of the Old Red Continent during the Siegenian. This landscape must have been separated from the source area by the more proximal Old Red Continent Foreland Belt (Fig. 18) that has not fully been delivered in the Rheinische Schiefergebirge.

The Northern Facies Belt and the northern part of the adjacent Central Facies Belt in the south were linked as parts of a huge deltaic system at the

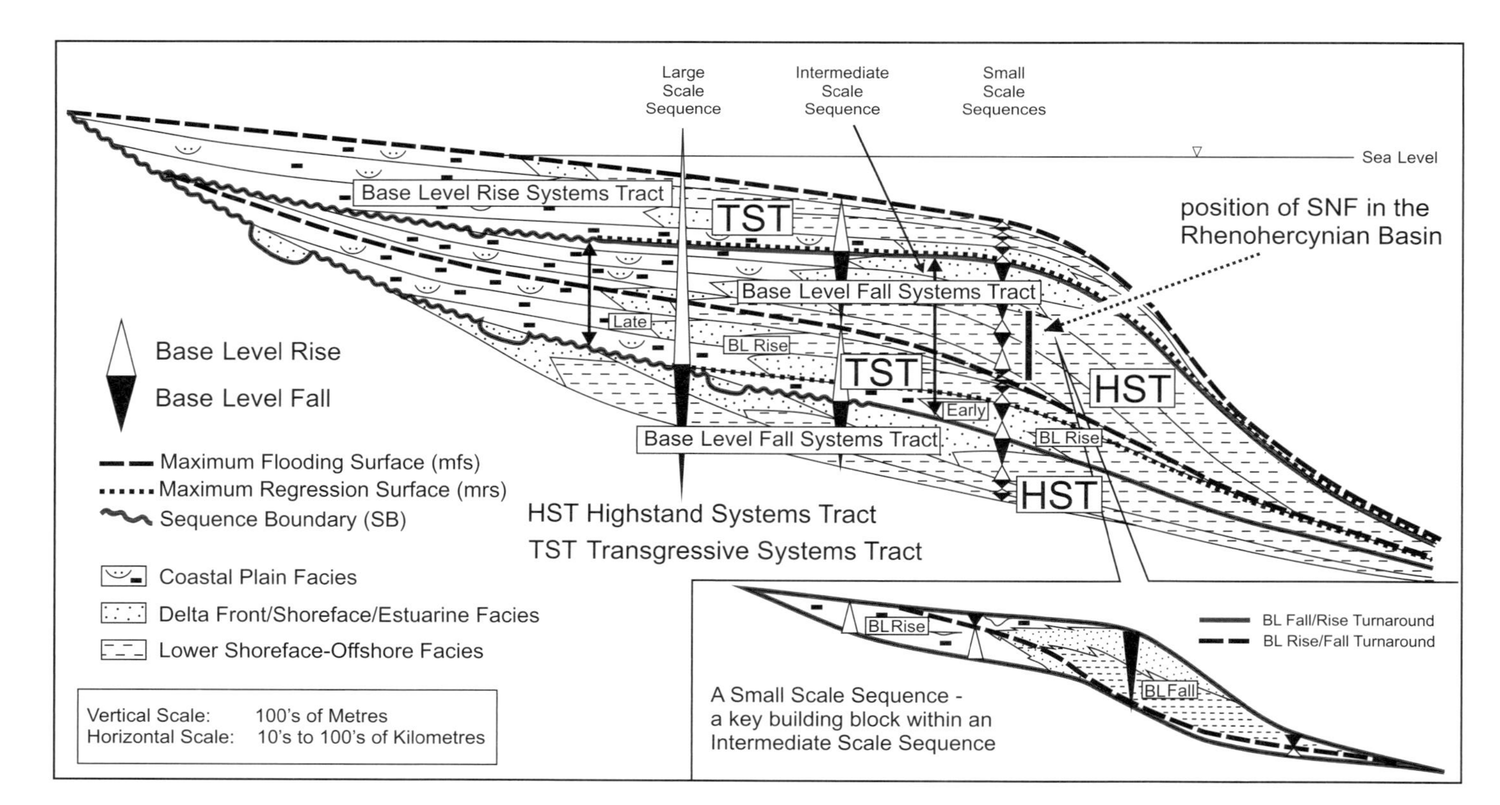

Fig. 17. Sequence cycles in a traditional sequence stratigraphy model interpreted in terms of base level cyclicity (Mjos *et al.* 1998). The Siegenian Normal Facies (SNF) of the Rhenohercynian Basin may be placed in the upper highstand systems tract above a maximum flooding surface, i.e. in the field with an intermediate-scale base level fall trend in the centre of the diagram. A small-scale cyclicity as indicated in the diagram has been shown in more detail in the columnar sections 1–7. A strong progradational trend of the Lower Devonian deltaic wedge towards the basin is signalled.

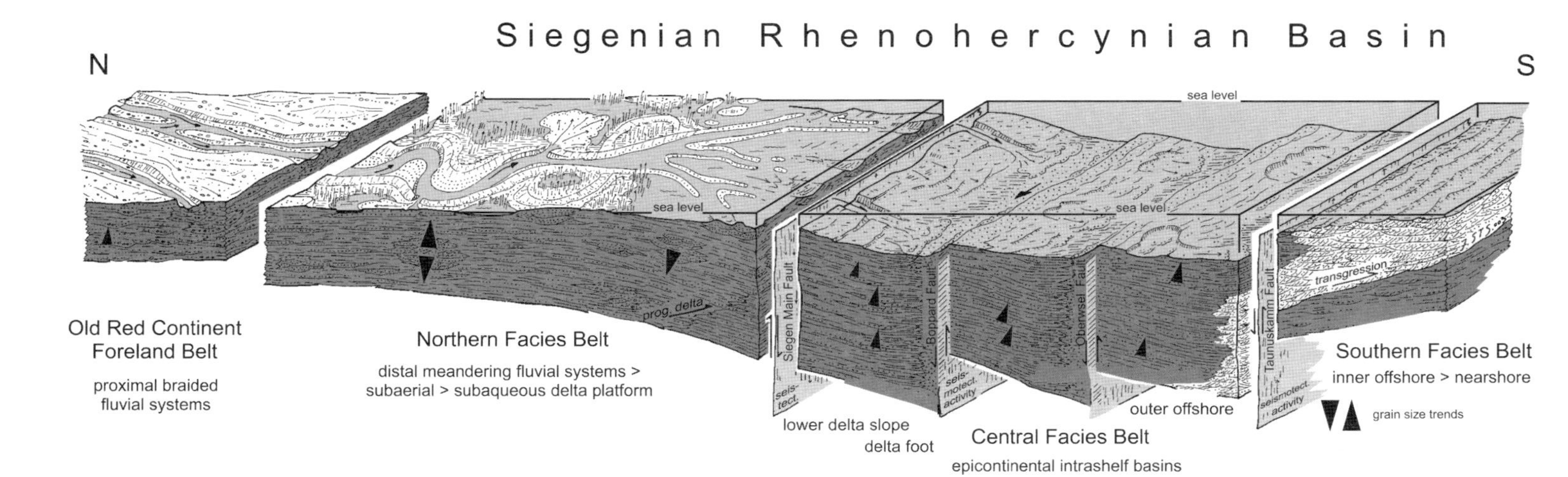

Fig. 18. Block diagram showing the different palaeoenvironments of the Siegenian Rhenohercynian Basin with the fluvial-dominated elongate bird-foot delta in the north (Stets & Schäfer 2003).

northern margin of the Rhenohercynian Basin (Fig. 18). The whole area was situated in the land–sea transition zone. The two facies belts were separated from each other by the Siegen Main Thrust. It inverted from a large syngenetic normal fault system during the Siegenian to a major thrust fault system during the Variscan (Hercynian) orogeny. Thus, parts of the delta, especially most of the lower delta slope, are missed or masked by the activity of this major structure today.

Caused by long-term steady and continuous processes, a rather homogeneous siliciclastic sedimentary wedge formed in the land–sea transition zone south of the Old Red Continent. This wedge reached a maximum thickness of up to 5000 m in the south. We imagine that large-scale, slowly acting synsedimentary tectonic tilting towards the south was a main factor of this evolution.

In the Northern Facies Belt, which was the eminent part of the land–sea transition zone, distal meandering fluvial systems existed in a low-relief mud and floodplain environment (Rheineck, Heimersheim, Schladern columnar sections). They gradually passed over to a subaerial delta platform towards the south. Here, large areas covered with lakes, swamps and marshlands existed. They were inhabited by the plant assemblages of a *'Psilophyton'* flora, and a *'Modiolopsis'* fauna. Meandering rivers passed through these plains. Toward the south, lagoons, flood basins and tidal flats existed, crossed by distributary channels in the seaward elongation of the river channels (Altenahr columnar section).

These Lower Devonian environments were widespread. They covered most of the area along the northern margin of the Rhenohercynian Basin. No sharp boundaries existed between the fluvial systems, the delta platform and the delta front. The transition of all these deltaic units was gradual.

On the delta platform, and near to its front, no major river systems or distributaries have been found yet. A wide network of smaller channels has to be assumed draining the vast subaerial lowland area. These rivers built up fluvial-dominated bird-foot deltas, each of them towards the south.

The rivers brought huge volumes of siliciclastic detritus to the sea that could not be fully incorporated in the fluvial landscape, or reworked near to the coast by wave activity. Thus, levees were built out into the shallow water of the delta front. From time to time, they were broken through with the outpouring detritus nourishing interdistributary bays and lagoons (Altenahr columnar section). Thus, fluvial activity was simulated in the nearshore marine environment. Indications of strict coastal environments are missing all over the Northern Facies Belt. But nearshore, sandy barrier systems in front of a foreshore tidal environment have been found (Unkel columnar section).

Seaward, the detritus spread out on the subaqueous delta platform. In places, marine biota had a chance to settle there (Bürresheim columnar section). Further towards the sea in the south, an upper delta slope existed that was characterized by suspension fall-out and transport down-slope. Lots of the silts and muds came to rest here. The water-saturated sediments formed slumps and slides. Submarine fans formed towards the bottom of the sea on the delta-foot plain (Augustenthal columnar section). We imagine that density flows driven by floods, slumps and suspension fall-out were the main mechanisms of sediment transport down-slope. They were probably triggered by seismotectonic shocks along syngenetic normal faults.

Indicative sedimentary features have shown that the water depth in front of the delta was rather shallow. There is no doubt that it was below storm wave base but surely not more than about 100 m below sea level just off the delta front. The 'Brachiopoda–Crinoidea palaeocommunities' containing anthozoans that lived there indicate fully oxygenated, well-suited open-marine conditions (Augustenthal columnar section), at least in places.

The river-dominated delta systems at the northern margin of the Rhenohercynian Basin covered a subsiding inclined epicontinental platform. As the water depth near the delta foot may not have exceeded 100 m, the delta slopes must have been only gently inclined and rather wide (Fig. 18). Therefore, sharp boundaries between the different subenvironments of the delta system, even down to the basin floor, are missed. Conditions of large deltas such as those in the land–sea transition zone of the Mississippi or Brahmaputra rivers cannot easily be applied to the Siegenian delta of the northern Rheinische Schiefergebirge.

Reviews were made by Klaus-Werner Tietze, University of Marburg (Germany) and by Ulrich Jansen, Senckenberg-Museum, Frankfurt/Main (Germany). We thank them for their constructive help and fruitful discussion. Also, the encouragement of Peter Königshof, Senckenberg-Museum, Frankfurt/Main (Germany), is appreciated for giving the initial impulse to prepare this paper.

References

BARTELS, C., BRIGGS, D. E. & BRASSEL, G. 1998. *The Fossils of the Hunsrück Slate: Marine Life in the Devonian*. Cambridge University Press, Cambridge, Cambridge Paleobiology Series, **31**.

BAUM, G. R. & VAIL, P. R. 1988. Sequence stratigraphic concepts applied to Paleogene outcrops, Gulf and Atlantic basins. *In*: WILGUS, C. K., HASTINGS, B. S., KENDALL, C. G. ST. C., POSAMENTIER, H. W., ROSS, C. A. & VAN WAGONER, J. C. *Sea-level Changes: An Integrated Approach*. Society for

Economic Palaeontology and Mineralogy Special Publication, **42**, 308–327.

BAUSCH VAN BERTSBERGH, J. W. 1940. Richtungen der Sedimentation in der Rheinischen Geosynklinale. *Geologische Rundschau*, **31**, 328–364.

BLIECK, A. & JAHNKE, H. 1980. Pteraspiden (Vertebrata, Heterostraci) aus den Unteren Siegener Schichten und ihre stratigraphischen Konsequenzen. *Neues Jahrbuch für Geologie Paläontologie, Abhandlungen*, **159**, 360–378.

BLUNDELL, D., FREEMAN, R. & MUELLER, ST. (eds) 1992. *A Continent Revealed – The European Geotraverse*. Cambridge University Press, Cambridge.

BRAUN, A. & MÖRS, T. (eds) 2001. Geologie und Paläontologie im Devon und Tertiär der ICE-Trasse im Siebengebirge. *Decheniana, Beihefte*, **39**.

BREIL-SCHOLLMAYER, A. 1990. *Zur Geologie des Siegenium und Emsium (Unterdevon) zwischen Aachen und Hellenthal (Nordeifel) unter besonderer Berücksichtigung der faziellen Entwicklung*. Thesis, Technical University of Aachen.

CARLS, P., JAHNKE, H., LUZNAT, M. & RACHEBOEUF, P. 1982. On the Siegenian Stage. *Courier Forschungsinstitut Senckenberg*, **55**, 181–198.

CROSS, T. A. & HOMEWOOD, P. W. 1997. AMANZ GRESSLY'S role in founding modern stratigraphy. *Geological Society of America, Bulletin*, **109**, 1617–1630.

CROSS, T. A. & LESSENGER, M. A. 1998. Sediment volume partitioning: rationale for stratigraphic model evaluation and high-resolution stratigraphic correlation. *In*: GRADSTEIN, F. M., SANDVIK, K. O. & MILTON, N. J. (eds) *Sequence Stratigraphy–Concepts and Applications*. Norwegian Petroleum Society (NPF) Special Publication, **8**.

CROSS, T. A., BAKER, M. R., CHAPIN, M. A. *ET AL*. 1993. Application of high-resolution sequence stratigraphy to reservoir analysis. *In*: ESCHAD, R. & DOLIGEZ, B. (eds) *Subsurface Reservoir Characterisation from Outcrop Observations*. Proceedings of the 7th Exploration and Production Research Conference, Technip, Paris, 11–33.

DAHMER, G. 1932. Fauna der belgischen "Quartzophyllades de Longlier" in Siegener Rauhflaserschichten auf Blatt Neuwied. *Jahrbuch preussisch geologische Landesanstalt*, **52**, 86–111.

DAHMER, G. 1934*a*. Die Fauna der Seifener Schichten. *Abhandlungen preussisch geologische Landesanstalt, N.F.*, **147**.

DAHMER, G. 1934*b*. Die Fauna der Siegener Schichten in der Umgebung des Laacher Sees. *Jahrbuch preussisch geologische Landesanstalt*, **55**, 122–141.

DAHMER, G. 1936. Die Fauna der Siegener Schichten von Unkel (Bl. Königswinter). *Jahrbuch preussisch geologische Landesanstalt*, **56**, 633–671.

DAHMER, G. 1937. Die Fauna der Siegener Schichten im Ahrgebiet. *Jahrbuch preussisch geologische Landesanstalt*, **57**, 435–464.

DAHMER, G. 1940. Die Fauna der Siegener Schichten (Unter-Devon) zwischen Bürresheim und Kirschesch in der Südost-Eifel. *Senckenbergiana*, **22**, 77–102.

DENNISON, J. M. 1985. Devonian eustatic fluctuations in Euramerica: Discussion. *Geological Society of America Bulletin*, **96**, 1595–1597.

DICKINSON, W. R., BEARD, L. S. *ET AL*. 1983. Provenance of North American Phanerozoic sandstones in relation to tectonic setting. *Geological Society of America Bulletin*, **94**, 222–235.

DSK 2002. *Stratigraphische Tabelle von Deutschland 2002*. Deutsche Stratigraphische Kommission, Potsdam.

EINSELE, G. 2000. *Sedimentary Basins – Evolution, Facies, and Sediment Budget* (2nd edn). Springer, Berlin.

ELKHOLY, H. & GAD, J. 2006. Die Wied-Gruppe (vormals Hunsrückschiefer): Eine neue lithostratigraphische Einheit am Nordrand der Moselmulde – Untersuchungen zu ihrer faziellen und stratigraphischen Einordnung. *Mainzer geowissenschaftliche. Mitteilungen*, **34**, 49–72.

ELLIOT, T. 1986. Deltas. *In*: READING, H. G. (ed.) *Sedimentary Environments and Facies*, Blackwell, London, 97–142.

ENGELS, B. 1956. Über die Fazies des Hunsrückschiefers. *Geologische Rundschau*, **45**, 143–149.

ERBEN, H. K. 1962. Zur Analyse und Interpretation der rheinischen und herzynischen Magnafazies des Devons. *2nd Int. Arbeitstagung Silur-Devon-Grenze Bonn/Bruxelles*, 42–61.

ERBEN, H. K. 1994. Das Meer des Hunsrückschiefers. *In*: KOENIGSWALD, W. & MEYER, W. (eds) *Erdgeschichte im Rheinland*. Pfeil, München, 49–56.

FRANKE, W. 1992. Phanerozoic structures and events in central Europe. *In*: BLUNDELL, D., FREEMAN, T. & MUELLER, S. (eds) *A Continent Revealed – The European Geotraverse*. Cambridge University Press, Cambridge, 164–180.

FRANKE, W. 2000. The mid-European segment of the Variscides: tectonostratigraphic units, terrane boundaries and plate tectonic evolution. *In*: FRANKE, W., HAAK, V., ONCKEN, O. & TANNER, D. (eds) *Orogenic Processes: Quantification and Modelling in the Variscan Belt*. Geological Society of London, Special Publication, **179**, 35–61.

FRANKE, W., EDER, W., ENGEL, W. & LANGENSTRASSEN, F. 1978. Main aspects of Geosynclinal sedimentation in the Rhenohercynian Zone. *Zeitschrift deutsche geologische Gesellschaft*, **129**, 201–216.

FUCHS, G. 1971. Faunengemeinschaften und Fazieszonen im Unterdevon der Osteifel als Schlüssel zur Paläogeographie. *Notizblatt hessisches Landesamt für Bodenforschung*, **99**, 78–105.

FUCHS, G. 1974. Das Unterdevon am Ostrand der Eifeler Nordsüd-Zone. *Beiträge naturkundliche Forschung Südwest-Deutschland, Beiheft*, **2**.

FUCHS, G. 1982. Upper Siegenian and Lower Emsian in the Eifel Hills. *Courier Forschungsinstitut Senckenberg*, **55**, 229–256.

FÜCHTBAUER, H. (ed.) 1988. *Sedimente und Sedimentgesteine. Sediment-Petrologie*, Teil II, (4th edn). Schweizerbart, Stuttgart.

GAD, J. 2005. Miosporen aus dem Hunsrückschiefer des Westerwaldes (Rheinisches Schiefergebirge, Unterdevon) und die stratigraphische Stellung der Mayen-Formation. *Mainzer geowissenschaftliche Mitteilung*, **33**, 167–218.

GAD, J. 2006. Was ist eigentlich Hunsrückschiefer? *Jahresberichte und Mitteilungen oberrheinischer geologischer Verein, N.F.*, **88**, 53–65.

GOLONKA, J., ROSS, M. I. & SCOTESE, C. R. 1994. Phanerozoic paleogeographic and paleoclimatic modeling maps. *In*: EMBRY, A. F., BEAUCHAMP, B. & GLASS, D. J. (eds) *Pangea: Global Environments and Resources*, Canadian Society Petroleum Geology, Memoir, **17**, 1–47.

GRABERT, H. 1979. *Erläuterungen zu Blatt 5111 Walbröl*. Geologische Karte Nordrhein-Westfalen 1:25 000, Erl. 5111, Krefeld.

HAVERKAMP, J. 1991. *Detritusanalyse unterdevonischer Sandsteine des Rheinisch-Ardennischen Schiefergebirges und ihre Bedeutung für die Rekonstruktion der sedimentliefernden Hinterländer*. Thesis, Technical University of Aachen.

HEINEN, V. 1996. Simulation der präorogenen devonisch-unterkarbonischen Beckenentwicklung und Krustenstruktur im linksrheinischen Schiefergebirge. *Aachener geowissenschaftliche Beiträge*, **15**.

HESEMANN, J. 1975. *Geologie Nordrhein-Westfalens*. Ferd. Schöningh, Paderborn.

HOLL, H. G. 1995. Die Siliziklastika des Unterdevon im Rheinischen Trog (Rheinisches Schiefergebirge) – Detritus-Eintrag und P, T-Geschichte. *Bonner geowissenschaftliche Schriften*, **18**.

HOMEWOOD, P. W., MAURIAUD, P. & LAFONT, F. 2000. *Best Practices in Sequence Stratigraphy*. Bulletin Centre Recherche Elf Exploration Production, Memoir, **25**.

JAHNKE, H. 1971. Fauna und Alter der Erbslochgrauwacke (Brachiopoden und Trilobiten, Unterdevon, Rheinisches Schiefergebirge und Harz). *Göttinger Arbeiten Geologie Paläontologie*, **9**.

JANKOWSKY, W. 1955. Schichtenfolge, Sedimentation und Tektonik im Unterdevon des Rheintales in der Gegend von Unkel – Remagen. *Geologische Rundschau*, **44**, 59–86.

JOHNSON, G. A., KLAPPER, G. & SANDBERG, C. A. 1985. Devonian eustatic fluctuations in Euramerica. *Geological Society of America, Bulletin*, **96**, 567–587.

KEGEL, W. 1950. Sedimentation und Tektonik in der rheinischen Geosynklinale. *Zeitschrift deutsche geologische Gesellschaft*, **100**, 267–289.

KIRNBAUER, T. 1991. Geologie, Petrographie und Geochemie der Pyroklastika des Unteren Ems/Unterdevon (Porphyroide) im südlichen Rheinischen Schiefergebirge. *Geologische Abhandlungen Hessen*, **92**.

KOCH, C. 1881. Über die Gliederung der rheinischen Unterdevon-Schichten zwischen Taunus und Westerwald. *Jahrbuch. Königlich Preußische Landesanstalt für 1880*, **1**, 190–242.

KOSSMAT, F. 1927. Gliederung des varistischen Gebirgsbaues. *Abhandlungen lls Sächsisch Geologisch Landesamt*, **1**, 1–39.

KRÄUSEL, R. 1936. Neue Untersuchungen zur paläozoischen Flora: Rheinische Devonfloren. *Bericht deutsche Botanische Gesellschaft*, **54**, 307–328.

KRÄUSEL, R. & WEYLAND, H. 1930. Die Flora des deutschen Unterdevons. *Abhandlungen preußisches geologisches Landesamt, N.F.*, **131**.

KUSZNIR, N. J. & ZIEGLER, P. A. 1992. The mechanics of continental extension and sedimentary basin formation: A simple shear/pure shear flexural cantilever model. *Tectonophysics*, **215**, 117–137.

KUTSCHER, F. 1931. Zur Entstehung des Hunsrückschiefers am Mittelrhein und auf dem Hunsrück. *Jahrbuch nassauischer Verein für Naturkunde*, **81**, 177–232.

LANGENSTRASSEN, F. 1983. Neritic sedimentation of the Lower and Middle Devonian in the Rheinisches Schiefergebirge east of River Rhine. *In*: MARTIN, H. & EDER, F. W. (eds) *Intracontinental Fold Belts*. Springer, Berlin, 43–77.

LANGENSTRASSEN, F., BECKER, G. & GROOS, H. 1979. Zur Fazies und Fauna der Brandenberg-Schichten bei Lasbeck (Eifel-Stufe, Rechtsrheinisches Schiefergebirge). *Neues Jahrbuch Geologie Paläontologie, Abhandlungen*, **158**, 64–99.

LGB 2005. *Geologie von Rheinland-Pfalz*. Landesamt für Geologie und Bergbau Rheinland-Pfalz, Schweizerbart, Stuttgart.

MEYER, W. 1958. Geologie der Siegener Schichten zwischen Ahr und Nette (Osteifel). *Zeitschrift deutsche geologische Gesellschaft*, **109**, 452–462.

MEYER, W. 1965. Gliederung und Altersstellung des Unterdevons südlich der Siegener Hauptaufschiebung in der Südost-Eifel und im Westerwald (Rheinisches Schiefergebirge). *Max-Richter-Festschrift*, 35–47.

MEYER, W. 1994. *Geologie der Eifel* (3rd edn). Schweizerbart, Stuttgart.

MEYER, W. & RAUFI, F. 1991. Das pflanzenführende Unterdevon in der Umgebung des Wahnbachtals im südwestlichen Bergischen Land. *Niederrheinische Landeskunde*, **10**, 99–107.

MEYER, W. & STETS, J. 1975. Das Rheinprofil zwischen Bonn und Bingen. *Zeitschrift deutsche geologische Gesellschaft*, **126**, 15–29.

MEYER, W. & STETS, J. 1980. Zur Paläogeographie von Unter- und Mitteldevon im westlichen und zentralen Rheinischen Schiefergebirge. *Zeitschrift deutsche geologische Gesellschaft*, **131**, 725–751.

MEYER, W. & STETS, J. 1994. Geologie des Rheinisch-Ardennischen Schiefergbirges. *In*: KOENIGSWALD, W. V. & MEYER, W. (eds) *Erdgeschichte im Rheinland*. Pfeil, München, 13–34.

MEYER, W. & STETS, J. 1996. Das Rheintal zwischen Bingen und Bonn. *Sammlung geologischer Führer*, **89**.

MITTMEYER, H.-G. 1974. Zur Neufassung der Rheinischen Unterdevon-Stufen. *Mainzer geowissenschaftliche Mitteilungen*, **3**, 69–79.

MITTMEYER, H.-G. 1980. Zur Geologie des Hunsrückschiefers. *Natur und Museum*, **110**, 148–155.

MITTMEYER, H.-G. 1982. Rhenish Lower Devonian Biostratigraphy. *Courier Forschungsinstitut Senckenberg*, **55**, 257–270.

MJOS, R., HADLER-JACOBSEN, F. & JOHANNESSEN, E. P. 1998. The distal sandstone pinchout of the Mesa Verde Group, San Juan Basin, and its relevance for sandstone prediction of the Brent Group, northern North Sea. *In*: GRADSTEIN, F. M., SANDVIK, K. O. & MILTONN, N. J. (eds) *Sequence Stratigraphy – Concepts and Applications*. Elsevier, Amsterdam, Norwegian Petroleum Society (NPF) Special Publication, **8**, 263–297.

MUNTZOS, TH. 1990. Terrestrische Sedimentation im Unterdevon (Ems) des Rheinischen Schiefergebirges. *Neues Jahrbuch Geologie Paläontologie, Abhandlung*, **181**, 1–3, 19–39.

NÖRING, F. K. 1939. Das Unterdevon im westlichen Hunsrück. *Abhandlung preußisch geologische Landesanstalt, N.F.*, **192**, 1–96.

ONCKEN, O., PLESCH, A., WEBER, K., RICKEN, W. & SCHRADER, S. 2000. Passive margin detachment during arc-continent collision (Central European Variscides). *In*: FRANKE, W., HAAK, V., ONCKEN, O. & TANNER, D. (eds) *Orogenic Processes: Qantification and Modelling in the Variscan Belt.* Geological Society, London, Special Publication, **179**, 199–216.

RAMÓN, J. C. & CROSS, T. A. 1997. Characterization and prediction of reservoir architecture and petrophysical properties in fluvial channel sandstones, middle Magdalena Basin, Colombia. *Ciencia, Technologia y Futuro*, **1**, 19–46.

REINECK, H.-E. & SINGH, I. B. 1980. *Depositional Sedimentary Environments* (2nd edn). Springer, Berlin.

RICHTER, R. 1931. Tierwelt und Umwelt im Hunsrückschiefer; zur Entwicklung eines schwarzen Schlammsteines. *Senckenbergiana*, **13**, 299–342.

SALVADOR, A. (ed.) 1994. *International Stratigraphic Guide* (2nd edn). GSA, Boulder, Colorado.

SCHÄFER, A. 2005. *Klastische Sedimente – Fazies und Sedimentstratigraphie.* Elsevier, Heidelberg.

SCHÄFER, A. & STETS, J. 1995. The Lower Devonian 'Emsquarzit' – tidal sedimentation in the Rhenish Basin (Rheinisches Schiefergebirge, Germany). *Zentralblatt Geologie Paläontologie, Teil I*, **1994**, 227–244.

SCHIEVENBUSCH, TH. 1992. Bilanzierte Profile, Profilabwicklung und Verformungsanalyse im westlichen Rheinischen Schiefergebirge zwischen Sötenicher Kalkmulde und Moselmulde. *Bonner geowissenschaftliche Schriften*, **3**.

SCHINDLER, T., AGHAI SOLTANI, L., BRAUN, A., ELKHOLY, H. & SCHMITZ, A. 2001. Geologie und Paläontologie des Großaufschlusses Aegidienberg-Tunnel und Logebachtal-Brücke der ICE-Neubaustrecke Köln-Rhein/Main (Unter-Devon, südliches Siebengebirge, Rheinland). *Decheniana, Beiheft*, **39**, 7–68.

SCHINDLER, T., AMLER, M. R. W. *ET AL.* 2004. Neue Erkenntnisse zur Paläontologie, Biofazies und Stratigraphie der Unterdevon-Ablagerungen (Siegen) an der ICE-Neubaustrecke bei Aegidienberg (Siebengebirge, W-Deutschland). *Decheniana*, **157**, 135–159.

SCHMIDT, WO. 1952. Die paläogeographische Entwicklung des linksrheinischen Schiefergebirges vom Kambrium bis zum Oberkarbon. *Zeitschrift deutsche geologische Gesellschaft*, **103**, 151–177.

SCHMIDT, WO. 1958. Die ersten Agnathen und Pflanzen aus dem Taunus-Gedinnium. *Notizblatt hessisches Landesamt Bodenforschung*, **86**, 31–49.

SCHWEITZER, H. J. 1983. *Die Unterdevonflora des Rheinlandes.* 1. Teil. Paläontographica, B 189. Schweiterbarth, Stuttgart.

SCHWEITZER, H. J. 1987. Introduction to the plant bearing beds and the flora of the Lower Devonian of the Rhineland. *Bonner Paläobotanische Mitteilungen*, **13**.

SCHWEITZER, H. J. 1994. Die ältesten Pflanzengesellschaften Deutschlands. *In*: KOENIGSWALD, W. V. & MEYER, W. (eds) *Erdgeschichte im Rheinland.* Pfeil, München, 57–70.

SCHWEITZER, H. J. 2003. Die Landnahme der Pflanzen. *Decheniana*, **156**, 177–215.

SEILACHER, A. 1970. Begriff und Bedeutung der Fossillagerstätten. *Neues Jahrbuch Geologie Paläontologie, Monatshefte*, **1970**, 34–39.

SEILACHER, A. & HEMLEBEN, CH. 1966. Spurenfauna und Bildungstiefe des Hunsrückschiefers (Unterdevon). *Notizblatt hessisches Landesamt Bodenforschung*, **94**, 40–53.

SIMPSON, S. 1940. Das Devon der Südost-Eifel zwischen Nette und Alf, Stratigraphie und Tektonik mit einem Beitrag zur Hunsrückschiefer-Frage. *Abhandlungen senckenbergische naturforschende Gesellschaft*, **447**.

SOLLE, G. 1951. Obere Siegener Schichten, Hunsrückschiefer, tiefstes Unterkoblenz und ihre Eingliederung ins Rheinische Unterdevon. *Geologisches Jahrbuch*, **65**, 299–380.

SOLLE, G. 1956. Die Wattfauna der unteren Klerfer Schichten von Greimerath (Unterdevon, Südost-Eifel). *Abhandlungen hessisches Landesamt Bodenforschung*, **17**.

SOLLE, G. 1972. Abgrenzung und Untergliederung der Ober-Ems-Stufe, mit Bemerkungen zur Unter-/Mitteldevon-Grenze. *Notizblatt hessisches Landesamt Bodenforschung*, **100**, 60–91.

STEINMANN, G. & ELBERSKIRCH, W. 1929. Neue bemerkenswerte Funde im ältesten Unterdevon des Wahnbachtales bei Siegburg. *Sitzungsbericht naturhistorischer Verein Preußen Rheinland und Westfalen*, **21/22**, 1–74.

STETS, J. & SCHÄFER, A. 1990. Unterdevon des Rheinischen Schiefergebirges. *Exkursionsführer SEDIMENT 90*, 5. Sedimentologen-Treffen, Juni 1990, Bonn.

STETS, J. & SCHÄFER, A. 2002. Depositional Environments in the Lower Devonian Siliciclastics of the Rhenohercynian Basin (Rheinisches Schiefergebirge, W-Germany) – Case Studies and a Model. *Contributions to Sedimentary and Geology*, **22**.

STETS, J. & SCHÄFER, A. 2003. Lower Devonian Depositional Environments of the Rhenohercynian Basin (Rheinisches Schiefergebirge, W-Germany). *Sediment 2003*, 10–13 June 2003, Wilhelmshaven (poster).

STETS, J. & SCHÄFER, A. 2004. Deltasedimentation im Nordabschnitt des Rhenoherzynischen Beckens (Unterdevon, Rheinisches Schiefergebirge). *In*: KUKLA, P., LITTKE, R., STOLLHOFEN, H. & SCHWARZER, D. (eds) *Sediment 2004. Schriftenreihe Deutsche Geologische Gesellschaft*, **33**, 256–273.

STRAHLER, A. H. & STRAHLER, A. 1997. *Physical Geography*. Wiley, New York.

STROETMANN-HEINEN, V. 1999. Simulation einer paläozoischen Plattenrandentwicklung in den mitteleuropäischen Varisziden – Das Linksrheinische Schiefergebirge. *Zeitschrift Deutsche Geologische Gesellschaft*, **150**, 451–470.

TAIT, J. A., BACHTADSE, V., FRANKE, W. & SOFFEL, H. 1997. Geodynamic evolution of the European Variscan fold belt: palaeomagnetic and geological constraints. *Geologische Rundschau*, **86**, 585–598.

WALLISER, O. H. & MICHELS, D. 1983. Der Ursprung des Rheinischen Schelfes im Devon. *Neues Jahrbuch Geologie Paläontologie, Abhandlungen*, **166**, 3–18.

WEHRMANN, A., HERTWECK, G. *ET AL.* 2005. Paleoenvironment of an Early Devonian land-sea transition: a case study from the southern margin of the Old Red Continent (Mosel valley, Germany). *Palaios*, **20**, 101–120.

WHEELER, H. E. 1964. Base level, lithosphere surface, and time-stratigraphy. *Geological Society of America Bulletin*, **75**, 599–610.

WIERICH, F. 1999. Orogene Prozesse im Spiegel synorogener Sedimente; korngefügekundliche Liefergebietsanalyse siliziklastischer Sedimente im Devon des Rheinischen Schiefergebirges. *Marburger Geowissenschaften*, **1**.

WINTERFELD, C. V. 1994. Variscische Deckentektonik und devonische Beckengeometrie der Nordeifel – ein quantitatives Modell (Profilbilanzierung und Strain-Analyse im linksrheinischen Schiefergebirge). *Aachener Geowissenschaftlichen Beiträge*, **2**.

WINTERFELD, C. V., BAYER, U., ONCKEN, O., LÜNENSCHLOSS, B. & SPRINGER, J. 1994*a*. Das westliche Rheinische Schiefergebirge. *Geowissenschaften*, **12**, 10–11, 320–324.

WINTERFELD, C. V., DITTMAR, U., MEYER, W., ONCKEN, O., SCHIEVENBUSCH, Th. & WALTER, R. 1994*b*. Crustal-scale geometry of the Rhenohercynian fold and thrust belt. *Journal Czech Geological Society*, **39**, 124–125; Praha.

WITZKE, R. 1990. Palaeoclimatic constraints for Palaeozoic Palaeolatitudes of Laurentia and Euramerica. *In*: MCKERROW, W. S. & SCOTESE, C. R. (eds) *Palaeozoic Palaeogeography and Biogeography*. Geological Society, London, Memoir, **12**, 57–73.

WUNDERLICH, H.-G. 1964. Maß, Ablauf und Ursachen orogener Einengung am Beispiel des Rheinischen Schiefergebirges, Ruhrkarbons und Harzes. *Geologische Rundschau*, **54**, 861–882.

ZIEGLER, P. 1990. *Geological Atlas of Western and Central Europe* (2nd edn). Shell International Petroleum Maatschappij, Den Haag.

ZIEGLER, P. A. 1992. Geodynamics of rifting and implications for hydrocarbon habitat. *Tectonophysics*, **215**, 221–253.

Frasnian reef evolution and palaeogeography, SE Lennard Shelf, Canning Basin, Australia

ANNETTE D. GEORGE[1]*, KATE M. TRINAJSTIC[1] & NANCY CHOW[2]

[1]*School of Earth and Environment, University of Western Australia, Crawley, WA 6009, Australia*

[2]*Department of Geological Sciences, University of Manitoba, Winnipeg, Manitoba R3T 2N2, Canada*

**Corresponding author (e-mail: ageorge@cyllene.uwa.edu.au)*

Abstract: Frasnian reef complexes of the southeastern Lennard Shelf (northern Canning Basin) developed on tilt-block highs and their evolution was controlled by fault-related subsidence. Tectonic control on relative sea-level changes was, therefore, a major factor influencing Early–Middle Frasnian palaeogeography of the Lennard Shelf. However, palaeogeographic reconstruction is not consistent with simple landward (northward) backstepping and younging of reef complexes in response to basin extension and subsidence of fault blocks. Using a sequence-stratigraphic approach, in conjunction with sedimentological and biostratigraphic data, we propose that two neighbouring fault blocks (Lawford area on the eastern side of Bugle Gap and the Hull platform to the north) record a similar history and that the reef complexes on those blocks were initiated at similar times. Seven phases of Early–Middle Frasnian platform growth (Fr2–8) are identified. All are bounded by third-order flooding surfaces associated with backstepping of platform margins and three surfaces (defined by conodont Zone 4, Zone 6 and late Zone 6) correlate across the two fault blocks. Only one sequence boundary has been clearly identified and a second relative sea-level fall is proposed based on a major collapse event following progradation and associated coarse siliciclastic facies. We propose that the correlation of flooding events across the SE Lennard Shelf is related to episodes of basin-margin faulting centred on the large, long (*c.* 25 km) faults which border these blocks (shelf-parallel faults for the Lawford block and an oblique north-trending transfer zone for the Hull block). There is limited evidence for relative sea-level falls and those recognized most likely resulted from eustatic events. The correlation between the Lawford block and Hull platform suggests linkage between major NW-trending shelf-parallel and oblique transfer faults and an evolved rift system by the Early Frasnian.

The well preserved facies relationships in the exhumed reef complexes of the northern Canning Basin (Fitzroy Trough) make them a world-class example of Late Devonian reef building (Playford 1980). The strike length of the reef complexes provides considerable exposure; however, the extensional tectonic history of the northern part of the basin during reef building, and post-Devonian deformation and exhumation, have created a fragmented record for much of the Frasnian strata. Correlations between separate platforms, as well as between platforms and their coeval fore-reef successions, have been hampered by difficulties in dating the shallow water successions although a number of Frasnian–Famennian fore-reef successions have been well documented and dated using goniatite and/or conodont biostratigraphy (e.g. Becker *et al.* 1993; Nicoll & Playford 1993; Becker & House 1997; George *et al.* 1997; George & Chow 2002).

Resolution of these problems is required to reconstruct the history of Late Devonian carbonate platform evolution in the northern Canning Basin. We have used a sequence-stratigraphic approach, underpinned by facies analysis and biostratigraphy, to aid in developing a better understanding of the Early–Middle Frasnian palaeogeography of the SE Lennard Shelf. Such an approach has been shown in field-based studies elsewhere to be very useful for unravelling the complexities of carbonate platform evolution (e.g. Chen *et al.* 2002; Whalen *et al.* 2002*b*; Bauer *et al.* 2003; Ruiz-Ortiz *et al.* 2004; Warrlich *et al.* 2005; MacNeil & Jones 2006).

We have combined the sedimentological–stratigraphic record with conodont biostratigraphy to provide the highest-resolution age control available, and microvertebrate (fish) remains to provide additional age control and for insights into palaeoenvironmental conditions. The conodont biostratigraphy has also been significant in this study to correlate proximal and distal facies within and between fore-reef successions as well as for constraining timing of events such as platform collapse

From: KÖNIGSHOF, P. (ed.) *Devonian Change: Case Studies in Palaeogeography and Palaeoecology*.
The Geological Society, London, Special Publications, **314**, 73–107.
DOI: 10.1144/SP314.4 0305-8719/09/$15.00

and backstepping which have been recognized for decades (e.g. Playford 1980; Kennard *et al.* 1992; Becker *et al.* 1993). This approach also highlights phases of progradation which have tended to be overshadowed by the overall and well-demonstrated transgressive character of the older Frasnian reef complexes.

An integrated approach also provides an opportunity to draw together a wide variety of sedimentological and biostratigraphic data and interpretations published on the Canning Basin reef complexes over many years (see below). In this paper we have focused on some of the older platforms which are widely distributed on the southeastern end of the Lennard Shelf (Fig. 1) as a basis for linking a number of well-known localities in larger synthesis. We present detailed work on a relatively poorly described area (Northern Lawford Range) because this area, with its well exposed and datable fore-reef succession (known as the McIntyre Knolls area), provides a record of platform evolution even though most of the coeval platform has been removed (Fig. 2). Elsewhere in the Canning Basin fore-reef successions have proven very useful for establishing reef complex history where platforms have been eroded or only a fragmentary record is preserved (e.g. George *et al.* 2002). The fore-reef succession of the Northern Lawford platform provides a critical link to some of the better known localities in the area (e.g. McWhae Ridge, Fig. 1). Platforms of (?)Middle to Late Frasnian age are exposed in the central and northwestern Lennard Shelf and their distribution and evolution have already been documented (Playford & Hocking 1998; Ward 1999; George *et al.* 2002).

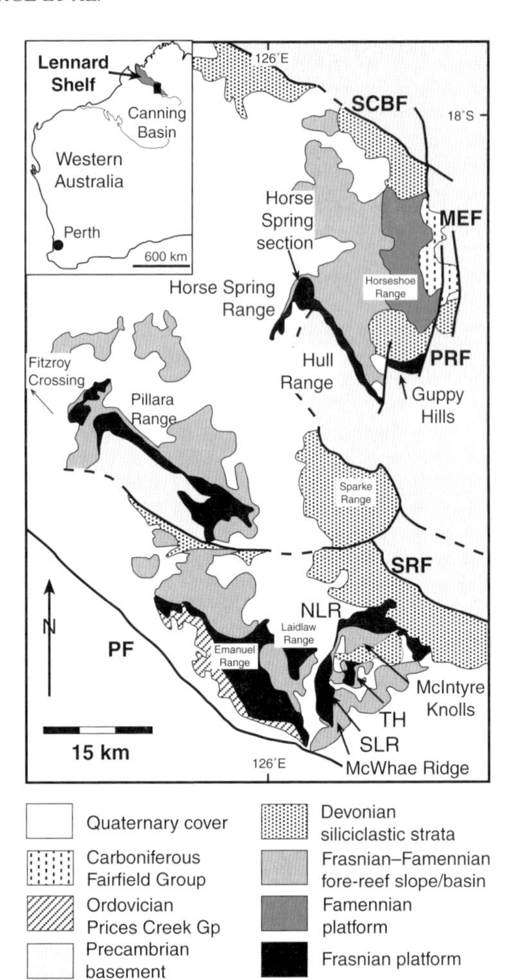

Fig. 1. Geological map of the southeastern Lennard Shelf, northern Canning Basin, showing distribution of major rock units (modified from Playford & Hocking 1998) and location of major limestone ranges, notably Southern and Northern Lawford Ranges (SLR and NLR respectively), Hull Range, Horse Spring Range and Guppy Hills. Abbreviations: MEF, Mount Elma Fault; PF, Pinnacles–Harvey–Beagle Bay fault system; PRF, Painted Rocks Fault; SCBF, Stony Creek–Barramundi fault system; SRF, Sparke Range Fault; TH, Teichert Hills.

Tectonic setting

The Australian continent formed part of Gondwana in the late Early to Late Devonian. The Canning Basin was located in an equatorial position at *c.* 12–14°S with an arid climate that had existed since at least the Late Silurian (Yeates *et al.* 1984; Li & Powell 2001). The Canning Basin represents an intracratonic–rift basin that developed in the Early Ordovician during the first of several extensional phases recorded in its long history (Shaw *et al.* 1994). The tectonic and depositional history of the basin can be subdivided into four megasequences representing phases of rifting, subsidence and deposition in the major depocentres (Kennard *et al.* 1994; Romine *et al.* 1994).

The major extensional phase in the late middle Devonian (mid-Givetian; Grey 1991) led to development of the NW-trending rift basin (Fitzroy Trough) along the northern margin of the Canning Basin (Fig. 1; Drummond *et al.* 1988). This phase has been linked to major shortening in central Australia (Alice Springs Orogeny) and transcurrent movement on faults of the Halls Creek Orogen which also opened the NE-trending Petrel Sub-basin in the neighbouring Bonaparte Basin (e.g. Baillie *et al.* 1994).

The Fitzroy Trough contains *c.* 15 km of predominantly Palaeozoic–Mesozoic fill with up to 5 km of strata on the flanking margins (e.g. Yeates *et al.* 1984). These margin successions include the Late Devonian reef complexes which are only exposed

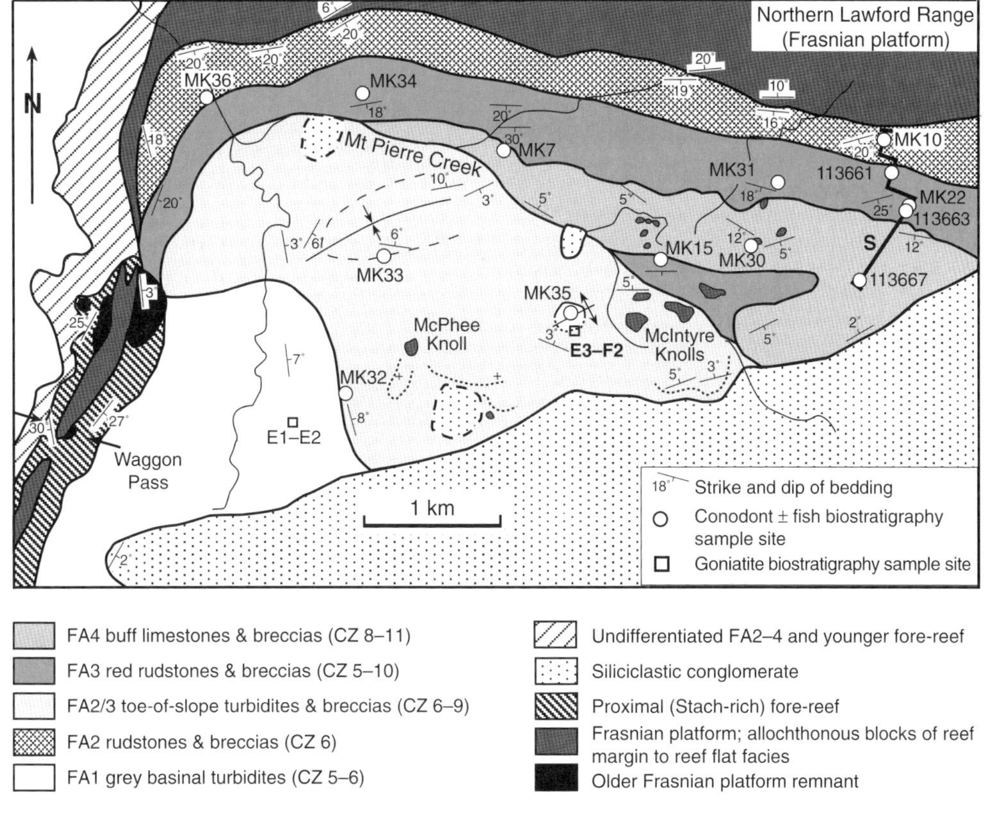

Fig. 2. Geological map of the Northern Lawford Range and associated fore-reef succession of the McIntyre Knolls area showing distribution of major slope facies associations (FA1–4). Lithostratigraphically, FA1 belongs to Gogo Formation and FA2–4 belong to Virgin Hills Formation. Measured section (S). Goniatite biostratigraphy from Becker *et al.* (1993).

on the inner Lennard Shelf (Fig. 1). The Harvey–Beagle Bay–Pinnacles Fault System defines the edge of the Lennard Shelf and has an estimated normal displacement of *c.* 6 km (Playford 1980). Post-Devonian tectonic history of the Fitzroy Trough includes two major transpressional phases. Mid-Carboniferous transpression caused regional inversion and uplift (Shaw *et al.* 1994) and marked erosion and karstification of the reef complexes (Playford 1980, 2002; Lehmann 1986; Wallace *et al.* 1991). Late Triassic–Early Jurassic dextral transpression also caused reactivation and inversion and development of new structures (folds and faults) in the trough (Yeates *et al.* 1984; Shaw *et al.* 1994).

Southeastern Lennard Shelf reef complexes

The SE Lennard Shelf is up to 75 km wide and features several NW-trending limestone ranges, including the Emanuel, Pillara, Home and Hull Ranges (Fig. 1), which represent exhumed parts of reef complexes. The orientation of these ranges is strongly controlled by major faults that define the northwesterly structural grain of the Lennard Shelf and are parallel to the major faults of the adjacent King Leopold Orogen. Extension produced an irregular topography with reef complexes forming on the tilt-block highs across the shelf (Playford 1980; Begg 1987; Kemp & Wilson 1990).

The Bugle Gap area also features north-trending and west-trending platforms (Laidlaw and Lawford Ranges, respectively), platform atolls (Teichert Hills, Lloyd Hill) and large major embayments (e.g. Paddy's Valley) and inter-reef basins (Fig. 1). To the north, the NW-trending Hull Range and NE-trending Horse Spring Range form part of a larger platform that also includes Guppy Hills to the east (Fig. 1). The present-day distribution of platforms, with variable preservation of reef-margin facies, and associated fore-reef deposits, have been summarized on the geological maps of Playford &

Hocking (1998). These platforms belong to the Givetian–Frasnian 'Pillara' phase of platform development. The overlying latest Frasnian–Famennian 'Nullara' phase is represented in the area by Famennian platform in the Horseshoe Range and Famennian fore-reef deposits which are distributed more widely in the broader area (Fig. 1; Playford 1980; Playford & Hocking 1998).

Previous studies undertaken in the Bugle Gap area have contributed significantly to understanding of the Late Devonian geology of the Canning Basin (e.g. Teichert 1943, 1949; Guppy *et al.* 1958; Rattigan & Veevers 1961; Playford & Lowry 1966) and recognition of facies relationships (Playford 1980, and additional references therein). Palaeontological and biostratigraphic studies were an important part of this early work (e.g. Teichert 1941; Glenister 1958; Veevers 1959; Glenister & Klapper 1966; Hill & Jell 1970; Seddon 1970; Druce 1976), and also provided a basis for development of international Devonian biozonations for dating and correlation using conodonts and goniatites (e.g. Klapper 1989; Becker *et al.* 1993; Becker & House 1997). Important Devonian macrovertebrate (fish) faunas have been recovered from the basinal Gogo Formation (the 'Gogo fish') over a number of years (e.g. Long 1988, 1995; Ahlberg 1989; Long & Trinajstic 2000) including *Mcnamaraspis kaprios* Long 1995 which is the fossil emblem of the state of Western Australia.

Read (1973*a*, *b*) provided new insights into the sedimentology of the Pillara platforms, particularly the back-reef facies, their cyclic arrangements and interpretation of palaeogeography based on constituent facies. This work has been followed by comprehensive descriptions and interpretations of reef-margin, reef-flat and back-reef associations (e.g. Kerans 1985; Copp 2002), re-examination of platform cyclicity (e.g. Hocking *et al.* 1995; Brownlaw *et al.* 1996, 1998), and comparisons with South China (Shen *et al.* 2008). Exploration for Mississippi Valley-type base-metal deposits generated detailed studies of the Pillara limestones of the northern and central Pillara Range (e.g. Hall 1984; Benn 1984; Cooper *et al.* 1984) which expanded understanding of early carbonate deposition on the Lennard Shelf. Base-metal deposits such as Cadjebut (now mined out) focused a great deal of attention on the Emanuel Range (Fig. 1) and subsurface on its western side, especially the evaporitic facies of the basal Pillara platform (Hocking *et al.* 1995; Warren & Kempton 1997) which were the likely sulphur source for mineralization (e.g. Tompkins *et al.* 1994; Wallace *et al.* 2002).

The ranges of the SE Lennard Shelf are separated from those to the NW (e.g. Napier and Oscar Ranges) by the Margaret Embayment which represents a major and complex accommodation zone known as the Colombo Fault Zone (Shaw *et al.* 1994). This zone has previously been referred to as the Fossil Downs Graben (e.g. Kemp & Wilson 1990; Vearncombe *et al.* 1996) and is considered a likely major conduit for the dispersal of siliciclastic sediment into the Fitzroy Trough (Holmes & Christie-Blick 1993; Shaw *et al.* 1994).

Overview of Northern Lawford Range

The Northern Lawford Range is composed of a broad west-trending platform, which dips 10–20° to the north, and a well exposed south-dipping fore-reef slope succession on its southern margin (Figs 1, 2). The northern margin of the platform is interbedded with coarse siliciclastic strata indicating that it was attached to the Precambrian hinterland. Several NNW-trending faults cross the platform (mapped as normal faults; Playford & Hocking 1998); on aerial photographs some of these faults show minor offsets in the platform strata. The Northern Lawford Range is joined by a narrow north-trending reefal belt to the Southern Lawford Range which has the well known Frasnian–Famennian section at McWhae Ridge at its southern end (Fig. 1).

The uppermost slope strata and platform margin have a dissected karst topography that is difficult to traverse in the eastern part of the margin except along streams that cross the platform margin. Platform margin facies are only locally preserved (see also Shen *et al.* 2008). They form a narrow zone (*c.* 20–30 m wide), and are dominantly massive, red to pink-coloured stromatoporoid boundstones with large hemispheroidal and massive stromatoporoids (up to 0.75 m across) with *Renalcis*, lithistid ('doughnut') sponges and minor solitary rugose corals, with abundant red fine sandy (peloidal) matrix. *Stachyodes* and other stromatoporoid debris are common in the matrix. This massive facies also features subhorizontal geopetal cements.

Platform margin facies grade landward into medium to very thickly bedded pale-coloured *Stachyodes*-rich boundstones, with common *Renalcis*, that characterize a zone up to *c.* 80–100 m wide essentially parallel to the platform edge. Bedding dips shallowly to the north (*c.* 6–10°) but locally as much as 20° and is cross-cut by numerous fibrous calcite cement- and sediment-filled neptunian dykes. The facies and diagenetic features are characteristic of the reef-flat environment (Kerans 1985). These limestones grade into bedded stromatoporoid rudstones with minor *Renalcis* which dip up to 20° and are considered more characteristic of the back-reef environment (e.g. Kerans 1985).

Most of the platform margin exposed in the Northern Lawford Range has a steep erosional surface which is typically abutted by coarse

(a)

(b)

Fig. 3. Field photographs of Northern Lawford platform margin. (**a**) View to the east showing steep eroded platform margin (M) abutted by very thick breccias and bedded limestones of FA2. (**b**) View to the west showing irregular low-angle erosional surface on thickly bedded, platform (reef-flat facies) onlapped by FA2 dipping *c*. 20° to the south.

fore-reef slope facies which dip up to *c*. 25° (Figs 2, 3). More rarely, steeply dipping *Renalcis*-bearing limestones abut the platform. Overall the fore-reef succession on the southern side of the Lawford platform forms a large wedge-shaped unit which preserves *c*. 200 m of strata proximally and up to *c*. 20 m distally (Fig. 2). The 'proximal' strata comprise the modern topographic slope whereas finer grained facies occupy the base-of-slope and basin-floor area between the Northern Lawford Range and Teichert Hills to the south (Fig. 1). The fore-reef slope succession is here described by facies associations rather than using the multiformational lithostratigraphic nomenclature for the southeastern Lennard Shelf (e.g. Playford 1980). However, for broader comparison with previous work the fore-reef succession in the McIntyre Knolls area has been mapped as Virgin Hills Formation (Playford & Lowry 1966; Playford & Hocking 1998). Although the Frasnian portion of this formation is composed of bioclastic limestones and carbonate breccias, with minor siliciclastic sandstone and conglomerate (Guppy *et al.* 1958), it is best known for bright red, fine fossiliferous thinly to very thinly bedded toe-of-slope facies exposed at some of the classic Lennard Shelf localities such as McWhae Ridge and Horse Spring Range (e.g. Nicoll & Playford 1993; Becker *et al.* 1991; Fig. 1).

The most famous feature of the lowermost slope is the set of very large (up to 200 m long) allochthonous blocks of platform limestone known as the 'McIntyre Knolls' (Playford 1981) which are located *c*. 1.5 km from the edge of the Lawford platform (Figs 2, 4a). These blocks, and the large isolated block 'McPhee Knoll' (Fig. 2), record a major platform-margin collapse event (Playford 1981; Mullins & Hine 1989). Smaller blocks (5 to 30 m across) are also embedded throughout the slope succession locally forming clusters of blocks (Figs 2, 4b).

The eastern and southern margins of the carbonate fore-reef succession are in contact with a large conglomeratic unit (Playford & Hocking 1998). The carbonate and siliciclastic facies are clearly

(a)

(b)

Fig. 4. Field photographs of allochthonous blocks. (**a**) View towards northeastern corner of southwestern McIntyre Knoll block underlain by 1.5 m thick, clast-supported debris-flow breccia. Note thickly bedded part of block (arrowed) which shows steep (*c*. 70°) dip. Geopetal cavity fills show that this block is upright. Trees in foreground are *c*. 3 m high. (**b**) View of three major blocks, some fragmented, on the lower (modern) slope to NE of McIntyre Knolls (Fig. 2) showing massive (locally crudely bedded) platform limestones of reef-margin affinity. Blocks on the left and centre have steep dips (70–85°). Vehicle track in foreground for scale.

interbedded along this contact. An overall westerly directed sediment transport direction is interpreted for these conglomerates which thin to the west and are contained along the northern edge of Teichert Hills (Figs 1, 2).

Fore-reef facies associations: sedimentology and biostratigraphy

The fore-reef slope succession can be subdivided into four facies associations (FA1–4) in ascending stratigraphic order (Fig. 5). The basal association (FA1) is only locally exposed at the platform margin whereas the other three associations (FA2–4) define discrete slope-parallel belts (Fig. 2). These slope-parallel belts are all characterized by carbonate breccias interbedded with medium to thickly bedded bioclastic rudstones and floatstones. In FA2 and 3 very thickly bedded breccias are spectacularly developed where they form the tops of coarsening-upward trends and bases of fining-upward trends respectively (Fig. 5). Near the platform, these breccias are as much as 8 m thick and typically ungraded. They are very similar to the cohesionless debris-flow breccias (debrites) described elsewhere on the Lennard Shelf (Playford 1984; George *et al.* 1997) and for clarity in the following sections they are referred to as 'debris-flow breccias'. The rudstones–floatstones are commonly associated with breccias that contain a significant proportion of slope-derived lithoclasts and minor to large amounts of platform-derived debris as described below (and are referred to as 'intraclastic

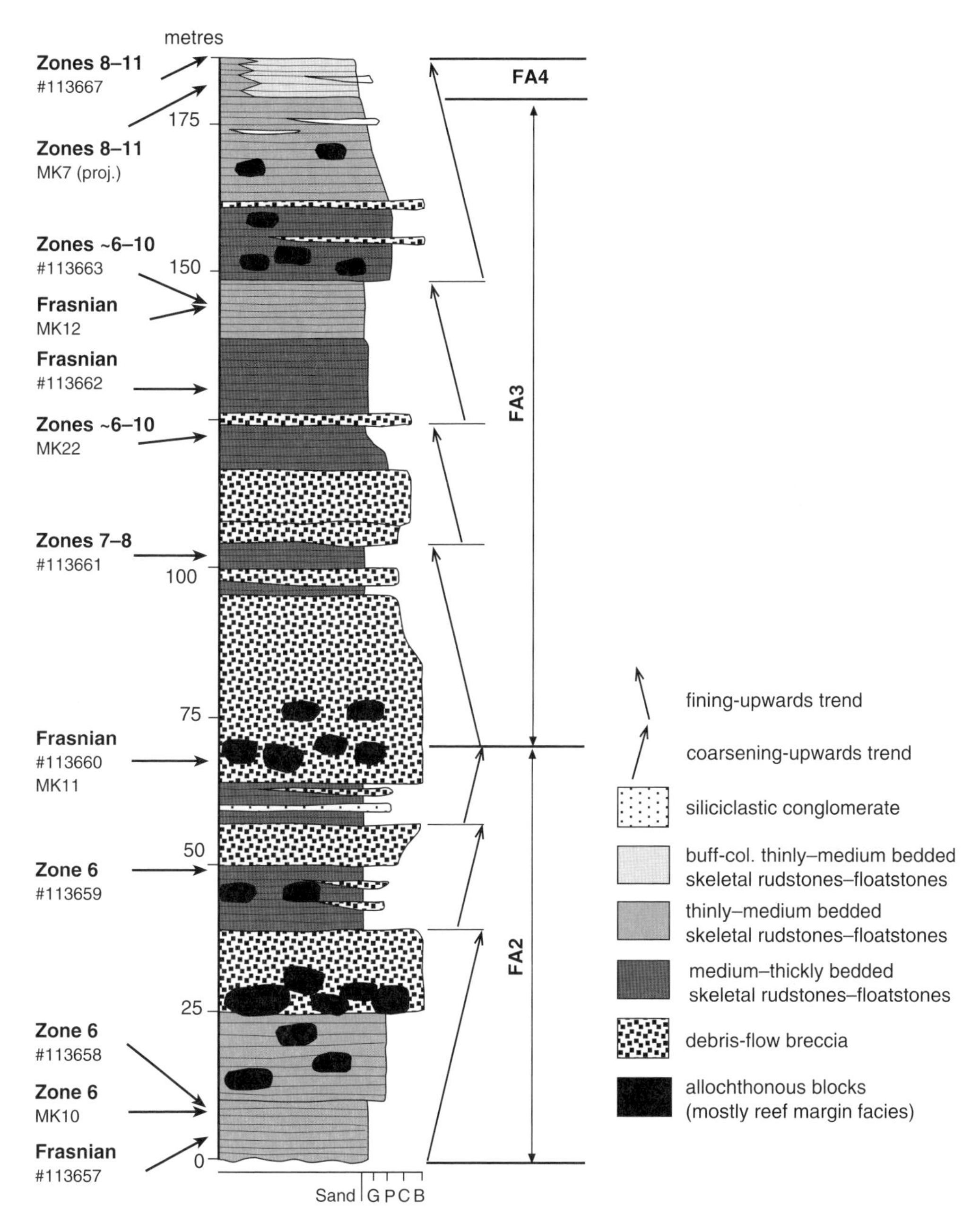

Fig. 5. McIntyre Knolls measured section (location shown on Fig. 2) showing major lithofacies, facies associations FA2–4, stacking patterns and location of conodont sample horizons. Modified from Trinajstic & George (2009). Specific zonal assignments summarized in Appendix. 'Frasnian' assignment is zonally undiagnostic. Note that MK7 sample site is projected along strike onto the section. Note that stacking patterns in FA2 are coarsening upward whereas those in FA3 are fining upward. Major block horizon at the boundary of FA2 and FA3 is considered coeval with the collapse event in which the McIntyre Knolls and associated clusters of blocks were emplaced. Grain size scale: G, granule; P, pebble; C, cobble; B, boulder.

breccias'). The presence of grading and mixed debris of shallow and deeper water origin in the rudstones–floatstones indicates deposition by sediment gravity flows (mostly turbidity currents and cohesionless debris flows/density-modified grain flows) on the slope. Siliciclastic facies are volumetrically minor within the fore-reef succession (Fig. 2); however, quartz sand is a common component in the bedded limestones and in the matrix of debris-flow breccias throughout the area.

Facies associations 1–3 comprise a conformable succession of thinly to medium-bedded grey, tan-grey and purplish-pink limestones (typically graded floatstones–rudstones to grainstones) and a variety of carbonate breccias on the basin floor. The grey limestones of FA1 are distinctive; however, the distal parts of FA2 and 3 are difficult to separate lithologically and for the most part we have not attempted to do so (Fig. 2). These facies are introduced in FA2 with further description and interpretation presented in FA3. Dips are invariably shallow to subhorizontal although the beds are commonly warped in some areas (Fig. 2).

The rudstones–floatstones of FA1–4 contain a diverse fossil assemblage including abundant stromatoporoid and skeletal debris indicative of a Frasnian age (Playford 1980). This is supported by Frasnian goniatite and conodont ages for the basinal and toe-of-slope facies (Becker *et al.* 1993) and dominance of stromatoporoid boundstones, and associated debris in the Lawford platform and as allochthonous blocks of platform limestone. Conodont biostratigraphy through the measured section (proximal facies; Fig. 3) and spot samples from key localities in the area (Fig. 2) confirm that the entire fore-reef succession is Frasnian (conodont Zones 5 to 11; Fig. 5; hereafter 'conodont zone' is shortened to 'Zone'). Yields are composed mostly of P_1 elements (notation of Purnell *et al.* 2000) of *Polygnathus* (*c.* 60%) and *Ancyrodella* (*c.* 20%). The remaining 20% are P_1 elements of *Palmatolepis* species (typically broken or juvenile specimens) and indeterminate P_2 and other ramiform elements. P_1 elements are typically over-represented in Late Palaeozoic faunas (von Bitter & Purnell 2005). We interpret the limited representation of other elements in our samples as reworking and sorting by downslope transport and/or differential preservation as a result of differences in size or robustness of elements of the conodont apparatus.

Conodont biostratigraphy has been the key for correlating toe-of-slope and basin-floor facies with their proximal equivalents. There is a noticeable downslope fining of bioclastic debris with lower to toe-of-slope facies dominated by thinly to medium-bedded turbiditic rudstones–floatstones–grainstones and a variety of breccias. These strata contain diverse microvertebrate fish remains, but of low abundance, composed of phoebodont teeth (providing the highest resolution biostratigraphic information) with thelodont and palaeoniscoid scales (see Appendix). Preservation of these remains is excellent with phoebodont teeth possessing tooth cusps and bases, and thelodont scales possessing spurs and ornament. Micro-ornament is preserved in both taxa which suggests limited abrasion. Other taxa include protacrodont and ctenacanthid scales (see Appendix); these remains are indeterminate at species level and are, therefore, of limited biostratigraphic use.

The four large McIntyre knoll allochthonous blocks, as well as the clusters of smaller blocks nearby (Fig. 2), display various fabrics both within and between blocks. Common fabrics are: massive to poorly bedded pale grey stromatoporoid framestones–rudstones (notably hemispheroidal and tabular forms, with abundant *Stachyodes* debris); massive to poorly bedded *Renalcis*-rich bindstones and *Renalcis*–stromatoporoid boundstones (Fig. 6a) with skeletal debris scattered or concentrated in rudstone–floatstone layers (brachiopods, *Stachyodes*, crinoids, goniatites, gastropods and *Amphipora*); massive sponge-rich bindstones, typically with *Renalcis* and skeletal debris (Fig. 6a); and thickly to very thickly to poorly bedded *Stachyodes*–*Renalcis* boundstones with skeletal rudstones. Minor but distinctive facies include well-laminated fenestral microbial limestones, quartz sandstones and, in one of the large blocks, well bedded *Amphipora* rudstones-floatstones. Siliciclastic sand is also observed in the carbonate facies suggesting that this sediment was routinely incorporated into facies forming on the platform. In one block quartz sandstone overlies a very irregular surface on massive *Renalcis*–*Stachyodes* limestone, also filling underlying vugs and fractures, suggesting subaerial exposure (Fig. 6b). The variety of fabrics is indicative of platform-margin facies on the Lennard Shelf (e.g. as described by Kerans 1985) with massive to poorly bedded stromatoporoid- and *Renalcis*-rich fabrics well known from the reef margin and the bedded *Stachyodes*-rich fabrics more commonly attributed to the reef flat. Very similar facies distributions and fossil assemblages have been documented from other Devonian reef complexes (e.g. Machel & Hunter 1994; Wendte 1994; Whalen *et al.* 2000*a*; Chen *et al.* 2001, 2002).

Common geopetal cavity fills featuring red- and ochre-coloured laminations (locally with quartz sand) and fibrous calcite cements (Fig. 6c) are also indicative of reef margin to reef flat (Kerans 1985) and show that the four major blocks are lying on their sides and upright (strikes trend WNW and dip steeply south or north). These cavity fills are

(a)

(b)

(c)

Fig. 6. Field photographs showing features of the allochthonous blocks. (**a**) Common reefal fabric of massive red *Renalcis*–stromatoporoid boundstone showing masses of *Renalcis* (R) and tabular stromatoporoids (S) with abundant red sediment. Lens cap 5 cm. (**b**) *Stachyodes* floatstone overlies quartz sandstone (dark) which overlies a very irregular surface developed on a *Renalcis–Stachyodes* boundstone. The highly irregular scalloped surface with angular limestone fragments embedded in the sandstone, and sandstone-filled vugs beneath the surface, potentially suggest a period of subaerial exposure during platform development. Hammer 35 cm. (**c**) Close up of large primary geopetal cavity showing quartz sandstone (Q) overlain by laminated ochre-red (dark) peloidal limestone (L) and fibrous early marine calcite (C).

cross-cut by post-emplacement geopetal cavity fills with very shallow to subhorizontal dips in all directions. The western two blocks rest on thick debris-flow breccias (Fig. 4a) whereas the strata beneath the other two blocks are not exposed. This relationship is discussed further below in relation to timing of emplacement of these blocks.

Facies association 1

Description. Adjacent to the platform, FA1 is composed of medium- to thickly bedded red bioclastic rudstones–floatstones. Bioclasts are commonly stromatoporoids, gastropods, crinoids and tabulate corals (e.g. *Thamnopora*). *Renalcis* in these beds most likely acted to bind and stabilize the slope given the steep dips observed (*c.* 55°). Direct conodont ages for these strata have not been obtained; however, their stratigraphic position beneath FA2 indicates that they are no younger than Frasnian Zone 6.

In the basin, very thinly to medium-bedded grey turbiditic facies (floatstones–rudstones to grainstones–packstones interbedded with mudstones and minor breccias) underlie FA2 (Fig. 2). Thinner beds are massive or normally graded (T_{abcd}, T_{bcd} and T_{bd} beds common) whereas thicker, coarser-grained beds may be normally or inversely graded. Simple sinuous subhorizontal burrows are locally present on bedding planes. Fossil debris in the turbidites includes crinoids, brachiopods, *Amphipora* (locally abundant), *Stachyodes*, *Alveolites* fragments, solitary rugose corals and angular pieces of stromatoporoid (locally silicified). Hematized straight nautiloids and goniatites (after pyrite which replaced the primary carbonate skeletons) are typically well preserved where they have weathered out of the muddier facies. Iron nodules (after pyrite) and hematized woody pieces are also commonly weathered out. Beds also commonly contain quartz sand. These facies belong to the Gogo Formation (Playford & Hocking 1998) and include medium to thickly bedded breccias in this area which contain clasts of reefal limestone (stromatoporoid–*Thamnopora* and *Stachyodes–Renalcis* fabrics observed). Much of the bioclastic material was transported into the basin by sediment gravity flows from higher up the slope, either from the east or north (Fig. 2).

In the western part of the inter-reef basin, goniatite ages (Frasnian E1–E2) equivalent to Frasnian Zone 5 have been obtained from facies which clearly underlie FA2 (Becker *et al.* 1993; their section 365; Fig. 2). These facies are also exposed in the centre of a small anticlinally folded area immediately west of McIntyre Knolls (Fig. 2) and have yielded goniatites with slightly younger Frasnian E3–F2 ages equivalent to lower part of Zone

6 (Becker *et al.* 1993; their section 367). Our conodont biostratigraphy from thinly to very thinly bedded grey-buff grainstones near the centre of the outcrop gives a consistent age assignment of Zones 2–6 (see Appendix).

Interpretation. The proximal part of this association represents the uppermost slope of a coarse debris apron around the platform margin. The distribution of the proximal facies near preserved reef-margin facies suggests that this debris apron formed contemporaneously with the platform that forms at least the southern part of the Northern Lawford Range. A similar fore-reef apron is seen around many parts of the platforms in the Bugle Gap area including good examples of gradational contacts that have been described around the southern part of the Lawford Range (e.g. Nadji Cave; Playford 1981, 1984). Although the lateral facies relationships in the McIntyre Knolls area are covered by younger fore-reef strata, the coarse upper slope facies most likely grade over a short distance into the toe-of-slope to basin-floor turbidites. A similar relationship has been described for the McWhae Ridge area (Playford 1980; Becker & House 1997, based on the earlier work in Playford 1981) and is well demonstrated by Paddy's Valley embayment (Fig. 2) which is largely filled with basinal facies (e.g. Playford 1980).

Facies association 2

Description. FA2 forms a major slope-parallel belt adjacent to the Northern Lawford platform where it typically abuts the steep margin (Figs 2, 3a) and locally onlaps FA1. Bedding dips consistently *c.* 18–20° with shallower or steeper dips observed locally (Fig. 2). For example, shallower dips (*c.* 10–12°) have been observed onlapping irregular platform margin contacts (Fig. 3b). These facies are characterized by very thick debris-flow breccias (individual beds 1.5–8 m thick, average *c.* 2–3 m), that have clast-supported, very poorly sorted fabrics (Fig. 7a). Grading is not developed although locally large clasts are embedded in breccia tops. Breccia clasts range up to 1.5 m and commonly display reefal fabrics; *Renalcis* boundstone is common along with stromatoporoid–*Renalcis*–*Stachyodes* framestones and red stromatactis limestone (locally sponge-rich; Fig. 7b). Bioclasts of stromatoporoids, corals and skeletal material are common. Minor clast types are red (slope-derived) rudstones and quartz sandstones. Most breccias contain a coarse red peloidal matrix (*c.* 10–15%; Fig. 7b); however, some have no primary matrix and are cemented with early marine calcite (Fig. 7c).

The rudstones–floatstones are massive to normally graded and contain diverse fossil assemblages (Fig. 7d). The most common fossils are angular pieces of massive and tabular stromatoporoids, *Stachyodes* and crinoids, with brachiopods, gastropods and corals, including tabulates (e.g. *Thamnopora, Alveolites*) and various rugosan genera (e.g. *Frechastrea* and *Kuangxiastraea*) and large and small solitary forms (e.g. *Temnophyllum*; Brownlaw 2000). *Amphipora* is locally common with minor oncoids, nautiloids and goniatites. Matrix-poor parts of beds typically feature abundant fibrous calcite cement. Medium to thickly bedded, clast-supported intraclastic breccias associated with the bedded limestones feature common slope-derived clasts (same composition as the adjacent facies) with various amounts of stromatoporoid debris included.

Allochthonous blocks of platform limestone (up to 20 m across) are common. Typical fabrics are pale-coloured to pinkish *Renalcis*–*Stachyodes* boundstone, reddish sponge–*Renalcis* boundstone, massive pale-coloured *Renalcis* boundstone and pink to red stromatoporoid–*Renalcis* boundstone. The last two are also observed in gradational contact within individual blocks.

Conodonts yielded by the rudstones–floatstones are consistently in Zone 6 below the conglomerate (Fig. 3, Appendix). In addition, the outermost buff to pink toe-of-slope turbidites are also in Zone 6 which shows that these strata (in the inter-reef basin) are the distal equivalents of the proximal FA2 (Fig. 2, Appendix). The inter-reef facies are dominated by thinly to medium-bedded, graded, pinkish buff limestones that display well-developed Bouma sequences (typically T_{abcd}, T_{abd} and T_{bd} beds; Fig. 8a). Sinuous simple or back-filled traces are present on bedding surfaces.

Clast-supported conglomerate is a minor facies near the top of FA2 in the measured section where it forms a very thick (*c.* 1 m), lenticular bed composed of rounded to subrounded Precambrian quartzite, metavolcanic schist, and vein quartz pebbles (up to 10 cm long) cemented by calcite (Fig. 8b, c). The carbonate facies below also contain quartz sand, i.e. in the matrix of the rudstones–floatstones and breccias, as much as *c.* 15–20%, and the overlying rudstones contain pebbles. Patches of Precambrian pebbles (of quartzite, vein quartz and schist), with some exposure of coherent conglomerate and pebbly sandstone, are also present on the basin floor, both at the base of the fore-reef slope and near the conglomerates to the south (Fig. 2). These conglomerates and sandstones (which locally include limestone clasts) are everywhere interbedded with quartzy or pebbly carbonate facies. The basin floor also features carbonate breccias, some of which have a siliciclastic sandstone matrix although typically this material is highly weathered and the breccias crop out as low

Fig. 7. Field photographs of FA2. (**a**) General view of very thickly bedded breccias overlain by thinly to medium-bedded rudstones–floatstones. Very thick breccia is *c.* 2.5 m thick. (**b**) Typical clast-supported debris-flow breccia showing closely packed angular clasts of reefal limestone with a red grainy (peloidal) matrix. Hammer 35 cm (arrowed). (**c**) Very thick breccia near base of association showing closely packed, matrix-poor fabric cemented with fibrous calcite cement (white). (**d**) Close-up of red skeletal rudstone showing large crinoidal stem fragments, reefal limestone (R) and slope limestone (S) clasts and small angular fragments of massive stromatoporoid (M). Hammer 35 cm.

ridges (*c.* 1–1.5 m high) of carbonate cobble–boulder rubble (Fig. 8d). These facies are located towards the outer edges of the fore-reef slope succession and west of McIntyre Knolls where they clearly overlie the turbidites of FA1 (Fig. 2).

Interpretation. FA2 represents a new phase of platform growth (although the resulting shallow-water facies have since been eroded) with debris shed onto a fore-reef slope that prograded into the inter-reef basin. Progradation is also indicated by coarsening-upward patterns in the proximal facies which potentially suggest several phases of progradation and collapse of the platform margin (Fig. 5). The thickest breccias are concentrated near the platform which suggests FA2 proximal facies represent upper fore-reef slope, and some of the cemented breccias (which lack primary matrix, e.g. Fig. 6c) could be talus deposits rather than the products of flows. The latter would also suggest some relief on the platform margin.

The uppermost coarsening-upward package has some distinctive features. Firstly, it contains a Precambrian-pebble conglomerate. Although this is a minor facies, its stratigraphic position suggests that it is potentially related to conglomeratic patches which crop out at the base of the modern slope and especially the sandstone-matrix rubbly carbonate breccias on the basin floor. Quartz sand is a component of a range of carbonate facies (e.g. platform facies as seen in allochthonous blocks and slope facies); however, what is distinctive about the conglomerates is that they are significantly coarser grained than the more widespread sand-sized sediment. This suggests that a different mechanism is required to bring coarse debris out onto the platform where it can be transported to and deposited on the slope and basin floor.

Secondly, the allochthonous blocks and breccia clasts are dominantly reef-margin varieties; however, there is a noticeable proportion of reef-flat facies. This is very similar to the composition of the large

Fig. 8. Field photographs of FA2. (**a**) Buff-coloured graded (T_{abcd}) bed of distal facies. Note inverse-normal grading in A division and low-angle cross-lamination of C division. Hammer 35 cm. (**b**) View of calcite-cemented conglomerate overlain by skeletal rudstones near top of association (Fig. 5). (**c**) Close-up of conglomerate showing rounded to subrounded, moderately sorted, clast-supported fabric. (**d**) Carbonate 'rubble' breccia which typically forms low ridges on the basin floor. Matrix is typically weathered out but where preserved is dominantly quartz sandstone. Notebook for scale (arrowed) 15 cm long.

McIntyre Knolls blocks and clusters of blocks nearby. The stratigraphic position of the McIntyre Knolls blocks on top of red matrix-bearing debris-flow breccias at the toe of the slope suggests emplacement during or at the end of FA2 deposition in response to a major platform collapse event which cut well back into the reef flat and locally even into back reef. In addition, the clusters of smaller blocks are onlapped by turbidites of FA3, which also constrains timing of emplacement. Given the similar composition and size of McPhee Knoll (Fig. 2) it was most likely emplaced during the same event.

Facies association 3

Description. FA3 overlies FA2 with similar dips (commonly 17–21°) and is characterized by red, coarse stromatoporoid (notably *Stachyodes*) rudstones–floatstones and carbonate breccias which form a major margin-parallel unit on the proximal slope (Fig. 2). The association fines up overall with common breccias and minor intercalated medium- to thickly bedded floatstones–rudstones in the lower part and thinner bedded rudstones and floatstones with fewer breccias in the upper part (Figs 5, 9a). Higher-order fining-upward patterns are well developed (Fig. 5).

The basal breccias (*c.* 1.5–4 m thick) invariably have poorly sorted, clast-supported fabrics with minor red matrix. Clasts are angular to subangular, up to 3 m across) and composed of various platform-margin facies with stromatoporoid and coral (e.g. *Alveolites*) bioclasts, and slope-derived clasts. Platform-derived clast types include red tabular stromatoporoid–*Renalcis* (±sponge) framestone, pale-coloured and reddish *Renalcis* boundstone, stromatoporoid–*Stachyodes*–*Renalcis* boundstone and *Stachyodes*–*Renalcis* boundstone (Fig. 9b). The clast types are mostly indicative of reef-margin facies. Slope-derived clasts are red bioclastic

(a)

(b)

(c)

Fig. 9. Field photographs of FA3. (**a**) General view to the east showing typical facies. Medium, planar-bedded facies are rudstones–floatstones whereas thick beds with a more massive appearance are breccias. Note allochthonous blocks embedded near the top of this package (arrowed). (**b**) Close-up of a very thick massive breccia featuring angular to subangular clasts of reefal limestone (pale), irregularly overlain by medium-bedded skeletal rudstones–floatstones. Top of breccia at upper end of hammer handle (35 cm). (**c**) Medium-bedded inversely graded rudstones showing abundant angular clasts of stromatoporoids and reefal clasts (pale coloured) with assorted skeletal debris. Hammer 35 cm.

rudstones–floatstones and rare quartz sandstone. The red matrix (usually *c.* 15% of the rock) is grainy (peloidal ± ooids) with crinoidal and brachiopod debris, and usually contains some fine-medium quartz sand.

The rudstones–floatstones are medium- to thickly bedded (*c.* 0.1–1 m) and massive, or display normal or inverse-normal grading (Fig. 9c). Coarser parts of beds are rudstones with abundant angular clasts of massive stromatoporoid, tabular stromatoporoid and *Stachyodes* debris. Finer parts display floatstone fabrics and smaller bioclasts, notably corals (e.g. solitary rugose), brachiopods, crinoids and *Amphipora*. Matrix in the rudstones is typically up to 15–20% and fibrous calcite cement is common. Medium to coarse quartz sand is a persistent but minor component (up to *c.* 10%) of these facies.

Clast-supported, lenticular, red intraclastic breccias are interbedded with the rudstones–floatstones. The breccias are composed of pebble- to cobble-sized clasts (slope facies ± reef margin varieties) and are commonly inversely graded. Clast-rich parts of beds are typically calcite-cemented (Fig. 10a) and remaining porosity may be filled with a secondary red, typically laminated matrix. The reef-margin clasts are mainly red tabular stromatoporoid–*Renalcis* framestones and massive pink-grey (probably microbial) boundstones similar to the clast types in the very thick breccias. Allochthonous blocks of reef-margin limestone (typically 5–10 m across) embedded in this facies association are composed of similar rock types. Blocks are particularly common near the top of the association (e.g. Fig. 9a).

Towards the top of the measured section, and elsewhere at this stratigraphic level, the dominant facies are very thinly to medium bedded (<0.1–0.3 m), normally graded floatstones to grainstones–rudstones and medium to thickly bedded, inverse-normally graded, intraclastic breccias (up to 0.6 m) with slope-derived and reefal clasts. The floatstones commonly contain red silty intraclasts. The fossil assemblage is commonly composed of crinoids, brachiopods, *Amphipora* with minor *Stachyodes*. One sample leached for microfossils yielded abundant glass microspherules ≤0.5 mm across (MK34; Fig. 2). The microspherules are pale brown, typically transparent (some have a frosted appearance) and spherical, resembling, in their physical characteristics, the Si-rich microspherules of Ma & Bai (2002).

Lower on the slope, these facies commonly abut the upslope side and lap onto the sides and downslope side of allochthonous blocks (Fig. 10b). Allochthonous blocks scattered along the top part of this association are composed of massive pinkish-red *Renalcis*-spar-rich limestones (± stromatoporoids, skeletal debris and sponges), massive to crudely bedded, distinctive red platy–tabular stromatoporoid–*Renalcis* framestones (Fig. 10c) and red stromatactis limestones. The fabrics, particularly in the *Renalcis*-bearing blocks, are cross-cut by neptunian dykes lined with fibrous calcite cement and filled with red sediment. Some of these blocks are draped by thick to very thick red-matrix breccias composed of common *Renalcis*-rich limestone clasts and other platform-margin varieties.

At the base of the modern slope and onto the basin floor, the distal part of this association is conformable with FA2 (Fig. 2). The facies are thinly to medium-bedded buff-grey-coloured turbidites (e.g. T_{ab}, T_{bc}, T_{bcd} beds) with common fossil debris, locally silicified (notably stromatoporoid bioclasts) with intraclastic breccias (Fig. 11a, b). Sinuous, unlined, simple burrows are visible on bedding planes.

The measured section yields conodonts in Frasnian Zones 6–11 (Fig. 3). One sample in the measured section provides a more specific assignment of Frasnian conodont Zone 7 to lower zone 8 at *c.* 100 m above the base of the section (*c.* 25 m above the base, of FA3; Fig. 3). One spot sample near the base of the modern slope is in Zone 8 (MK15; Fig. 2, Appendix) and a sample from very close to the top of the association is in Zones 9–11 (MK34; Fig. 2, Appendix). Samples generally have low conodont abundance and feature predominantly shallow-water genera, notably *Polygnathus* and *Icriodus*, although there are rare *Palmatolepis* and *Ancyrodella* of deeper-water affinity. The palmatolepids are more common in the finer-grained samples towards the top of the section. The well-dated underlying FA2 (Zone 6) and recognition of Zones 9–11 for

(a)

(b)

(c)

Fig. 10. Field photographs of FA3 features. (**a**) Clast-supported intraclastic breccia composed of abundant slope-derived clasts of red limestone (dark) and some reefal debris (pale coloured) cemented by fibrous calcite. Hammer 35 cm. (**b**) Allochthonous block of massive *Renalcis* boundstone (B) overlain by bedded red skeletal rudstones– floatstones on the downslope side of the block (locality MK15, Fig. 2). (**c**) Good example of tabular–platy stromatoporoids–*Renalcis* boundstone (framestone) with *Renalcis* encrusting the undersides of the stromatoporoids. Abundant red sediment, in places laminated, fills the cavities.

(a)

(b)

Fig. 11. Field photographs of FA3. (**a**) Thinly- to medium-bedded buff-coloured limestones on the basin floor. Note subhorizontal dips and massive thick lenticular breccia on left side of photograph. (**b**) Close-up of graded (T_{ab}) bed showing coarse base (A division) with planar laminated top (B division). Coarse debris is mostly fossil material (silicified pieces are dark) and some slope limestone clasts (C division). Lens cap 5 cm.

uppermost samples suggest that the depositional age range of FA3 is upper Zone 6 to Zone 9.

The spot samples also yielded microvertebrate (fish) remains including well-preserved scales (i.e. not abraded or broken) of the thelodont *Australolepis seddoni* and palaeoniscoid fragments (Fig. 2, Appendix). *A. seddoni* is a Frasnian taxon and has been shown by Turner (1997) to range from the base of the Frasnian to the top of Zone 4 although this range can be extended to the top of Zone 10 on the Lennard Shelf (Trinajstic & George 2009). The palaeoniscoids are long-ranging (Frasnian to lower Carboniferous; Basden *et al.* 2006).

Interpretation. The marked change to fining-upward patterns in this association is interpreted to reflect a change to an overall retrogradational platform phase during deposition of FA3. The prominence of massive *Renalcis*-rich fabrics in the blocks and breccia clasts suggests the platform debris was largely derived from the reef margin and potentially also uppermost slope, with diversity of clast material also suggesting some reworking of older strata. The broader-ranging conodont ages for the association in general supports reworking and deposition of debris further downslope. In addition, the presence of silty limestone clasts in breccias and the observation that the distal part of FA3 does not appear to have extended as far to the south as the underlying FA2 slope, supports a change to retrogradation.

The thelodont *A. seddoni* has generally been used to interpret nearshore or shallow marine conditions; however, it has also been found in deeper, lower-energy settings (Trinajstic 2001; Trinajstic & George 2009). Moreover, the well-preserved material in FA3 and the expected proportions of scales from different parts of the fish (e.g. Trinajstic 2001) suggest *in situ* accumulation of fish remains on the toe-of-slope rather than transportation from shallower slope or platform settings.

Sample MK34 (Fig. 2) with the microspherules is *c.* mid-Frasnian and probably Zone 9 based on the age of the overlying association (Fig. 5). This age is clearly older than the Frasnian–Famennian boundary (and potentially older allowing for reworking) so no relationship between a meteorite impact and the latest Frasnian extinction events can be shown for this sample. Moreover, this time interval is regarded as relatively stable with no observed rise in species level extinctions (Becker 2000).

Facies association 4

Description. The thin uppermost package of fore-reef strata exposed in the area typically shows a marked colour change from the underlying red FA3 to buff-coloured bioclastic limestones and associated breccias (FA4). The contact with FA3 is conformable although locally may appear to onlap because dips in FA4 are noticeably lower ($\leq 10^\circ$) than FA3 (Figs 2, 12a). The colour change varies from sharp to transitional over a few metres. The bioclastic limestones are very thinly to medium bedded ranging from a few centimetres up to 0.25 m; they are commonly normally graded and display T_{abd}, T_{abc} or T_{acd} sequences. The thicker beds have rudstone bases with angular pebbles–cobbles of rugose corals, locally very abundant, and massive stromatoporoids (also seen encrusting corals) grading into floatstone or grainstone tops (Fig. 12b). The finer-grained material features common crinoids with peloids and intraclasts. Additional fossil debris includes bivalves, brachiopods, *Stachyodes*, *Amphipora*, *Thamnopora* and tabulate corals. Tabulate and massive corals are commonly encrusted by massive stromatoporoids.

The turbiditic limestones are interbedded with medium-bedded, massive to inversely graded, clast-supported breccias (up to 0.25 m thick) composed of clasts of the turbiditic facies with some reefal limestone (*Renalcis* or *Renalcis–Stachyodes* boundstones). Crinoid and brachiopod debris is common in finer-grained parts and in the matrix. At the eastern end of the study area, these facies are intercalated with siliciclastic conglomerates and sandstones and are typically red (Fig. 2). Elsewhere quartz sand or, locally, subrounded pebbles of Precambrian rock types are a conspicuous although minor component in rudstones and breccias. Allochthonous blocks of reefal limestone are common within this association; however, the turbidites of LFA4 appear to wrap around the lower parts of the blocks and their deposition most likely post-dated block emplacement. Dips are commonly shallow and mostly to the south or SW where measured on the lower part of the present-day slope. At the base of this slope and in the vicinity of allochthonous blocks, bedding is very shallow dipping (typically $<5^\circ$) and in variable directions (Fig. 2).

Conodont biostratigraphy shows that this association in the measured section is in Zones 8–11 (Fig. 5) which is consistent with spot samples along the slope to the west which are in Zones 9–11 (see Appendix). This suggests that the depositional age range of FA4 is likely to be no older than Zone 9 and certainly no younger than Zone 11. FA4 also features a change in the composition of the conodont fauna with an increase in the species of *Palmatolepis* as well as a greater diversity of conodont genera consistent with the presence of phoebedont shark teeth (see Appendix).

(a)

(b)

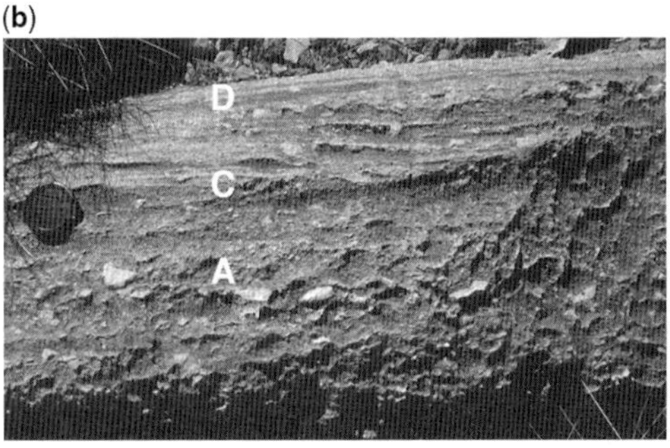

Fig. 12. Field photographs of FA4. (**a**) View of lower part of modern slope showing sharp conformable contact between bedded limestones and breccias of FA4 and FA3. Top of red limestones of FA4 arrowed. Note lenticular debris-flow breccia and small allochthonous block debris beneath arrow. (**b**) Close-up of T_{acd} bed with coarse pebbly base and finer-grained cross-laminated to planar laminated top. Dark speckly appearance shows significant component of medium-coarse quartz sand. Lens cap 5 cm.

Interpretation. FA4 represents the youngest depositional phase within the inter-reef basin on the southern margin of the Lawford Range. The change to buff-coloured crinoidal limestones for much of the association, coupled with the deeper-water affinity of the conodonts and phoebedont sharks, is interpreted as continuation of deeper-water, open-marine conditions. A deeper-water interpretation is also consistent with the downslope increase in crinoidal debris observed in FA2 and 3. Compositionally, these facies contain clasts of mixed origins, including locally abundant coarse bioclasts of coral and stromatoporoid debris of a shallow, platform-margin origin, and crinoids and intraclasts suggestive of a deeper-water fore-reef slope origin. The intercalation of siliciclastic material and facies indicates contemporaneous deposition of carbonate and siliciclastic sediments at the eastern end of the inter-reef basin where conditions were probably also shallower.

The angularity of the coral and stromatoporoid debris suggest erosion of some kind of resistant buildup, rather than simply reworking of older material, especially as stromatoporoid encrusted coral debris is also present. Similar buff-coloured limestones (although younger) have been recognized elsewhere on the Lennard Shelf and have been attributed to deeper-water conditions, e.g. a crinoidal–intraclastic fore-reef slope unit in the Napier Range (e.g. George *et al.* 1997; George & Chow 2002).

Evolution of the Northern Lawford reef complex

The abundant siliciclastic facies and sediment in all parts of the Northern Lawford platform reflect proximity of this reef complex to Precambrian basement. The platform typically has a steep, scarp-like margin eroded into thickly bedded *Stachyodes–Renalcis*-rich limestones of the reef flat and against which younger fore-reef strata (FA2) onlap. Rarely preserved reef-margin facies are associated with steeply dipping fore-reef strata (FA1) which abut the platform margin and are likely to be coeval. A similar steep debris apron fringing platforms is seen around Waggon Pass (Fig. 2) and elsewhere in Bugle Gap (e.g. Playford 1980). FA1 (Frasnian Zones 5–6) is more widely represented by basin-floor turbidites and siltstones which underlie the main fore-reef succession. The grey colour of the FA1 turbidites on the basin floor and presence of hematized goniatites led Becker *et al.* (1993) to interpret hypoxic conditions. The presence of trace fossils suggests semi-restricted conditions on the sea floor most likely due to poor circulation in the small inter-reef basin. Trace fossils are common on the basin-floor turbidites of the overlying associations, suggesting that conditions improved.

The three major facies associations of the McIntyre Knolls fore-reef to basinal succession (FA2–4) are interpreted to record three younger and discrete phases of reef-complex evolution during the late Early to Middle Frasnian. FA2 (Frasnian Zone 6) records an important phase of progradation following backstepping of the platform margin relative to the position of the platform during FA1 time. This is indicated by mid-upper slope facies (proximal FA2) abutting the older platform with the correlative uppermost slope deposits no longer preserved. Conodont biostratigraphy from FA1 and FA2 indicates that this backstep and flooding event occurred in the early part of Zone 6.

An overall progradational trend for FA2 is shown by the breccias and rudstones being deposited much further into the basin than similar facies in FA1. Higher-order coarsening-upward trends within FA2 highlight progradational pulses culminating in a major phase of debris deposition and platform collapse (Fig. 3). The coeval platform and reef margin is no longer preserved except as abundant debris in the form of allochthonous blocks (reef margin and reef flat facies) which crop out along the lowermost slope and toe-of-slope, and as clasts in the very thick debris-flow breccias which feature prominently in this association.

The major collapse phase is represented by the McIntyre Knolls and associated blocks embedded higher up the present-day slope (Figs 2, 4b). The knoll blocks slid over 1 km into the basin and came to rest on very thick debris-flow breccias and associated finer-grained facies. In the measured section, a prominent zone of reef margin and reef flat allochthonous blocks is associated with debris-flow breccias at the boundary between FA2 and FA3 (Fig. 5). The position of the blocks on top of debris-flow breccias deposited on the basin floor, and the stratigraphic position of similar blocks in the measured section, constrain the timing of collapse to late in Frasnian Zone 6. This timing is similar to but slightly younger than the timing of early in Zone 6 suggested by Becker *et al.* (1993). However, it is more consistent with an important event subsequently recognized by Becker (2000) as discussed further below. Whittam *et al.* (1994) proposed that the McIntyre Knolls blocks were part of a lowstand systems tract that correlated with the platform succession in Windjana Gorge. However, although the latter is also Frasnian it is clearly younger in age (late Middle or Late Frasnian; Ward 1999; George *et al.* 2002).

Progradation of the platform (during highstand deposition of FA2) suggests that collapse may have occurred in response to major relative sea-level fall in conjunction with instability due to oversteepening (triggered by storms, tsunamis or seismic events; e.g. Cook & Mullins 1983; Hine *et al.* 1992). It is interesting to note that siliciclastic conglomerate is present just below the collapse debris in the measured section (top of FA2; Fig. 3). It is likely that was also a time of major influx of siliciclastic debris from the eastern side of the inter-reef basin (Fig. 2) but not necessarily initiation of deposition. The association of siliciclastic sedimentation may also support a relative sea-level lowstand as seen elsewhere on the Lennard Shelf (e.g. George & Powell 1997) and inferred from seismic data in the deeper parts of the Fitzroy Trough (e.g. Southgate *et al.* 1993). However, the abundance of siliciclastic sediment in this carbonate system overall suggests this needs to be considered within the stratigraphic context of platform-margin collapse and platform demise, rather than simply attributing a lowstand origin.

FA3 records flooding and backstepping of the next platform phase as indicated by the marked change to an overall fining-upward trend and higher order retrogradational stacking patterns. The base of FA3 is likely to be late Zone 6 (based on the ages above and below; Fig. 5) and this is the age we assign to the second backstepping event in the Northern Lawford platform following its initial formation. Retrogradation is also indicated by the broad ranges in conodont ages in individual samples (e.g. Zone 6–10) suggesting reworking of previously deposited slope sediments (particularly FA2 age material). Overall retrogradation is also consistent with the distribution of this association close to the present-day platform (Fig. 2) and the appearance of thelodonts indicative of deeper water. Collapse of the platform margin and likely reworking of platform debris during this time are also suggested by the common allochthonous blocks of reef-margin facies embedded within FA3 strata (Fig. 5).

FA4 is a relatively thin unit that shows the most limited distribution in the area. It is interpreted to record continued deeper-water conditions based on the change in colour and dominance of crinoidal–intraclastic debris suggesting reworking of slope sediments. Stromatoporoid and coral debris were derived from erosion of shallower water buildups upslope. The overall platform style may have changed from the types of platform margins observed and/or interpreted for FA1–3 to a lower relief or possibly ramp-style setting; the abundance of massive coral debris may be significant in this regard. Conodont biostratigraphy gives a minimum depositional age in Zone 9 for the top of FA3 and a maximum depositional age in upper Zone 8 for FA4 (Fig. 3) which suggests an age for the base of FA4 within Zone 9. Shallower conditions were probably maintained along the contact with the siliciclastic sediments on the eastern side of the inter-reef basin.

Bugle Gap platforms

The Northern Lawford platform history now needs to be considered in the broader context of the evolution of the platforms on the eastern side of Bugle Gap and the neighbouring platforms to the west. The western end of the Northern Lawford platform narrows and bends south into a thin belt of massive reefal limestone flanked on the eastern side by coarse rudstones and breccias of FA2 which are overlain by coarse *Stachyodes*-rich rudstones of FA3 (Fig. 2). On the western side, coarse fore-reef deposits equivalent to FA2 and younger have been mapped (Playford & Hocking 1998).

Immediately to the south (still north of Waggon Pass) this narrow belt is matched by a similar belt of massive to crudely bedded limestones composed of large tabular and hemispheroidal stromatoporoid framestones and *Stachyodes* rudstones–floatstones (Fig. 2). These facies are underlain by a subhorizontal remnant of platform (back-reef to reef-flat facies) which crops out on the eastern side (Fig. 2) and is also shown in Playford & Hocking (1998). A similar but small remnant of well-bedded platform is also preserved in Waggon Pass (southern side of track, beneath the reef facies; see Shen *et al.* 2008) onto which fore-reef *Stachyodes*-rich facies downlap (Fig. 13a). These relationships show that the platform remnant: (1) was drowned resulting in development of narrow elongate reef on top (a relationship well described from other parts of Bugle Gap, e.g. Playford *et al.* 1989); and (2) is older than the Northern Lawford platform.

Teichert Hills has been interpreted as a platform atoll that is flanked by a narrow zone of fore-reef along almost all sides (Fig. 1; Playford & Lowry 1966; Playford & Hocking 1998). This platform atoll is best known for the chute and buttress structures at its southwestern corner (Playford 1981; Figs 1, 13b). The ENE-trending buttresses were constructed by domal growths of tabulate corals, stromatoporoids, calcimicrobes and early marine cements (Wood & Oppenheimer 2000). Their origin has been contentious, with Playford (1981) and Wood & Oppenheimer (2000) favouring a growth origin and Southgate *et al.* (1993) suggesting that the chute and buttress morphology is erosional. Nonetheless, these authors all agree that drowning of the platform atoll (with pinnacle reefs developing on the older platform; Playford *et al.* 1989) caused its ultimate demise. Phosphatic crusts mantle the drowned surface (Southgate *et al.* 1993) and conodont biostratigraphy of fore-reef strata (associated with subsequent reef development) that lie just above this surface indicate Zone 3–lower Zone 4 (G. Klapper pers. comm. 2006). It is also clear that Teichert Hills atoll existed prior to development of the Northern Lawford platform because the northern margin of the atoll prevented siliciclastic sediment from the eastern side of the basin being dispersed to the south and SE (Figs 1, 2).

The Southern Lawford Range is a broad platform which is everywhere flanked by a unit of proximal fore-reef (Playford & Hocking 1998). This fore-reef is noticeably wider where the pinnacle and elongate reefs were established on older platform (e.g. western side of Fig. 2) including the Southern Lawford Range and Teichert Hills. A good example of this relationship has been described from Kelly's Pass near McWhae Ridge (Fig. 13c). Early conodont biostratigraphy by Seddon (1970) established that this fore-reef succession, at least around the southern part of the Lawford platform, is Early Frasnian (in the equivalent range Zones 2–6). The presence of the goniatite *Timanites* (Seddon 1970) is consistent with this age assignment (equivalent Zone 4; Becker *et al.* 1993).

The top of the proximal fore-reef along the southeastern platform margin at McWhae Ridge, with its well-known oncoid rudstones and nautiloid-rich bed top (Playford 1980), has been dated as Frasnian Zone 4 (Nicoll 1984; Becker & House 1997). Significant deepening and drowning of the fore-reef and platform at this locality were recognized many years ago (Playford & Lowry 1966; Playford 1981; Nicoll 1984) with basinal muds deposited on top of the platform by Zone 5 time (Becker & House 1997). Deep-water stromatolites accumulated on the slope with coeval muds deposited on the basin floor (Playford 1980, 1981). Overlying very thinly bedded red limestones are in Frasnian conodont Zone 7 on the basin floor (Becker & House 1997, section no. 372) and Zone 12 higher up the slope (Becker & House 1997, section no. 371B; G. Klapper pers. comm.). In the Northern Lawford Range the Late Frasnian part of this depositional phase is recognized on the western side of the range where conglomerates are overlain by thinly bedded red limestones; the basal beds are dated as conodont Zone 11 (Becker & House's (1997) Windy Knolls locality). These relationships suggest that the conglomerates eventually spilled westwards over the limited carbonate reef and potentially provided a base for a younger (Late Frasnian) platform phase. The presence of Zone 11 strata in both localities suggests that carbonate production, potentially from relatively limited systems, was occurring to the north (at McWhae Ridge) and east (at Windy Knolls) and that both depositional sites were distal to source. Becker & House (1997) suggested that these

(a)

(b)

(c)

Fig. 13. Frasnian Fr2 platform features. (**a**) Platform remnant in Waggon Pass (south side) distinguished by thinly bedded *Amphipora*-skeletal floatstones (P) overlain by forereef *Stachyodes*-rich rudstones dipping to the east (into photo). Dips steepen up-section with topmost beds currently dipping at *c*. 27°. (**b**) Chute and buttress structures on the SW corner of Teichert Hills platform atoll showing morphology interpreted in other studies as growth features. (**c**) Platform relationships at Kelly's Pass near McWhae Ridge showing thinly bedded subhorizontal platform limestones (P) of Fr2 overlain by a younger reefal phase characterized by massive limestones (M) and bedded, dipping fore-reef facies (F) of Fr3.

strata represent a time of maximum flooding in the SE Lennard Shelf. We have not included these strata in FA4 because potentially they record a new phase of platform production which ultimately led to marked progradation, as seen in the downlapping thinly bedded Frasnian–Famennian fore-reef succession that has been well described at McWhae Ridge (Playford 1980; Nicoll & Playford 1993). An iridium anomaly discovered in this section heightened interest in a potential meteorite impact at the Frasnian–Famennian boundary (e.g. Playford *et al.* 1984); however, conodont biostratigraphy showed that this anomaly is Early Famennian (*crepida* Zone) with iridium microbially concentrated in a *Frutexites* layer (Nicoll & Playford 1993).

A comparison of these platforms suggests that three main phases (Fr2–4) are represented in the western Bugle Gap area (Fig. 14). The oldest phase (Fr2) includes the Southern Lawford platform, Teichert Hills platform atoll, and platform remnants exposed at and north of Waggon Pass. All show evidence for drowning which is suggested to be *c.* Zone 3 from the basal strata overlying the phosphatic surface at Teichert Hills. Drowning caused backstepping and development of elongate and pinnacle reefs (Fr3) with deposition of coeval flanking fore-reef debris (Figs 13a, c, 14). The shift in the locus of carbonate production landward (i.e. northward) to form the Northern Lawford platform (Fr4 at McIntyre Knolls) suggests that drowning was most likely widespread to the south. The style of platform margin with a narrow flanking fore-reef slope is very similar to the fore-reef described from around the Southern Lawford platform and Teichert Hills atoll, but represents a younger phase (Zone 5) as shown by FA1 age relationships. Similarly, Seddon (1970) suggested that the fore-reef around the Lawford Range, although similar to and mapped as part of the same formation (Sadler Limestone; Playford & Hocking 1998), was younger than the type section at Sadler Ridge (Fig. 1). Backstepping promoted development of the Northern Lawford platform so we think it is unlikely that an extensive older platform phase is present beneath it (cf. Playford & Hocking 1998) and that, instead, the platform was probably built on siliciclastic sediment eroded from

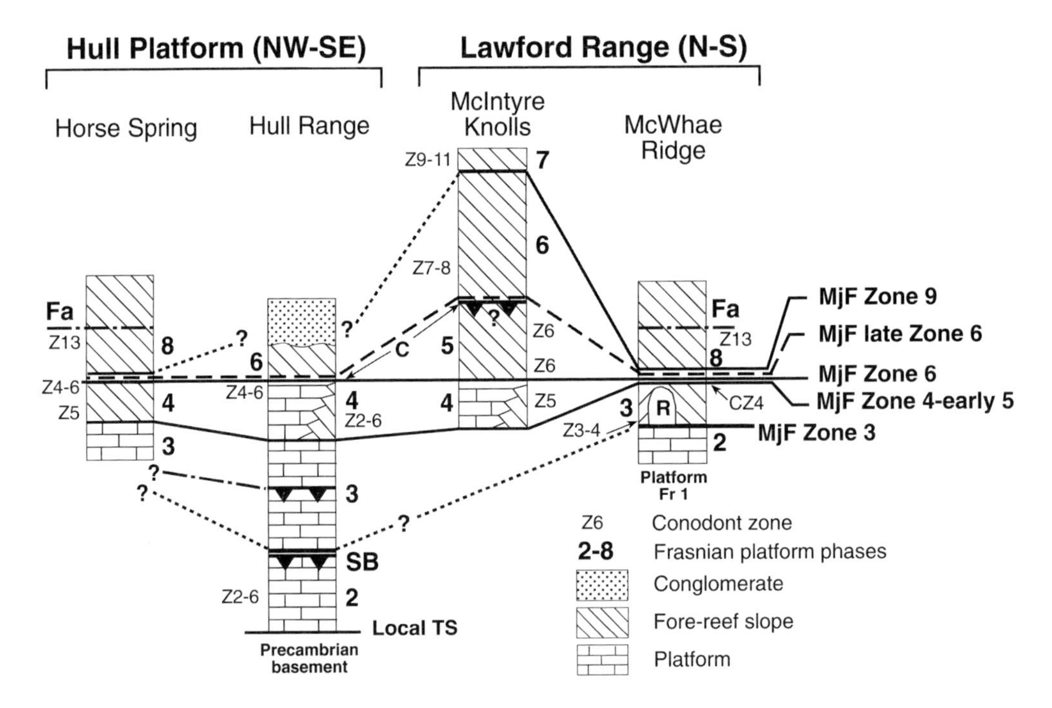

Fig. 14. Summary of Hull Platform and Lawford Range areas showing Frasnian platform phases (represented by platform facies and/or fore-reef facies) labelled 2–8. Platform phases are separated by major flooding surfaces (MjF) which are commonly accompanied by backstepping. Samples with diagnostic conodont biostratigraphy are shown in their stratigraphic positions (Zone 6 = Z6). The McIntyre Knolls section provides an expanded history of phases Fr5–7 which are largely condensed carbonate sections at the well-known McWhae Ridge and Horse Spring sections. Hull Range section provides an expanded view of Early Frasnian platform phases (Fr2–4). Sequence boundaries representing relative sea-level falls are labelled SB. McIntyre Knolls and Rattigan's Rocks collapse events are labelled 'C' and considered most likely coeval. Deposition of Late Frasnian strata is shown as platform phase Fr8 for simplicity.

Precambrian basement uplifted along the Sparke Range Fault (Fig. 1).

Playford & Hocking (1998) interpreted the Southern Lawford platform to overlie an older platform phase which is well exposed along the western Emanuel Range and is proposed to be Givetian at its base (Fig. 1; Hocking *et al.* 1995; referred to in our scheme as Fr1). The platform unconformably overlies a veneer of Ordovician rocks on Precambrian basement and is also interpreted to underlie Paddy's Valley, Lloyd Hill atoll and the Laidlaw Range (Fig. 1; Playford 1980, 1981; Playford & Hocking 1998). The latter two show similarities in platform morphology and fore-reef association and are mapped as coeval with the older platforms on the eastern side of Bugle Gap (Playford & Hocking 1998). Moreover, pinnacle development along the northern Laidlaw Range has been attributed to backstepping and drowning, similar to the platforms across the gap (Playford *et al.* 1989).

The phases of platform development preserved in the McIntyre Knolls fore-reef section are subsequently labelled Fr5, 6 and 7 (Fig. 14). Each is underlain by a flooding surface (Zone 6, late Zone 6 and Zone 9 respectively, as outlined above) with only one platform (Fr5) showing evidence for a relative sea-level fall and major platform collapse prior to flooding. There is no clear evidence of palaeokarst but parts of the platform margin may have been exposed (Fig. 14). The dominance of flooding surfaces in these Frasnian platforms is not surprising given the overall transgressive character of Frasnian reef development on the Lennard Shelf. We agree with Holmes & Christie-Blick (1993) that platforms which develop during overall transgressive conditions are most easily differentiated by flooding surfaces. This is especially in contrast to sequence boundaries (with evidence of subaerial exposure) separating latest Frasnian through Famennian platforms on the northwestern Lennard Shelf (George *et al.* 2002; Chow *et al.* 2004) when overall regressive conditions ensued (Playford 1980; Southgate *et al.* 1993).

Hull Platform (Hull Range, Guppy Hills and Horse Spring Range)

The Hull Range is located NW of the Lawford Range and is composed of platform strata with some flanking fore-reef strata preserved (Fig. 1; Playford & Hocking 1998). Precambrian basement is exposed on the western side of the range (Fig. 1). The platform is continuous with the NE-trending Horse Spring Range which is also composed of platform strata with a thicker fore-reef slope succession. Guppy Hills is considered the easternmost extension of the Hull Range and is composed of platform strata. The limestone ranges expose mostly back-reef facies with some preservation of leeward platform margin towards the top.

The Hull platform (our collective term for Hull Range, Horse Spring Range and Guppy Hills) is important because it contains clear evidence of subaerial exposure and flooding events for which we have biostratigraphic control (microvertebrate and conodont biostratigraphy) to show that the entire platform succession is within Frasnian Zones 2–6 (George & Chow 1999; George *et al.* 2009). This is in contrast to Hocking & Playford (2000) who considered Horse Spring Range and Guppy Hills to be largely Givetian.

A detailed analysis of the lateral and vertical facies relationships in the Hull platform will be presented in a separate publication. A key observation from that work is that the northwestern end of the platform is dominated throughout by aggradational shallow subtidal to intertidal facies (e.g. fenestral limestones) whereas the southeastern end is characterized by deep subtidal to shallow subtidal facies which are aggradational to retrogradational upsection. We interpret this broad pattern to reflect the control on platform development by the bounding normal faults to the east (Fig. 1), both in producing the original tilt block basement geometry and continually generating accommodation towards the fault by tilting and subsidence (George *et al.* 2009).

In this study we have defined three discrete growth phases in the platform strata based on our surfaces and biostratigraphy (Fig. 14). The oldest phase (Fr2) was initiated after flooding of the Precambrian basement and was terminated by subaerial exposure and local karstification (George & Chow 1999). The second phase (Fr3) was initiated by flooding of the platform which also initiated carbonate production and deposition in the eastern part of Guppy Hills onto siliciclastic facies (Fig. 1). Within this phase there is evidence for subaerial exposure, most notably at the northeastern end of the platform; however, as it is not clear this was a major event we have not subdivided this phase. Termination of Fr2 was by flooding and a shift of the platform margin towards the platform interior.

This third phase (Fr4) is represented by reef-margin facies in the subsurface in the northern part of the Hull Range (George *et al.* 2009). Exposed pale-coloured fore-reef strata deposited on the eastern (leeward) side of the Hull Range are likely to be coeval with the third phase (Fig. 14). Conodont biostratigraphy shows that the top of this fore-reef package is in the range Frasnian Zone 4 to early Zone 6. The Horse Spring Range also features a package of buff-coloured, partially dolomitic, proximal fore-reef facies (breccias and bioclastic rudstones) that abut the eroded edge of the platform indicating that they were deposited after collapse of this older platform. Limited published dates on this package indicate deposition during Zone 5 (Becker *et al.* 1993) and our conodont biostratigraphy confirms that the top of the proximal package is also Zones 4–6.

Both areas are sharply overlain by red slope limestones. These are well described from the Horse Spring section which is one of the best dated Frasnian–Famennian sections on the Lennard Shelf (Becker *et al.* 1991). The strata at the base of this section are in Zone 6 (Becker *et al.* 1991) and thus the top of the youngest exposed Hull platform is defined by a Zone 6 flooding surface. The basal *c.* 2.5 m of the Horse Spring section is highly condensed as shown by conodont and goniatite biostratigraphy (Zones 6–11; Klapper 2007; Fig. 14). The section does not show evidence for significant coarsening of the facies until the earliest Famennian.

Synthesis and discussion

Sedimentological and biostratigraphic examination of the fore-reef succession in the McIntyre Knolls area provides an expanded history for the evolution of the Lawford Range with four discrete phases of platform development separated by flooding events associated with backstepping. Conodont biostratigraphy has been used to constrain formation of these surfaces to *c.* Zone 3, Zone 6, late Zone 6 and Zone 9 (Fig. 14). Only one flooding event (late Zone 6) appears to have definitely followed a relative sea-level fall ($\pm$subaerial exposure of the platform margin) suggested by major collapse of the prograding platform margin and associated coarse siliciclastic deposition. Comparison with the McWhae Ridge locality to the south shows how highly condensed the geological record is at this southern locality (as is well known from previous studies cited above), with all four flooding surfaces located within a few metres stratigraphically (Fig. 14).

Comparison of the Lawford Range area with the Hull platform has provided some surprising results in terms of the degree of correlation between these two major areas and implications for palaeogeography and the interplay between active tectonism (rifting) and eustatic changes. The Zone 6 flooding surface has proven a very useful datum beneath which we have placed the three platform phases recognized in the Hull platform (Fr2–4). Each is separated by a major flooding surface associated with backstepping, with one phase (Fr2) providing clear evidence for subaerial exposure at its top prior to flooding that initiated deposition of phase Fr3 (Fig. 14).

Basal platform phases

The oldest phase of platform development recognized in the Lawford and Hull areas appears to be broadly coeval although this timing is not tightly constrained biostratigraphically. In the Southern Lawford Range the exhumed platform (Fr2; Fig. 14) has been interpreted by other workers to overlie an older platform that is Givetian at its base. In the Hull Range, basal strata are a mixture of siliciclastic and carbonate facies which were deposited directly on Precambrian basement (Fig. 15a). To date, the basal Frasnian conodont Zone 1 has not been identified in outcrop in the Canning Basin; the estimate of Zones 2–3 is reasonable given the overlying flooding surface. A feature which potentially strengthens this correlation is that Hull Fr2 is capped by a subaerial exposure surface (George & Chow 1999) that is variably represented along the *c.* 20 km strike-length of the platform as it is currently exposed (George *et al.* 2009). Playford (1981) described mild karstification at the top of Lawford Fr2 near McWhae Ridge which suggests regional control on relative sea-level fall. This is probably most easily achieved by a eustatic fall; however, the alternative tectonic control (e.g. by footwall uplift) is discussed further below.

A simple extensional model predicts that, during basin evolution, younger reef complexes develop in progressively more landward positions. The Hull platform was initiated later, given the interpretation of an older, Givetian platform (FR1) in the Emanuel Range. It is interesting that initiation of the Hull platform, although located *c.* 30 km north of the Northern Lawford Range (Fig. 15) and the most northerly of all the platforms on the SE Lennard Shelf, is most likely early Early Frasnian (*c.* Zone 2) and, therefore, not markedly younger.

Zone 3 flooding

The subaerial exposure surface in the Hull platform was flooded enabling development of a thick, aggradational platform package (Hull Fr3). A significant feature of this flooding event is that it resulted in expansion of the platform towards the east and initiation of deposition in the eastern Guppy Hills area which had previously been siliciclastic (probably shallow marine–terrestrial) deposition adjacent to the bounding faults (George *et al.* 2009; Fig. 15b). This suggests that flooding was also related to development of accommodation close to the fault and that it was tectonically controlled. Eustatic sea level may still have been low or rising at this time but its signature would most likely have been overwhelmed by fault-related subsidence. Widespread flooding of the Lawford Fr2 platforms has been widely documented and

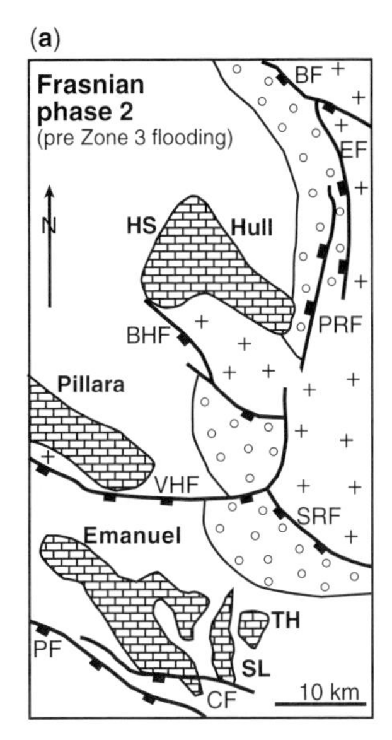

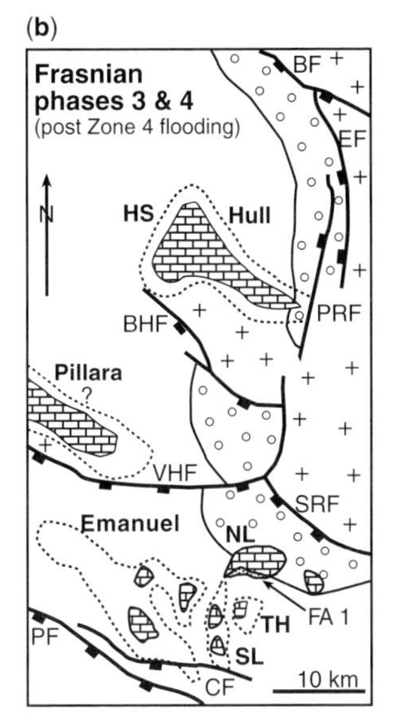

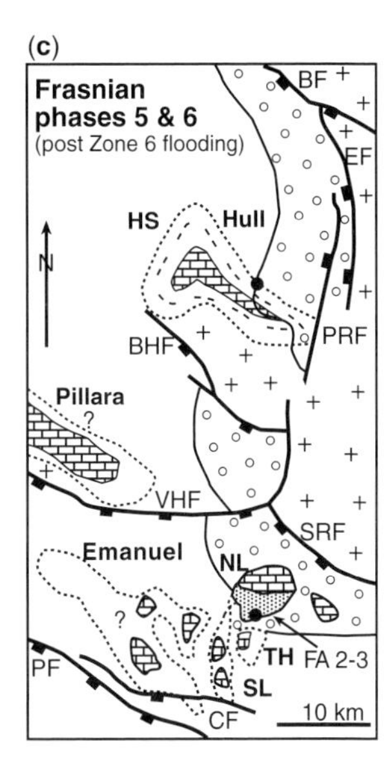

Fig. 15. Palaeogeographic interpretations for the Lawford Range area and Hull platform area. Major normal faults (symbol on hanging wall) and distribution of platforms in (A) are based on geological map (Playford & Hocking 1998). Original shapes of platforms are schematic especially as the windward (western) margin of the Hull platform has been eroded. Lawford block is bounded by the Sparke Range Fault (SRF) to the north and Pinnacles Fault (PF) to the south. Hull platform is bounded by north-trending fault zone featuring the Mt Elma Fault (EF) and Painted Rocks Fault (PRF) to the east and north-trending faults to the west which have no obvious surface expression. (**a**) Frasnian phase Fr2 prior to flooding in Zone 3 time (SL, Southern Lawford Platform; TH, Teichert Hills platform atoll). Emanuel and Pillara Ranges are included although they were not specifically included in this study. (**b**) Effects of flooding events (Zone 3 and Zone 4) have flooded large parts of the Southern Lawford Range and Teichert Hills (and presumably Emanuel Range) and promoted backstepping and formation of the Northern Lawford Platform on a siliciclastic base (*c*. Zones 4–5). Flooding of the Hull platform causes expansion (backstepping) of the platform onto siliciclastics (Zone 3) and of the platform margins. Subsequent flooding causes further backstepping although the northwestern end of the platform (Horse Spring area) continues to aggrade subtidal–intertidal facies. (**c**) Platform phases 5 and 6 in which Hull platform has undergone further flooding and backstepping. Northern Lawford Platform progrades during phase 5 and backsteps to form phase 6. Megabreccias formed by collapse are shown as solid circles.

summarized above. It resulted in development of backstepped reefs and pinnacles around the southern end of the Lawford Range, Teichert Hills and on the platforms on the western side of Bugle Gap ('Kelly's Pass Event' of Playford 2002) and shown schematically on Figure 15b.

Zone 4 flooding

The age of this event in the Hull platform is not directly constrained biostratigraphically (only a single long-ranging Frasnian conodont taxon was recovered from dark mudstones, and no microvertebrate remains). At Horse Spring the vertical eroded margin of Hull Fr3 is currently exposed and is abutted by coarse proximal fore-reef of *c*. Zone 5 (belonging to Fr4 although the actual platform has been eroded at this locality). A similar relationship is recognized at McIntyre Knolls with the presently exposed platform of the Northern Lawford Range and associated fore-reef package (FA1) suggesting deposition was underway by Zone 5 (Fig. 15b). This platform records a significant backstep in response to drowning of many parts of the Lawford Range to the south, most notably represented at McWhae Ridge where the top of the associated proximal fore-reef is well-dated to Zone 4 (Becker & House 1997). As a result of drowning, platform Fr4 is not recognised at McWhae Ridge (Fig. 14). Instead, much of the Bugle Gap area may have been characterized by patch reefs, pinnacles or small atolls at this time (Fig. 15b).

Zone 6 flooding

The McIntyre Knolls section is very useful for delineating two important flooding events. The first of these events is associated with backstepping of the Northern Lawford platform shown more obviously by subsequent deposition of a fore-reef package of Zone 6 age during progradation of Lawford Fr5 (Figs 14, 15c). It is difficult to assess how widespread Fr5 platform was in the Hull and Lawford areas because of erosion. In addition, most of the key sections appear to have been distal, with limited sediment delivered to those areas (Figs 14, 15c). Comparison of our results with goniatite biodiversity synthesized by Becker (2000) from his considerable earlier work in the Bugle Gap area (much of which is cited above) shows some very good agreement. Becker (2000) recognized regional diversification within Zone 6 (*Gogoceras nicolli* goniatite Zone, UD IF2) which he correlated with a major phase of hypoxic deposition in Bugle Gap (hematized goniatites in basinal facies; Becker & House 1997). This transgressive event fits very well with the Zone 6 flooding event recognized in this study. It is an important correlatable surface in the region because it is marked in fore-reef successions by a major change from coarse proximal (typically steeply dipping) facies associated with the underlying platform to shallower dipping coarse to fine, bright red facies (depending on position relative to platform Fr5). This Zone 6 flooding event is probably the same as the 'Shady Bore Event' shown, but not described, by Playford (2002, fig. 2).

Late Zone 6 collapse

The McIntyre Knolls section also provides evidence for a major platform collapse event following progradation of platform Fr5 in the late part of Zone 6. Becker (2000) named the 'Bugle Gap Event' for elevated regional species-level extinctions which he associated with regression at the end of UD IF2 (within Zone 6). In addition, he proposed that the Bugle Gap Event was potentially of regional significance. This regression fits very well with our sedimentological evidence for relative sea-level fall and collapse of the Northern Lawford platform towards the end of Zone 6 (Fig. 14). The increased flux of coarse siliciclastic sediments into the broader area at this time (e.g. as recognized at McIntyre Knolls) with potential for increased nutrients from land surface runoff, may have played a role in these extinctions.

This phase of lowered relative sea level was possibly in response to eustatic fall (as suggested by Becker 2000). This northern area would be the most likely area to preserve evidence of a eustatic event because the areas to the south were mostly too deep at this time. An interesting additional factor to consider is that the basal red strata on the eastern side of the Hull Range contain numerous large allochthonous blocks (known as 'Rattigan's Rocks') interpreted by Playford & Lowry (1966) as a megabreccia most likely formed by platform-margin collapse. We suggest that this collapse event is very similar in stratigraphic position and age to the platform-margin collapse event represented by McIntyre Knolls. This correlation may provide further weight to an interpretation of eustatic fall affecting the region. The collapse event overlies a progradational package at the McIntyre Knolls (FA2) which is essentially absent at the Hull Range locality (Fig. 14). The fine-grained facies deposited in the southern Lawford and Horse Spring localities at that time indicate they were probably too distal from platforms to receive coarse carbonate sediment (Figs 14, 15c). The McIntyre Knolls blocks slid *c.* 1 km from their origin and Rattigan's Rocks may also represent outrunner blocks deposited a considerable distance from their platform (Fig. 15c).

An alternative to eustatic fall is footwall uplift during faulting – a mechanism that we and other workers have invoked in other parts of the Lennard Shelf (e.g. Shaw *et al.* 1994; Chow *et al.* 2004). In the McIntyre Knolls area there is no clear evidence for a fault along the southern side of the Northern Lawford Range in terms of linear features and/or dolomitized zones (although one may exist at depth). The NNW-trending faults offset the platform which suggests they are post-Devonian or at least post-Frasnian age. Likewise the eastern side of the Hull platform also has no obvious platform-parallel structures but there are some large north-trending (normal and reverse) faults mapped (Playford & Hocking 1998) which may have been active at the time. Given the apparent synchronicity of the collapse events at both locations we favour eustatic control on this relative sea-level fall.

Late Zone 6 and Zone 9 flooding events, and post-FA4 deposition

Development of deeper-water conditions in the late part of Zone 6 is represented by FA3 at McIntyre Knolls and interpreted as overall retrogradation of the Northern Lawford platform. Again, the McIntyre Knolls area preserves the thickest example of this phase (Fig. 14). Elsewhere we think that conditions were relatively deep with patch reefs or other aerially confined buildups persisting (Fig. 15c). The youngest fore-reef package (FA4) in the Southern Lawford fore-reef succession indicates continued development of deep water conditions away from the siliciclastic areas. The base of FA4 is constrained by conodont biostratigraphy to *c.* Zone 9; elsewhere at this time the areas focused on in this study were distal to any carbonate factories and sedimentation was condensed (e.g. Horse Spring, McWhae Ridge; Figs 14, 15c).

Distal and condensed sedimentation is also recorded in Zone 11 which Becker & House

(1997) proposed was the peak of a major transgression. This record, although condensed in many places, does suggest that carbonate production was occurring locally and that these areas would ultimately provide the base across which platforms of the earliest Famennian would prograde. In some areas it is likely that the large deposits of conglomerate and sandstone provided a base on which the Famennian platforms could develop and/or prograde. In some areas these platforms are preserved in down-faulted areas, e.g. Horseshoe Range (north of Hull Range; Fig. 1).

Transgressions during Zones 9 and 11 on the SE Lennard Shelf broadly coincide with the development of carbonate platforms on the NW Lennard Shelf which are now exhumed. Of the four major Frasnian platform phases recognized (A, B, C^{lower}, C^{upper}). Platform C^{lower} is in Zone 12 to potentially the lower part of Zone 13 based on coeval fore-reef deposits, and was considered to have developed during maximum flooding of the central and NW Lennard Shelf (George *et al.* 2002). The underlying Platforms A and B, which also exhibit backstepping geometry (Ward 1999), could potentially represent deposition at least during Zone 11 and potentially as old as Zone 9. Independent biostratigraphic dating of these older platforms is clearly required; this proposal represents a first step towards a biostratigraphically based correlation of the reef complexes SE and NW of the Margaret Embayment.

Fault-controlled subsidence

Initiation of platforms on tilt blocks has been long established for the Lennard Shelf reef complexes (e.g. Playford 1980; Kemp & Wilson 1990). The backstepping geometry of the Frasnian platforms was also recognized several decades ago (Playford & Lowry 1966) and was attributed to accommodation generated by basin subsidence rather than solely by eustasy (Playford *et al.* 1989). The overall architecture of the Lennard Shelf reef complexes was most likely controlled by relative sea-level changes in which both tectonics and eustasy played a part. However, during the Givetian–Frasnian synrift phase of basin development (Kennard *et al.* 1992; Southgate *et al.* 1993) it is highly likely that major stratigraphic patterns reflect fault-controlled subsidence as earlier suggested by Playford *et al.* (1989) and, in particular, tilt block evolution (e.g. Brachert *et al.* 2002 and references therein). Eustatic transgressive phases may have been coincident and, therefore, enhanced relative sea-level highstands created primarily by tectonism.

We consider that development of the platform phases presently exposed in and close to the Lawford Range was strongly controlled by extension along the Sparke Range Fault to the north and Pinnacles and Cadjebut Faults to the south (Figs 1, 14). Additional faults such as the west-trending Virgin Hills Fault were probably important in controlling local subsidence on the northwestern part of this block, particularly given the earlier platform development that has been proposed by others for that area. The Hull platform grew on a fault block bounded on the east by the major north-trending Mt Elma Fault and associated north-trending faults (Fig. 1). Facies distributions indicate that the Mt Elma fault system fundamentally controlled evolution of the Hull platform; it involved tilting (by rotation of the tilt block) and subsidence (George *et al.* 2009) in a way clearly described from carbonate systems elsewhere (e.g. Bosence *et al.* 1998).

An interesting outcome of our study of platform phases of the Lawford block and comparisons with the Hull block to the north is the coincidence of third-order flooding events (within the resolution of conodont biostratigraphy) in the region. As discussed above, it is unlikely that these events are simply eustatically controlled because different fault blocks in an active rift could also have differing relative sea-level histories as eustasy and local tectonics combine to control their evolution. Indeed, the two areas focused on in this study do show differences in platform history (Fig. 15).

The Frasnian Zone 6 flooding event is the most recognizable in the Lawford and Hull platforms and the most correlatable based on conodont biostratigraphy, goniatite biodiversity (Becker 2000) and the marked change in fore-reef facies in both platforms. The thickest package of overlying fore-reef facies is seen in the Southern Lawford platform (FA2) and indicates progradation of the platform following backstepping. Elsewhere Zone 6 limestones are very thin (e.g. Horse Spring section is <1 m; Klapper 2007). Progradation led to collapse and formation of a megabreccia at McIntyre Knolls (Southern Lawford Range) and, we suggest, also at Rattigan's Rocks (Hull Range; Fig. 15c). Both areas are associated with siliciclastic facies. In the case of McIntyre Knolls, the siliciclastic facies were deposited mainly as a large body in the inter-reef basin and as smaller bodies on and at the toe of the slope. At Rattigan's Rocks, the carbonate package is unconformably overlain by conglomerates (Playford & Lowry 1966; Playford & Hocking 1998). Retrogradational FA3 (Fig. 2) is well represented in the Southern Lawford fore-reef but is part of condensed sections elsewhere (Fig. 13). The carbonate facies associated with the Rattigan's Rocks allochthonous blocks are assigned an age equivalent to FA3 (Hull Fr6) although we currently have no independent biostratigraphic control.

Lawford Fr2 was partially drowned during an Early Frasnian event (*c.* Zone 3) coinciding with flooding following a relative sea-level fall (and subaerial exposure) of Hull Fr2 and eastward expansion of the Hull platform onto siliciclastic sediment shed from the footwall of the Mt Elma Fault (during phase Fr3; Fig. 15b). In the Lawford Range (and on platforms on the western side of Bugle Gap) this flooding event promoted the growth of elongate and pinnacle reefs whereas the Hull platform was able to keep up during this time. The expansion of the Hull platform onto siliciclastic sediments adjacent to the Mt Elma Fault is a good analogue for initiation of the Northern Lawford Platform on a siliciclastic pile adjacent to the Sparke Range Fault without there being any older carbonate platform beneath. We suggest, moreover, that the relief on which the Northern Lawford Platform developed could have been provided by the siliciclastic pile rather than a smaller tilt block.

An interesting aspect to consider is that with deepening over the Southern Lawford Range, more open conditions on the windward (western) side of Teichert Hills atoll may have promoted development of the chute and buttress structures. Both Playford (1981) and Wood & Oppenheimer (2000) highlighted the high wave energy required to produce these types of structures, and furthermore, Playford (1981) proposed that reef-margin blocks had been transported to the platform interior by waves. Wood & Oppenheimer (2000) proposed that the orientation of the buttresses may record refracted waves approaching from the SW, presumably with more open water to the south. However, in comparison with well-studied modern reefal systems (e.g. Blanchon & Jones 1997), the robust style of the Teichert Hills buttresses is also consistent with growth coincident with oncoming waves able to traverse parts of the Southern Lawford Platform. Such a scenario is speculative and remains to be tested.

The event which resulted in initiation of the Northern Lawford Platform (Zone 4) is interpreted as fault-controlled subsidence adjacent to the Sparke Range Fault accompanied by basinward tilting to create different amounts of accommodation across the fault block. Most accommodation was generated in the southern part of the block leading to drowning of platform Lawford Fr3 most notably seen at McWhae Ridge (Fig. 14). We have correlated this flooding event to the marked flooding event on the Hull platform that is associated with collapse of the windward margin (Horse Spring area) of Hull Fr3 and backstepping of the margins towards the interior. The surface underlies fore-reef strata in Zones 5–6 and is reasonably interpreted as forming in Zone 4 to potentially early Zone 5.

Our correlation leads us to propose that the correspondence in flooding events across two major fault blocks on the SE Lennard Shelf is related to basin-margin tectonic activity, given the interpretation that has existed for many years that the backstepping platforms are fault-controlled. The synchronous development of some surfaces in two large neighbouring fault blocks, in particular the early Zone 6 flooding surface and late Zone 6 surface following a relative sea-level lowstand, and the potential for synchronous or near-synchronous development of several other surfaces (e.g. *c.* Zone 3 and Zone 4), strongly suggests major tectonic events affecting the bounding faults.

For the Lawford block, the Sparke Range Fault is a major basin-bounding fault and the Pinnacles fault system to the south defines the basinward edge of the Lennard Shelf (Fig. 1). The preservation of deepest water facies in the Southern Lawford Range and northward backstepping of the reef complexes in this block indicates basinward tilting during the Early Frasnian. The north-trending Mt Elma and Painted Rocks faults bounding the eastern side of the Hull block are oblique to the NW trend of the shelf; we propose that these faults form a transfer zone. A potential bounding fault is the Black Hills Fault on the southern side (Fig. 15) and north- to NNE-trending faults on the western side of the Horse Spring Range which are recognized in the subsurface and related to the eastern margin of the Colombo Fault Zone (Kemp & Wilson 1990). Pillara-age platform atolls were also developed on fault blocks along this eastern margin as imaged on seismic profiles and interpreted from well logs (e.g. Margaret 1 which is located *c.* 10 km SE of the Horse Spring Range; Kemp & Wilson 1990). Elsewhere on the Lennard Shelf accommodation/transfer zones are characterized by oblique-slip movement (Shaw *et al.* 1994). Continued oblique-slip movement may also explain the ?post-Devonian offset of the Guppy Hills from the southeastern end of the Hull Platform.

The correlation of major flooding events suggests that the Mt Elma–Painted Rocks fault system was linked to the WNW-trending Sparke Range Fault at least by the Early Frasnian. The Mt Elma–Painted Rocks fault system could have linked the Sparke Range Fault with the NW-trending Stony Creek–Barramundi fault system to the north; evidence for this is beyond the scope of this study. Nonetheless, linkage of the faults bounding the Lawford and Hull blocks is indicative of an evolved rift system rather than one in which normal faults are essentially isolated segments (e.g. Gawthorpe & Leeder 2000). An important aspect of our study has been to include areas located relatively close or distal to the bounding faults to gain a broader view of the responses of the underlying basement fault blocks during the Early–Middle Frasnian on the Lennard Shelf. We do not expect that episodes of fault growth and linkage during evolution of the Fitzroy Trough were absolutely synchronous because that is generally considered unlikely based on studies elsewhere (e.g. Gawthorpe *et al.* 1994; Gupta *et al.* 1999). However, we do recognize broadly coeval events at third-order scale (<1 Ma and potentially <0.5 Ma); these suggest fundamental links between the adjacent fault blocks. We propose that basin-margin tectonic deformation may have been largely centred on these major, long (*c.* 25 km) border faults as described from other rift basins characterized by strong tilt-block geometries (e.g. Jackson *et al.* 2005). Moreover, studies such as Gawthorpe *et al.* (1994) and Jackson *et al.* (2005) have highlighted the dominance of transgressive and highstand systems tracts near fault zones with limited development of eustatic lowstands because of fault-controlled subsidence. Gawthorpe *et al.* (1994) also pointed out that during greenhouse times, such as the Frasnian, the magnitude of eustatic falls is reduced.

Conclusions

The Frasnian reef complexes of the SE Lennard Shelf are distributed across the *c.* 75 km wide northern margin of the Fitzroy Trough. They record mixed carbonate–siliciclastic sedimentation on tilt-block highs during active extension of this sub-basin. A sequence-stratigraphic approach, underpinned by facies analysis and biostratigraphy, has enabled us to develop a temporal framework in which to establish the history of reef complex evolution, in particular the palaeogeography and role of tectonism (notably fault-related subsidence) in controlling relative sea-level changes. Application of biostratigraphy has enabled as high resolution dating as possible to constrain event ages and also to provide insights into palaeoenviromental conditions during the Early–Middle Frasnian on the Lennard Shelf. The conodont biostratigraphy has been particularly important for identifying coeval proximal and distal facies.

This study has focused on the Lawford Range area on the eastern side of Bugle Gap, specifically the Northern Lawford Range, to gain insight into Frasnian platforms that are no longer preserved at the better known localities, e.g. McWhae Ridge, in the Southern Lawford Range. This area is the eastern part of a large fault block bounded by two major NW-trending faults which define the basinward and landward edges of the shelf. Seven discrete platform phases are identified (Lawford Fr2–8; Fig. 14) and all are associated with flooding and backstepping. Major flooding in Early Frasnian Zone 3 had a significant effect on the palaeogeography of this block; we interpret a large reduction in the area of these platforms as a result (Fig. 15b). The demise of Fr3 reefs, particularly in the south, through flooding (Early Frasnian Zone 4) initiated formation of the Northern Lawford Platform to the north. Only one phase (Fr5) is clearly associated with a relative sea-level fall prior to flooding. The relative fall, which we interpret as eustatically controlled, is associated with major platform margin collapse and an influx of coarse siliciclastic sediments.

The Hull platform (exposed in three adjacent limestone ranges) occupies the fault block to the north and is controlled by a major north-trending fault system on its eastern margin that we interpret as a transfer zone. The Hull platform is useful because it contains a good record of Early Frasnian platform facies. Three discrete platform phases (Hull Fr2–5) are exposed with much of the overlying phases represented by condensed sections (Fig. 14). These phases are separated by flooding surfaces and are associated with backstepping. One major sequence boundary is recognized (top Fr2) and evidence for a younger relative sea-level fall (Fr5) is also preserved.

Comparison of platform development on the Lawford Block with that of the Hull platform on the neighbouring block to the north shows that at least two of the flooding surfaces can be recognized (within the <1 Ma resolution of conodont biostratigraphy). There are notable differences in platform evolution of these two blocks which suggests that attributing flooding to eustatic sea-level rises is too simplistic, although coincident highstands may well have amplified some of the events. Moreover, the well-known backstepping geometries of the Pillara platforms have been attributed in large part to fault-controlled subsidence for many years (Playford *et al.* 1989; Ward 1999).

We propose that the correspondence in flooding events across two major fault blocks on the SE Lennard Shelf is related to basin-margin tectonic activity which created accommodation via fault-controlled subsidence. Basal platforms were initiated on tilt blocks and continued tilting and subsidence were important controls on the generation of accommodation, particularly on the hanging-walls of the main border faults. The synchronous development of surfaces in two large neighbouring fault blocks, in particular the Zone 4 and Zone 6 flooding surfaces, and late Zone 6 surface following a relative sea-level lowstand, and the potential for synchronous or near-synchronous development of other surfaces (e.g. Zone 3), strongly suggest major tectonic events affecting the border faults. In addition, the correlation between the Lawford block and Hull platform, which is largely controlled by a north-trending transfer zone, suggests linkage between major NW-trending shelf-parallel faults and an evolved rift system by the Early Frasnian. This study helps illustrate the importance of investigating the local and regional tectonic history of a basin to help clarify potential second- and third-order eustatic signals.

Fieldwork in the McIntyre Knolls area was initially undertaken by A.D.G. in 1994 supported by an Australian Research Council Large Grant (A39232819) and subsequent sedimentological and biostratigraphic work was supported by small ARC, NSERC (Canada), UWA and UoM grants to A.D.G. and/or N.C. More recent support from a National Geographic Grant (7860-05) and ARC Discovery Grant (DP0664703) is also acknowledged. It is our pleasure to work with Gil Klapper and we thank him sincerely for the earlier conodont biostratigraphy of the McIntyre Knolls area, his assistance and discussions with K. M. T. regarding subsequent identifications, and considerable editorial patience. K. M. T. also wishes to thank Sue Turner, Mikael Ginter and Alexander Ivanov for helpful discussions. We are grateful for the ongoing support and encouragement of our work from colleagues who have inspired us through their significant contributions to understanding carbonate systems and/or things Devonian, in particular: John Talent, Noel James, Eric Mountjoy, Jack Wendte, Fred Read, Maurice Tucker, Brian Jones, Rick Sarg, and Brian Williams. Our thanks to the owners/managers of the stations – Mt Pierre (Louie and Marianne Dolby), Fossil Downs (Annette and John Henwood) and Gogo (Dan Grant) – for permission to work on their property. Field assistance in the McIntyre Knolls area during 1994–95 field seasons from Barbara Bilo, Linghua Zheng, Carolyn Green and Scott Brownlaw, and helpful discussions at various times with Scott Brownlaw, Peter Southgate, Phil Playford, Roger Hocking, Malcolm Wallace, Gregg Webb, Rachel Wood, Gil Klapper, Thomas Becker, Ken McNamara, Raimund Feist, John Jell and the late Michael House and Chris Powell, are gratefully acknowledged by A.D.G. We thank Peter Koenigshof for the opportunity to contribute to this special publication and for his editorial input, and John Talent and Peter Koenigshof for kindly reviewing the manuscript.

Appendix: conodont biostratigraphy by G. Klapper & K. Trinajstic

Table A1. *Measured section samples (in stratigraphic order; position shown on Fig. 5)*

GSWA no. (Grid ref.*)	UWA no. (Grid ref.*)	Conodont taxa	Microvert. taxa	Age determination†
113657 (976386)		*Ancyrodella gigas* form 1? (juv. spec.), *Polygnathus webbi, P.* sp., *Icriodus symmetricus, Pandorinellina* sp., *Belodella* sp.		Frasnian, zonally undiagnostic
	MK10 (976386)	*Palmatolepis bohemica, P. transitans, Polygnathus alatus*		Frasnian, Zone 6 based on known overlap of first two species
113658 (976386)		*Palmatolepis bohemica, P. transitans, Ancyrodella* sp. indet. (1 fragment), *Polygnathus alatus, P.* n. sp., *Belodella* sp.		Frasnian, Zone 6 based on known overlap of first two species
113659 (977384)		*Palmatolepis transitans P. punctata, Ozarkodina trepta, Ancyrognathus ancryognathoideus, Ancyrodella curvata*, early form *Polygnathus webbi, P.* n. sp.		Frasnian, Zone 6 based on overlap in the range of the first and third species
113660 (978383)		*Pandorinellina* sp., *Belodella* sp.		Frasnian, zonally undiagnostic
	MK11 (978383)	*Polygnathus webbi*		Frasnian, zonally undiagnostic
			Palaeoniscoid indet.	
113661 (978382)		*Palmatolepis punctata, P.* sp., *Ozarkodina nonaginta, Ancyrodella curvata*, early form *Polygnathus* n. sp., *P. alatus, P. angustidiscus, Icriodus symmetricus, Belodella* sp.		Frasnian, base Zone 7 to within Zone 8 based on known range of the third species

(*Continued*)

Table A1. *Continued*

GSWA no. (Grid ref.*)	UWA no. (Grid ref.*)	Conodont taxa	Microvert. taxa	Age determination[†]
	MK22 (979380)	*Palmatolepis punctata, Ancyrodella curvata*		Frasnian, ~Zone 6[‡] to Zone 10 based on known range of first species and stratigraphic position
				Frasnian, zonally undiagnostic
113662 (978381)		*Polygnathus* n. sp., *Icriodus symmetricus*		
	MK12 (978380)	*Polygnathus webbi*		Frasnian, zonally undiagnostic
113663 (977380)		*Palmatolepis punctata, Ancyrodella curvata*, early form? *A.* sp. indet., *Polygnathus* sp. [specimen lost], *Icriodus symmetricus*		Frasnian, ~Zone 6[‡] to Zone 10 based on known range of first species and stratigraphic position
113664 (978378)		*Ancyrodella curvata*, early form *Polygnathus webbi P.* sp. (cf. *P. decorosus*)		Frasnian, zonally undiagnostic
		Icriodus symmetricus		
113665 (978378)		*Palmatolepis* sp. indet. (P_2 element only possibly *P. plana*), *Ancyrodella* sp. indet. (fragments), *Polygnathus* spp. (cf. *P. webbi* and cf. *P. decorosus*), *Icriodus symmetricus*		Frasnian, zonally undiagnostic
113667 (977374)		*Palmatolepis ljaschenkoae, P. kireevae, P. plana?* (1 atypical specimen), *Ancyrognathus coeni, Ancyrodella curvata*, early form *Polygnathus* spp.		Frasnian, maximum range from upper Zone 8 to lower Zone 11 based on overlap in ranges of the first, second and fourth species.
		Icriodus symmetricus		

Determinations by G. Klapper (GSWA samples) and K. Trinajstic (UWA samples). For footnotes see Table A2.

Table A2. *Spot samples by facies association (location shown on Fig. 2)*

UWA no. (Grid ref.*)	Facies association	Conodont taxa	Microvert. taxa	Age determination
MK35 (954372)	FA1	*Mesotaxis ovalis, Polygnathus alatus*		Frasnian, Zones 3 to 6 based on known range of first species
			Palaeoniscoid indet.	
MK36 (928388)	FA2	*Palmatolepis transitans, Ancyrodella gigas* form 2, *Mesotaxis ovalis,*		Frasnian, Zone 6 based on the known range of the second species

(*Continued*)

Table A2. *Continued*

UWA no. (Grid ref.*)	Facies association	Conodont taxa	Microvert. taxa	Age determination
		Polygnathus webbi, Icriodus symmetricus		
			Palaeoniscoid indet.	
MK32 (937366)	FA2 (distal)	*Palmatolepis bohemica, P. spinata, Ancyrodella curvata*, early form *Polygnathus webbi, Icriodus symmetricus*		Frasnian, Zone 6 to lower part of Zone 8 based on range of first species
			Phoebodus fastigatus Ctenacanthus Protocrodus indet.	Frasnian, Zones 6 to 11 based on known range of first species
		Placoderm indet. Palaeoniscoid indet.		
MK33 (940375)	FA2/3 (distal)	*Palmatolepis punctata, Ancyrodella curvata, Polygnathus webbi, Icriodus symmetricus*		Frasnian, ~Zones 6[‡] to 10 based on known range of first species
		Australolepis seddoni, Moythomasia n. sp. *Ctenacanthus*, Placoderm indet.		Frasnian, up to Zone 8 based on known range of first species[§]
MK15 (961376)	FA3	*Palmatolepis ljaschenkoae?, P. punctata, Ozarkodina nonaginta, Ancyrognathus tsiensi, Ancyrodella curvata*		Frasnian, Zone 8 based on overlap in ranges of third and fourth species
			Protocordus indet., *Ctenacanthus, Moythomasia* n. sp.	Frasnian, zonally undiagnostic
MK31 (970382)	FA3	*Ancyrodella curvata* late form		Frasnian, zonally undiagnostic
			Ctenacanthus	Frasnian, zonally undiagnostic
MK34 (938389)	FA3	*Palmatolepis ljaschenkoae, P. proversa, Ancyrodella curvata*, late form *Polygnathus decorosus, P. webbi, Icriodus symmetricus*		Frasnian, Zones 9 to lower part of Zone 11 based on known range of second species
			Phoebodus fastigatus, Moythomasia durgaringa, M. n. sp.	Frasnian, Zone 6 to 11 based on known range of first species
MK7 (951384)	FA4	*Ancyrognathus tsiensi, Ancyrodella curvata, A. lobata, A. gigas,*		Frasnian, Zones 8 to 11 based on known range of first species

(*Continued*)

Table A2. *Continued*

UWA no. (Grid ref.*)	Facies association	Conodont taxa	Microvert. taxa	Age determination
		Polygnathus decorosus, Icriodus symmetricus		
			Phoebodus fastigatus, Moythomasia durgaringa, Ctancanthus sp. scales, Palaeoniscoid indet.	Frasnian, Zones 6 to 11 based on known range of first species
MK30 (967377)	FA4	*Palmatolepis ljaschenkoae, P. proversa, Ancyrodella curvata*, late form *Polygnathus webbi, Icriodus symmetricus*		Frasnian, Zones 9 to lower part of Zone 11 based on known range of second species
			Moythomasia durgaringa	Frasnian, zonally undiagnostic

Determinations by K. Trinajstic.
*Grid references pertain to the Australian Map Grid Zone 52.
†Zonal nomenclature for the Frasnian follows Klapper & Foster (1993) and Klapper *et al.* (1996). Equivalent conodont Zone assignment for microvertebrates based on published ranges of Turner (1997), Basden *et al.* (2006) and K.Trinajstic (unpubl. data).
‡The full range of *P. punctata* is Frasnian Zone 5 to Zone 10; however, these samples overlie older samples which suggests a depositional age no older than Zone 7 for MK 22 and no. 113663 and a depositional age of no older than Zone 6 for MK33.
§Range of *A. seddoni* known up to Zone 10 on the Lennard Shelf (Trinajstic & George 2009).

References

AHLBERG, P. 1989. Fossil Fish from Gogo. *Nature*, **337**, 511–512.

BAILLIE, P. W., POWELL, C. McA., LI, Z. X. & RYALL, A. M. 1994. The tectonic framework of Western Australia's Neoproterozoic to Recent sedimentary basins. *In*: PURCELL, P. G. & PURCELL, R. R. (eds) *The Sedimentary Basins of Western Australia*. Proceedings of Petroleum Exploration Society of Australia Symposium, Perth, 1994, 45–62.

BASDEN, A., TRINAJSTIC, K. & MERRICK, J. 2006. Eons of fishy fossils. *In*: MERRICK, J. R., ARCHER, M., HICKEY, G. M. & LEE, M. S. Y. *Evolution and Biogeography of Australasian Vertebrates*. Australian Scientific Publishing.

BAUER, J., KUSS, J. & STEUBER, T. 2003. Sequence architecture and carbonate platform configuration (Late Cenomanian–Santonian), Sinai, Egypt. *Sedimentology*, **50**, 387–414.

BECKER, R. T. 2000. Palaeobiogeographic relationships and diversity of Upper Devonian ammonoids from Western Australia. *Records of the WA Museum, Supplement*, **58** 385–401.

BECKER, R. T. & HOUSE, M. R. 1997. Sea-level changes in the Upper Devonian of the Canning Basin, Western Australia. *Courier Forschung-Inst. Senckenberg*, **199**, 129–146.

BECKER, R. T., HOUSE, M. R., KIRCHGASSER, W. T. & PLAYFORD, P. E. 1991. Sedimentary and faunal changes across the Frasnian/Famennian boundary in the Canning Basin of Western Australia. *Historical Biology*, **5**, 183–196.

BECKER, R. T., HOUSE, M. R. & KIRCHGASSER, W. T. 1993. Devonian goniatite biostratigraphy and timing of facies movements in the Frasnian of the Canning Basin, Western Australia. *In*: HAILWOOD, E. A. & KIDD, R. B. (eds) *High Resolution Stratigraphy*. Geological Society, London, Special Publication, **70**, 293–321.

BEGG, J. 1987. Structuring and controls on Devonian reef development on the northwest Barbwire and adjacent terraces, Canning Basin. *The APEA Journal*, **28**, 137–151.

BENN, C. J. 1984. Facies changes and development of a carbonate platform, east Pillara Range. *In*: PURCELL, P. G. (ed.) *The Canning Basin WA*. Proceedings of Geological Society of Australia and Petroleum Exploration Society of Australia, Symposium, Perth, 221–228.

BLANCHON, P. & JONES, B. 1997. Hurricane control on shelf-edge-reef architecture around Grand Cayman. *Sedimentology*, **44**, 479–506.

BOSENCE, D., CROSS, N. & HARDY, S. 1998. Architecture and depositional sequences of Tertiary fault-block carbonate platforms: An analysis from outcrop (Miocene, Gulf of Suez) and computer modelling. *Marine and Petroleum Geology*, **15**, 203–221.

BRACHERT, T. C., KRAUTWORST, U. M. R. & STUECKRAD, O. M. 2002. Tectono-climatic evolution of a Neogene intermontane basin (Late Miocene Carboneras subbasin, southeast Spain): Revelations from basin mapping and biofacies analysis. *Basin Research*, **14**, 503–521.

BROWNLAW, R. L. S. 2000. *Rugose coral biostratigraphy and cyclostratigraphy of the Middle and Upper Devonian carbonate complexes, Lennard Shelf, Canning Basin, Western Australia*. PhD thesis, University of Queensland.

BROWNLAW, R. L. S., HOCKING, R. M. & JELL, J. S. 1996. High frequency sea-level fluctuations in the Pillara Limestone, Guppy Hills, Lennard Shelf, northwestern Australia. *Historical Biology*, **11**, 187–212.

BROWNLAW, R. S., HEARN, S. J. & JELL, J. S. 1998. Spectral analysis of the back-reef limestones of the 'Devonian Great Barrier Reef', Western Australia. *Proceedings of the Royal Society of Queensland*, **107**, 99–107.

CHEN, D., TUCKER, M. E., ZHU, J. & JIANG, M. 2001. Carbonate sedimentation in a starved pullapart basin, Middle to Late Devonian, southern Guilin, South China. *Basin Research*, **13**, 141–167.

CHEN, D., TUCKER, M. E., ZHU, J. & JIANG, M. 2002. Carbonate platform evolution: From a bioconstructed platform margin to a sand-shoal system (Devonian, Guilin, South China). *Sedimentology*, **49**, 737–764.

CHOW, N., GEORGE, A. D. & TRINAJSTIC, K. M. 2004. Tectonic control on Tectonic control on development of a Frasnian–Famennian (Late Devonian) palaeokarst surface, Canning Basin reef complexes, northwestern Australia. *Australian Journal of Earth Science*, **51**, 911–917.

COOK, H. E. & MULLINS, H. T. 1983. Basin margin environment. *In*: SCHOLLE, P. A. *ET AL*. (eds) *Carbonate Depositional Environments*. American Association of Petroleum Geologists Memoir, **33**, 539–617.

COOPER, R. W., HALL, W. D. M. & STYLES, G. R. 1984. The Devonian stratigraphy of the central Pillara Range. *In*: PURCELL, P. G. (ed.) *The Canning Basin WA*. Proceedings of Geological Society of Australia and Petroleum Exploration Society of Australia, Symposium, Perth, 229–234.

COPP, I. A. 2002. *Subsurface facies analysis of Devonian reef complexes, Lennard Shelf, Canning Basin Western Australia*. Geological Survey of Western Australia Report, **58**.

DRUCE, E. C. 1976. *Conodont biostratigraphy of the Upper Devonian reef complexes of the Canning Basin Western Australia*. Bureau of Mineral Resources Bulletin, **158**.

DRUMMOND, B. J., SEXTON, M. J., BARTON, T. J. & SHAW, R. D. 1988. The nature of faulting along the margins of the Fitzroy Sub-basin Canning Basin, and implications for the tectonic development of the Sub-basin. *Exploration Geophysics*, **22**, 111–115.

GAWTHORPE, R. L. & LEEDER, M. R. 2000. Tectono-sedimentary evolution of active extensional basins. *Basin Research*, **12**, 195–218.

GAWTHORPE, R. L., FRASER, A. J. & COLLIER, R. E. L. 1994. Sequence stratigraphy in active extensional basins: Implications for the interpretation of ancient basin-fills. *Marine and Petroleum Geology*, **11**, 642–658.

GEORGE, A. D. & CHOW, N. 1999. Palaeokarst development in a lower Frasnian (Devonian) platform succession, Canning Basin, northwestern Australia. *Australian Journal of Earth Sciences*, **46**, 905–913.

GEORGE, A. D. & CHOW, N. 2002. The depositional record of the Frasnian/Famennian boundary interval in a fore-reef succession, Canning Basin, Western Australia. *Palaeogeography, Palaeoclimatology, Palaeoecology*, **181**, 347–374.

GEORGE, A. D. & POWELL, C. M. 1997. Paleokarst in an Upper Devonian reef complex of the Canning Basin, Western Australia. *Journal of Sedimentary Research*, **67**, 935–944.

GEORGE, A. D., PLAYFORD, P. E., POWELL, C. M. & TORNATORA, P. M. 1997. Lithofacies and sequence development on an Upper Devonian mixed carbonate-siliciclastic fore-reef slope, Canning Basin, Western Australia. *Sedimentology*, **44**, 843–867.

GEORGE, A. D., CHOW, N. & TRINAJSTIC, K. M. 2002. Integrated approach to platform–basin correlation and deciphering the evolution of Devonian reefs, northern Canning Basin, Western Australia. *In*: KEEP, S. & MOSS, S. *The Sedimentary Basins of Western Australia 3*. Proceedings of the Petroleum Exploration Society of Australia Symposium, Perth, October 2002.

GEORGE, A. D., CHOW, N. & TRINAJSTIC, K. M. 2009. Syndepositional fault control on Lower Frasnian platform evolution, Lennard Shelf, Canning Basin, Australia. *Geology*, **37**, 331–334.

GLENISTER, B. F. 1958. Upper Devonian ammonoids from the *Mantioceras* Zone, Fitzroy basin, Western Australia. *Journal of Paleontology*, **32**, 58–96.

GLENISTER, B. F. & KLAPPER, G. J. 1966. Upper Devonian conodonts from the Canning Basin, Western Australia. *Journal of Paleontology*, **40**, 777–842.

GREY, K. 1991. A mid-Givetian miospore age for the onset of reef development on the Lennard Shelf, Canning Basin, Western Australia. *Review of Palaeobotany and Palynology*, **68**, 37–48.

GUPPY, D. J., LINDNER, A. W., RATTIGAN, J. H. & CASEY, J. N. 1958. *The geology of the Fitzroy Basin, Western Australia*. Bureau of Mineral Resources Bulletin, **36**.

GUPTA, S., UNDERHILL, J. R., SHARP, I. R. & GAWTHORPE, R. L. 1999. Role of fault interactions in controlling synrift sediment dispersal patterns: Miocene, Abu Alaqa Group, Suez Rift, Sinai, Egypt. *Basin Research*, **11**, 167–189.

HALL, W. D. M. 1984. The stratigraphy and structural development of the Givetian–Frasnian reef complex, Limestone Billy Hills, western Pillara Range. *In*: PURCELL, P.G. (ed.) *The Canning Basin WA*. Proceedings of Geological Society of Australia and Petroleum Exploration Society of Australia, Symposium, Perth, 235–244.

HILL, D. & JELL, J. S. 1970. *Devonian corals from the Canning Basin, Western Australia*. Geological Survey of Western Australia Bulletin, **123**.

HINE, A. C. *ET AL*. 1992. Megabreccia shedding from modern, low-relief carbonate platforms, Nicaraguan

Rise. *Geological Society of America Bulletin*, **104**, 928–943.

HOCKING, R. M. & PLAYFORD, P. E. 2000. Cycle types in carbonate platform facies, Devonian reef complexes, Canning Basin, Western Australia. *Geological Survey of Western Australia Annual Review 2000–2001*, 74–80.

HOCKING, R. M., COPP, I. A., PLAYFORD, P. E. & KEMPTON, R. H. 1995. The Cadjebut Formation: A Givetian evaporitic precursor to Devonian reef complexes of the Lennard Shelf, Canning Basin, Western Australia. *Geological Survey of Western Australia Annual Review 1995–1996*, 48–54.

HOLMES, A. E. & CHRISTIE-BLICK, 1993. Origin of sedimentary cycles in mixed carbonate-siliciclastic systems: An example from the Canning Basin, Western Australia. *In*: LOUCKS, R. G. & SARG, J. F. (eds) *Carbonate Sequence Stratigraphy: Recent Developments and Applications*. American Association of Petroleum Geologist Memoir, **57**, 181–212.

JACKSON, C. A. L., GAWTHORPE, R. L., CARR, I. D. & SHARP, I. R. 2005. Normal faulting as a control on the stratigraphic development of shallow marine syn-rift sequences: The Nukhul and Lower Rudeis Formations, Hammam Faraun fault block, Suez Rift, Egypt. *Sedimentology*, **52**, 313–338.

KEMP, G. J. & WILSON, B. L. 1990. The seismic expression of Middle to Upper Devonian reef complexes, Canning Basin. *APEA Journal*, **30**, 280–289.

KENNARD, J. M., SOUTHGATE, P. N. *ET AL*. 1992. New sequence perspective on the Devonian reef complex and the Frasnian–Famennian boundary, Canning Basin, Australia. *Geology*, **20**, 1135–1138.

KENNARD, J. M., JACKSON, M. J., ROMINE, K. K., SHAW, R. D. & SOUTHGATE, P. N. 1994. Depositional sequences and associated petroleum systems of the Canning Basin, WA. *In*: PURCELL, P. G. & PURCELL, R. R. (eds) *The Sedimentary Basins of Western Australia*. Proceedings of Petroleum Exploration Society of Australia Symposium, Perth, 657–676.

KERANS, C. 1985. *Petrology of Devonian and Carboniferous Carbonates of the Canning Basin*. Minerals and Energy Research Institute of Western Australia, Perth, Report, **12**.

KLAPPER, G. 1989. The Montagne Noire Frasnian (Upper Devonian) conodont succession. *In*: MCMILLAN, N. J., EMBRY, A. F. & GLASS, D. J. (eds) *Devonian of the World, Vol. III, Canadian*. Society of Petroleum Geologists Memoir, **14**, 449–468.

KLAPPER, G. 2007. Frasnian (Upper Devonian) conodont succession at Horse Spring and correlative sections, Canning Basin, Western Australia. *Journal of Paleontology*, **81**, 513–537.

KLAPPER, G. & FOSTER, J. T. 1993. Shape analysis of Frasnian species of the Late Devonian conodont genus *Palmatolepis*. *Paleontological Society Memoir*, **32**, 1–35.

KLAPPER, G., KUZ'MIN, A. V. & OVNATANOVA, N. S. 1996. Upper Devonian conodonts from the Timan-Pechora region, Russia, and correlation with a Frasnian composite standard. *Journal of Paleontology*, **60**, 131–152.

LEHMANN, P. R. 1986. The geology and hydrocarbon potential of the EP 104 Permit, northwest Canning Basin, Western Australia. *APEA Journal*, **26**, 261–284.

LI, Z. X. & POWELL, C. McA. 2001. An outline of the palaeogeographic evolution of the Australasian region since the beginning of the Neoproterozoic. *Earth-Science Reviews*, **53**, 237–277.

LONG, J. A. 1988. Late Devonian fishes from the Gogo Formation, Western Australia. *National Geographic Research and Exploration*, **4**, 436–450.

LONG, J. A. 1995. A new plourdosteid Arthrodire from the Upper Devonian Gogo Formation of Western Australia. *Palaeontology*, **38**, 39–62.

LONG, J. A. & TRINAJSTIC, K. M. 2000. Devonian microvertebrate faunas of Western Australia. *Courier Forschung-Inst. Senckenberg*, **223**, 471–485.

MA, X. P. & BAI, S. L. 2002. Biological, depositional, microspherule, and geochemical records of the Frasnian/Famennian boundary beds, South China. *Palaeogeography, Palaeoclimatology, Palaeoecology*, **181**, 325–346.

MACNEIL, A. J. & JONES, B. 2006. Sequence stratigraphy of a Late Devonian ramp-situated reef system in the Western Canada Sedimentary Basin: Dynamic responses to sea-level change and regressive reef development. *Sedimentology*, **53**, 321–359.

MACHEL, H. G. & HUNTER, I. G. 1994. Facies models for Middle to Late Devonian shallow-marine carbonates, with comparisons to modern reefs: A guide for facies analysis. *Facies*, **30**, 155–176.

MULLINS, H. T. & HINE, A. C. 1989. Scalloped bank margins: beginning of the end for carbonate platforms? *Geology*, **17**, 30–33.

NICOLL, R. S. 1984. Conodont distribution in the marginal-slope facies of the Upper Devonian reef complex, Canning Basin, Western Australia. *In*: CLARK, D. L. (ed.) *Conodont Biofacies and Provincialism*. Geological Society of America, Special Paper, **196**, 127–141.

NICOLL, R. S. & PLAYFORD, P. E. 1993. Upper Devonian iridium anomalies, conodont zonation and the Frasnian–Famennian boundary in the Canning Basin, Western Australia. *Palaeogeography, Palaeoclimatology, Palaeoecology*, **104**, 105–113.

PLAYFORD, P. E. 1980. Devonian 'Great Barrier Reef' of Canning Basin, Western Australia. *American Association of Petroleum Geologists Bulletin*, **64**, 814–840.

PLAYFORD, P. E. 1981. *Devonian reef complexes of the Canning Basin, Western Australia*. Australian Geological Convention, Field Excursion Guidebook.

PLAYFORD, P. E. 1984. Platform-margin and marginal slope relationships in Devonian reef complexes of the Canning Basin. *In*: PURCELL, P. G. (ed.) *The Canning Basin WA*. Symposium Proceedings, Geological Society of Australia and Petroleum Exploration Society of Australia, 189–214.

PLAYFORD, P. E. 2002. Palaeokarst, pseudokarst and sequence stratigraphy in Devonian reef complexes of the Canning Basin, Western Australia. *In*: KEEP, M. & MOSS, S. J. (eds) *The Sedimentary Basins of Western Australia 3*. Proceedings of the Petroleum Exploration Society of Australia Symposium, Perth, 763–793.

PLAYFORD, P. E. & HOCKING, R. M. 1998. Geological maps of the Lennard Shelf, Plates 1, 4 and 5, *GSWA Bulletin*, 145.

PLAYFORD, P. E. & LOWRY, D. C. 1966. Devonian reef complexes of the Canning Basin, Western Australia. *GSWA Bulletin*, **18**.

PLAYFORD, P. E., MCLAREN, D. J., ORTH, C. J., GILMORE, J. S. & GOODFELLOW, W. D. 1984. Iridium anomaly in the Upper Devonian of the Canning Basin, Western Australia. *Science*, **226**, 437–439.

PLAYFORD, P. E., HURLEY, N. F., KERANS, C. & MIDDLETON, M. F. 1989. Reefal platform development, Devonian of the Canning Basin, Western Australia. *In*: CREVELLO, P. D., WILSON, J. L., SARG, J. F. & READ, J. R. (eds) *Controls on Carbonate Platform and Basin Development*. SEPM Special Publication, **44**, 187–202.

PURNELL, M. A., DONOGHUE, P. C. J. & ALDRIDGE, R. J. 2000. Orientation and anatomical notation in conodonts. *Journal of Paleontology*, **74**, 113–122.

RATTIGAN, J. H. & VEEVERS, J. J. 1961. Devonian, in The Geology of the Canning Basin, Western Australia. *Bureau of Mineral Resources Bulletin*, **60**, 22–61.

READ, J. F. 1973*a*. Carbonate cycles, Pillara Formation (Devonian), Canning Basin, Western Australia. *Canadian Petroleum Geology Bulletin*, **21**, 38–51.

READ, J. F. 1973*b*. Paleo-environments and paleogeography, Pillara Formation (Devonian), Western Australia. *Canadian Petroleum Geology Bulletin*, **21**, 344–394.

ROMINE, K. K., SOUTHGATE, P. N., KENNARD, J. M. & JACKSON, M. J. 1994. The Ordovician to Silurian phase of the Canning Basin WA: Structure and sequence evolution. *In*: PURCELL, P. G. & PURCELL, R. R. (eds) *The Sedimentary Basins of Western Australia*. Proceedings of Petroleum Exploration Society of Australia Symposium, Perth, 677–702.

RUIZ-ORTIZ, P. A., BOSENCE, D. W. J., REY, J., NIETO, L. M., CASTRO, J. M. & MOLINA, J. M. 2004. Tectonic control of facies architecture, sequence stratigraphy an drowning of Liassic carbonate platform (Betic Cordillera, Southern Spain). *Basin Research*, **16**, 235–257.

SEDDON, G. 1970. Frasnian conodonts from the Sadler Ridge – Bugle Gap area, Canning Basin, Western Australia. *Geological Society of Australia Journal*, **16**, 723–753.

SHAW, R. D., SEXTON, M. J. & SEEILINGER, I. 1994. *The tectonic framework of the Canning Basin, WA. including the 1:2 million structural elements map of the Canning Basin*. Australian Geological Survey Organisation Record, **1994/48**.

SHEN, J. W., WEBB, G. E. & JELL, J. S. 2008. Platform margins, reef facies, and microbial carbonates; a comparison of Devonian reef complexes in the Canning Basin, Western Australia, and the Guilin region, South China. *Earth-Science Reviews*, **88**, 33–59.

SOUTHGATE, P. N., KENNARD, J. M., JACKSON, M. J., O'BRIEN, P. E. & SEXTON, M. J. 1993. Reciprocal lowstand clastic and highstand carbonate sedimentation, subsurface Devonian reef complex, Canning Basin, Western Australia. *In*: LOUCKS, R. G. & SARG, J. F. (eds) *Carbonate Sequence Stratigraphy: Recent Developements and Applications*. American Association of Petroleum Geologists, Memoirs, **57**, 157–179.

TEICHERT, C. 1941. Upper Devonian goniatite succession of Western Australia. *American Journal of Science*, **239**, 148–153.

TEICHERT, C. 1943. The Devonian of Western Australia, a preliminary review. *American Journal of Science*, **241**, 69–94.

TEICHERT, C. 1949. Stratigraphy and palaeontology of Devonian portion of the Kimberley division, Western Australia. Australia Bureau of Mineral Resources, Reports, **2**.

TOMPKINS, L. A., RAYNER, M. J., GROVES, D. I. & ROCHE, M. T. 1994. Evaporites: In situ sulfur source for rhythmically banded ore in the Cadjebut MVT Zn-Pb deposit, Western Australia. *Economic Geology*, **89**, 467–492.

TRINAJSTIC, K. 2001. A description of additional variation seen in the scale morphology of the Frasnian thelodont *Australolepis seddoni* Turner and Dring, 1981. *Records of the Western Australian Museum*, **20**, 237–246.

TRINAJSTIC, K. M. & GEORGE, A. D. 2009. Microvertebrate biostratigraphy of Upper Devonian (Frasnian) carbonate rocks in the Canning and Carnarvon Basins of Western Australia. *Palaeontology*, **52**, in press.

TURNER, S. 1997. Sequence of Devonian thelodont scale assemblages in East Gondwana. Geological Society of America, Special Papers, **321**, 295–315.

VEARNCOMBE, J. R., CHISNALL, A. W., DENTITH, M. C., DÖRLING, S. L., RAYNER, M. J. & HOLYLAND, P. W. 1996. Structural controls on Mississippi Valley-type mineralization, the southeast Lennard Shelf, Western Australia. *In*: SANGSTER, D. F. (ed.) *Carbonate-hosted Lead-Zinc Deposits 75th Anniversary Volume*. Society of Economic Geology, Special Publications, **4**, 74–95.

VEEVERS, J. J. 1959. *Devonian brachiopods from the Fitzroy Basin, Western Australia*. Australia Bureau of Mineral Resources, Bulletin, **45**.

VON BITTER, P. H. & PURNELL, M. A. 2005. An experimental investigation of post-depositional taphonomic bias in conodonts. *In*: PURNELL, M. A. & DONOGHUE, P. C. J. (eds) *Conodont Biology and Phylogeny: Interpreting the Fossil Record*. The Palaeontological Association, London, Special Papers in Palaeontology, **73**, 39–56.

WALLACE, M. W., KERANS, C., PLAYFORD, P. E. & MCMANUS, A. 1991. Burial diagenesis in the Upper Devonian reef complexes of the Geikie Gorge region, Canning Basin, Western Australia. *American Association of Petroleum Geologists Bulletin*, **75**, 1018–1038.

WALLACE, M. W., MIDDLETON, H. A., JOHNS, B. & MARSHALLSEA, S. 2002. Hydrocarbons and Mississippi Valley-type sulfides in the Devonian reef complexes of the eastern Lennard Shelf, Canning Basin, Western Australia. *In*: KEEP, M. & MOSS, S. J. (eds) *The Sedimentary Basins of Western Australia 3*. Proceedings of the Petroleum Exploration Society of Australia Symposium, Perth, 795–816.

WARD, B. W. 1999. Tectonic control on backstepping sequences revealed by mapping of Frasnian backstepped platforms, Devonian reef complexes, Napier Range, Canning Basin, Western Australia. *In*: HARRIS, P. M., SALLER, A. H. & SIMO, J. A. T. (eds)

Advances in Carbonate Sequence Stratigraphy: Applications to Reservoirs, Outcrops, and Models. SEPM, Special Publications, **63**, 47–74.

WARREN, J. K. & KEMPTON, R. H. 1997. Evaporite sedimentology and the origin of evaporiteassociated Mississippi Valley-Type sulphides in the Cadjebut mine area, Lennard Shelf, Canning Basin, Western Australia. *In*: MONTANEZ, P., GREGG, J. M. & SHELTON, K. L. (eds) *Basin-wide Diagenetic Patterns: Integrated Petrologic, Geochemical and Hydrologic Considerations*. SEPM, Special Publications, **57**, 183–205.

WARRLICH, G., BOSENCE, D. & WALTHAM, D. 2005. 3D and 4D controls on carbonate depositional systems: Sedimentological and sequence stratigraphic analysis of an attached carbonate platform and atoll (Miocene, Nijar Basin, SE Spain). *Sedimentology*, **52**, 363–389.

WENDTE, J. C. 1994. Cooking Lake platform evolution and its control on Late Devonian Leduc reef inception and localization, Redwater, Alberta. *Bulletin of Canadian Petroleum Geology*, **42**, 499–528.

WHALEN, M. T., EBERLI, G. P., VAN BUCHEM, F. S. P. & MOUNTJOY, E. W. 2000*a*. Facies models and architecture of Upper Devonian carbonate platforms (Miette and Ancient Wall), Alberta, Canada. *In*: HOMEWOOD, P. W. & EVERLI, G. P. (eds) *Genetic Stratigraphy on the Exploration and Production Scales—Case Studies from the Pennsylvanian of the Paradox Basin and the Upper Devonian of Alberta*. Bulletin, Centre Recherche Elf Exploration-Production, Memoire, **24**, 139–178.

WHALEN, M. T., EBERLI, G. P., VAN BUCHEM, F. S. P., MOUNTJOY, E. W. & HOMEWOOD, P. W. 2000*b*. Bypass margins, basin-restricted wedges, and platform-to-basin correlation, Upper Devonian, Canadian Rocky Mountains: Implications for sequence stratigraphy of carbonate platform systems. *Journal of Sedimentary Research*, **70**, 913–936.

WHITTAM, D. B., KENNARD, J. M., KIRK, R. B., SARG, J. F. & SOUTHGATE, P. N. 1994. A proposed third-order sequence framework for the Upper Devonian outcrops of the Northern Canning Basin. *In*: PURCELL, P. G. & PURCELL, R. R. (eds) *The Sedimentary Basins of Western Australia*. Proceedings of Petroleum Exploration Society of Australia Symposium, Perth, 697–701.

WOOD, R. & OPPENHEIMER, C. 2000. Spur and groove morphology from a Late Devonian reef. *Sedimentary Geology*, **133**, 185–193.

YEATES, A. N., GIBSON, D. L., TOWNER, R. R. & CROWE, R. W. A. 1984. Regional geology of the onshore Canning Basin, WA. *In*: PURCELL, P. G. (ed.) *The Canning Basin WA*. Proceedings of Geological Society of Australia and Petroleum Exploration Society of Australia Symposium, 23–55.

Distribution, geometry and palaeogeography of the Frasnian (Late Devonian) reef complexes of Banks Island, NWT, western arctic, Canada

P. COPPER[1]* & E. EDINGER[2]

[1]*Department of Earth Sciences, Laurentian University, Sudbury, P3E 2C6, Canada*

[2]*Geography Department, Memorial University, St John's, A1B 3X9, Canada*

**Corresponding author (e-mail: pcopper@laurentian.ca)*

Abstract: Following the collapse of the >2000 km long Givetian (Middle Devonian) Inuitian/Ellesmere carbonate platform factory in arctic Canada, within the 0° to 10° equatorial palaeolatitudes north, the only Frasnian reefs in high arctic Canada retreated westwards, confined to northeastern Banks Island. These reefs, numbering well over 130, and dominated by corals and stromatoporoid sponges, were spread over *c*. 5000 km^2, within the 220 m thick Mercy Bay Formation. Reefs were developed at four different stratigraphic levels (termed the A, B, C and D levels) during early and middle Frasnian time, periodically smothered by intervening siliciclastics during sea-level lowstands, and were finally buried by thick siliciclastic sands, silts and muds derived from the east during the late Frasnian. The Frasnian–Famennian boundary is masked within continental, deltaic facies bearing plant remains. The Banks reef and carbonate succession is preserved as horizontally stratified, *in situ* limestones, with the succeeding reefs backsteppping towards the east, in response to cyclic Frasnian sea-level rises. The Banks reefs were developed on the distal lobes of a megadelta, periodically covered by extensive lowland forests that stretched more than 2000 km west from Greenland and Ellesmere Island. Reefs are three-dimensionally preserved, unaffected by diagenesis, dolomitization, major tectonics, vegetative cover or glaciation, and thus display some of the finest, pristine Late Devonian reef complexes known. The reefs represent a range of geometries, from small circular to oval patch reefs a few tens of metres in diameter and <5–10 m thick, to larger prominent tabular ('reef platform' style) and domal reef structures >2 km across, or 300–400 m diameter, and 40–60 m thick. Coral and stromatoporoid-rich biostromes also formed a significant part of the reef seascape. Morphology of the reefs leads to an estimation of penecontemporaneous relief of 10–20 m above the prevailing sea floor, with reef-core facies generally dominated by stromatoporoid sponges, and flank facies by a variable mix of colonial tabulate and rugose corals, as well as stromatoporoids. Bryozoans and calcimicrobialites were rare, in contrast to other Late Devonian reefs such as those of the Canning Basin (Australia), or the Guangxi platform (South China); mudmound reef facies, such as seen in Belgium were limited.

The Middle Devonian (Eifelian–Givetian) was a time of unprecedented, globally warm Phanerozoic climates (noted as the 'supergreenhouse': Copper 2002*b*; Copper & Scotese 2003) marked by the most widespread reef development of the past 540 million years, with tropical barrier reef complexes in latitudes of $\geq$45° north and south, equivalent to modern temperate latitudes. The stable isotope record of this time demonstrates very high concentrations of CO_2 at levels 16× to 24× those of today (well over 5000 ppm; Berner 2001, 2004), corroborated by extensive low- to mid-latitude barrier reef and marine carbonate factories worldwide (Copper 1994, 2002*a*, *b*). By the end of Givetian time, tropical rainforests became established for the first time, with large rooted trunks, and leaf canopies as high as 8 m, as demonstrated by the famous giant cladoxylopid fossil forest at Gilboa, New York state, just south of the palaeoequator (Stein *et al.* 2007). Similar large tree trunks, with diameters up to 10–15 cm, are present in the Frasnian siliciclastic sediments of Banks Island, mostly adjacent to the reefs; they are presumably waterlogged trees that floated down-delta. Wilder (1989, 1994) has previously suggested that the establishment of ancient tropical lowland forests in the Late Devonian led to dramatic changes along shorelines, and the production of coal beds, as well as a sharp rise in atmospheric photosynthetic icehouse-forcing oxygen. Banks Island (Fig. 1), an example for the Wilder model during the Frasnian, was located within 10° latitude north of the equator, in a distal delta-margin siliciclastic setting (Embry & Klovan 1976). The Frasnian–Famennian (F-F) boundary on Banks Island is defined by the sharp introduction of a new

From: KÖNIGSHOF, P. (ed.) *Devonian Change: Case Studies in Palaeogeography and Palaeoecology*.
The Geological Society, London, Special Publications, **314**, 109–124.
DOI: 10.1144/SP314.5 0305-8719/09/$15.00

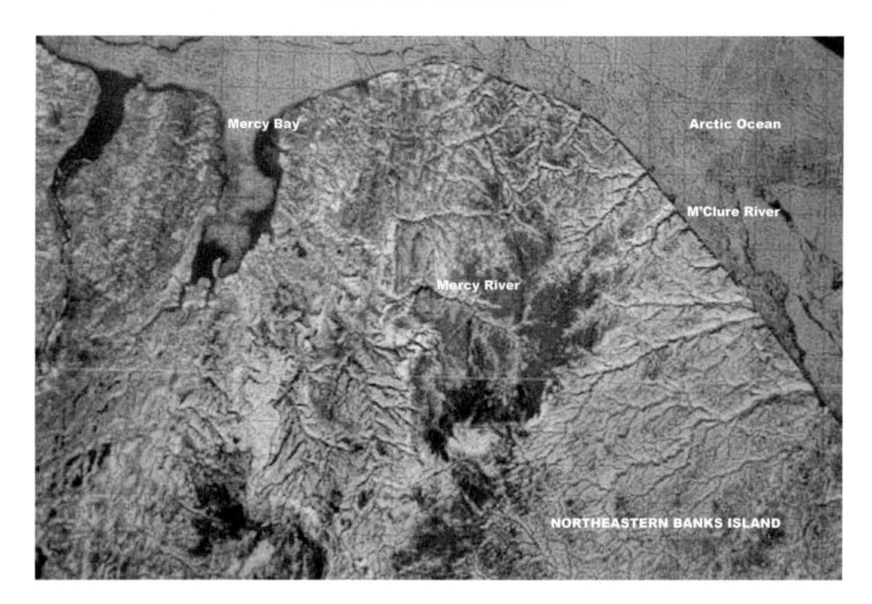

Fig. 1. False-colour radar satellite imagery of northeastern Banks Island, focusing on about 5000 km^2 of the Mercy carbonate outcrop belt, representing high topographic terrain. The Late Devonian carbonate and reef outcrop is outlined in yellow and orange colours, as these are an approximation of radar reflection data, e.g. permafrost soil moisture, elevation and lithology. The Arctic Ocean (blue), mostly ice covered, is on the north side. Important rivers that incised the Mercy carbonates and siliciclastics are marked.

Lophozonotriletes floral assemblage within such a terrestrial setting (Hills *et al.* 1971); however, the F-F contact is not seen in the marine reef setting, as reefs were buried by delta sands before the extinctions. Coal beds in the Banks Late Devonian siliciclastics are as thick as 20 cm (Manning 1956).

Copper (1977, 1986) suggested that Late Devonian cooling was produced by the tectonic collision of the Laurentia, Baltica and Gondwana plates, closing off the Iapetus ocean and rerouting ocean circulation to produce southern hemisphere glaciation, as well as equatorial cooling and shrinking of the tropical reef ecosystem. Berner *et al.* (2007) marked this transition to higher oxygen states as beginning approximately 360 million years ago, during the Late Devonian, and reaching a peak in the icehouse Carboniferous. Royer *et al.* (2004) and Berner *et al.* (2007) have promoted CO_2 change as the primary climate driver. In contrast, Veizer *et al.* (2000) proposed a decoupling of CO_2 and climate during the Phanerozoic, and later Steuber & Veizer (2002) suggested plate tectonic forcing as a cause, corroborating earlier ideas of the cold climatic push–pull of ocean circulation and sea-level change. Sandberg (1983) marks the Late Devonian as the transition state from a calcite to an aragonite ocean system, thus affecting skeletal mineralogy of reef-builders. It is in this changing, cyclical, climatic greenhouse to icehouse transitional setting, and ocean mineralogy, that the reefs of Banks Island grew near the palaeoequator. The four Banks reef levels are interpreted by us as responding to similar sea-level and reef-building cycles present in North America and Europe (Johnson *et al.* 1985; Sorauf & Pedder 1986; McLean & Klapper 1998; Boulvain 2007).

The critical timing in the Late Devonian is a major factor to consider: the turnover from a super-greenhouse Middle Devonian to the start of the icehouse Famennian occurred during and after the F-F mass extinction events (MEEs) that coincided with multiple continental glaciations in South America (Isaacson *et al.* 1999), and corresponding sea-level drawdowns. In marine settings, the transition to a more oxygenated atmosphere coincided with the switch from a more stratified calcite warm ocean to aragonite-dominated, ventilated cool ocean states around the F-F boundary (Sandberg 1983; Copper 1986; Stanley 1988). Stanley & Hardie (1999) proposed that the mineralogical change in the transition from calcite to aragonite oceans was forced by mid-ocean sea-floor spreading and a concomitant chemistry switch in Ca and Mg saturation. This reversal from a calcite to aragonite ocean from the Frasnian into the Famennian, and the F-F mass extinction, does not appear to be merely a random, coincident phenomenon of sea-water chemistry, climate, atmospheric greenhouse gas content and biodiversity losses and changes. The Devonian tropical stromatoporoid sponges (which commonly dominated the Banks reefs) suffered a 90% loss or more during the F-F extinctions, and died out completely at the end of the Famennian: they had a relatively soluble aragonite (or high Mg calcite) skeleton. One of the keystone mid-Palaeozoic reef-building and skeleton-constructing taxa thus vanished, alongside virtually all the colonial rugosans and most of the tabulate corals. Whether this was due to cooler climates or the switch to aragonite oceans (or other causes such as the loss of photosynthetic symbionts, or a temperature switch that shut off the gene(s) for biocalcification; e.g. Jackson *et al.* 2007), is still unclear. Calcite oceans (with calcitic tabulate and rugose corals, common reef elements on Banks Island) were favoured in global greenhouse episodes such as the Cambrian to Devonian, and the Mesozoic. Aragonite oceans (with aragonite skeletal secretion) dominated in the Famennian to Permian, and in the Cenozoic (Ries *et al.* 2006). Corals and carbonate platforms likewise have a strong temperature dependence as we do not find shallow-water reefs or thick carbonate platforms in the cooler temperate regions (Isern *et al.* 1996; Schrag & Linsley 2002; Hart & Kench 2007).

In addition to the evolution of the earliest tropical, and possibly warm temperate rainforests, the Late Devonian was also a time of other conquests of the terrestrial biome, e.g. the development of extensive soil horizons (Retallack 1985) that would have had an impact on the erosion and supply of siliciclastic sediments, such as those on Banks Island, especially at sea-level lowstands. Linked to the first forest ecosystem was the expansion of the terrestrial insect fauna (Grimaldi & Engel 2005; Glenner *et al.* 2006), Insects may have provided part of a nutrient base for the evolving fresh-water fishes and other consumers, and accelerated the transition from lobe-finned fishes to the first limbed and fully terrestrial amphibians, with such taxa as *Ichthyostega, Acanthostega* and *Tiktaalik* (Shubin *et al.* 2006; armoured fish remains are not uncommon in the coaly non-marine siliciclastics on Banks Island). This periodically altered the input of phosphorus into the distal reef ecosystem of Banks Island (P inhibits carbonate production and reef growth). Thus the Frasnian reef horizons of Banks Island, located in their unusual delta setting at a time of oscillating sea levels, provide a key glimpse into the Late Devonian world at a time of multiple changes and crises in the ecosystems of ancient shallow tropical seas and their adjacent land areas.

Banks Island: background and interpretation

Presently Banks Island is part of the Canadian cold arctic desert and permafrost, broadly bisecting latitude 74° north (Fig. 1). The elevated northeastern corner of Banks Island (sometimes labelled the 'Mercy Platform'), where the reefs are located, was effectively unglaciated during the Pleistocene, as noted by Craig & Fyles (1960; Fyles in Thorsteinsson & Tozer 1962: 8), who stated that 'this area probably lies beyond the northwestern boundary of the last ice sheet'. As a result, the Frasnian reefs were not buried in glacial tills, nor moraines, nor peneplaned, nor destroyed by subglacial ice push, though in places there is 'thin, non-descript glacial debris' (Fyles in Thorsteinsson & Tozer 1962: 12). Holocene river erosion, mass wasting, frost shattering, permafrost melt and downslope slip have removed some parts of the reef complexes, or allowed them to collapse or calve off the reefs or reef platforms. The Mercy Bay carbonates and reefs occupied some 5000 km^2 of area, 80 km east–west and 70 km north–south, on the northeastern corner of Banks Island, located some 400 km NE from its only settlement, Sachs Harbour.

We initially mapped reefs from stereo air photos at a scale of 1:40 000; these were subsequently checked on the ground with GPS co-ordinates at some 117 localities. Others were located, photographed and viewed from the air by helicopter (Copper & Edinger 2004). More than 800 digital and film photographs were used to record the reefs. Reefs were readily identifiable, as reefal limestones stand out as topographic relief, and in white or light grey, contrasting with the dark grey surrounding siliciclastics (Fig. 1). Most of the reefs occur within the vicinity of the Kanikshar, Kamik, Mercy, Manning and M'Clure rivers, east and NE of the Gyrfalcon River. The strata are nearly flat-lying, with the highest-level reefs occurring at elevations of *c.* 400 m above sea level, and the lowest reefs exposed along the Arctic Ocean shoreline at the south end of Mercy Bay, e.g. at the mouth of the Gyrfalcon River. We measure the cumulative thickness of the Mercy Bay carbonates as *c.* 220 m (Fig. 2); this is reasonably close to the 230–280 m estimate of Thorsteinsson & Tozer (1962, 1970), who calculated a maximum of 500–600 feet for the reefs and intervening clastics. Only minor post-Devonian tectonic displacement and local faulting have displaced some reef structures (Thorsteinsson & Tozer 1962, 1970). In their interpretation, Klovan & Embry (1971) and Embry & Klovan (1971) postulated a number of anticlinal, synclinal and fault displacements on Banks Island, as well as normal faults, and used this explanation to suggest a single reefal carbonate horizon. We show, based on abundant field evidence, that this would not account for the superposiiton of the various reef horizons we have measured *in situ* at a number of sites, nor would it account for the horizontal disposition of the isolated reefs (see below).

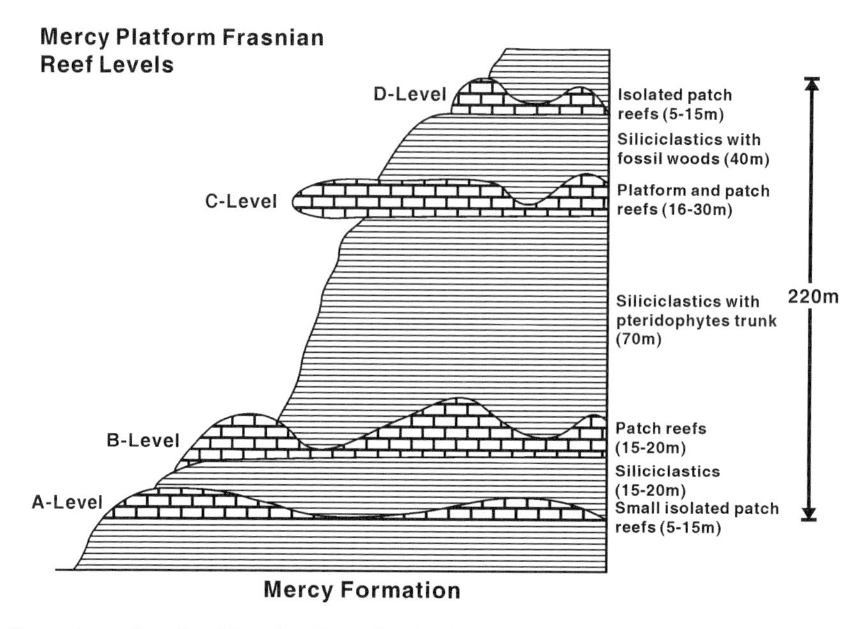

Fig. 2. Composite section of the Mercy Bay Formation showing the four Frasnian reef levels from A to D, with total thickness estimated at 220 m from the base of the lower carbonate A horizon, to the top of the upper unit D (average, not maximal thicknesses are indicated). Lower unit A reefs may attain a variable thickness of *c.* 5–20 m, e.g. at the mouth of the Gyrfalcon River, and lower Mercy River and lower Manning and Vesey rivers. B-level reefs occur only on the northwestern fringes, as far south and east as the forks of the West and East Mercy River. The upper unit D reefs, e.g. at the M'Clure River localites, also reach a thickness of 5–30 m. Tabular or platform reefs, of variable plan views, typical of the C level, reached optimal thicknesses of *c.* 30–60 m, and several kilometres across areally.

The first person to describe the limestones and fossils on Banks Island was Armstrong in 1857, whose ship was ice-locked on the west side of Mercy Bay, in search of the famed Franklin Expedition. Armstrong undoubtedly traversed to Gyrfalcon Bluff, the most prominent reef outcrop visible along Mercy Bay, from where the ship anchored. Because of its remoteness flanking the Arctic Ocean, Thorsteinsson & Tozer (1962, 1970) were the first to provide a brief geological description and age for the Banks reefs, describing three reef horizons, while on a mapping survey using small aircraft and ground traverses. This was followed in 1971 by Embry & Klovan who studied reefs at four localities – the Manning River, upper East Mercy River, Gyrfalcon Bluff, and M'Clure River – and measured them as '200 feet' (*c.* 60 m) thick. They detailed several of the extensive reefs, and used a descriptive terminology for reef sediments that has been extensively used in the geological literature, e.g. bafflestone, bindstone, rudstone, etc. Klovan & Embry (1971) attributed all the reef development to a single 30–60 m thick unit, the Mercy Bay Member, within the siliciclastic non-marine Devonian weatherall Formation, whose type locality lies on Melville Island to the north (it is Givetian in age; McGregor & Uyeno 1972). Embry (1988, 1991) and Embry & Klovan (1976) suggested that the Banks Island reefs were developed on distal lobes of the giant Ellesmere delta during the Late Devonian, with continental sediments over 2 km thick in the Okse Bay area of Ellesmere Island, thinning westwards towards Banks Island to *c.* 800 m. Miall (1976, 1979) added more data on the clastic sediments of the area, using borehole petroleum exploration and seismic data.

The Banks reefs and reef platforms are dominated by colonial corals and the dense $CaCO_3$ skeletons of the stromatoporoid sponges, whereas modern coral reefs and platforms feature >90% aragonitic scleractinian corals and calcitic coralline algae (Hart & Kench 2007). The Holocene global carbonate budget also contains a substantial carbonate contribution by aragonitic green algae such as *Halimeda*, as these have high seasonal $CaCO_3$ production rates, since they die back each year (Rees *et al.* 2007); green algae were not recorded in the Banks sequence.

We report on two extensive surveys of the Mercy Bay Platform in the summers of 2000 and 2003, in which 117 localities were sampled *in situ* using Polar Shelf helicopter support. Some 20 sections were measured, and over 130 reef localities discovered, many in areas well beyond those noted previously. Reefs were identified at four different intervals spanning a total thickness of some 220 m. In this paper we focus on the geometry, distribution and palaeogeographic location of these reefs. We here employ the term 'reef' in the broad biological sense as structures produced by skeletal reef-builders, elevated above the sea floor; this includes such frequently synonymous terms as organic buildup, mudmound, reef bank and bioherm. Biostromes are considered as thickets or meadows of carbonate producers and are excluded in this preview, though they may be incorporated in reef cores or flanks. Most of the structures studied were patch reefs varying from 5 to 500 m in diameter, and from 2 to 60 m in height. At level C, large reef platforms similar to those on the modern Great Barrier Reef ('bank' or 'tabular' reefs) extended over distances of several kilometres and were up to 60 m thick, with sloping reef flanks tapering off distally (we suggest that some have been misinterpreted as anticlinal structures). Some patch reefs were incorporated into, or overgrown by, platform reefs.

The lowest reef level has been labelled as the 'A-level', and the highest, and latest, the 'D-level' reefs (see Fig. 2). Reefs were mapped with GPS using latitudes and longitudes, and plotted on grids that match the regional topographic maps, at scales of 1:50 000, identified by six UTM map sheets. Reef photographs are identified by mercator grid references. The Mercy River, the largest in the area, is divided into sections (Figs 1, 3) and has five branches: one to the north (the North Mercy River, exposing A- and B-level reefs), the West Mercy River (level C), the East Mercy River (levels B, C and D), Middle Mercy River (levels C and D) and the lower Mercy River (exposing A- and B-level reefs). We estimate at present (the conodonts are being evaluated by T. Uyeno) that the Banks reef levels span the Swan Hills through Leduc reef horizons of western Canada, i.e. approximately conodont zones 1–11 (see McLean & Klapper 1998).

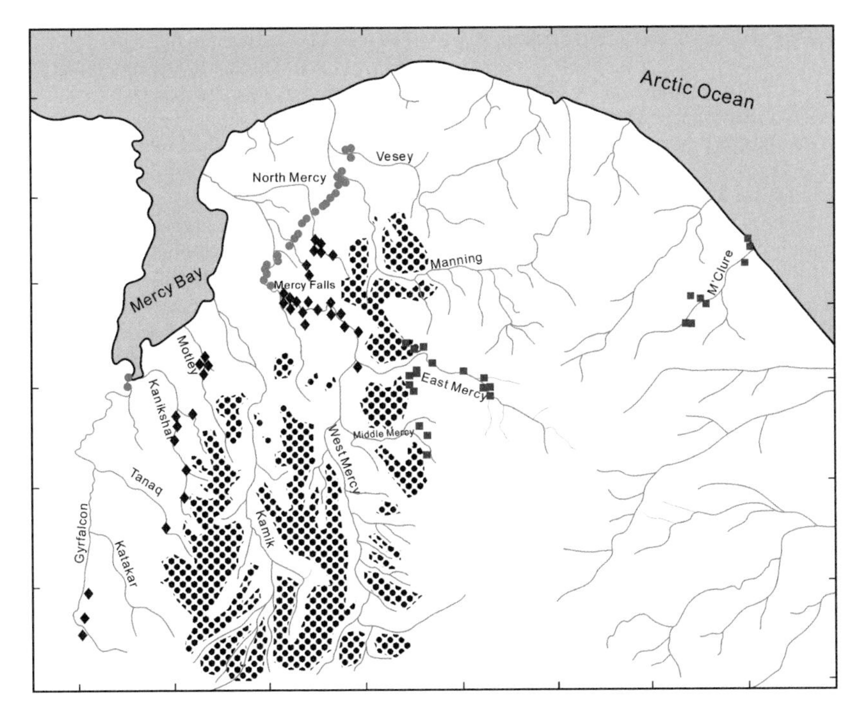

Fig. 3. Location map of the major reef levels of northeastern Banks Island with the Mercy carbonate tract at between 73 and 74° north (see Fig. 1). The A-level patch reefs (green circles) fringe the northern and western margins. The B-level patch reef clusters (blue diamonds), some 70 m lower, also on the northern fringes, are found on the river valleys of the lower Mercy, Motley, Kamik, Kanikshar and Tanaq rivers. The C-level reefs (dotted terrain) are the most widespread, making up most of the tabular and platform reefs up to some 5 km across, as well as isolated reefs on the eastern and southern margins of the complex, such as the double reef at Gyrfalcon Bluff (not shown, but NE of the 'K' of 'Kanikshar'). The dotted pattern marks the estimated extent of the C horizon, not outcrop. The upper D-level patch reefs (red squares) are well exposed in the M'Clure River area (ten prominent reefs), and the eastern flanks of the carbonate platform reefs around the East and Middle Mercy River. The D level is a reef platform only on the East Mercy and Middle Mercy tract.

The coral fauna is similar to that of the Hay River sequence in western Canada, spanning conodont zones 1–8. Neither the coral nor brachiopod evidence we found indicate the latest Frasnian conodont zones 12–13.

We base our interpretation regarding the four early and middle Frasnian reef levels, and the regional structure, on the following solid, hard evidence.

(1) The direct superposition, with continuous, measurable, stepped outcrop, of three of the reef levels (B, C and D) that occur along the East Mercy River is clearly evident in air photo coverage and in panoramic views from a helicopter (covered by topographic map 88C/14, co-ordinates 59000–63000:09000–11000). This is locality 25 of Thorsteinsson & Tozer (1962: 58), who stated 'three biostromal reefs, separated by 300 to 400 feet of clastic sediments were observed in sequence on one valley wall'. We corroborate that the lowest, B-level reef B27 (60000:10700), which lies on a promontory at the junction of the West and East Mercy River branches, is overlain by a gap of *c.* 70 m to reef level C (localities B9, B41, B42), and another 40 m to level D (localities B43, B9, 8C/14, 60000–66000:08000–10000), capped at elevations of 340–350 m. We measured continuous sections on both banks of the East Mercy River at this site. Levels C and D are present at the Middle Mercy River branch due south of the East Mercy. Level-D patch reefs also crop out 10–12 km further upstream on the East Mercy River as isolated reefs.

(2) Levels A and B were measured in continuous, superposed, unfaulted sections on the lower Mercy River (88F/3, 49000–5000:18000–20000, reef localities B13, B14–B19, B20, B21), with the lower A-level reef exposed in the river bed, and directly traceable to the A-level reefs downstream at Mercy Falls, and other reefs 2–4 km downstream from Mercy Falls (see Fig. 4a, c, d). These A-level reefs lie well beyond the main reef tracts; they were unknown to Thorsteinsson & Tozer (1962), and were neither discovered nor plotted by Klovan & Embry (1971).

(3) None of the continuous shallow-dipping, completely exposed, 300 m high coastal sections from the M'Clure River, extending more than 30 km westwards, show any faults in either our photographs or visits, or in photographs taken by Klovan & Embry (1971: pls. 1–2), or Miall (1976). The only structural feature evident on the north coast is a gentle syncline centred around Rodd Head on the north coast, with a regional dip of $<5^\circ$ (see Thorsteinsson & Tozer 1962: pl. 23). This gentle syncline satisfactorily explains the up-dip D-level reefs along the M'Clure River. It is inexplicable why inland sections should be faulted, but not coastal sections, if the down-faulted reefs indeed represent a single reef level, as suggested by Klovan & Embry (1971). No graben structures have been outlined that might explain the reefs at the base of the Kanikshar and Tanaq river valleys.

(4) We measured and sampled the three reefs along the Kanikshar River said to have 'slumped' down the valley (Klovan & Embry 1971: pl. 4, fig. 3; see also photographs in Thorsteinsson & Tozer 1962: pl. 24; D. A. Ehman & J. C. Wise unpubl. work, 1971), as well as the reefs at the mouth of the Gyrfalcon River (reef B6) and east of Gyrfalcon Bluff, along Gyr Creek (reefs B4a–d located some 70–80 m below the top of the north Gyrfalcon Reef). None of these reefs were visited by Embry & Klovan (1971): we examined all. These reefs are horizontally emplaced, untilted and are intact three-dimensionally (no faults mark their boundaries, nor are platform carbonates broken off at the sides). If these had slumped and become detached from the C-level platform, we expect that it would be highly unlikely, indeed structurally impossible, that all eight such reefs in this area east and south of Gyrfalcon Bluff, some as much as 600 m in diameter, came to rest horizontally and remained completely intact, with slope facies visible on all sides (see Fig. 5b, c, d, f). Neither are any of the reefs in the North Mercy tract (reefs B7a–q) tilted, nor out of place.

(5) All the A-level reefs occur at the northern and western margins of the Banks reef tract (Fig. 3). They were measured and sampled *in situ* for a distance of some 40 km along strike, from the Vesey River in the NE to the mouth of the Gyrfalcon River at the west end. These reefs are directly correlatable with the reefs exposed at Mercy Falls and downstream. None of these reefs show evidence for displacement (none of these sites were visited or sampled by Klovan & Embry (1971), though one site was plotted on the lower reaches of the Manning River (their fig. 1; see Fig. 4e, f). It is also difficult to visualize how reef B6, which we correlate with level A (see Fig. 4b), and is more than 400 m in diameter with reef flanks exposed, could have become detached from the Gyrfalcon Bluff C-level platform and 'slumped' some 100 m lower topographically, and 1 km to the west, while maintaining a perfectly horizontal position, with all flanks preserved.

(6) None of the faults on the Banks platform, shown by D. A. Ehman & J. C. Wise (unpubl. work, 1971) and Miall (1979), would explain the detachment and displacement of any of the A-level or B-level reefs; indeed, none are shown at the locations needed to explain tectonic slumping or glacial displacement. To explain the reefs located on the valley floors of the Kanikshar Gyr and Motley rivers, two separate and parallel grabens, normal to regional strike, would have to be demonstrated.

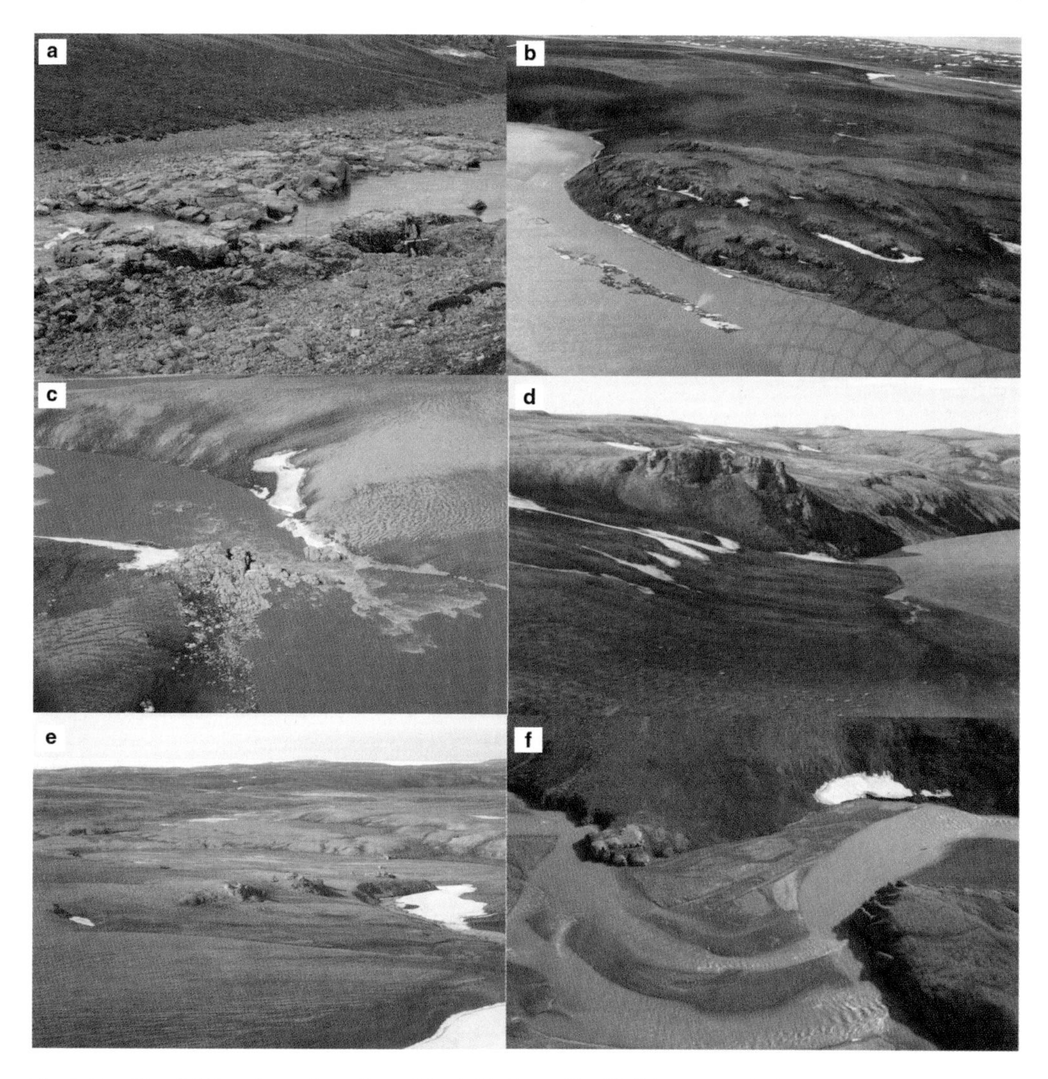

Fig. 4. Basal level A Frasnian reefs on Banks Island. **(a)** Core of Reef B17 located on the lower Mercy River *c.* 10 km downstream from the main reef tract (NTS 8/F4, 54800–55300:18200–300; E. Edinger for scale). This low, domal patch reef is *c.* 100 m in diameter and 5–7 m thick. **(b)** Reef B6, an elongated patch reef, orientated roughly north–south, on the west side of the mouth of the Gyrfalcon River (NTS 88C/13, 35800–36000:11700–12500), *c.* 800 m by 200 m in plan view, *c.* 20 m thick. **(c)** Mercy Falls patch Reef B13 downstream on the lower Mercy River (88F/3, 49700:20800, reef centre), a low domal patch reef *c.* 5 m thick, 120 m diameter. **(d)** Reef B44, a high domal patch reef on the lower Mercy River below Mercy Falls, *c.* 30 m thick and 200 m in diameter. **(e)** Patch reef tract on the lower reaches of the Manning River, reefs B67–B70 (and unnumbered reefs along the middle and east banks of the river, 88F/13, 57000–8000:31000–2000); some eight patch reefs <80 m in diameter are visible. **(f)** Lower Manning River Patch reef B67 on the west bank (upper left of photo), facing matching patch reef of the east bank, <50 m in diameter, 4–5 m thick.

Klovan & Embry (1971: 720) discuss only 'faults parallel to the folds'. In addition, there is little evidence that the seven folds mapped by Klovan & Embry (1971: fig. 10) exist, as these reefs, examined by us, represent dipping reef flanks, or margins of separated tabular or platform reefs, or reefs dissected by river erosion with the reef thinning to the east and west.

(7) We regard it as unlikely that the Banks Island Frasnian reef tract, the only one in the Canadian arctic at this time, was unaffected by frequent Late Devonian global sea-level cycles, *sensu* Johnson *et al.* (1985). We do not believe that the 220 m thick Banks complex would fail to match multiple sea-level Frasnian reef cycles as shown in western

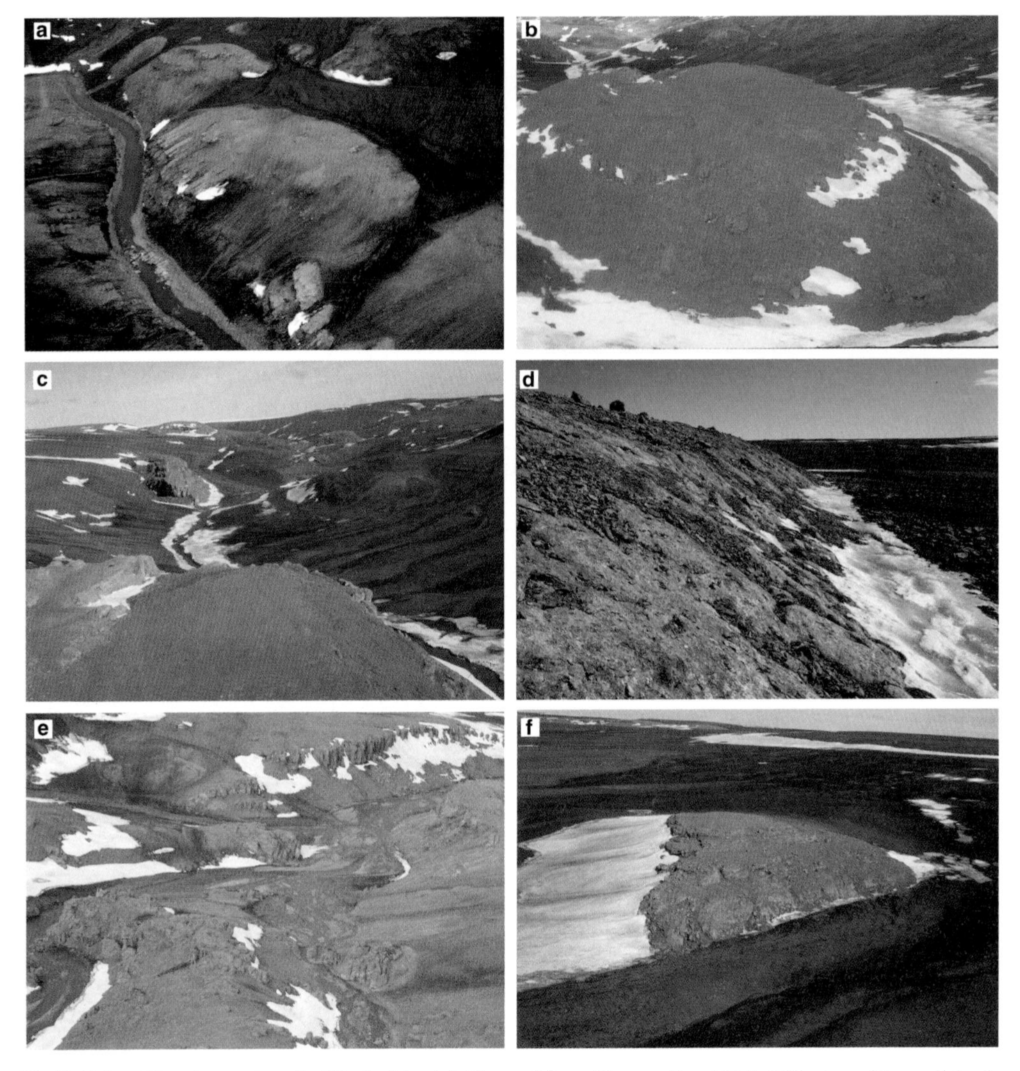

Fig. 5. B-level Frasnian patch reefs of Banks Island. **(a)** Lower Mercy River reef tract B14–B18, view of the south bank (right, with lowermost A reef B17 from Fig. 4 in river bed), centred around patch reef B16, *c.* 30 m thick, cut by the river (88F/3, centred *c.* 54000:48200), composite reef *c.* 30 m thick, built in four stages. **(b)** West bank Kanikshar River Reef B33 (see Thorsteinsson & Tozer 1962: fig. 24) circular patch reef *c.* 250 m diameter, 30 m thick, with a cap of platy stromatoporoids (NTS 88C/14, 40800–1000:97800–98100). **(c)** From Reef B33 in foreground looking downstream at Reef B34 (88C/14, 41000:9880–99000) *c.* 30 m thick, 200 m diameter, cap of stromatoporoid platestone. **(d, f)** East bank Kanikshar River patch Reef B38, with view of 20° reef flank *c.* 250 m diameter, 10 m thick, and panoramic view of same (88C/14, centred at 44200:04700). **(e)** North branch Mercy River patch Reef complex B7a–t, a cluster of *c.* 12 reefs covering about 1.5 km^2, ranging from *c.* 50 m to 400 m diameter, 20–30 m thick (8F/3, 54600–5600: 23000–4000).

Canada (e.g. McLean & Klapper 1998), and Europe (Boulvain 2007).

A-level reefs (Fig. 4)

These are the oldest Frasnian reefs exposed on Banks Island. More than 40 of these isolated patch reefs occur on the western and northern fringes of the Mercy Platform, not connected by any carbonate platform between, extending from the Vesey River to the NE to the mouth of the Gyrfalcon River to the west for *c.* 40 km. Typically these are relatively small patch reefs, compared to later reef levels on Banks Island. The largest and thickest A-level reef is exposed at the mouth of the Gyrfalcon River that drains into Mercy Bay (Fig. 5, reef locality B6, 88C/13,

35800–36000:11700–12500: this reef was not cited or mapped by Embry & Klovan 1971). It is oval, about 550 m long and 250 m wide, some 30 m thick, trending roughly north–south. The single reef is capped by thick, platy stromatoporoids overlying 5–7 m of smaller, platy stromatoporoids and alveolitid corals, and in turn is founded on a 2–3 m thick biostromal framestone of stromatoporoids and tabulate corals. The last also make up the flank beds, where brachiopod shells occur. The top of Gyrfalcon Bluff (level C) is stratigraphically and elevationally *c.* 130 m higher.

Mercy Falls on the lower Mercy River (locality B13, 88F/3, 49700:20800), is about 20 m high, and the primary clue to the lower reef horizons (Fig. 4c). The Mercy Falls reef is *c.* 150 m in radius, *c.* 12 m in thickness, has a low syndepositional relief of <3–5 m, low angled flanks of <5°, a 'mudmound' reef core and a cap of dense micrite, and rare phaceloid rugose corals overlying a thamnoporid coral bafflestone biostrome. Upstream, about 1–2 km from Mercy Falls, are three additional patch reefs (B20, B21a, B21b, 88F/3, 53100–700:18200–800: see Fig. 5a), all with low relief and with relatively sparse faunas of amphiporid stromatoporoids, phaceloid rugosans, and dominated by carbonate mudstones. A small A-level patch reef, 30 m in diameter, with domal stromatoporoids and phaceloid rugose corals is also exposed upstream from Mercy Falls along the river bed, and at the base of locality B18, below the B-level reefs (88F/3, 56600:18700; see close-up Fig. 4a, distal view Fig. 5b). An alternative explanation for the A-level reef horizon on Banks Island, as seen from the lower Mercy tract, is that this marks the beginning of the B-cycle, with a disruption and burial by siliciclastics prior to B-level reef growth. If the A- and B-level reefs are considered as one cycle, only three reef levels, as originally postulated by Thorsteinsson & Tozer (1962), would be present.

Downstream from Mercy Falls, about five A-level patch reefs crop out, with the largest at locality B44 (88F/3, 49440:22267; see foreground Fig. 4d). The large and prominent patch reef B44 was measured at *c.* 32 m thick, oval and *c.* 100 m in diameter. Framebuilders are digitate amphiporid stromatoporoids and alveolitids, astreoid rugosans cap and flank the reef, as well as the reef core, and platy stromatoporoids occur at the base (we did not find any amphiporids at the C and D levels). Other smaller A-level reefs are exposed on the opposite river bank of the lower Mercy River, and inland west of the river, and further patch reefs occur some 1–2 km downstream (background Fig. 4d).

The northernmost A-level reefs (reefs B67–B69, 88F/3, 57200–500:31300–400, here termed the Manning–Vesey reef belt) were discovered mostly along the lower stretches of the Manning and Vesey rivers, several kilometres downstream from the Manning River C-level mesa figured by Embry & Klovan (1971). These are about 40 km NNE from the southwestern A-level reef exposed at the mouth of the Gyrfalcon River (locality B6), and lie 10–15 km north and west of the main C-level reef tract on the Manning and East Mercy rivers. These are all small, circular to oval patch reefs usually <10 m in diameter and with relief of <3–5 m, some cropping out in the middle, and on the east bank of the wide Manning River valley (Fig. 4e, f). Stromatoporoids, branching thamnoporoids and platy alveolitids dominate the cores and flanks; many are broken, indicating storm damage. The Vesey River itself exposes three patch reefs, the most northerly of the area (locality B80a–c, 88F/3, 56400:34200). These small patch reefs have a diverse fauna of thamnoporoids, cerioid rugose corals and stromatoporoids, forming a lower biostromal unit and upper reef core, capped by 3–4 m of massively bedded micrites with stromatactis. Overall, the outer A-reef belt consists of isolated patch reefs, cropping out at virtually every river and creek draining NE over a distance of more than 40 km on the north and west flanks of the Mercy outcrop belt. All of these reefs are *in situ*, with a horizontal base: there is no evidence for faulting and transport of these reefs downslope, either individually or en masse, from the B or C level (this would require a major fault parallel to the reef belt to the south, which is neither postulated nor identified by previous authors, e.g. D. A. Ehman & J. C. Wise unpubl. work, 1971; Miall 1979). There is no evidence that these are distal reef remnants, pinchouts, or anticlinal extensions of the Manning and East Mercy River tracts at the 320–340 m topographic elevation levels 10–25 km to the south.

B-level reefs (Fig. 5)

This is an extensive reef tract, with roughly twice as many reefs, and with larger reefs, backstepped eastwards from the A-level reefs by 1 to 10 km. A carbonate platform only rarely connect these B-level reefs to each other (e.g. North Mercy River, reef localities B7a–r: see Fig. 5e). The north Mercy River reef cluster, with some 18 reefs from B7a to B7r (88F/3, 54000–51000:23000–500), stretch for about 1.5 km along the North Mercy River, partly connected by platform limestones. These reefs vary in size from <50 m to some 200–300 m in diameter; some are elongated, and thicknesses range to 30 m. The larger reefs can be seen with three or four internal discontinuities, and reef regrowth. A number of these patch reefs are small mounds made up, sometimes almost entirely, of branching thamnoporids that formed dense baffling thickets (centre Fig. 5e), somewhat like modern *Porites* patch reefs in the Caribbean, or *Acropora* thickets in the Pacific ocean. Other patch reefs here, especially the larger ones, are dominated by platy stromatoporoids, with thamnnoporid coral flanking beds. Reef B7q (88F/3, 55400:23300) appears to be a mudmound, as skeletal material seems proportionally rare, and micrite dominates.

Other excellent outcrops of clustered B-level patch reefs occur along the lower Mercy River (localities B14–B20, B56–B66) as a large reef cluster that extends for about 2 km upstream, in part connected by platform limestones (Fig. 5a). Striking here is that some of the reefs show three or four stages of development, with disconformities between that may represent storm breaks, or changes in sedimentation rates and siliciclastic supply (see reef B17, UTM 88F/3, 54800:18200–300, as in Fig. 5a; reef B62, 56770:17900; reef B63, 55579:18237). Upstream along the main Mercy River valley north of the East Mercy, there are several spectacular, three-dimensionally exposed, prominent isolated mounds that are readily visible from a helicopter or air photos (reefs B56 to B62). These are tracked downstream from the East Mercy branch (reef B27 section), curving some 8–9 km south of the East Mercy River towards the Lower Mercy tract on both sides of the river. Most are 500 to 600 m in diameter, and up to 30 m thick; some of these internally also show three or four growth cycles. These reefs are dominated by tabulate and rugose corals, though stromatoporoids are also common. Thus for both the isolated mounds and the patch reef tracts, reef growth was periodically disrupted and restarted. This is much less evident in C-level platform reefs.

Less than 1 to 1.5 km south and east of Gyrfalcon Bluff, along the Gyr River, the short river east of Gyrfalcon bluff (localities B1–B4, 88C/14, 41500–2500:98000–12000), several isolated small B-level patch reefs occur about 50–70 m below the reef level of Gyrfalcon Bluff. Upstream along the Motley River further to the east (reef locality B28, 88F/3, 494700–45000:05300–400), some 8–10 km SSE of Gyrfalcon Bluff, five isolated B-level reefs crop out on both sides of the river. These are up to about 300 m in diameter, *c.* 20–30 m thick, and are constructed of domal cerioid rugosans and platy stromatoporoids. Reef core blocks in the stream show calcimicrobes such as *Renalcis* and *Girvanella*. About 15–18 km south of Gyrfalcon Bluff isolated B-level patch reefs crop out on the Katakar, Tanak and Kanikshar rivers at stratigraphic levels some 50–70 m below the C-level platform complex. Three of these reefs, south of Gyrfalcon Bluff, are shown in a panoramic photo of Thorsteinsson & Tozer (1962: pl. 24; see also Klovan & Embry 1971: pl. 4, fig, 3: in the foreground is our reef locality B33, 88C/14, 40800–4100:97800–98000), a 200–300 m diameter, 20 m thick reef dominated by platy stromatoporoids. Other reefs in this area are reefs B34 to B39 on branches of the Kanikshar River. Embry & Klovan (1971) suggested, from aerial views and data in D. A. Ehman & J. C. Wise (unpubl. work, 1971), that these reefs had slumped down from the main C-level platform. Close examination, mapping and outcrop evidence of these *in situ* reefs showed no slumping, tilting or sediment deformation, nor surrounding brecciated blocks that would indicate detachment from adjacent reef platforms.

We note the complete absence of B-level (and lower A-level) reefs in the Kamik River valley, in the West Mercy Valley (south of the East Mercy branch), in the Middle Mercy, and on the upper reaches of the Manning River. If slumping were a common feature in the Mercy area (as proposed by D. A. Ehman & J. C. Wise unpubl. work; 1971 Klovan & Embry 1971), the absence of lower reef levels in river valleys to the east and south can only be explained by the confinement of the A- and B-level reefs to the northern and western margins of the Banks reef tracts, and by backstepping eastwards of the C and D levels during sea-level highstands.

The most westerly (and some of the most southerly) Mercy reefs are two prominent, isolated pinnacle structures (reefs B23–B24), at a short tributary on the east bank of the Gyrfalcon River, that flows almost due north towards Mercy Bay: these two reefs, facing each other across a short tributary (88C/13, 31500–2300:89000–700) are located about 22 km SSW of Gyrfalcon Bluff, and about 5–6 km west of C-level platform reefs B29–B30. Their location and topographic level, covered by siliciclastics to the east, indicate they are B-level reefs (see Fig. 3).

C-level reefs (Fig. 6)

This represents the maximal Banks reef and platform development in terms of area covered, and thickness, in the central portion of the Mercy reef belt (the reef level explored by Embry & Klovan (1971) at the East Mercy and Manning rivers). The reefs here include Gyrfalcon Bluff, the Middle and East Mercy reef tracts, the tracts along the upper Manning River, as well as reef platforms south and upstream along the Kamik, West Mercy and Katakar rivers to the south. Gyrfalcon Bluff

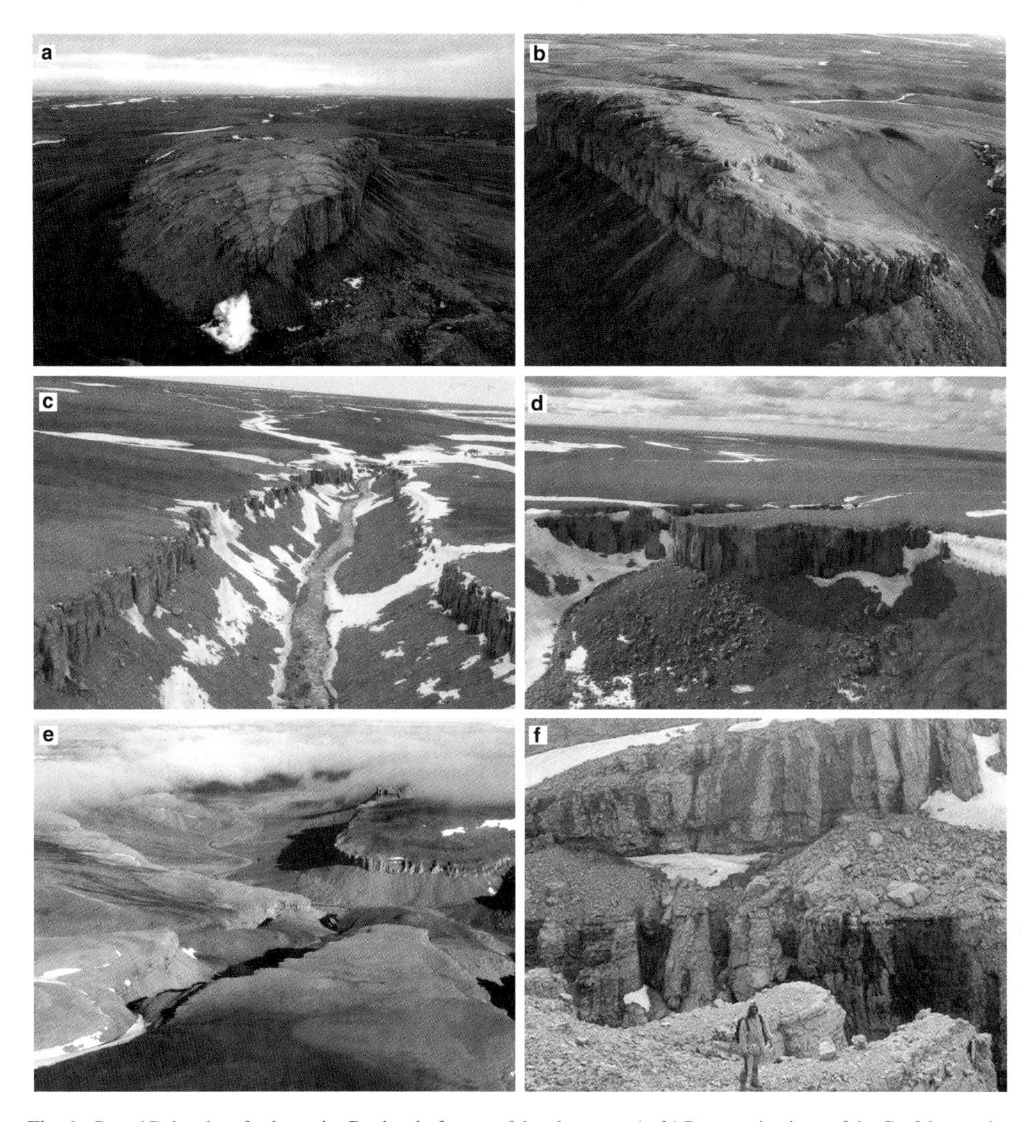

Fig. 6. C- and D-level reefs, the major Banks platform reef development. **(a, b)** Panoramic views of the Gyrfalcon twin reef (Gyrfalcon reef B5a–b): (a) view from the north looking at the north reef with the south reef obscured in the background, and (b) view from the SW looking at the tapering reef flanks of the north reef with intervening siliciclastics with pteridophyte wood trunks and roots. **(c)** Kamik River B22 platform reef, view due south, dissected by the river upstream, in the southernmost outcrop belt of the Banks reef tract (88C/11, centred *c.* 50000:81000–2000). The platform thins southwards, as well as east and west, and is locally thickened by eight patch reef within the platform shelf reef. **(d)** SW fork Kamik River, west bank patch Reef B48, view northwards (88C/14, centred at 44700:96800–900), isolated low domal reef, *c.* 600 m diameter, *c.* 35 m thick, with creek cutting south side. **(e)** East Mercy River, Reefs B9a–d, B41–B43, view northwards downstream Mercy River, with south bank D-level reef platform foreground centre (B42), some 40 m above the C-level platform in the centre, and on the opposite north bank (B43), and B-level reef B29, upper left down the valley. Panorama covers *c.* 4 km field of view laterally (88C/14, 60000–4000:10000). **(f)** Upper D-level reef south bank East Mercy River, Reef B9a, view looking SE (88C/14, centred at 62000–3000:09000). These reefs overlie the lower level B42 reef platform.

(reefs B5a, b, Fig. 6a, b), a north–south elongated, prominent high about 800 m in length, exposes a double reef, with steep flanks exposed on the east side: the bluff stands out at about 170–180 m above the surrounding terrain, i.e. about 120–130 m above the A-level reef at the mouth of the Gyrfalcon River (reef B6, Fig. 4b). Gyrfalcon Bluff is not a single reef, but consists of two adjacent reefs, thinning out towards each other, separated laterally by *c.* 100–150 m of siiliciclastic flagstones, with coal and plant remains (B5, 88C/14, northern reef 40200–500:12300–800; southern reef, 40400–500:11900–12280). The west side of both reefs has been partially removed by erosion, but the eastern side has preserved the steeply dipping flank beds. The two reefs are primarily composed of thick, platy boundstones of stromatoporoids, but the early growth stages of the reefs are at the same level as the bafflestone, biostromal platforms of other reefs to the east and south.

The south branch of the Manning River (Embry & Klovan 1971) displays an extensive reef platform partly dissected by the river and its tributaries, with two waterfalls dropping over the resistant carbonate platform (reefs B8a–f, 88F/3, 61500–62500: 20600– 22700). Platform carbonates are commonly dark grey micrites with isolated phaceloid rugosan colonies up to 1 m or more in diameter, and some 40–50 cm thick. This is similar to those seen in the southern margins of the outcrop area (below Fig. 6c, d: reef localities B22, B48). At locality B8d along the Manning River (88F/3, 61500:21500) pteridophyte fossil trunks 3–10 cm in diameter, and a metre or more long, are common, some encrusted by corals or stromatoporoids that occur in peri-reefal strata alongside crinoid ossicles. The reef platform is buried by siliciclastics on both sides of the river and upstream, but on the west flank the platform continues, with breaks, thinning southwards to connect with the East Mercy section. We were unable to find either the lower B or higher D reef levels on the upper Manning River, unlike the section at the East Mercy River.

The southwestern portion of the West Mercy River (*c.* 15–20 km south of most other Mercy Platform outcrops at the East Mercy), displays the most southerly reefs of the Mercy River drainage basin. This area has platform carbonates similar to those of the Middle Mercy and East Mercy, though thinner than those to the north and disappearing southwards. More than 20 patch reefs occur as part of a platform unit in the southernmost West Mercy tract (localities B12a–h, 88C/14, 63100–65600:89200–500). These patch reefs are generally small, usually <50 m diameter, and 3–8 m thick, dominated by digitate thamnoporid and platy alveolitid corals (locally making up 90% or more of such structures), phaceloid rugosans, or platy stromatoporoids, Some reefs are up to 400 m in diameter, but not more than 25 m thick. South of this platform tract there are few isolated patch reefs. One of these is locality B46 (88C/14, 64300–400:86700–86800), which marks an isolated outcrop of a small patch reef, the southernmost reef along the West Mercy River. This isolated B46 reef (*c.* 3 km south of the B12 cluster), is 100–150 m in diameter and 15 m thick, including a lower rubbly weathering biostromal, bafflestone layer of alveolitids, thamnoporids and phaceloid rugosans, and a cap of massive, platy stromatoporoids. Other isolated reefs crop out over *c.* 15 km between the northern Middle Mercy B10-f reef tract, and southern West Mercy locality B46 (i88C/14, 64000:89000–03000).

Our measured sections of the East Mercy River branch (localities B9, B27, B41–B43, 88C/14, *c.* 59000:10700–11500) expose reefs at the B, C and D levels, with the C level containing the main Mercy Platform reef outcrops (locality 25; Thorsteinsson & Tozer 1962). The main reef C platform is up to *c.* 20–30 m thick but local patch reefs, 30–500 m in diameter, thicken portions of the platform. Hardgrounds occur within some of the reefs and platform. Locally the reefs are capped by giant, continuous, flat stromatoporoids more than 8–10 m in diameter and several metres thick, perhaps the largest coenostea of stromatoporoids ever reported. The C platform stretches to several kilometres long, with flat mesas on both sides of the Mercy River. To the east, the uppermost D-level reefs occur along the river bed (e.g. Fig. 6f).

The Kamik River drainage basin (named after Kamik, or 'Boot' Lake, near the mouth of the river), runs west of and parallel to the West Mercy River and contains the most southerly reefs of the Banks carbonate section, extending as far south as localities marked by platform reefs B22a–d (UTM 88C/11, 500000:81000, roughly matching locality 9 of Embry & Klovan 1971). These are about 5–6 km further south than the southernmost West Mercy patch reefs at B46 (noted above). On the west bank of the Kamik River, stretching due north for nearly 45 km from its headwaters, is a long series of mesas whose cap is the C-level reef platform and which terminate at a reef locally called 'The Sentinel' on the east bank, as it guards the narrow canyon leading upstream (reef B26, 88C/14, 48500:07000). Within the tabular reef platforms on the west bank are smaller to larger patch reefs, e.g. reef B48 (Fig. 6d: 88C/14, 44700: 97000), and reefs facing 'The Sentinel' on the west bank downstream.

On the downstream tract of the Kamik River (east bank), there are four very large, irregularly shaped, elongate, tabular platform reefs, between 1 and 1.2 km in maximal length, and stretching for about 7 km, adjacent to 'The Sentinel' (extending south and west from locality B25, 88C/14, 48000–52000:05000–10000). These platform reef outcrops are spread over some 15 km^2. Whether these were connected to the carbonate platform on the west bank is uncertain. Patch reefs jut out about 10–15 m above the local reef platform on both sides of the river; 'Sentinel' reef is a castellated, tower-like structure (reef B26). There is not a single 'slumped' reef in the long Kamik River valley, which would be expected if slumping was the explanation for reefs occurring below the C reef horizon. One small patch reef (88C/14, 46800:17000) on the west bank of the Kamik River extends away from the main reef platform but is not slumped: only the west side of the reef with its sloping flanks, is preserved. D. A. Ehman & J. C. Wise (unpubl. work, 1971) interpreted the platform outcrops between the Kamik and Tanaq–Kanikshar rivers to the west as the 'Hamilton anticline', presumably because reef flank strata dip and thin away on both sides.

C-level reefs and reef platforms also occur prominently on the upper reaches of the Tanaq and Kanikshar rivers (localities B32, B36–B38, B51–B53). These represent the western side of the same reef platform that occurs across the water divide on the west bank of the Kamik River (described above). Most of these reefs are part of a reef platform structure and can be walked out along the mesas on the east bank of the Tanaq and Kanikshar rivers. It is along the Kanikshar River that the C-level reef platform overlies the isolated B-level patch reefs along the river bed (reefs B33–B35), first illustrated by Thorsteinsson & Tozer (1962: pl. 24). To the north, the C-level platform along the Tanaq and Kanikshar rivers gives way to isolated patch reefs (B37–B39), the isolated and eroded reef core called 'The Monument' (3–4 km south of Gyrfalcon), and the two reefs which form Gyrfalcon Bluff itself. There are no further reefs to the west of this tract except the two B-level reefs along the east side of the Gyrfalcon River (Fig. 3).

D-level reefs (Figs 6 and 7)

The final phase of Banks Frasnian reef development was less spectacular, with stratigraphic levels about 30–40 m higher than the C platform level. They are evident at three main areas: the eastern reaches of the East Mercy River tract above the C platform (e.g. locality B9: Figs 6e, f, 7f), the Middle Mercy River branch (localities B10a–f, B42, centred around 88C/14, 64000:04000), and the ten reefs mapped along the coast around the M'Clure river bordering the Arctic Ocean (Figs 1, 3: reefs B11a, B11b, B49, B50, B92–B98). Nearly all D-level reefs are isolated, small patch reefs, with only minor platform development around the East Mercy River.

The M'Clure River reefs (B11a, b, B49, B50, B92–B98), about 40 km east of the main Mercy Platform, first noted by Embry & Klovan (1971), are capped by dark siliciclastic flagstones, in turn overlain by outcrops of pure white quartz sands (these may be mistaken for reef carbonates on air photos). The most prominent of the ten reefs in this area is a striking patch reef exposed 4–5 km upstream on the M'Clure River (locality B11b), bisected by the river, thus exposing the reef on both sides (Fig. 7a, b). The reef core is 40 m thick, and *c.* 100 m in diameter, with a rich coral fauna at the top, with brachiopods and crinoids on the flanks. Another small reef (<10 m thick) crops out on the bluff of a NW branch of the M'Clure River (reef B49, Fig. 7d). Reef B11a is exposed downstream on the upper ledges of the south bank, about 500 m to the east (Fig. 7a, foreground, left). Other patch reefs are exposed near the mouth of the river (Fig. 7e), and two occur northwestwards from the M'Clure mouth, along the unfaulted coastal bluffs. Other reefs, on the SW fork of the upper M'Clure River (B50, B92–B94, Fig. 7c) show fossil wood overgrown by the colonial rugose coral *Argutastraea*. Locally, conchoidally fracturing bitumen seeps out of the micritic reef core (reef B50).

The East Mercy River upstream area has a series of circular to oval isolated patch reefs well exposed on both river banks, and evident further upstream eastwards in helicopter overflights (localities B85–B91: Fig. 7f). One of the most easterly on the East Mercy River, some 8 km from outcrops B9a–f, is reef outcrop B90 on the south wall of the canyon (88C/15, 72468:09718). Reef B91 (Fig. 7f), on the south bank of an East Mercy river fork (88C/15, 73300:86000), represents the most easterly occurrence along the Mercy River. At the B9 localities, which include both C- and D-level reefs, the higher D-level reefs are usually patch reef structures, mostly 2–5 m thick and <10 m diameter, incorporated into a reef platform.

Other areas of development of the D-level reefs are the localities centred around B10, along the Middle Mercy River canyon that flows westwards into the main part of the Mercy River (B10a–f, 88C/14, *c.* 64000:04000). A prominent bank of thin reef platform is present on both river banks, thinning eastwards and southwards; parts of the limestone banks are bituminous, black micrites (organic content appears to be higher in the D level than lower reef levels, either a facies change to lagoonal sediments, or upward migration of bitumen). Reefs on the Middle Mercy (see Fig. 3) tend to be <10–15 m in diameter, and <5–8 m thick, some showing overgrowth of platform over patch reef. They tend to be dominated by tabulate and rugose corals, and are relatively separate and isolated, with steep flanks at *c.* 15–25°. None form prominent reef mounds such as those seen in earlier A-, B- and C-level reefs, and they may be easily missed on overflights.

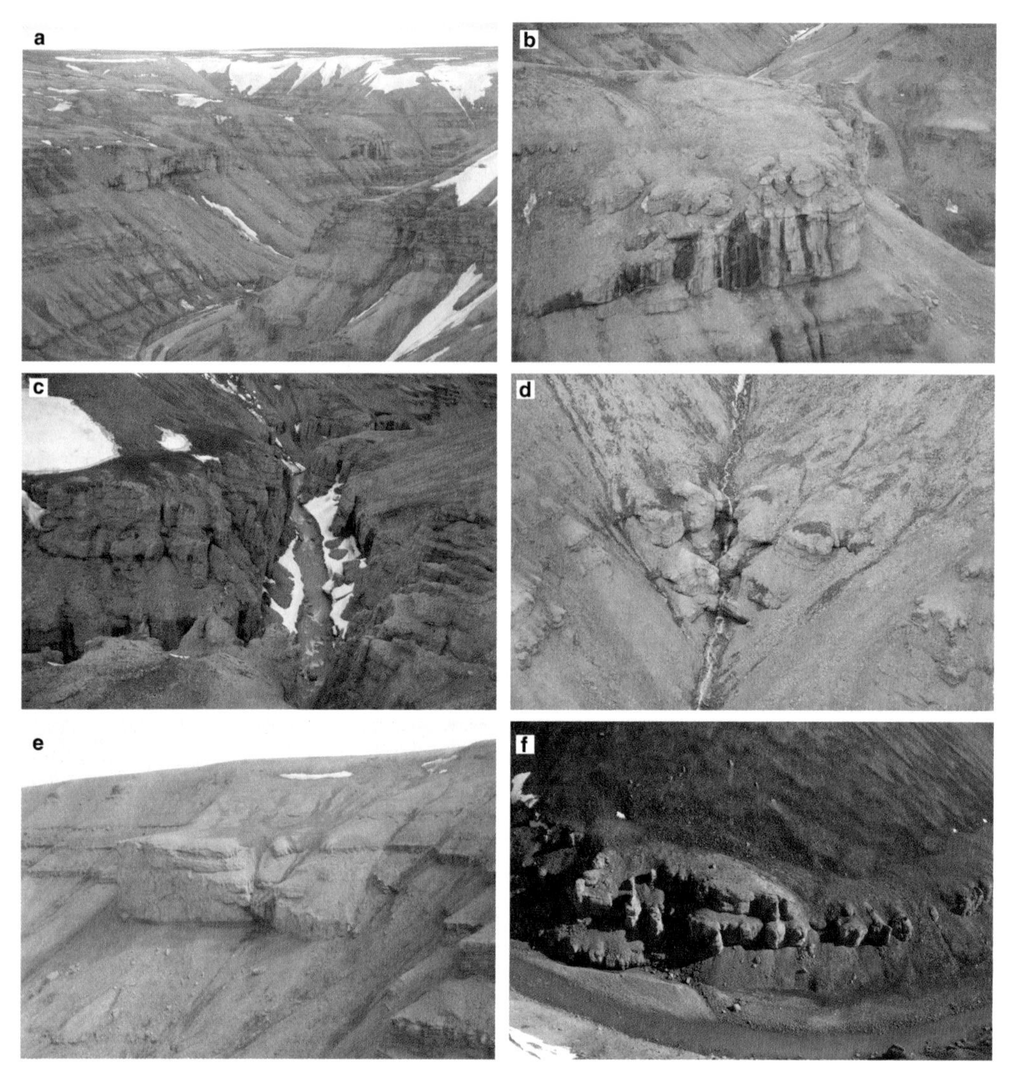

Fig. 7. D-level reefs along the M'Clure and East Mercy River. (**a**) View westwards upstream, exposing M'Clure patch reefs B11a–b on the south bank (*c*. 80–100 m diameter, 30 m thick, isolated by siliciclastics (88F/2, 95000–7000: 19000). (**b**) M'Clure River Patch reef B11b, north bank, the river bisects the reef; note the capping structure, and steeply dipping flank beds (88F/2, 95000:19400; see Embry & Klovan 1971: pl. 13, fig. 2). (**c**) M'Clure patch reef B92–94, cluster of four reefs, largest of the M'Clure reefs, *c*. 2 km upstream from B11b, cut by the SW fork of the river (88F/2, 93700–4000:7200–500; see Embry & Klovan 1971: pl. 13, fig. 1). (**d**) North fork M'Clure River B49, small patch reef, *c*. 500 m NW of B11b, north bank (88F/2, 94600:19400). (**e**) SE bank M'Clure River patch reef B94 (see Embry & Klovan 1971: pl. 14), eroded portion emerging from cliff face, surrounded by siliciclastics (88F/2, 00200:22900). (**f**) South bank East Mercy River, D-level reef B91, *c*. 150 m in diameter, upstream, *c*. 15 km east of fork West and East Mercy rivers (88C/15, 73300:86000). Note the small patch reef foundation, and reef margins tapering out to the east as isolated patches.

Conclusions

The Banks Island reefs are the only reefs of Frasnian age in the high Canadian arctic, and were palaeogeographically located at palaeolatitudes $<10°$ north of the equator at the time, covering *c*. 5000 km^2. The reefs occur in a distal megadelta setting, and ultimately appear to have been buried by westward prograding siliciclastics prior to the late Frasnian extinctions, during sea-level lowstands (one of the oldest known reef records on delta lobes). The reefs grew during early and middle

Frasnian sea-level highstands, as they indicate backstepping eastwards from the oldest to youngest reef horizons. Faunally, the reefs have a high abundance of corals and/or stromatoporoids, but their diversity is low, with usually fewer than a dozen reef-building species, similar to other early and middle Frasnian reefs worldwide. Other reef components such as brachiopods, crinoids and bryozoans were generally scarce, even in off-reef facies, presumably because of the prevailing siliciclastic facies between reefs. Some corals and stromatoporoids were seen to be directly encrusted on submerged or sunken fossil tree trunks from some of the debris swept downstream along the delta lobes, the oldest known such occurrence in the fossil record. Reef geometries varied at the four stratigraphic reef levels, from widespread small (<5–10 m diameter) to very large patch reef mounds (up to 700–800 m diameter), to modest carbonate platform development several kilometres across. No single large carbonate platform appears to have been present.

We thank the Polar Continental Shelf Project (Canada Department of Energy, Mines and Resources) for their help over two field seasons (2000, 2003), with logistics to and from, and on Banks Island, and chartered aircraft from Inuvik, Twin Otter cargo flights to Polar Bear Cabin field base from Sachs Harbour, and daily pilot and helicopter support on the rugged upland Mercy carbonates. The Natural Sciences and Engineering Research Council of Canada were generous in providing the basic research support, flights to Inuvik, field supplies, student support, and laboratory funding. Parks Canada in Inuvik and Sachs Harbour provided additional help and advice, as some of the area is located within the new Aulavik National Park. We thank our assistants in the field, Jennifer Neild and Joe Kudlak, who endured long traverses across rock and permafrost, and loads of samples. This paper is a preliminary assessment of overall Banks reef development, and will be followed by papers dealing with reef petrography, reef communities and palaeoecology. We thank Dr Ross McLean (Calgary) for identification of the rugose corals. Specimens are stored at the Geological Survey of Canada, Ottawa, in the charge of curator Jean Dougherty. Some prepared and polished slabs are on display at the Sachs Harbour and Inuvik offices of Parks Canada. We thank Dr Eberhard Gischler and Dr Ashton Embry for their comments on the manuscript: we remain responsible for the ideas generated.

References

ARMSTRONG, A. 1857. *A Personal Narrative of the Discovery of the Northwest Passage*. Hurst & Blackett, London.

BERNER, R. A. 2001. Modeling atmospheric O_2 over Phanerozoic time. *Geochimica et Cosmochimica Acta*, **65**, 658–694.

BERNER, R. A. 2004. *The Phanerozoic Carbon Cycle: CO_2 and O_2*. Yale University Press, New Haven.

BERNER, R. A., VAN DEN BROOKS, J. M. & WARD, P. D. 2007. Oxygen and evolution. *Science*, **316**, 557–558.

BOULVAIN, F. 2007. Frasnian carbonate mounds from Belgium: Sedimentology and paleooceanography. *In*: ALVARO, J. J., ARETZ, M., BOULVAIN, F., MUNNECKE, A., VACHARD, D. & VENNIN, E. (eds) *Palaeozoic Reefs and Bioaccumulations: Climatic and Evolutionary Controls*. Geological Society, London, Special Publications, **275**, 125–142.

COPPER, P. 1977. Paleolatitudes in the Devonian of Brazil and the Frasnian–Famennian mass extinctions. *Palaeogeography, Palaeoclimatology, Palaeoecology*, **21**, 165–207.

COPPER, P. 1986. The Frasnian–Famennian mass extinction event and cold water oceans. *Geology*, **14**, 835–839.

COPPER, P. 1994. Ancient reef ecosystem expansion and collapse. *Coral Reefs*, **13**, 3–11.

COPPER, P. 2002*a*. Reef development at the Frasnian/Famennian mass extinction boundary. *Palaeogeography, Palaeoclimatology, Palaeoecology*, **181**, 27–65.

COPPER, P. 2002*b*. Silurian and Devonian reefs: 80 million years of global greenhouse between two ice ages. *In*: KIESSLING, W., FLÜGEL, E. & GOLONKA, J. (eds) *Phanerozoic Reef Patterns. SEPM, Special Publications*, **72**, 181–238.

COPPER, P. & EDINGER, E. 2004. Fossil reefs on Banks Island. *Annual Report Research Monitoring National Parks Western Arctic*, **2003**, 21–22.

COPPER, P. & SCOTESE, C. R. 2003. Megareefs in Middle Devonian supergreenhouse climates. *In*: CHAN, M. A. & ARCHER, A. W. (eds) *Extreme Depositional Environments: Mega End Members in Geologic Time*. Geological Society of America, Special Paper, **370**, 209–230.

CRAIG, B. G. & FYLES, J. G. 1960. *Pleistocene Geology of Arctic Canada*. Geological Survey of Canada, Paper, **60-10**.

EMBRY, A. F. 1988. Middle–Upper Devonian sedimentation in the Canadian arctic islands and the Ellesmerian Orogeny. *In*: MCMILLAN, N. J., EMBRY, A. F. & GLASS, D. J. (eds) *Devonian of the World*. Canadian Society of Petroleum Geologists, Memoir, **14**(2), 15–28.

EMBRY, A. F. 1991. Middle-Upper Devonian clastic wedge of the arctic islands. *In*: TRETTIN, H. P. (ed.) *Geology of the Innuitian Orogen and Arctic Platform of Canada and Greenland*. Geological Survey of Canada, Geology of Canada, **3**, 263–279.

EMBRY, A. F. & KLOVAN, J. E. 1971. A Late Devonian reef tract on northeastern Banks Island, N. W. T. *Bulletin of Canadian Petroleum Geology*, **19**, 730–781.

EMBRY, A. F. & KLOVAN, J. E. 1976. The Middle-Upper Devonian clastic wedge of the Franklinian Geosyncline. *Bulletin of Canadian Petroleum Geology*, **24**, 485–639.

GLENNER, H., THOMSEN, P. F., HEBSGAARD, M. B., SORENSEN, M. V. & WILLERSLEV, E. 2006. The origin of insects. *Science*, **314**, 1883–1884.

GRIMALDI, D. & ENGEL, M. S. 2005. *Evolution of the Insects*. Cambridge University Press, Cambridge.

HART, D. E. & KENCH, P. S. 2007. Carbonate production of an emergent platform, Warraber Island, Torres Strait, Australia. *Coral Reefs*, **26**, 53–68.

HILLS, L. V., SMITH, R. E. & SWEET, A. R. 1971. Upper Devonian megaspores, northeastern Banks Island, N. W. T. *Bulletin of Canadian Petroleum Geology*, **19**, 799–817.

ISAACSON, P. E., HLADIL, J., SHEN, J. W., KALVODA, J. & GRADER, G. 1999. Late Devonian (Famennian) glaciation in South America and marine offlap on other continents. *Abhandlungen der Geologischen Bundesanstalt Wien* **54**, 239–257.

ISERN, A. R., MCKENZIE, J. A. & FEARY, D. A. 1996. The role of sea-surface temperature as a control on carbonate platform development in the western Coral Sea. *Palaeogeography, Palaeoclimatology, Palaeoecology*, **124**, 247–272.

JACKSON, D. J., MACIS, L., REITNER, J., DEGNAN, B. M. & WOERHEIDE, G. 2007. Sponge paleogenomics reveals an ancient role for carbonic anhydrase in skeletogenesis, *Science*, **316**, 1893–1894.

JOHNSON, J. G., KLAPPER, G. & SANDBERG, C. A. 1985. Devonian eustatic sealevel fluctuations in Euramerica. *Geological Society of America Bulletin*, **96**, 567–587.

KLOVAN, J. E. & EMBRY, A. F. 1971. Upper Devonian stratigraphy, northeastern Banks Island, N. W. T. *Bulletin of Canadian Petroleum Geology*, **19**, 705–729.

MCGREGOR, D. C. & UYENO, T. T. 1972. *Devonian spores and conodonts of Melville and Bathurst islands, District of Franklin*. Geological Survey of Canada, Paper, **71-13**.

MCLEAN, R. A. & KLAPPER, G. A. 1998. Biostratigraphy of Frasnian (Upper Devonian) strata in western Canada, based on conodonts and rugose corals. *Bulletin of Canadian Petroleum Geology*, **46**, 515–563.

MANNING, T. H. 1956. Narrative of a second Defense Research Board Expedition to Banks Island, with notes on the country and history. *Journal of the Arctic Institute of North America*, **9**, 3–77.

MIALL, A. D. 1976. Proterozoic and Paleozoic geology of Banks Island, arctic Canada. *Geological Survey of Canada Bulletin*, **258**, 1–77.

MIALL, A. D. 1979. *Mesozoic and Tertiary geology of Banks Island, Arctic Canada: History of an unstable cratonic margin*. Geological Survey of Canada, Memoir, **387** (Mercy map sheet, 1455A).

RETALLACK, G. J. 1985. Fossils soils as grounds for interpreting the advent of large plants and animals on land. *Philosophical Transactions of the Royal Society of London, Biological Sciences*, **B 309**, 105–142.

REES, S. A., OPDYKE, B. N., WILSON, P. A. & HENSTOCK, T. J. 2007. Significance of *Halimeda* bioherms to the global carbonate budget based on a geological sediment budget for the northern Great Barrier Reef, Australia. *Coral Reefs*, **26**, 177–188.

RIES, J. B., STANLEY, S. M. & HARDIE, L. A. 2006. Scleractinian corals produce calcite, and grow more slowly, in artificlal Cretaceous seawater. *Geology*, **34**(7), 525–528.

ROYER, D. L., BERNER, R. A., MONTANEZ, I. P., TABOR, N. J. & BEERLING, D. J. 2004. CO_2 as a primary driver of Phanerozoic climate. *GSA Today*, **14**(3), 4–10.

SANDBERG, P. A. 1983. Oscillating trend in Phanerozoic nonskeletal carbonate mineralogy. *Nature*, **305**, 198–222.

SCHRAG, D. P. & LINSLEY, B. K. 2002. Corals, chemistry and climate. *Science*, **296**, 277–278.

SHUBIN, N. H., DAESCHLER, E. B. & JENKINS, F. A. 2006. The pectoral fin of *Tiktaalik roseae* and the origin of the tetrapod limb. *Nature*, **440**, 747–749.

SORAUF, J. E. & PEDDER, A. E. H. 1986. Late Devonian rugoase corals and the Frasnian-Famennian crisis. *Canadian Journal of Earth Sciences*, **23**, 1265–1287.

STANLEY, S. M. 1988. Paleozoic mass extinctions: Shared patterns suggest global cooling as a common cause. *American Journal of Science*, **288**, 344–352.

STANLEY, S. M. & HARDIE, L. A. 1999. Hypercalcification: Paleontology links plate tectonics and geochemistry to sedimentology, *GSA Today*, 9–17.

STEIN, W. E., MANNOLINI, F., HERNICK, L. V., LANDING, E. & BERRY, C. M. 2007. Giant cladoxylopsid trees resolve the enigma of the Earth's earliest forest stumps. *Nature*, **446**, 904–907.

STEUBER, T. & VEIZER, J. 2002. Phanerozoic record of plate tectonic control of seawater chemistry and carbonate sedimentation. *Geology*, **30**, 1123–1126.

THORSTEINSSON, R. & TOZER, T. E. 1962. *Banks, Victoria and Stefansson islands, arctic archipelago*. Geological Survey of Canada, Memoir, **22**, 1–85.

THORSTEINSSON, R. & TOZER, T. E. 1970. Geology of the arctic archipelago. *In*: DOUGLAS, R. J. W. (ed.) *Geology and Economic Minerals of Canada*. Economic Geology Report, **1**, Geological Survey of Canada, Ottawa, 549–590.

VEIZER, J., GODDERIS, Y. & FRANCOIS, L. M. 2000. Evidence for decoupling of atmospheric CO_2 and climate during the Phanerozoic eon. *Nature*, **408**, 698–701.

WILDER, R. 1989. Neue Ergebnisse zum oberdevonischen Riffsterben am Nordrand des mitteleuropaeischen Variscikums. *Fortschritte Geologie Rheinland Westfalen* **35**, 57–74.

WILDER, R. 1994. Death of Devonian reefs – implications and further investigations. *Courier Forschungsinstitut Senckenberg*, **172**, 241–247.

Parasites in Emsian–Eifelian *Favosites* (Anthozoa, Tabulata) from the Holy Cross Mountains (Poland): changes of distribution within colony

M. K. ZAPALSKI

Faculty of Geology, Warsaw University Żwirki i Wigury 93, 02–089 Warszawa, Poland and Laboratoire de Paléontologie stratigraphique FLST and ISA, UMR 8157 'Géosystèmes' du CNRS. 41, rue du Port, 59046 Lille cedex, France

Present address: Institute of Paleobiology, PAS, Twarda 51/55, 00-818 Warszawa, Poland (e-mail: m.zapalski@twarda.pan.pl)

Abstract: Organisms of unknown biological affinities, assigned to the genus *Chaetosalpinx*, are known to infest Palaeozoic tabulate corals and stromatoporoids. Analysis of distribution of these parasites, performed on Emsian–Eifelian material of *Favosites goldfussi* (Anthozoa, Tabulata) from the Northern Region of the Holy Cross Mountains (Poland), shows that parasites were absent in the early astogenetical stages, and that during astogeny both the absolute number of parasites per colony and the number of parasites per polyp were increasing. The latter can reach 2.7 parasites per polyp. Preferred settling places are in corallite corners (junction of three individuals), but dense infestation also produced settlement in the corallite walls (between two individuals). Probable causes of the increase are insufficient protection by host's cnidae, insufficient immune system response, and parasite ability to adapt to the host's defences.

Endobionts belonging to the genus *Chaetosalpinx* (and other closely related genera such as *Helicosalpinx*) inhabited coralla of various tabulate corals (Tapanila 2005). They were originally considered as their commensals (e.g. Oekentorp 1969), but Stel (1976) suggested that they may be parasites. A recent analysis (Zapalski 2004, 2007) has shown that the relation between them and their hosts was parasitic rather than commensal. Howell (1962) proposed placing them among serpulids, but the biological affinity of these endobionts remains unknown.

Although these organisms were often described and illustrated (e.g. Oekentorp 1969; Plusquellec 1968*a*, *b*; Tapanila 2002), very little attention was paid to their distribution within the host's corallum. Tapanila (2004) analysed Ordovician *Helicosalpinx* infesting *Calapoecia* and *Columnopora* tabulates, and stated that there is no particular pattern in the distribution of the endobionts within the host's corallum.

This paper attempts to answer the question of whether the number of parasites increased in relation to the number of individuals during astogeny, or remained constant. Such a study has never been undertaken before. The increase of the number of parasites in relation to the number of polyps could be an effect of insufficient protection by host nematocysts, and/or insufficient immune system response. This analysis uses Emsian–Eifelian material of *Favosites goldfussi* d'Orbigny infested by *Cheatosalpinx ferganensis* Sokolov from the northern part of the Holy Cross Mountains, Poland.

Material and methods

The analysed material consists of three coralla of *Favosites goldfussi* infested by *Chaetosalpinx* parasites (infested coralla are rare and therefore the material is limited). They come from the vicinity of Grzegorzowice in the Łysogóry Region (Northern Region) of the Holy Cross Mountains (for a map see Zapalski 2005; for description of this material of *Chaetosalpinx* see Zapalski 2007). The Emsian–Eifelian mudstones of Grzegorzowice Beds (*serotinus–patulus* conodont zones according to Malec & Turnau 1997; see also Halamski & Racki 2005) contain numerous rugose and tabulate corals (Stasińska 1954, 1958; Zapalski 2005) and other macrofossils (Pajchlowa 1957).

Three specimens are analysed here. Two were collected in the late 1940s to early 1950s and used by Stasińska (1958) for taxonomic work on favositids; the third was collected by the present author in summer 2006. The collection of A. Stasińska is dispersed, and therefore it is not possible to conclude how many coralla are infested and what is the ratio between infested/uninfested coralla. The material is partly housed at the Faculté libre des Sciences et Technologies, Lille (GFCL), and partly at the Institute of Palaeobiology, Polish Academy of Sciences, Warsaw (ZPAL).

Corallа were sectioned perpendicularly to the axis of corallum growth. Each corallum provided several thin sections, spaced 5–12 mm (altogether

From: KÖNIGSHOF, P. (ed.) *Devonian Change: Case Studies in Palaeogeography and Palaeoecology*. The Geological Society, London, Special Publications, **314**, 125–129.
DOI: 10.1144/SP314.6 0305-8719/09/$15.00

19 thin sections); additionally acetate peels were made in order to record changes in morphology (altogether seven peels). On each thin section the number of corallites and parasites (*Chaetosalpinx*) was counted; parasite prevalence (P) was calculated as follows (calculation method introduced here):

$$P = T_p/C$$

where C stands for number of complete cross-sections of corallites in the thin section and T_p for the total number of parasites in the thin section. The 'prevalence' in epidemiology is a term indicating ratio of infested individuals to total number of individuals (e.g. Palm & Klimpel 2008; Vathsala *et al.* 2008). The formula presented here is adapted to colonial animals, where several individuals are parasitized at once and parasites are shared between individuals.

The analysed material is rather poorly preserved: central parts of coralla are calcified, and precise observations cannot always be performed. Counted numbers of parasites should be treated as approximate, and probably the true P values are slightly higher (in recrystallized parts of coralla the corallites can be counted, but the *Chaetosalpinx* tubes are invisible).

Results

Serial thin sections

The T_p values calculated for each thin section for two coralla are given in Tables 1 and 2. The third corallum (specimen ZPAL T.II/54) is hemispherical, measuring 25 (height) × 40 × 46 mm, and displayed only two parasites (*Chaetosalpinx*) in the late stages of growth. The gradual increase in the number of parasites during astogeny in the two specimens is illustrated in Figures 1 and 2.

Table 1. *Parasite infestation in specimen GFCL 2174**

Thin section no.	Distance from bottom of corallum (mm)	C	T_p	P
GFCL 2174a	10	–	0	0
GFCL 2174b	23	107	11	0.103
GFCL 2174c	31	171	52	0.304
GFCL 2174d	40	255	188	0.737
GFCL 2174e	47	250	160	0.640
GFCL 2174f	54	321	192	0.598
GFCL 2174g	58	414	242	0.585
GFCL 2174h	65	369	419	1.136

*Morphology: bulbo-columnar; size: 71 (height) × 52 × 66 mm.

Table 2. *Parasite infestation in specimen ZPAL T.II/17**

Thin section no.	Distance from bottom of corallum (mm)	C	T_p	P
ZPAL T.II/17a	13	36	7	0.194
ZPAL T.II/17b	19	161	21	0.130
ZPAL T.II/17c	25	317	192	0.606
ZPAL T.II/17d	29	248	345	1.391
ZPAL T.II/17e	32	312	332	1.064
ZPAL T.II/17f	37	83	227	2.735

*Morphology: discoidal; size: 41 (height) × 56 × 82 mm.

Patterns of parasite distribution with in the host's corallum

In these cross-sections the parasites do not display any general preference for location in either the centre or peripheries of the corallum, but there is variation of the location of parasites relative to individuals of the host. In the early astogenetic stages parasites prefer settlement in the corners of corallites (at which three individuals join). Conversely, in the late astogenetic stages, they seem to occupy both corners and walls between adjacent individuals.

Discussion

Any comparison of the parasite infestation in fossil and modern cnidarians will remain more or less speculative, as the interactions between colonial hosts and their parasites are poorly understood (Hill & Okamura 2007). Nonetheless it is worth trying to understand what might be the factors controlling the infestation.

The distribution of the *Chaetosalpinx* parasites within both of the analysed coralla shows three important features: (1) absence or scarcity of parasites during the early stages of astogeny; (2) increase in the total number of parasites during astogeny; (3) increase in number of parasites per corallite (T_p) during astogeny.

The increase in the number of parasites does not seem to be unusual, as long as the number of individuals in the colony increases. Newly growing coral individuals created further space for the parasites to infest the colony. The decreased number of parasites in certain thin sections is caused by sectioning a smaller area of corallum (as is the case in the sample ZPAL T.II/17 in sections close to the corallum surface, see Fig. 2) or recrystallization, excluding some areas of the thin section from counting (where it was possible to count the corallites, but not the parasites, as in the case of sample GFCL 2174, see Fig. 2). Therefore, as stated above, the

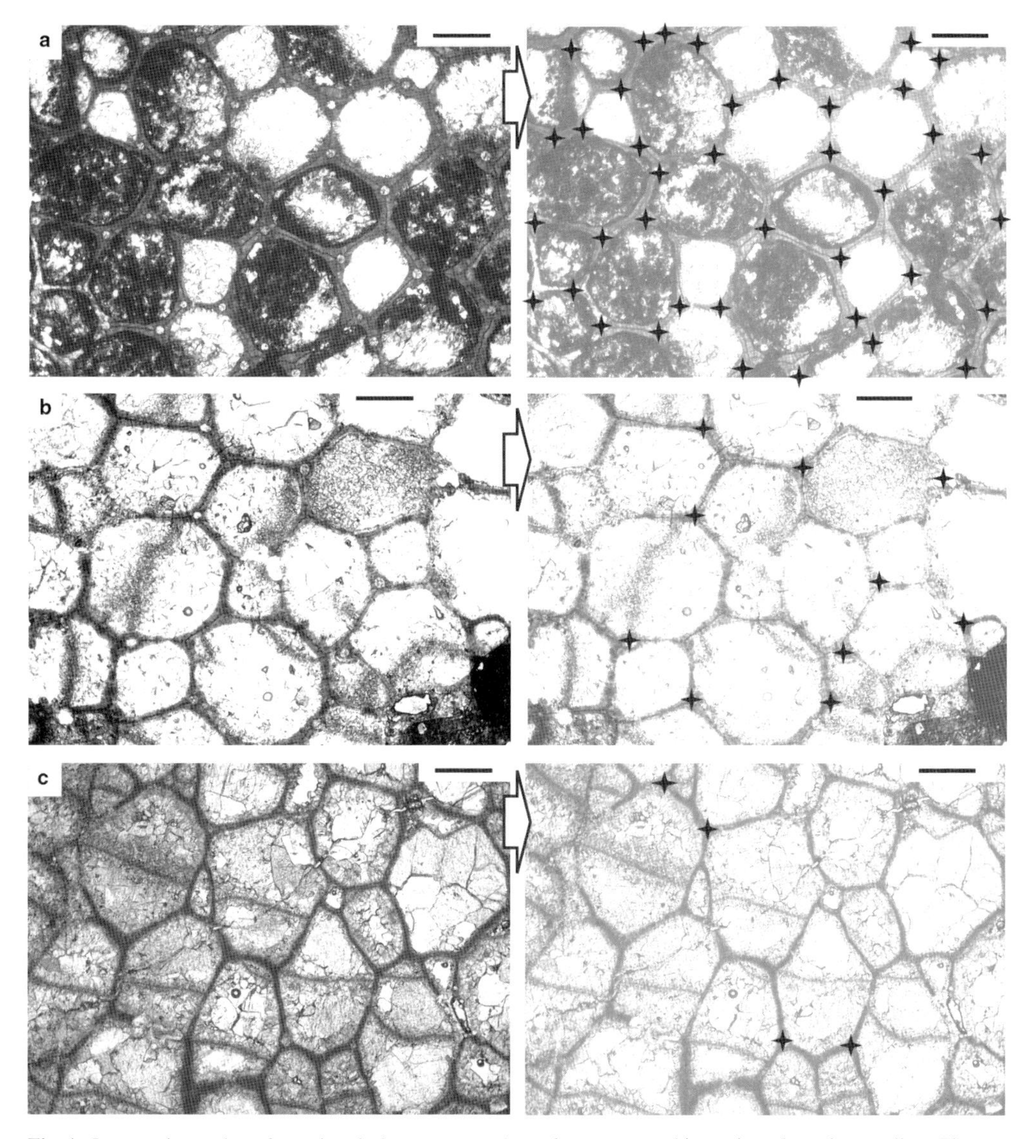

Fig. 1. Increase in number of parasites during astogeny, shown by transverse thin sections through a corallum. Pictures on the left show thin sections, and on the right the same thin sections with crosses marking the positions of *Chaetosalpinx* parasites. (**a**) Thin section GFCL 2174h, 65 mm from the bottom of the corallum (6 mm from the top); (**b**) thin section GFCL 2174d, 40 mm from the bottom of the corallum; (**c**) thin section GFCL 2174a, 10 mm from the bottom of the corallum. Note the strong increase in number of parasites. Grzegorzowice Beds, Emsian–Eifelian; Grzegorzowice, Holy Cross Mountains, Poland. Scale bar 1 mm.

true P values may be higher than the 2.7 calculated in the analysis. Research on modern bryozoans has shown that the high number of parasites in colonial animals is correlated with their low virulence (Hill & Okamura 2007); this may be similar to the discussed case.

The increasing number of parasites per individual (parasite prevalence) during astogeny may be a sign of two different processes: (1) insufficient protection by cnidae; (2a) insufficient response from the coral immune system or (2b) ability of the parasite to adapt to the immune response of the host, and therefore (2a). The insufficient protection by cnidae was probably the first factor allowing infestation. Cnidarians have no specialized immune cells (Kuznetsov & Bosch 2003), therefore the insufficient immune response seems to be highly possible; however, some modern corals do appear

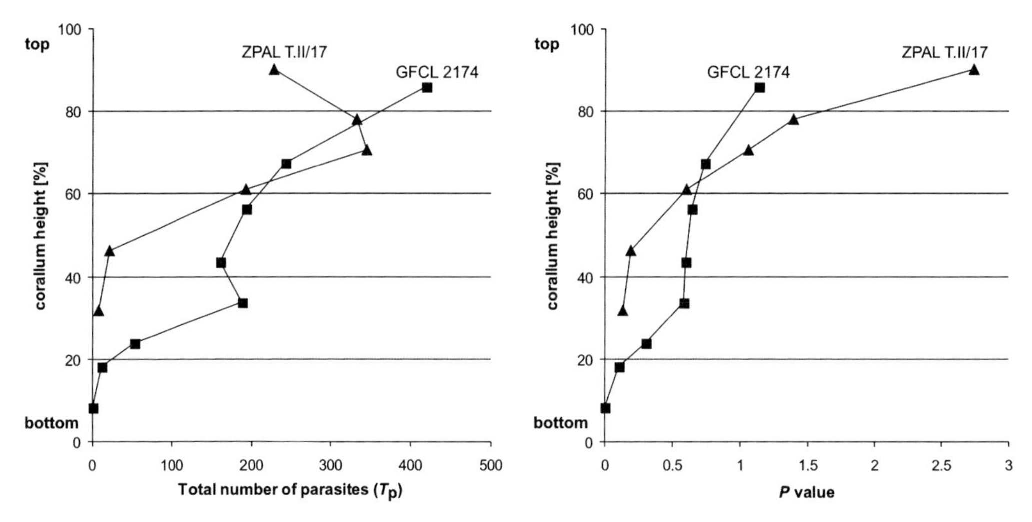

Fig. 2. Diagrams showing the increase of the total number of parasites (left) and the parasite prevalence (right) plotted over the corallum height (standardized). The decrease in the total number of parasites in specimen ZPAL T.II/17 is caused by the smaller area of thin sections, while the decrease in their number in specimen GFCL 2174 is caused by local recrystallization of specimen. Note the systematic increase of parasite prevalence during astogeny (right diagram).

to have a well developed self-defence system (Chadwick-Furman & Rinkevich 1994), at least in the terms of self/non-self recognition. The increase in the number of parasites per individual might indicate that in the investigated *Favosites* the immune system was undeveloped. Parasite adaptation to the general host defences is not uncommon among modern organisms (Schulenburg *et al.* 2007), but it is correlated with immune systems unable to create a new immune answer to the changing strategy of the parasite.

Comparisons with parasites of modern corals are problematic. Digenean meacercariae, copepods and cirripedians are the most common parasites of recent scleractinians (e.g. Cheng & Wang 1974; Humes 1986; Ross & Newman 1995; Simon-Blecher & Achituv 1997; Aeby 2007). These parasites are usually less numerous (several individuals within a colony) and much larger (Aeby 2003), and their distribution cannot be compared with that of *Chaetosalpinx*. The only modern analogue are parasites of *Favia* occurring between corallites; their dynamics of infestation remain, however, unknown (Rosen 1968).

The preferred location of the parasite seems to be in the corners of corallites. On the other hand, when the infestation attains high T_p values, parasites seem to occupy more frequently the walls between adjacent individuals. This was probably an effect of prior occupation of the corners: the newcomers had to settle in the corallum parts that were not occupied yet. The absence of preferred zones of settlement within the corallum (e.g. in the centre or periphery) is similar to that observed by Tapanila (2004).

Conclusions

The absolute number of *Chaetosalpinx* parasites within a single corallum increased during astogeny.

The number of *Chaetosalpinx* parasites per individual in the host colony increased during astogeny and could reach 2.7 parasites per polyp; such a high infestation may indicate insufficient protection by cnidae and insufficient immune system response.

Chaetosalpinx parasites did not prefer any particular location within the corallum (neither centre nor periphery). On the other hand, their placement in relation to individuals is usually in corallite corners (where three individuals join), but they also occur in the walls between individuals (where two individuals join); the latter situation seems to be more common in late astogeny stages.

I wish to express gratitude to Bruno Mistiaen (Lille) and Stefan Schröder (Köln) for valuable comments on the manuscript. Adam T. Halamski (Warszawa) and Benoît Hubert (Lille) discussed an early draft of the text. I am very grateful to John Brenner (Wokingham) who improved my English. The Foundation for Polish Science is thanked for the scholarship funding.

References

AEBY, G. S. 2003. Corals in the genus *Porites* are susceptible to infection by a larval trematode. *Coral Reefs*, **22**, 216.

AEBY, G. S. 2007. Spatial and temporal patterns of *Porites trematodiasis* on the reefs of Kaneohe Bay, Oahu, Hawaii. *Bulletin of Marine Science*, **80**, 209–218.

CHADWICK-FURMAN, N. & RINKEVICH, B. 1994. A complex allorecognition system in a reef-building coral: Delayed responses, reversals and nontransitive hierarchies. *Coral Reefs*, **13**, 57–63.

CHENG, T. C. & WONG, A. K. L. 1973. Chemical, histochemical, and histopathological studies on corals, *Porites* spp., parasitized by tremato de metacercariae. *Journal of Invertebrate Pathology*, **23**, 303–317.

HALAMSKI, A. T. & RACKI, G. 2005. [R 220 di 05, R 220 dm 05]. *In*: WEDDIGE, K. (ed.) Devonian Correlation Table. *Senckenbergiana lethaea*, **85**, 192–195.

HILL, S. L. L. & OKAMURA, B. 2007. Endoparasitism in colonial hosts: Patterns and processes. *Parasitology*, **134**, 841–852.

HOWELL, B. F. 1962. Worms. *In*: MOORE, R. C. (ed.) *Treatise on Invertebrate Paleontology. Part W. Miscellanea*. Geological Society of America and University of Kansas Press, Lawrence, W144–W177.

HUMES, A. G. 1986. Two new species of *Cerioxynus* (Copepoda: Poecilostomatoida) parasitic in corals (Scleractinia: Faviidae) in the South Pacific. *Systematic Parasitology*, **8**, 187–198.

KUZNETSOV, S. G. & BOSCH, T. C. G. 2003. Self/nonself recognition in Cnidaria: Contact to allogenetic tissue does not result in elimination of nonself cells in *Hydra vulgaris*. *Zoology*, **106**, 109–116.

MALEC, J. & TURNAU, E. 1997. Middle Devonian conodont, ostracod and miospore stratigraphy of the Grzegorzowice-Skały section, Holy Cross Mountains, Poland. *Bulletin of the Polish Academy of Sciences, Earth Sciences*, **45**, 67–86.

OEKENTORP, K. 1969. Kommensalismus bei Favositiden. *Münstersche Forschungen zur Geologie und Paläontologie*, **12**, 165–217.

PAJCHLOWA, M. 1957. Dewon w profilu Grzegorzowice-Skały. *Biuletyn Instytutu Geologicznego*, **122**, 145–254.

PALM, H. W. & KLIMPEL, S. 2008. Metazoan fish parasites of *Macrourus berglax* Lacepède, 1801 and other macrourids of the North Atlantic: Invasion of the deep sea from the continental shelf. *Deep Sea Research II*, **55**, 236–242.

PLUSQUELLEC, Y. 1968*a*. Commensaux des Tabulés et Stromatoporoïdes du Dévonien armoricain. *Annales de la Société Géologique du Nord*, **88**, 47–56.

PLUSQUELLEC, Y. 1968*b*. De quelques commensaux de Coelentérés paléozoïques. *Annales de la Société Géologique du Nord*, **88**, 163–171.

ROSEN, B. R. 1968. An account of a pathologic structure in the Faviidae (Anthozoa): a revision of *Favia valenciennesii* (Edward & Haime) and its allies. *Bulletin of the British Museum (Natural History), Zoology*, **16**, 325–362.

ROSS, A. & NEWMAN, W. A. 1995. A coral-eating barnacle, revisited (Cirripedia, Pyrgomatidae). *Contributions to Zoology*, **65**, 129–175.

SCHULENBURG, H., BOEHNISCH, C. & MICHIELS, N. K. 2007. How do invertebrates generate a highly specific innate immune response? *Molecular Immunology*, **44**, 3338–3344.

SIMON-BLECHER, N. & ACHITUV, Y. 1997. Relationship between coral pit crab *Cryptochirus coralliodytes* Heller and its host coral. *Journal of Experimental Marine Biology and Ecology*, **215**, 93–102.

STASIŃSKA, A. 1954. Koralowce Tabulata z dewonu Grzegorzowic (badania wstępne). *Acta Geologica Polonica*, **4**, 277–290.

STASIŃSKA, A. 1958. Tabulata, Chaetetida et Heliolitida du Dévonien Moyen des Monts de Sainte-Croix. *Acta Palaeontologica Polonica*, **3**, 161–282.

STEL, J. H. 1976. The Palaeozoic hard substrate trace fossils *Helicosalpinx*, *Chaetosalpinx* and *Torquaysalpinx*. *Neues Jahrbuch für Geologie und Paläontologie, Monatshefte*, **1976**, 726–744.

TAPANILA, L. 2002. A new endosymbiont in Late Ordovician tabulate corals from Anticosti Island, eastern Canada. *Ichnos*, **9**, 109–116.

TAPANILA, L. 2004. The earliest *Helicosalpinx* from Canada and the global expansion of commensalism in Late Ordovician sarcinulid corals (Tabulata). *Palaeogeography Palaeoclimatology, Palaeoecology*, **215**, 99–110.

TAPANILA, L. 2005. Palaeoecology and diversity of endosymbionts in Palaeozoic marine invertebrates: Trace fossil evidence. *Lethaia*, **38**, 89–99.

VATHSALA, M., MOHAN, P., SACIKUMAR, & RAMESSH, S. 2008. Survey of tick species distribution in sheep and goats in Tamil Nadu, India. *Small Ruminant Research*, **74**, 238–242.

ZAPALSKI, M. K. 2004. Parasitism on favositids (Tabulata). *Palaeontology Newsletter*, **57**, 194.

ZAPALSKI, M. K. 2005. A new species of Tabulata from the Emsian of the Holy Cross Mts., Poland. *Neues Jahrbuch für Geologie und Paläontologie, Monatshefte*, **2005**, 248–256.

ZAPALSKI, M. K. 2007. Parasitism versus commensalism – the case of tabulate endobionts. *Palaeontology*, **50**, 1375–1380.

Devonian (Emsian to Frasnian) crinoids of the Dra Valley, western Anti-Atlas Mountains, Morocco

G. D. WEBSTER[1]* & R. T. BECKER[2]

[1]*Department of Geology, Washington State University, Pullman, WA, USA 99164-2812*

[2]*Geologisch-Paläontologisches Institut, Westfälische Wilhelms-Universität, Corrensstr. 24, D-48149 Münster, Germany*

**Corresponding author (e-mail: webster@wsu.edu)*

Abstract: Emsian to Frasnian crinoids are described from pelagic facies of the eastern part of the Dra Valley (Tata area), western Anti-Atlas Mountains, southern Morocco. The crinoids show only minor relationship with previously described crinoids from the Tafilalt and Ma'der areas of the eastern Anti-Atlas. The differences are judged to reflect the different environments of these areas. The Dra Valley hexacrinids show greater affinity with European faunas, whereas the amabilicrinids show greater affinity with North American taxa. New taxa described are *Hexacrinites chenae* sp. nov., *Dracrinus crenulatus* gen. and sp. nov., *Coquinacrinus revimentus* gen. and sp. nov, and *Embolocrinus quadruus* gen. and sp. nov.

The Dra Valley of southern Morocco is a major drainage extending almost 600 km; it runs parallel to the south side of the western part of the Anti-Atlas Mountains from near Zagora in the NE to Tan-Tan near the Atlantic Ocean in the SW. Devonian strata are exposed along the Dra Valley over much of this area and form spectacular outcrops especially in the northeastern half of the valley (Fig. 1). Early stratigraphic studies of the Dra Valley began in the 1920s (Gentil 1929) and continued into the 1940s (Descossy & Roch 1934; Bondon & Clariond 1934; Bourcart 1938; Choubert & Marcais 1948). A more detailed and systematic survey, including formal lithostratigraphy and international correlations, were subsequently based on numerous studies by the famous French geologist H. Hollard (e.g. Hollard 1963, 1967, 1978, 1981*a*, *b*; Hollard & Jaquemont 1956). The geographic position along the disputed Moroccan–Algerian border prevented scientific progress for several decades. But the area, with its laterally interfingering and vertically alternating neritic and pelagic sediments, is of prime importance for the correlation of Devonian nearshore and offshore facies and faunas, one of the prime targets of IGCP 499 on 'Devonian Land–Sea Interaction: Evolution of Ecosystems and Climates'. Based on detailed investigations by several working groups, adavances in the Devonian stratigraphy of the Dra Valley were presented at a joint IGCP 499 and SDS (International Subcommission on Devonian Stratigraphy) field trip (El Hassani 2004). The most recent overviews of Dra Valley Devonian facies development, litho- and biostratigraphy, were given by Becker *et al.* (2004*a*) and Jansen *et al.* (2007).

Termier & Termier (1950) reported the first unquestioned crinoids from Morocco. Subsequently, they have been reported from all series of the Devonian (Le Maitre 1958; Termier & Termier 1974; Le Menn 1992; Le Menn & Regnault 1993; Klug *et al.* 2003; Webster *et al.* 2005). Although most of the reports described disarticulated cup plates and stem ossicles, cup and crowns were described from pelagic sediments of the Ma'der and Tafilalt areas of southeastern Morocco by Webster *et al.* (2005). Most of the Ma'der and Tafilalt Famennian crinoids were thought to have been attached to logs. It is not yet known which taxa produced locally very extensive and thick Upper Givetian, Middle Frasnian or Middle Famennian encrinites of the Tafilalt.

Preceding the SDS field conference in 2004, field investigations in 2001 and 2003 by research teams of the Westfälische Wilhelms-Universität, Münster, discovered crinoid cups in encrinitic beds of the Anoû Smaira Formation of the Tata region in the eastern Dra Valley. In addition, numerous crinoid stem ossicles and pluricolumnals were found in various horizons within the Frasnian succession, as reported in the Devonian field guide (El Hassani 2004). During the field conference additional crinoid cups and crowns were discovered in older horizons in the Devonian as well as additional specimens in the Anoû Smaira Formation. The present paper is based on all of these specimens. They are described and related to previously described crinoids from the Ma'der and Tafilalt areas of southeastern Morocco as well as other world occurrences. The abundance of mostly disarticulated crinoid remains in many neritic or

From: KÖNIGSHOF, P. (ed.) *Devonian Change: Case Studies in Palaeogeography and Palaeoecology*.
The Geological Society, London, Special Publications, **314**, 131–148.
DOI: 10.1144/SP314.7 0305-8719/09/$15.00

(a)

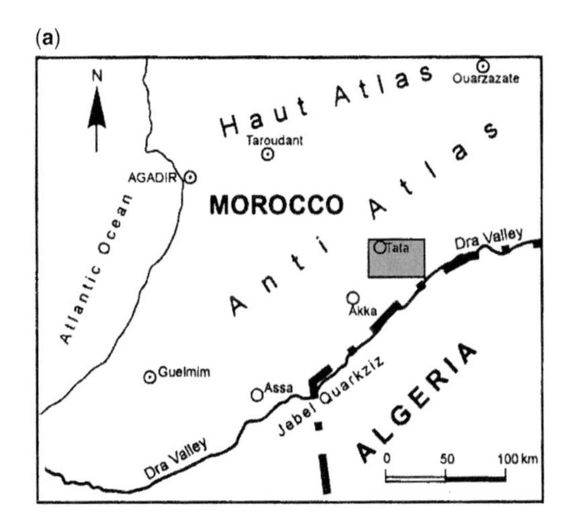

(b)

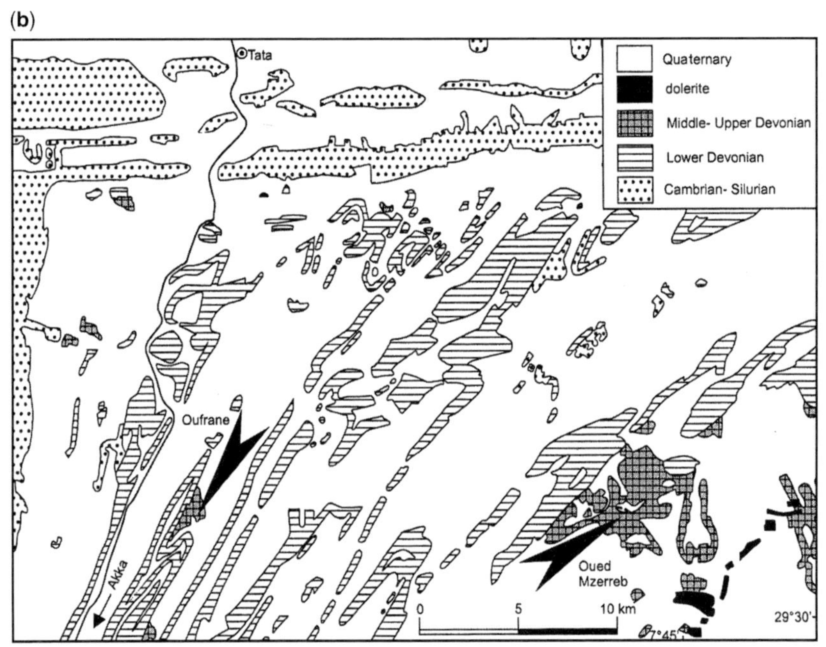

Fig. 1. (**a**) Location of the Tata region in the Anti-Atlas. (**b**) Location of studied Givetian and Frasnian successions at Oued Mzerreb.

pelagic settings shows that knowledge about them could provide an important tool for comparisons and correlations across facies belts. In this respect, it is hoped that this contribution will improve the background for future crinoid-based palaeoecological interpretation of shelf habitats.

Stratigraphy

All new crinoid material comes from the eastern Dra Valley succession which crops out between south of Foum Zguid and south of Tata (Fig. 1). Specimens which can be assigned to new genera and species come from three sections: section Foum Zguid III of Jansen *et al.* (2004) and two sections at Oued Mzerreb (Becker *et al.* 2004*b*).

Foum Zguid III

A group of three Lower Devonian sections lies close to the main road from Agadir to Foum Zguid, *c.* 25 km SW of the latter, and has been investigated in detail by Jansen *et al.* (2004). Brachiopods, dacryoconarids, ostracods, trilobites, conodonts and subordinate goniatites allow a rather fine biostratigraphic dating and a precise placing of the Lower–Middle Devonian boundary. The massive sandstones of the Rich 3 Sandstone Member, at the top of the Mdâour-el-Kbír Formation, forms a widely recognizable marker unit that embraces the Lower–Upper Emsian transition. A massive trilobite limestone forms the transgressive base of the lower Member of the Timrhanrhart Formation, which is overlain by a condensed, massive *Sellanarcestes* Limestone with abundant goniatites and conodonts of the *serotinus* Biozone. Subsequent limestone nodules yielded abundant trilobites (*Phacops, Hollardops*), the tabulate coral *Pleurodictyum* and the new cladid *Dracrinus crenulatus* gen. and sp. nov. The Emsian–Eifelian boundary beds follow slightly higher on the steep slope. Although no typical latest Emsian goniatites have been found locally, it can be assumed that the new crinoid falls in the upper part of the upper Emsian (Lower Devonian IV-D *sensu* Becker & House 2000; *c.* upper part of *serotinus* to *patulus* Conodont Biozones).

Oued Mzerreb West

The Givetian sedimentary and faunal successions at Oued Mzerreb West were first discovered by Bensaid (1974) and more recently have been studied in detail by Aboussalam (2003) and Becker *et al.* (2004*b*). There is a rather wide outcrop area of Devonian rocks at 27 km SE of Tata (Fig. 1). Section West ranges through most of the Givetian within a very shallow anticline structure. All of the succession belongs to the Ahrererouch Formation, which is mostly a pelagic succession of deeply weathered marles, often with pyritic faunas, and intercalated thin limestones and siltstones (Fig. 2). Bed 8a is a characteristic marker unit that resembles the reddish nodular limestones of the Montagne Noire (France), which are called 'griotte'. The 'Lower Red Griotte' differs from under- and overlying beds since it contains an admixture of typical neritic (proetid trilobites, large-eyed phacopids, diverse gastropods, bivalves, brachiopods, rugose and thamnoporid corals) with pelagic elements, such as styliolinids and goniatites. The marker species *Maenioceras decheni* gives a precise age within the Middle Givetian (*decheni* Biozone, Middle Devonian II-C2), just above the Upper Pumilio Limestone, which falls in the basal part of the *ansatus* (= Middle *varcus*) Conodont Biozone. The rather common crinoid remains include a basal circlet attributed to the family Periechocrinidae as well as *Hexacrinites chenae* sp. nov.

A thin siltstone forms the top of Bed 9, which is characterized by a goniatite fauna with common *Agoniatites* but lacking *Maenioceras*. This justified naming it the *Agoniatites* Bed. Only few crinoid remains were found in the main marl but the siltstone (Bed 9b) yielded one melocrinitid basal circlet. Correlating the Oued Mzerreb section and Tafilalt Givetian (Becker *et al.* 2004*b*) it can be suggested that Bed 9b is the last unit within the *ansatus* Biozone before the sea-level rise driving the Taghanic Onlap.

Oued Mzerreb, Frasnian section

Knowledge of the Frasnian of the Dra Valley is still very poor. The section measured in detail several hundred metres above section West (Becker *et al.* 2004*c*) is thus far the only studied Frasnian succession of this vast region. All units are assigned to the Anoû Smaira Formation, which consists of alternating calcareous/marly layers, often with pyritic faunas, thin siltstones, nodular limestones, and levels of large, often rather irregular marl and limestone concretions (Fig. 3). As in the Givetian, a reddish griotte unit, the Upper Red Griotte (Bed 4), forms a local marker level. It yielded the conodont *Avignathus decorosus*, which gives an age not older than Montagne Noire Biozone 10. Overlying dark grey marls are very pyrite-rich sediments with pyretic goniatite faunas (secondarily transformed into hematite) (secondarily hematitic), especially above a level of unfossiliferous limestone concretions (Bed 6). Intercalated are two or three levels of encrinite plates, especially in the higher part of Bed 6, where goniatites are much less frequent than at the base. There are very few other benthic fossils, rare bivalves (*Buchiola* and others), gastropods, rhynchonellids and thamnoporid tabulates. The predominating goniatite is a new species of *Carinoceras*, which gives the name *Carinoceras* Beds. Correlation with the international cephalopod zonation (Becker & House 2000) is better based on the co-occurring genus *Trimanticoceras*, on rare *Aulatornoceras auris*, and on advanced specimens of *Beloceras tenuistriatum* with eight ventral lobes at median size. *Trimanticoceras* first occurs in Germany or Australia in Upper Devonian I-I, at the base of the Upper Frasnian. *A. auris* appears at the same level in Germany, Australia, New York, or in the Timan of Polar Russia. In the well- documented beloceratid succession of the Canning Basin of NW Australia (Becker & House 2009), the first specimens of *Beloceras* with eight ventral lobes were found in the upper part of I-I (regional *Playfordites tripartitus* Biozone). Upper Devonian I-I correlates with the Montagne Noire conodont Biozone 11. Most likely, the *Carinoceras* Beds were deposited during the deepening pulse related to the global *semichatovae* transgression (Becker *et al.* 2004*b*), slightly above the base of MN Biozone 11.

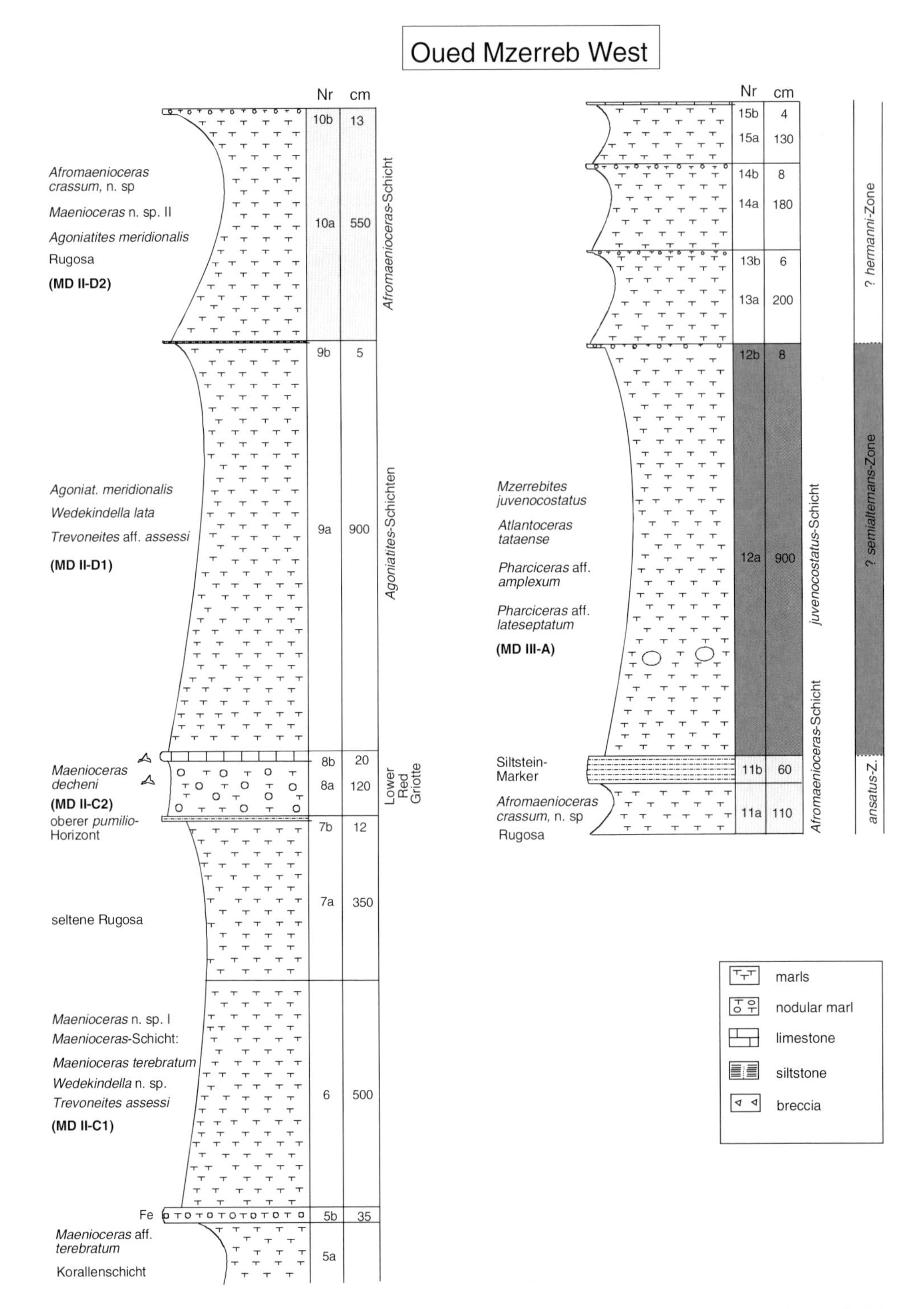

Fig. 2. Lithological log at Oued Mzerreb West, showing the position of crinoid-bearing units (Red Griotte = Bed 8a and thin siltstone unit = Bed 9b).

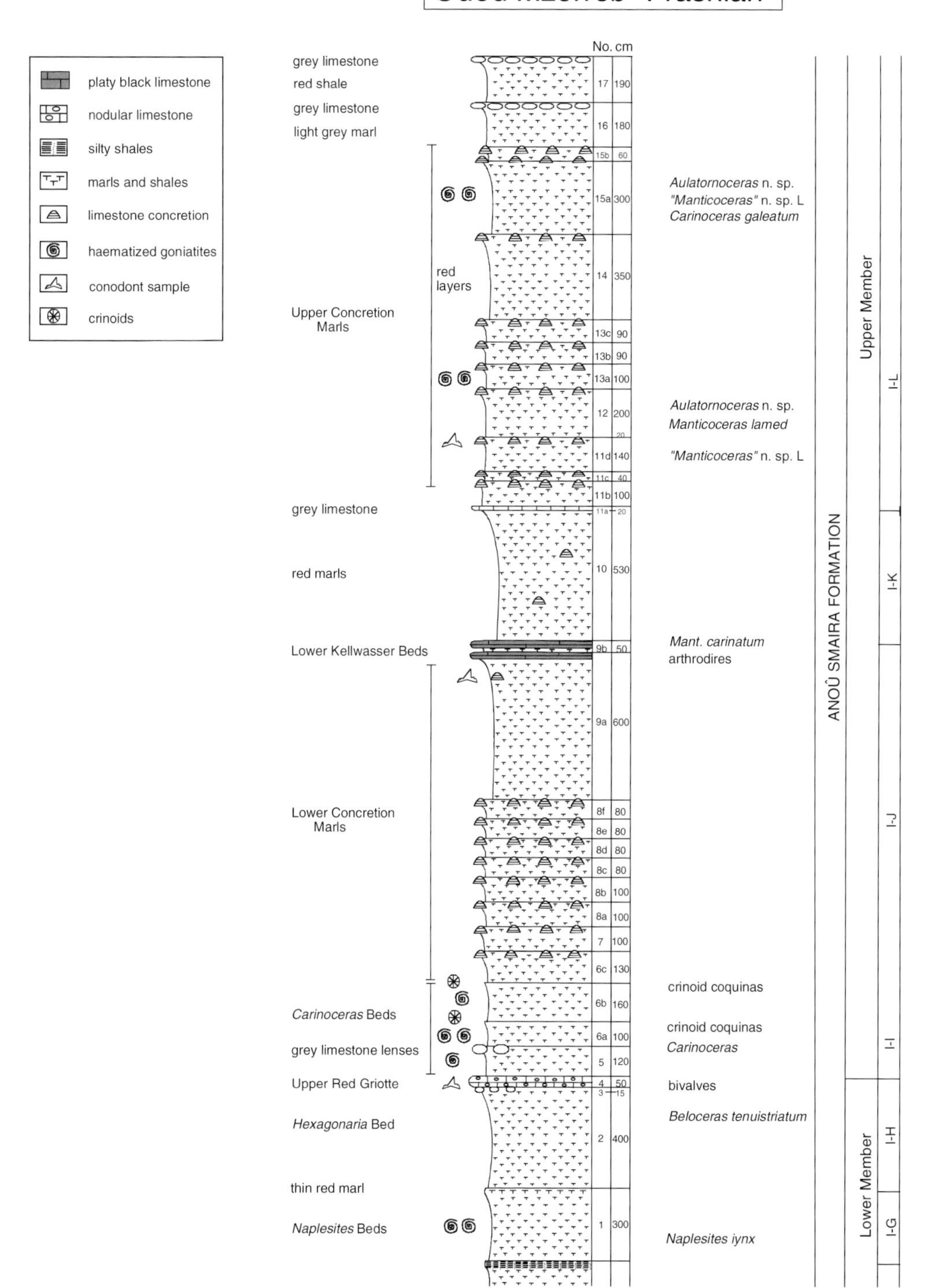

Fig. 3. Lithological log for the Frasnian at Oued Mzerreb, showing the position of crinoid coquinas in the *Carinoceras* Beds (Beds 5 to 6b) of the Anoû Smaira Formation.

Fig. 4. (**a, b**) Melocrinitoidea, genus uncertain, figured specimen GPIM B3B-1/2, oral and aboral views of basal circlet with proximal columnals. (**c**) Melocrinitidae, genus uncertain, figured specimen GPIM B3B-1/3, partial set of arms lacking ramus. (**d**) Melocrinitidae, genus uncertain, figured specimen GPIM B3B-1/4 partial set of arms with ramus. (**e**) *Dracrinus crenulatus* gen. and sp. nov., posterior view, holotype GPIM B3B-1/7. (**f–i**) *Hexacrinites chenae*

Crinoid taphonomy and palaeobiogeography

In the Emsian to Frasnian parts of the Devonian sections in the Dra Valley the crinoids occur in most of the carbonate beds, but mainly as disarticulated stem ossicles and rare cup plates. The few cups and crowns are isolated occurrences within these deposits. They are considerd to have been buried in or nearly *in situ*, with some disaggregation which results from uncertain causes, including wave action, scavengers, bioturbation and compaction.

Cups and crowns in the *Carinoceras* Beds in the basal part of the Anou Smaira Formation occur in two crinoid coquina horizons. Each of these horizons consists of a thin layer (generally between 1 and 2 cm thick) of pluricolumnals (having a length usually less than 4 cm) and deposited in nearly parallel arrangement or forming a jumbled mass of smaller pluricolumnals with a length of less than 1 cm. Cups, crowns and arm fragments were deposited aligned with the elongate pluricolumnals and amongst the jumbled pluricolumnals. The aligned specimens suggest that during deposition the current was unidirectional and jumbled specimens suggest a sudden burying related to a major storm or tectonic event.

The fragmentary remains here referred to the periechocrinids and melocrinitids cannot be attributed to known genera without question. All of the specimens are within the known stratigraphic range of these groups. Nevertheless, this is the first report of either of the periechocrinids or melocrinitids in Morocco, thus extending their palaeogeographic range to the western part of the Devonian Prototethyan seas.

Hexacrinites Austin & Austin (1843) ranges from Middle Silurian into the Early Mississippian in North America (but is relatively rare) and is cosmopolitan in the equatorial belt during its acme from Emsian to Frasnian (Webster 2003). *Hexacrinites* had its highest diversity and abundance during its acme in Germany and Belgium (Hauser 1997, 1999, 2001) but it was relatively rare in the Famennian. Previous occurrences in the Devonian of the Prototethys are from the Lochovian (Jell 1999) and Frasnian (Jell & Jell 1999) of Australia, the Famennian of China (Lane *et al.* 1997; Waters *et al.* 2003), and the Famennian of Iran (Webster *et al.* 2007). The present report of *Hexacrinites chenae* sp. nov. in Morocco extends the genus stratigraphic range downward into the Givetian, the palaeogeographic range to the western part of the Prototethys, and shows greater affinity to European faunas than North American faunas.

Dracrinus crenulatus nov. gen. and sp. is an advanced dendrocrinid with biserial or uniserial wedge-shaped brachials and pinnules (derived characters), but retaining primitive (plesiomorphic) characters in the cup and tegmen. Thus, it displays a mosaic of derived and plesiomorphic characters. Advanced dendrocrinids here refers to cladids previously assigned to the Poteriocrinida. The primitive dendrocrinids reached their acme in the Devonian and are cosmopolitan in the equatorial belt. *Dracrinus* occurs in nodular limestones within mixed neritic–pelagic deposits suggesting a deeper environment.

Amabilicrinids Webster *et al.* (2003) are dendrocrinids that have cuneate or rectilinear brachials bearing pinnules (i.e. a mosaic of derived and plesiomorphic characters). They were reported from Middle Devonian deposits of the eastern United States (Webster *et al.* 2003), Famennian deposits in the Ma'der (Webster *et al.* 2005) and western United States (Webster *et al.* 2003), and early Mississippian of southern Iran and the United States (Webster *et al.* 2003). The occurrence of *Coquinacrinus* gen. nov. and *Embolocrinus* gen. nov. in the Upper Frasnian pelagic deposits of the Dra Valley extends the palaeogeographic range of the amabilicrinids within Morocco and shows that they were present in the region earlier than previously known. The Morocco amabilicrinids may have been the progenitors of some of the Early Mississippian amabilicrinids of Iran, suggesting a seaway existed along the northern coast of Africa during the Middle to Late Devonian. They also show greater affinity with North American faunas than European faunas.

Flexibles were widespread in Palaeozoic crinoid faunas, but usually in small numbers and low diversity. The fragmentary arms found in the *Carinoceras* Beds document their occurrence in the Dra Valley, but provide no significant biostratigraphic or palaeoecologic information.

Amabilicrinids occur both in the Dra Valley and Ma'der pelagic deposits, but these two regions in Morocco share no other common taxonomic groups. This undoubtedly reflects the different environments of these regions. Most of the crinoids

Fig. 4. (*Continued*) sp. nov., holotype GPIM B3B-1/6, oral, A-ray, posterior, and aboral views. (**j**) Periechocrinidae, genus uncertain, aboral view of basal circlet, figured specimen GPIM B3B-1/1. (**k–o**) *Embolocrinus quadruus* gen. and sp. nov: (k) paratype 5, GPIM B3B-1/23 posterior view; (l) paratype 6, GPIM B3B-1/24, posterior view of crown showing quadrate pinnulars; (m) paratype 8 GPIM B3B-1/26, EA or AB interray view; (n) paratype 3, GPIM B3B-1/21, lateral view of uncertain ray orientation showing proximal stem; (o) holotype, GPIM B3B-1/18, lateral view of uncertain orientation showing proximal stem, biserial brachials and quadrate pinnulars.

from the Ma'der and Tafilalt area were presumed to have been attached to logs. Most of the crinoids found in the Dra Valley sediments were probably attached to firm objects in the bottom sediments and were buried in or nearly *in situ*. Only one of the holdfasts of *Coquinacrinus revimentus* gen. and sp. nov. has a grainy texture suggesting that it was attached to a log.

Disarticulated crinoidal ossicles and pluricolumnals are common elements in the carbonate deposits of the Dra Valley and added to the bulk of the sediments in the region. Future studies of these sediments should yield a greater number of taxa from other horizons within the Emsian to Frasnian as well as possibly older and younger sediments. Especially promising in the western Dra Valley (Assa to Torkoz regions) are the upper Emsian *Sellanarcestes* Limestone (middle part of the Khebchia Formation) and lower Eifelian Crinoid Marl Member of the lower part of the Yeraifa Formation (Becker *et al.* 2004*c*).

Systematics

Specimens are reposited in the collections of the Geologisch-Paläontologisches Institut Münster and bear numbers preceded by GPIM. Morphologic terms follow Moore & Teichert (1978), with measurement terminology after Webster & Jell (1999), anal terminology after Webster & Maples (2006), and noditaxis patterns after Webster (1974).

Class CRINOIDEA J. S. Miller 1821
Subclass CAMERATA Wachsmuth & Springer 1885
Order MONOBATHRIDA Moore & Laudon 1943
Superfamily PERIECHOCRINOIDEA Bronn 1849
Family PERIECHOCRINIDAE Bronn 1849
Genus indet.
Text Figure 4J

Figured specimen. GPIM B3B-1/1, basal circlet discovered by Xiu-Qin Chen.

Type locality and horizon. Bed 8a, Red Griotte Beds, Oued Mzerreb Member, Ahrerouch Formation, Middle Givetian, Oued Mzerreb West Section (Becker *et al.* in El Hassani 2004), Tata region, eastern Dra Valley; GPS: N29°33′12″, W7°46′19″.

Description. Basal circlet equally tripartite, hexagonal in basal view, diameter 11.2 mm, length 3.9 mm, proximally thickened base subhorizontal for stem attachment, no flange, distally widely outflaring and gently upflaring. Basals pentagonal, length 5.7 mm, width 10.3 mm, ornament of coarse granulations and three linear ridges: one medial, two lateral each paralleling basal/basal suture, making a double ridge stellate ornament parallel with the sutures and between plates at plate boundaries. Stem facet circular, diameter 4 mm. Axial canal pentastellate, diameter 0.8 mm.

Remarks. The specimen is weathered and has lost some plate edges and surfaces from breakage along cleavage planes. Surface features of the articular facet are lost to solution or abrasion weathering. The combination of the equally tripartite basals and linear ridge ornament suggests relationship to the periechocrinids to which the specimen is tentatively referred. The linear ridge ornament, however, is one found in a number of the camerate genera, including diplobathids and monobathrids as illustrated in the *Treatise on Invertebrate Paleontology* (Moore & Teichert 1978). However, in the Devonian the other camerates with a tripartite basal circlet lack the linear ridge ornament present on the periechocrinids.

Superfamily MELOCRINITOIDEA d'Orbigny 1852
Incertae familiae
Genus indet., basal circlet
Text Figure 4A, B

Figured specimen. GPIM B3B-1/2, a single basal circlet.

Type locality and horizon. Thin siltstone top of *Agoniatites* Beds, Bed 9b, Tigusselt Member, Ahrerouch Formation, Middle Givetian, Oued Mzerreb West Section (Becker *et al.* in El Hassani 2004), Tata region, eastern Dra Valley; GPS: N29°33′12″, W07°46′19″.

Description. Basal circlet low, proximally horizontal, truncated for stem attachment, up- and outflaring distally, length 5.3 mm, diameter at truncation 10 mm, distal diameter 13 mm. Formed of four plates, moderately thick, posterior basal extended distally and truncated for supporting primanal; sutures with basals covered with fine anastomosing ridges. Proximal most three columnals incompletely developed, not fully visible in lateral or linear views. Fourth columnal complete, holomeric, round, 10 mm diameter, latus scalloped laterally, roundly convex linearly. Crenularium wide, bearing fine culmina extending full width of crenularium. Areola wide, smooth. Lumen narrow, pentalobate.

Remarks. Lacking the rest of the theca, generic attribution of this basal circlet is uncertain because the four-plated basal circlet is recognized in several genera of the melocrinitids as well as a few other non-melocrinitids in the Devonian such as *Liomolgocrinus* Strimple (1963). However, it is possible that the specimen is a melocrinitid because many species of *Melocrinites* have ornamentation similar to that of the specimen. Furthermore, Hauser

(2002) illustrated nearly 100 species of *Melocrinites* Goldfuss 1831, which is the most common melocrinitid in the Devonian. The number of species of *Melocrinites* is more than twice the number of species recognized in all other melocrinitid genera described in the Devonian. Such diversity could be overestimated and probably due to wide intraspecific variations.

Family MELOCRINITIDAE d'Orbigny 1852
Genus indet., partial sets of arms
Text Figure 4C, D

Specimens. Figured specimens GPIM B3B-1/3 and GPIM B3B-1/4, partial sets of arms; GPIM B3B-1/5, lot of 24 arms fragments.

Type locality and horizon. Crinoid coquina in *Carinoceras* Beds, basal part of Anou Smaira Formation, Late Frasnian, Qued Mzerreb Section; approximately 0.5 km north of GPS N29°33′12″, W07°46′19″.

Description. Arms formed of two semifused rami giving off pinnulate branches every fourth brachial in each ramule. Estimated 20 to 40 arms per ray. Brachials rectilinear, strongly convexly rounded transversely, gently convex longitudinally, much shorter than wide. Pinnules one per brachial on same side of arm. Pinnulars rectilinear, elongate, approximately twice as long as wide, taper gently distally, minimum five per pinnule.

Remarks. The larger of two fragmentary partial sets of arms of an indeterminate melocrinitid has the distal part of the semifused rami and arm branches with pinnules. The other specimen has only the pinnulate branches. Both specimens are well preserved and easily distinguished from the cuneate brachials with nearly square pinnulars of *Embolocrinus quadruus* gen. and sp. nov. The rectilinear brachials are easily distinguished from the brachials of *Coquinacrinus revimentus* gen. and sp. nov. which have lateral lappets.

The generic identification is uncertain because the two semifused rami in each ray are known in both *Melocrinites* and *Ctenocrinus* Bronn 1840. Other genera of the Melocrinitidae have four rami (*Alisocrinus* Kirk 1929) or give off additional large rami in each ray (*Promelocrinus* Jaekel 1902; *Trichotocrinus* Olsson 1912; *Mongolocrinus* Rozhnov 1992).

Superfamily HEXACRINITOIDEA Wachsmuth & Springer 1885
Family HEXACRINITIDAE Wachsmuth & Springer 1885
Genus HEXACRINITES Austin & Austin 1843
Hexacrinites chenae sp. nov.
Text Figure 4F–I

Derivation of name. Named for Xiu-qin Chen.

Holotype. GPIM B3B-1/6, a thecae, discovered by Xiu-Qin Chen.

Type locality and horizon. Bed 8a, Red Griotte Beds, Oued Mzerreb Member, Ahrerouch Formation, Middle Givetian, Oued Mzerreb West Section (Becker *et al.* in El Hassani 2004), Tata region, eastern Dra Valley; GPS: N29°33′12″, W7°46′19″.

Diagnosis. Recognized by the combination of a medium bowl shape, pronounced ray ridges, and granulose ornament.

Description. Cup small, length 3.8 mm, width maximum 6.3 mm, minimun 5.6 mm, average 5.95 mm, medium bowl shape, hexagonal in oral view; ornament of granular to short anastomosing ridges and rounded to slightly angular ray ridges increasing in development distally to merge with brachials. Basal circlet tripartite, shallow bowl shape, diameter 3.8 mm, slightly protruded truncated base for stem attachment. Basals 3, subequal, pentagonal, slightly wider (2.8 mm) than long (2 mm), proximally horizontal, moderately upflared distally. Radials 5, hexagonal, length approximately equal to width, but variable: A-radial length 3.1 mm, width 2.9 mm; B-radial length and width 3 mm; C-radial length 3 mm, width 2.9 mm; D-radial length and width 3.2 mm; E-radial length 3.1 mm, width 2.7 mm; widest near but not at distal ends of mutual sutures. Radial facets horseshoe shaped, wider than deep, width 1.5 mm, angustary, facet width/radial width 0.52, declivate gently, weakly developed broad transverse ridge not extending to lateral sides of facet, wide V-shaped ambulacral groove. Primanal pentagonal, large, length 2.6 mm, width 2.4 mm, in line of radials, bears longitudinal medial ridge. First primibrachial rectilinear, wider (1.5 mm) than long (0.7 mm), transversely horseshoe shaped, strongly rounded externally. Stem facet round, diameter 1.7 mm; axial canal pentagonal, diameter 0.3 mm. Distal arms, tegmen, and stem unknown.

Remarks. The cup of *Hexacrinites chenae* sp. nov. is slightly distorted and encrusted with attached micro-organisms and some matrix. It is most similar to *H. sartenaeri* Hauser (1999) and *H. ardennicus* Hauser (1999), both reported from the Frasnian of the southern margin of the Dinant Basin of Belgium. However, *H. chenae* differs from both of these species by having the pronounced ray ridges and granulose ornament. All other species of *Hexacrinites* differ from *H. chenae* by having grossly different cup shapes, ornament, or basal flanges.

Subclass CLADIDA Moore & Laudon 1943
Order DENDROCRINIDA Bather 1899
?Family DENDROCRINIDAE Wachsmuth & Springer 1886
Genus DRACRINUS gen. nov.

Type species. Dracrinus crenulatus gen. and sp. nov.

Derivation of name. Named for the Dra Valley.

Diagnosis. Distinguished by the presence of symplexy articular facets on both side of each brachial.

Description. As for *Dracrinus crenulatus* sp. nov.

Remarks. The presence of muscular articular facets on the brachials (except branching brachials and the attached distal brachial) is not recognized in Palaeozoic crinoids except in the articulates and flexibiles. In the extant Articulata the synarthrial facets are on the two sides of brachials in series of successive muscular articulations, and only on one side of each brachial in brachial pairs united by ligamentary articulation (synostosis or syzygy), whereas in the flexibles the muscular articular facets usually have a patelloid process on the proximal exterior end, have grossly different morphology of the facet, and lack pinnules. Lacking the articular facets on the brachials *Dracrinus* gen. nov. would be considered as a dendrocrinid retaining a primitive (plesiomorphic) anal arrangement and unquestionably assigned to the family Dendrocrinidae.

Occurrence. Devonian, Late Emsian, Western Anti-Atlas Mountains, eastern part of Dra Valley.

Dracrinus crenulatus sp. nov.
Text Figure 4E

Derivation of name. From the Latin *crenula* referring to the coarse culmina on the radial facets and brachial facets.

Holotype. GPIM B3B-1/7, a crushed theca, found by Jurgen Bockwinkle.

Type locality and horizon. Nodular limestone beds above *Sellanarcestes* Limestone in basal part of Timrhanrhart Formation, Late Emsian, Foum Zguid Section (unit 16a of Jansen *et al.* in El Hassani 2004), Foum Zguid Section, northeastern Dra Valley. GPS: N29°55.676′ W07°02.960′.

Diagnosis. Cup high truncated cone, coarse granular ornament, impressed sutures, moderately deep apical pits. Infrabasals forming basal third of cup. Anals 4. Radial facets peneplenary. Tegmen stout, formed of columns of laterally interlocking hexagonal plates. Arm number and branching pattern unknown. Brachials rectilinear, bearing symplexy muscular articular facets on both sides of each brachial. Stem holomeric, round, heteromorphic; crenularium narrow, wide lumen.

Description. Partial crown crushed on bedding surface with posterior view exposed, length 24.2 mm, width 15.3 mm. Cup high truncated cone, length 13.2 mm, width 15.3 mm; ornamentation of coarse granules, moderately impress sutures, and moderately deep apical pits on all triple plate junctions. Infrabasal circlet forming basal third of cup. Infrabasals 5(?), quadrangular, subhorizontal proximally for stem attachment, distally highly upflared, gently convex longitudinally and transversely, visible length 6.3 mm, width 5 mm. Basals 5(?), form nearly half of cup walls, BC and CD septagonal, DE hexagonal, longer than wide, gently convex longitudinally, moderately convex transversely; CD-basal length 6.3 mm, width 5.7 mm. Radials 5(?), form upper fifth of cup walls, pentagonal, wider than long (D-radial length 4.8 mm, width 6.3 mm), gently convex longitudinally, moderately convex transversely. Radial facets peneplenary, concave, with wide crenularium on outer margin bearing coarse crenulae, inner surface not exposed; facet width/radial width $5.7/6.3 = 0.9$ (D-radial). D-radial bears a narrow interradial notch on the mutual shoulder with the E radial. Anals, 4 below radial summit, Menoplax 1 subcondition. Primanal largest, length and width 4.4 mm, hexagonal, partly under secundanal, bearing two subequal anals distally. Secundanal above posterior basal with distal edge at radial summit, length and width 4.1 mm. Tertanal and right tube plate rest on primanal, proximal one-third below radial summit. Tegmen stout, consists of vertical columns of hexagonal plates staggered to interlock laterally. First primibrachial (D-ray) rectilinear, shorter (length 1.2 mm) than wide (width 4.5 mm, incomplete), straight longitudinally, roundly convex transversely, bear coarse crenulae on outer margin of proximal facet, crenulae coarser than those of columnal facet. Arm number and branching pattern unknown. Associated brachials have articular facets with coarse crenulae on outer margin, coarse transverse fulcral ridge centrally, and small muscle fields bordering moderately wide ambulacral groove ventrally on one side. Opposite side has coarse crenulae on outer margin, small muscle fields bordering moderately wide ambulacral groove ventrally and a transverse groove centrally for interlocking with transverse ridge of adjacent brachial. Crenulae with symplexy articulation. Stem holomeric, round, heteromorphic proximally, noditaxis N1. Latus strongly rounded on nodals and internodals, crenularium narrow, narrow areola, wide lumen.

Remarks. The partial crown of *Dracrinus crenulatus* gen. and sp. nov. is crushed with the anal interray centred, lacking distal parts of the arms, tegmen and stem. It is embedded in a coquina of rectilinear brachials and other ossicles judged to belong to the specimen or other specimens of the same species as disarticulated brachials show the roundly convex transverse outer surface and transverse fulcral ridge and wide crenulated outer margin on the facets. One disarticulated brachial at the base of the cup shows an irregularly denticulated transverse ridge, the wide crenulated outer margin, small muscle fields and deep ambulacral groove. The derived crenulated margin of the brachial facets appears on most of the exposed brachials and apparently extended up the arms for most if not all of the brachials. This is a character not recognized in other primitive cladids. The abundance and size diversity of disarticulated brachials in the coquina suggests that the arms branched at least once and perhaps multiple times. However, the arms are unbranched on the first or second primibrachial that are slightly disarticulated in the D-ray. The anals are a primitive menoplax arrangement, whereas the radial facets are an intermediate evolutionary feature between the smooth plesiomorphic facets of most primitive cladids and the complex derived facets of advanced cladids. Tegmen plates continue directly above the anals and the right most upper anal has a more rounded rib-like surface extending onto the overlying tegmen plates suggesting relationship with the glossocrinids.

Family AMABILICRINIDAE Webster *et al.* 2003
COQUINACRINUS gen. nov.

Derivation of name. From the crinoidal coquina wherein the specimens were found.

Type species. Coquinacrinus revimentus gen. and sp. nov., here designated.

Diagnosis. Distinguished by the presence of the flanges or lappets on the lateral edges of the brachials.

Remarks. Coquinacrinus gen. nov. retains primitive (plesiomorphic) features of the amabilicrinids in the conical cup, rectilinear brachials, and three or four anals in the cup.

Occurrence. Devonian, Late Frasnian, Western Anti-Atlas Mountains, eastern part of Dra Valley.

Coquinacrinus revimentus sp. nov.
Text Figures 5A–N

Derivation of name. From Latin, *revimentum*, meaning lappet or flange.

Type series. Holotype GPIM B3B-1/8; paratypes 1–4, GPIM B3B-1/9—B3B-1/12; lot of eight arm fragments, GPIM B3B-1/13; and holdfasts 1–4, GPIM B3B-1/14—B3B-1/17.

Type locality and horizon. Holotype, paratypes 1 and 4, lot of eight arm fragments, and holdfasts from lower crinoidal coquina and paratypes 2 and 3 from the upper crinoidal coquina in *Carinoceras* Beds, Anou Smaira Formation, Late Frasnian, Oued Mzerreb Section West; approximately 0.5 km north of GPS N29°33′12″, W07°46′19″.

Diagnosis. Cup low conicfal, truncate base, three or four anals, deep apical pits; radial facets peneplenary; arms 10, uniserial, isotomous branching on axillary second primibrachial, brachials rectilinear with lateral flanges or lappets, tegmen formed of hexagonal plates in vertical columns, alternating interlocking plicate and non-plicate; stem holomeric, round transversely; lumen pentagonal; crenularium wide with coarse culmina.

Description. Cup low conical, wider (12.2 mm maximum, 8.3 mm minimum, 10.2 mm average) than long (7.8 mm in posterior interray, 5.8 mm in A-ray, 6.8 mm average), with deep apical pits, deepest in anal interray, plates smooth. Infrabasal circlet low, diameter up to 5.4 mm, 4.2 mm minimum, 4.8 mm average, forming lower fifth of cup or less. Infrabasals 5, subhorizontal proximally for stem attachment, distally outflared, visible part wider (2.6 mm) than long (0.9 mm). Basals 5, hexagonal (DE, EA and AB) or septagonal (BC and CD), wider (4.4 mm) than long (4 mm), slightly concave to straight longitudinally, moderately convex transversely. Radials 5, pentagonal except C-radial hexagonal, wider (5.2 mm) than long (3.5 mm), slightly concave longitudinally, strongly convex transversely, outflaring. Radial facets peneplenary (width radial facet/width radial $4/5.2 = 0.76$), concave with transverse fulcral ridge and growth rings on outer margin, slightly declivate. Anals 3 or 4, Menoplax 2 or 1 subconditions. Primanal largest, hexagonal, approximately equidimensional (length 3.5 mm, width 3.7 mm), slightly undercutting secundanal, supports tertanal and quartanal (when present) distally. Secundanal in contact with and directly above CD-basal, hexagonal, slightly longer (3.7 mm) than wide (3.5 mm), distal third above radial summit. Tertanal and quartanal hexagonal, small, distal half to two-thirds above radial summit. Tegmen slender elongate, formed of alternating vertical columns of non-ribbed wider plicate and ribbed narrower hexagonal plates interlocking laterally. First primibrachial length 1.5 mm, width 5.2 mm. Second primibrachial axillary, isotomous branching. Brachials rectilinear, much wider than long, distal ends flat, centrally elevated, bearing a single pinnule on

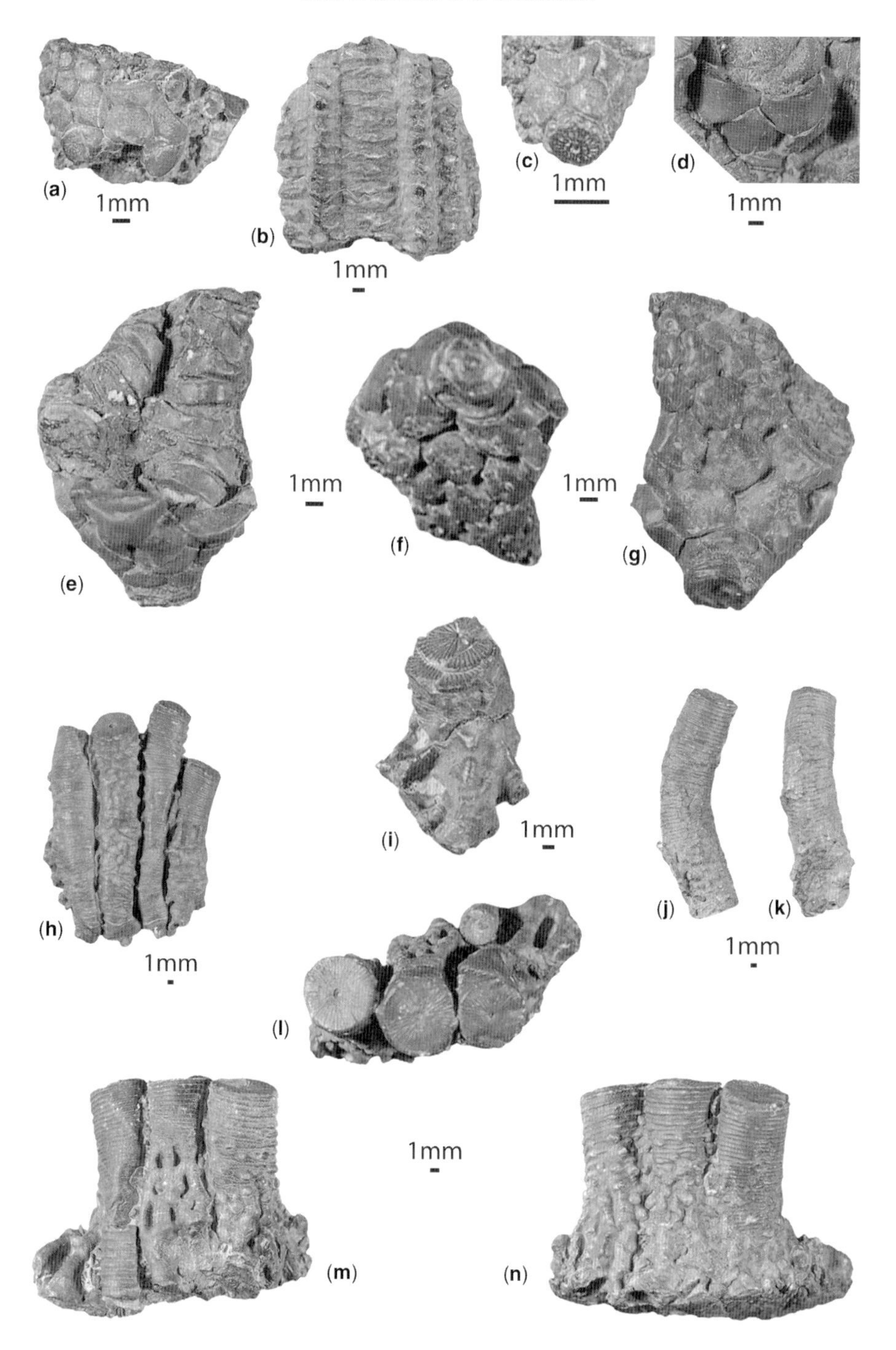

Fig. 5. (**a–n**) *Coquinacrinus revimentus* gen. and sp. nov. (a) C-ray view of paratype 1 GPIM B3B-1/9. (b) Lateral view of distal part of tegmen, paratype 4 GPIM B3B-1/12. (c) Oblique view of cup, ray uncertain, paratype 2 GPIM B3B-1/10. (d) Oblique lateral view of cup, ray uncertain, paratype 3, GPIM B3B-1/11. (e–g) A-ray, basal and posterior views, holotype GPIM B3B-1/8. (h) Lateral view of four distal pluricolumnals showing some short rootlets, all above a cementation pad, holdfast 3 GPIM B3B-1/16. (i) Lateral view of holdfast 1 showing branching rootlets and articular facet, GPIM B3B-1/14. (j, k) Lateral views of holdfast 2 GPIM B3B-1/15. (l–n) Oral and two lateral views of three holdfasts showing rootlets above a cementation pad, GPIM B3B-1/17.

alternate sides of arm. Pinnules slender, elongate, formed of more than five pinnulars. Pinnulars elongate, length twice width, transversely convex. Stem holomeric, round, symplexy articulation, proximal diameter 4.5 mm (weathered), homeomorphic proximally. Columnals short with gently convex latus; crenularium wide with coarse culmina; areola narrow. Lumen roundly pentagonal.

Holdfast 1: distalmost pluricolumnal with cementation surface, length 19 mm, width 12.8 mm, diameter of proximal columnals 7.8 mm. Columnals short, round in transverse section, with wide crenularium, narrow areola, pentagonal lumen, and gently convex latus; syzygial articulation. Culmina straight, moderately coarse, may branch proximally to distally, remain unbranched, or rarely inserted in distal half of crenularium. Distalmost columnals of variable length, develop nodes on latus, branch into rootlets which may have lateral connections, and terminate at attachment surface. Attachment surface with grainy texture in part, irregular otherwise.

Holdfast 2: specimen weathered and proximal- and distalmost parts not preserved; length 37.2 mm, diameter 8.5 mm. Column homeomorphic with wide crenularium, narrow areola, small lumen of uncertain shape. Surface with increasing nodes distally. Distally the column is parallel to the cementation surface, proximally curving away from the cementation surface at approximately 30°. Cementation surface concave.

Holdfast 3: distalmost parts of four columns with cementation surface parallel to columnals, specimen length 38.5 mm, width 27.8 mm. All columns tapering, branching into rootlets, and interlocking laterally distally. Facets like that of Holdfast 1. Specimen second from left has many more coarse nodes and the individual columnal length is highly irregular laterally. Proximal ends of three specimens on right beginning to curve away from cementation surface. Attachment surface has a thin calcite layer which shows a grainy texture transverse to the column lengths.

Holdfast 4: distalmost parts of three columns, specimen length 23.9 mm, width 31.7 mm. Proximal diameters of columns 8 mm, 8.6 mm, 8.5 mm (right to left). Left two columns homeomorphic; columnals transverse section round, short, syzygal articulation, latus mildly to strongly convex with numerous coarse nodes becoming more abundant near base; crenularium nearly full width of facet, with very narrow areola; crenulae coarse medially, or distally or remain unbranched, rarely inserted in distal half; lumen roundly pentagonal. Right specimen heteromorphic, N1 noditaxis, nodals with strongly convex latus with coarse nodes increasing in abundance distally; crenularium and lumen as in two left specimens. This specimen may not belong to *C. revimentus* sp. nov. Some columnals of all specimens of irregular lengths. Some rootlets with interconnecting lateral ridges. All columns branch into rootlets distally and positioned approximately normal to cementation surface.

Remarks. The description of *Coquinacrinus revimentus* sp. nov. is based on the holotype, paratypes, and four pluricolumnal holdfasts. The holotype is a cup with the anals, base of tegmen, proximal brachials of the D- and E-rays, and proximal three columnals moderately well presereved. The radial facets are weathered, accentuating the transverse ridge with loss of the growth rings on the outer marginal area. It is distorted with the anal interray partly dislocated into the cup cavity. Paratype 1 is crushed parallel to the A-ray/posterior plane of symmetray; it lacks the stem, infrabasal circlet, arms and distal tegmen. It is embedded in a coquina of dissociated brachials. Paratype 2 is a partly exposed cup with the cup plates slightly disarticulated embedded in a coquina of pluricolumnals. It retains the transverse fulcral ridge and growth rings on the outer margin of the radial facets. Paratype 3, a cup, is on the other side of the block with paratype 2. Paratype 4 is a weathered distal tip of a tegmen showing vertical ribbing on the interlocking hexagonal ossicles and the terminus below the summit. Eight other specimens consisting of arm fragments showing the rectilinear brachials and elongate pinnulars are also assigned to the taxon.

The above pluricolumnals have the same type of columnal facets, including the roundly pentagonal lumen and coarse culmina. They show considerable variation of the attachment structures and the curvature suggests the crown was elevated or suspended in the current away from the attachment. The grainy texture of holdfast 3 suggests that it was attached to a log.

?Family AMABILICRINIDAE Webster *et al.* 2003
Genus EMBOLOCRINUS gen. nov.

Derivation of name. From the Greek, *embolos*, meaning wedge and referring to wedge-shaped brachials.

Type species. Embolocrinus quadruus gen. and sp. nov., here designated.

Diagnosis. Distinguished by uniserial or biserial wedge-shaped brachials and short square or nearly square pinnulars.

Remarks. Classification of *Embolocrinus* gen. nov. is problematic because it displays a mosaic of plesiomorphic and derived characters. The low conical shape of the cup and presence of four anals are both primitive plesiomorphic characters of the cladids. The pinnulate biserial or uniserial arms are advanced derived characters in the Dendrocrinida. The cup of *Embolocrinus* is characteristic of genera of both the Amabilicrinidae and Scytalocrinidae. However, the biserial arms are not characteristic of either family as both have uniserial arms. Biserial arms are an evolutionary convergent trend appearing in different cladid families. Without the biserial arms *Embolocrinus* would be assigned to the Amabilicrinidae without question because the other characters of the genus fit the characters of the family. *Embolocrinus* is probably derived from *Hallocrinus* Goldring, 1923 by modification of the brachials to the wedge shape and lowering of the arm branching to the second primibrachial.

Occurrence. Late Devonian, Frasnian, western part of Anti-Atlas Mountains, eastern part of Dra Valley.

Embolocrinus quadruus sp. nov.
Text Figures 4K–O

Derivation of name. From the Latin, *quadruus*, meaning square and referring to the shape of the pinnulars.

Type series. Holotype GPIM B3B-1/18, a crown; paratypes 1–8 GPIM B3B-1/19—B3B-1/26; one crown and a crushed cup on a single slab GPIM B3B-1/27; three cups GPIM B3B-1/28—B3B-1/30, and three lots of arm fragments GPIM B3B-1/31—B3B-1/33.

Type locality and horizon. Lower crinoidal coquina in *Carinoceras* Beds, Anou Smaira Formation, Late Frasnian, Oued Mzerreb Section West; approximately 0.5 km north of GPS N29°33′12″, W07°46′19″.

Description. Crown small, slender, flaring distally, pear-shaped enclosed. Cup small, low truncate cone. Ornamentation lacking in juvenile specimens, developing shallow apical pits at junction of basals and radials, most prominent in adult forms. Infrabasal circlet low, proximally subhorizontal for stem attachment, distal tips upturned, forming small part of cup wall. Infrabasals 5, dart-shaped, distal tips barely upturned and visible in lateral view. Basals 5, DE, EA and AB hexagonal, BC and CD septagonal, outflared, slightly concave longitudinally, moderately convex transversely, form approximately one-half of cup wall. Radials 5, E, A and B pentagonal, C and D hexagonal, wider than long, gently convex longitudinally, moderately convex transversely, outflaring, forming upper half of cup wall. Radial facets peneplenary, moderately concave transversely, gently declivate, bear weak transverse fulcral ridge, shallow narrow ligament pit, narrow outer margin and narrow muscle areas. Interradial notches narrow, but obvious. Anals 3 or 4, Menoplax 2 or 1 subcondition. Primanal large, hexagonal, nearly separating smaller secunanal from CD-basal and supporting tertanal and quartanal directly above. Secundanal barely in contact with CD-basal, probably projects slightly above radial summit. Tertanal and quartanal small with proximal tips below radial summit. Arms 10, 2 per ray, uniserial or biserial, isotomous branching on second primibrachial. Brachials cuneate, much wider than long, wedge-shaped in transverse view, slightly convex longitudinally, strongly convex transversely. Primibrachials 2, rectilinear. Secundibrachials cuneate, bear a single pinnule on wide side. Pinnules elongate, formed of more than 15 pinnulars. Pinnulars short, nearly square proximallly, slighty wider than long distally, gently convex transversely, straight longitudinally. Anal tube robust, distal extent unknown. Stem holomeric, round transversely, homeomorphic proximally becoming heteromorphic distally with noditaxis of N1. Columnals short, transversely round, symplexy articulation; crenularium approximately one-half of facet radius; culmina moderately coarse, branching at midlength to distal tips; areola approximately one-half of crenularium depth, slightly concave; lumen pentagonal; latus irregular, convex longitudinally. Proximal columnal irregular on proximal surface adjoining basals and infrabasals. Measurements are given in Table 1.

Remarks. Embolocrinus quadruus gen. and sp. nov. is based on the holotype and eight paratypes. The holotype is a crown retaining the proximal part of the stem. Paratype 1 is on the same slab as the holotype and is a partially exposed cup with proximal parts of the arms attached or slightly dislocated. Paratype 2 is a cup with the A-ray and C-ray proximal parts of the arms slightly disassociated and a large primanal and secundanal, with the secundanal directly above and widely in contact with the CD-basal. Paratype 3 is a partial crown with only a single anal and the base of the arms of the B-, C- and D-rays are slightly dislocated. The single anal is in the primanal position in contact with the BC- and CD-basals and C- and D-radials. Paratype 4 is a cup on the same slab with paratype 3. Paratype 5 is a juvenile cup with the oral surface tilted and embedded in a coquina of crinoid ossicles with the anal interray clearly exposed; only the primanal is preserved, but the facets for the secundanal, tertanal and quartanal are preserved. Paratype 6 is a weathered crown showing the base of the anal tube and uniserial arms. It is on the same slab with three other non-type cups. Paratype 7 is a set of uniserial arms with the basal anal tube plates and paratype 8 is a cup with associated proximal parts of two arms showing the biserial brachials. Brachials are present on all of the crowns or partial crown specimens. Other specimens are two crowns, three cups, and five partial arms and pinnules on uncertain pluricolumnals.

Table 1. *Measurements (in mm) of four specimens of* Embolocrinus quadruus *n. gen. and sp.*

Types	Holotype	Paratype 2	Paratype 3	Paratype 5
Specimen length	32.0	2.6	6.1	
Specimen width	17.5	4.1	6.2	
Crown length	20.2		5.9	
Crown width	17.5		6.2	
Cup length	2.7	2.0	2.0	0.9
Cup width	3.7	4.1	4.0	2.0
Infrabasal circlet diam.		2.0		1.0
Infrabasal length (vis)	0.3	0.1		0.2
Infrabasal width (vis)		0.3		0.4
Basal length	1.3	1.2		
Basal width	1.4	1/3		
CD basal length			0.9	0.6
CD basal width		1.7		0.6
Radial length	1.5	1.4	1.0	0.5
Radial width	2.1	1.9	1.8	
Radial facet width	1.6	1.5	1.4	
Facet width/radial width	.76	.79	.78	
First primibrach length	0.5		0.5	
First primibrach width	1.6			
Primanal length			0.8	
Primanal width			1.0	
Length proximal stem	11.1			
Diameter stem proximal	2.1	0.5		
Diameter stem distal	2.1	2.2		

Variation in the arms is observed in the uniserial or biserial condition. The uniserial condition is continuous throughout the length of the arm, but individual brachials may approach the biserial condition. The biserial condition starts with the second or third secundibrachial and continues throughout the length of the arm. The pinnulars of both the uniserial and biserial arms are identical as are the cups. Variation in the number of anals is noted by the presence or absence of the fourth anal below the radial summit. The specimen with only a single anal in the primitive plesiomorphic or paedomorphic primanal position is considered an aberrant specimen, although it could be an extreme pattern in the intraspecific field of variation.

Subclass FLEXIBILIA Zittel 1895
Order TAXOCRINIDA Springer 1913
Superfamily TAXOCRINOIDEA Angelin 1878
Genus indet.

Mentioned specimen. GPIM B3B-1/34, disarticulated brachial.

Type locality and horizon. Upper crinoidal coquina in *Carinoceras* Beds, Anou Smaira Formation, Late Frasnian, Oued Mzerreb Section West; approximately 0.5 km north of GPS N29°33′12″, W07°46′19″.

Description. Arm fragment with brachials having well developed patelloid process, straight longitudinally, strongly convex transversely. Radial facet complex, with straight transverse ridges extending width of facet; wide outer margin hemicircle or hemiellipsoid shape, with elevated outer rim; muscle areas deep bordering ventral ambulacal groove.

Remarks. These indeterminate flexible brachials occur associated with a pluricolumnal suggesting they are part of a crown that was disarticulated by scavenging or current action. There are three axillary brachials of differing sizes suggesting that the arms branched a minimum of three times. The complex morphology of the facets is similar to that of some taxocrinids as illustrated by Springer (1920: pl. 54, fig. 19*c*) but lacks the intricate ridges of some forbesiocrinids as illustrated by Springer (1920: pl. 24, figs. 10*a*, 14, among others). This suggests possible affinity with the taxocrinids rather than the sagenocrinids. The brachials are mentioned to show the diversity in the fauna of the *Carinoceras* beds.

The organizational work and leadership of the 2004 International Field Conference by A. El Hassani is gratefully acknowledged. Our appreciation is extended to the participants of the Field Conference who collected part of the specimens, with special thanks to Jürgen Bockwinukel,

Xiu-Qin Chen, John Marshall and Ulrich Jansen. This investigation is a contribution of the IGCP 499 project. The manuscript was improved significantly by two anonymous reviews.

References

ABOUSSALAM, Z. S. 2003. Das Taghanic-Event im höheren Mittel-Devon von West-Europa und Marokko. *Münstersche Forschungen zur Geologie und Paläontologie*, **97**, 1–332.

ANGELIN, N. P. 1878. *Iconographia Crinoideorum: in stratis Sueciae Siluricis fossilium.* Samson and Wallin, Holmiae, 1–62.

AUSTIN, T. & AUSTIN, T. 1843. XXXIII. Description of several new genera and species of Crinoidea. *Annals and Magazine of Natural History, series 1*, **11**(69), 195–207.

BATHER, F. A. 1899. *A phylogenetic classification of the Pelmatozoa.* British Association for the Advancement of Science, 916–923.

BECKER, R. T. & HOUSE, M. R. 2000. Devonian ammonoid zones and their correlation with established series and stage boundaries. *Courier Forschungsinstitut Senckenberg*, **220**, 113–151.

BECKER, R. T. & HOUSE, M. R. 2009. Devonian ammonoid biostratigraphy of the Canning Basin. *Geological Survey of Western Australia, Bulletin*, **145**, in press.

BECKER, R. T., JANSEN, U., PLODOWSKI, G., SCHINDLER, E., ABOUSSALAM, Z. S. & WEDDIGE, K. 2004*a*. Devonian litho- and biostratigraphy of the Dra Valley – an overview. *In*: EL HASSANI, A. (ed.) *Devonian Neritic–Pelagic Correlation and Events in the Dra Valley (Western Anti-Atlas, Morocco).* Documents de l'Institut Scientifique, **19**, 3–20.

BECKER, R. T., ABOUSSALAM, Z. S., BOCKWINKEL, J., EBBINGHAUSEN, V., EL HASSANI, A. & NÜBEL, H. 2004*b*. The Givetian and Frasnian at Oued Mzerreb (Tata region, eastern Dra Valley). *In*: EL HASSANI, A. (ed.) *Devonian Neritic–Pelagic Correlation and Events in the Dra Valley (Western Anti-Atlas, Morocco).* Documents de l'Institut Scientifique, **19**, 37–55.

BECKER, R. T., BOCKWINKEL, J., EBBIGHAUSEN, V., ABOUSSALAM, Z. S., EL HASSANI, A. & NÜBEL, H. 2004*c*. Lower and Middle Devonian stratigraphy and faunas at Bou Tserfine near Assa (Dra Valley, SW Morocco). *In*: EL HASSANI, A. (ed.) *Devonian Neritic–Pelagic Correlation, and Events in the Dra Valley (Western Anti-Atlas Morocco).* Documents de l'Institut Scientifique, **19**, 125–139.

BENSAID, M. 1974. Etude sur des Goniatites a la limite du Dévonien Moyen et Supérieur, du Sud Marocain. *Notes du Service Geologique du Maroc*, **36**(274), 81–140.

BONDON, J. & CLARIOND, E. 1934. Itinéraire géologique d'Aqqa à Tindouf (Sahara marocain). *Compte Rendu hebdomadaires des séances, de l'Academie des Sciences, Paris*, **199**(sér. II), 45–58.

BOURCART, J. 1938. Une coupe géologique de l'Oued Noun a l'Quarkziz par Torkoz (Sahara marocain). *Comptes rendus sommaires des séances, Société géologique de France*, 12–14.

BRONN, H. G. 1840. *Ctenocrinus*, ein neues Krinoiden-Geschlecht der Grauwacke. *Neues Jahrbuch für Mineralogie, Geologie, und Paläontologie*, 542–548.

BRONN, H. G. 1848–49. *Index palaeontologicus*, unter Mitwirking der Herren Prof. H. R. Göppert und H. von Meyer. Handbuch einer Geschichte der Natur, 5, Abt. 1, (1, 2), pt. 3, A. Nomenclator Palaeontologicus, Stuttgart.

CHOUBERT, G. & MARCAIS, J. 1948. *La géologie marocaine.* Éditions de l'Encyclopédie coloniale et maritime, Paris, 1–12.

DESCOSSY, G. & ROCH, E. 1934. Sur quelques fossiles de la basse vallée du Dra de la région de Tindouf. *Comptes Rendus sommaires des séancse, Société géologique de France*, **4**(série 5), 104–105.

D'ORBIGNY, A. 1852. *Prodrome du paléontologie stratigraphique universelle des animaux mollusques et rayonnés faisant suite au cours élémentaire de plaéontologie et de géologie stratigraphique.* Masson, Paris.

EL HASSANI, A. (ed.) 2004. *Devonian Neritic–Pelagic Correlation and Events in the Dra Valley (Western Anti-Atlas, Morocco).* Documents de l'Institut Scientifique, **19**.

GENTIL, L. 1929. L'Anti-Atlas et de djebel Bani (Exploration de 1923). *Revue de Géographie Physique et Géologie Dynamique*, **2**(1), 5–66, 68–69.

GOLDFUSS, G. A. 1831. *Petrefacta Germaniae, tam ea, Quae in Museo Universitatis Regiae Borussicae Fridericiae Wilhelmiae Rhenanea, serventur, quam alia quaecunque in Museis Hoeninghusiano Muensteriano aliisque, extant, iconibus et descriiptionns illustrata.* Abbildungen und Beschreibungen der Petrefacten Deutschlands und der Angränzende Länder, unter Mitwirkung des Hern Grafen Georg zu Münster, herausgegeben von August Goldfuss. v. 1 Divisio secunda. Radiariorum reliquiae, [Echinodermata], 115–221.

GOLDRING, W. 1923. *The Devonian Crinoids of the State of New York.* New York State Museum, Memoir, **16**.

HAUSER, J. 1997. *Die Crinoiden des Mittel-Devon der Eifler Kalkmulden.* Bonn.

HAUSER, J. 1999. *Die Crinoiden der Frasnes-Stufe (Oberdevon) vom Südrand der Dinant Mulde (Belgische und französische Ardennen).* Bonn.

HAUSER, J. 2001. *Neubeschreibung mitteldevonischer Eifelcrinoiden aus der Sammlung Schultze (Museum of Comparative Zoology, The Agassiz Museum, Harvard University, Massachusetts, USA) nebst einer Zusammenstellung der Eifelcrinoiden (Holotypen) der Goldfuss-Sammlung.* Bonn.

HAUSER, J. 2002. *Oberdevonische Echinodermen aus den Dolomit-Steinbrüchen von Wallersheim-Loch (Rheinisches Schiefergebirge, Prümer Mulde; Eifel).* Bonn.

HOLLARD, H. 1963. Un tableau stratigraphique du Dévonien du Sud de l'Anti-Atlas. *Notes et Mémoires du Service Géologique du Maroc*, **23**, 105–109.

HOLLARD, H. 1967. Le Dévonien du Maroc et du Sahara nord-occidental. *In*: OSAWALD, D. H. (ed.) *International Symposium on the Devonian System*, Calgary, **1**, 203–244.

HOLLARD, H. 1978. Corrélations entre niveaux à brachiopods et à goniatites au voisinage de la limite Dévonien inférieur – Dévonien moyen dans les plaines du Dra (Maroc présaharien). *Newsletter on Stratigraphy*, **7**(1), 8–25.

HOLLARD, H. 1981*a*. Principaux caractères des formations dévoniennes de l'Anti-Atlas. *Notes et Mémoires du Service géologique du Maroc*, **42**(308), 15–22.

HOLLARD, H. 1981*b*. Tableaux de correlations du Silurien et do Dévonien de l'Anti-Atlas. *Notes et Mémoires du Service Géologique du Maroc*, **42**(308), 23.

HOLLARD, H. & JAQUEMONT, P. 1956. Le Gothlandien, le Dévonien et le Carbonifère des régions du Dra et du Zemout (confines algéro-marocains du Sud). *Notes et Mémoires du Service géologique du Maroc*, **15**(135), 7–33.

JAEKEL, O. 1902. *Über verschiedene Wege phylogenetischer Entwicklung*. 5. Verhandlungen des Internationalen Zoologischen Congress, Berlin, 1901, 1058–1117.

JANSEN, U., BECKER, G., PLODOWSKI, G., SCHINDLER, E., VOGAL, O. & WEDDIGE, K. 2004. The Emsian to Eifelian near Foum Zguid (NE Dra Valley, Morocco). *In*: EL HASSANI, A. (ed.) *Devonian Neritic–Pelagic Correlation and Events in the Dra Valley (Western Anti-Atlas, Morocco)*. Documents de l'Institut Scientifique, **19**, 21–35.

JANSEN, U., LAZREQ, N., PLODOWSKI, G., SCHEMM-GREGORY, M., SCHINDLER, E. & WEDDIGE, K. 2007. Neritic–pelagic correlation in the Lower and basal Middle Devonian of the Dra Valley (Southern Anti-Atlas, Moroccan Pre.Sahara). *In*: BECKER, R. T. & KIRCHGASSER, W. T. (eds) *Devonian Events and Correlations*. Geological Society, London, Special Publications, **278**, 9–37.

JELL, P. A. 1999. *Silurian and Devonian Crinoids from Central Victoria*. Memoirs of the Queensland Museum, **43**(1), 1–114.

JELL, P. A. & JELL, J. S. 1999. *Crinoids, a blastoid and a cyclocystoid from the Upper Devonian reef complex of the Canning Basin, Western Australia*. Memoirs of the Queensland Museum, **43**(1), 201–236.

KIRK, E. 1929. The status of the genus *Mariacrinus* Hall. *American Journal of Science*, **18**, 337–346.

KLUG, C., RÜCKLIN, M., MEYER-BERTHAUD, B., SORIA, A., KORN, D. & WENDT, J. 2003. Late Devonian pseudoplanktonic crinoids from Morocco. *Neues Jahrbuch für Geologie Paläontologie Monatschafte*, **3**, 153–163.

LANE, N. G., WATERS, J. A. & MAPLES, C. G. 1997. Echinoderm faunas of the Hongguleleng Formation, Late Devonian (Famennian), Xinjiang-Uygur Autonomous Region, People's Republic of China. *Journal of Paleontology, Memoir*, (Supplement to **71**, 2), 1–43.

LE MAITRE, D. 1958. Crinoides du Dèvonien d'Afrique du Nord. *Comptes Rendus sommaires des séances. Société géologique de France*, **14**, 344–346.

LE MENN, J. 1992. Evolution du genre *Haplocrinites*: crinoide Inadunata atypique du Devonien moyen. *In*: GAYET, M. (ed.) *Premier Congrés National de Paléontologie*, Paris, 17–19 Mai 1990, Geobios, Lyon, Mémoire Speciale, **14**, 105–112.

LE MENN, J. & REGNAULT, D. 1993. Découverte de microcrinoïdes dans le Dévonien inférieur de Maroc. *Comptes Rendus de l'Académie des Sciences, Paris*, série 2, **316**, 251–256.

MILLER, J. S. 1821. *A natural history of the Crinoidea, or lily-shaped animals; with observations on the genera, Asteria, Euryale, Comatula and Marsupites*. Bryon & Co., Bristol.

MOORE, R. C. & LAUDON, L. R. 1943. *Evolution and Classification of Paleozoic Crinoids*. Geological Society of America, Special Paper, **46**.

MOORE, R. C. & TEICHERT, C. (eds) 1978. *Treatise on Invertebrate Paleontology*, Part T, Echinodermata 2, Crinoidea. Geological Society of America and University of Kansas, Lawrence, T1–T1027.

OLSSON, A. 1912. New and interesting fossils from the Devonian of New York. *Bulletins of American Paleontology*, **5**(23), 27–38.

ROZHNOV, S. V. 1992. *Crinoids from the Silurian of Mongolia*. Sovmestnaya Sovetsko-Mongol 'Skaya Paleontologicheskaya Ekspeditsiya, Trudy, **41**, 84–89 [in Russian].

SPRINGER, F. 1913. Crinoidea. *In*: ZITTEL, K. A. VON. (ed.) *Text-book of Paleontology* 2nd edition, (translated and edited by C. R. Eastman) Macmillan, London, **1**, 173–243.

SPRINGER, F. 1920. *The Crinoidea Flexibilia*. Smithsonian Institution, Washington, Publication, **2501**.

STRIMPLE, H. L. 1963. Crinoids of the Hunton Group. *Oklahoma Geological Survey Bulletin*, **100**, 1–169.

TERMIER, G. & TERMIER, H. 1950. Paléontologie Marocaine II. Invertébres de l'Ere Primaire. 4. Annélides, Arthropodes, Echinodermes, Conularides et Graptolithes. *Service de la Carte géologique du Maroc, Notes et Mémoires*, **79**(4), 1–279.

TERMIER, H. & TERMIER, G. 1974. Une méthode nouvelle: l'utilisation des fragments d'échinodermes contenus dans le sédiments dévoniens et carbonifères du Maroc. *Service Géologique du Maroc*, Notes, **35**(255), 27–53.

WACHSMUTH, C. & SPRINGER, F. 1885–1886. Revision of the Palaeocrinoidea. *Proceedings of the Academy of Natural Sciences of Philadelphia*. **III** (Sec. 1.) Discussion of the classification and relations of the brachiate crinoids, and conclusion of the generic descriptions (1885), 225–364, pl. 4–9. III (Sec. 2.) Discussion of the classification and relations of the brachiate crinoids, and conclusion of the generic descriptions (1886), 64–226.

WATERS, J. A., MAPLES, C. G., LANE, N. G., MARCUS, S., LIAO, Z.-T., LIU, L., HOU, H.-F. & WANG, J.-X. 2003. A quadrupling of Famennian pelmatozoan diversity: new Late Devonian blastoids and crinoids from northwest China. *Journal of Paleontology*, **77**, 922–948.

WEBSTER, G. D. 1974. Crinoid pluricolumnal noditaxis patterns. *Journal of Paleontology*, **48**, 1283–1288.

WEBSTER, G. D. 2003. *Bibliography and index of Paleozoic crinoids, coronates, and hemistreptocrinoids, 1758–1999*. Geological Society of America, Special Paper, **363**.

WEBSTER, G. D. & JELL, P. A. 1999. *New Carboniferous Crinoids from Eastern Australia*. Memoirs of the Queensland Museum, **43**(1), 237–278.

WEBSTER, G. D. & MAPLES, C. G. 2006. Cladid crinoid (Echinodermata) anal conditions: a terminology problem and proposed solution. *Palaeontology*, **49**(1), 187–212.

WEBSTER, G. D., MAPLES, C. G., MAWSON, R. & DASTANPOUR, M. 2003. A cladid-dominated Early Mississippian crinoid and conodont fauna from Kerman Province, Iran and revision of the glossocrinids and rhenocrinids. *Journal of Paleontology*, Memoir **60** (supplement to **77**(3)), 1–35.

WEBSTER, G. D., BECKER, R. T. & MAPLES, C. G. 2005. Biostratigraphy, paleoecology, and taxonomy of Devonian (Emsian and Famennian) crinoids from southeastern Morocco. *Journal of Paleontology*, **79**, 1052–1071.

WEBSTER, G. D., MAPLES, C. G. & YAZDI, M. 2007. Late Devonian and Early Mississippian echinoderms from central and northern Iran. *Journal of Paleontology*, **81**, 1105–1118.

ZITTEL, K. A. VON. 1895. *Grundzüge der Palaeontologie (Palaeozoologie)*, R. Oldenbourg, München.

A re-evaluation of Famennian echinoderm diversity: implications for patterns of extinction and rebound in the Late Devonian

J. A. WATERS[1]* & G. D. WEBSTER[2]

[1]*Department of Geology, Appalachian State University, Boone, NC 28608, USA*

[2]*School of Earth and Environmental Sciences, Washington State University, Pullman, WA 99164, USA*

**Corresponding author (e-mail: watersja@appstate.edu)*

Abstract: Critical to understanding long-term trends in diversity is a dataset that is both worldwide in scope and based on a sound taxonomic foundation. In this paper we re-evaluate the Famennian (Late Devonian) echinoderm dataset, which has changed radically in the past decade, and reinterpret patterns of Late Devonian echinoderm extinction and rebound based on these new data. Historically, Famennian (Late Devonian) and earliest Carboniferous echinoderms have been poorly known on a global basis leading to interpretations of prolonged rebound from the Devonian extinction events. Recent discoveries of abundant and diverse Famennian echinoderm faunas from northwestern China, Colorado, Australia and Iran, together with re-examination of previously known echinoderm faunas from Germany and England, have altered drastically our understanding of the patterns of extinction and rebound of Famennian and earliest Carboniferous echinoderm communities. Overall, Famennian echinoderm diversity at the generic level is nearly five times greater than reported in the 2002 Sepkoski compilation, and familial level diversity is more than seven times greater than previously thought. Despite the increases in diversity, Famennian echinoderm faunas show a reduced diversity of camerate crinoids that typify both Middle Devonian and Lower Mississippian faunas and portend the rise of cladid crinoid diversity later in the Carboniferous. Individual Famennian faunas are numerically dominated by blastoids, which also portends trends seen at various times later in the Palaeozoic. In general, we are able to recognize the following trends. Rebound from the Late Devonian extinction events in echinoderms was more rapid than previously thought, but seems to be concentrated in Asia. Palaeogeographically Famennian echinoderms can be grouped into two broad regions: one includes China, Australia and Iran, all of which bordered the Palaeotethys; the other includes regions from Laurussia (Europe and North America) and northern Africa (Morocco).

Patterns of extinction and diversification in the fossil record typically depend on large-scale database compilations such as that of Sepkoski (2002) which contained the stratigraphic ranges of more than 37 000 genera taken mostly from the primary taxonomic literature published prior to 1998. One of our goals in this paper is to critically evaluate the data on Famennian crinoid and blastoid echinoderms in Sepkoski's compilation and update available information for inclusion into database successors such as the Palaeobiology Database Project.

Famennian echinoderm faunas are rare in North America and Europe and have been considered rare on a global basis. However, in recent years discoveries of new localities and re-evaluation of museum collections have shed new light on this important interval in echinoderm evolution. Lane *et al.* (1997), Waters *et al.* (2003) and Webster & Waters (2009) described an abundant and diverse fauna of Famennian blastoids and crinoids from the Hongguleleng Formation, Xinjiang Province, PRC. Additional Famennian echinoderm faunas either have been redescribed recently or are described for the first time. Lane *et al.* (2001*a*, *b*) re-evaluated the taxonomy of crinoids and blastoids from the type Devonian in England and from the Etroeungt of Germany. Thomas & Haude (2006) compiled data on Famennian crinoids from Germany incorporating additional taxa recently discovered from new outcrops. In addition, Jell & Jell (1999) reported a modestly diverse echinoderm fauna from the Canning Basin, Australia. Webster *et al.* (1999) reported on a Famennian echinoderm community from Colorado and Webster *et al.* (2007) described a Famennian fauna from Iran. Webster *et al.* (2005) described Famennian crinoids from Morocco and G. D. Webster *et al.* (unpubl. data) are currently investigating a Famennian crinoid fauna from Saalfeld, Germany. Lastly, Webster & Waters (unpubl. data) are revising some of the Famennian crinoids reported as Middle Devonian by Chen & Yao (1993) from Yunnan Province, China.

From: KÖNIGSHOF, P. (ed.) *Devonian Change: Case Studies in Palaeogeography and Palaeoecology*.
The Geological Society, London, Special Publications, **314**, 149–161.
DOI: 10.1144/SP314.8 0305-8719/09/$15.00

The results of these studies are presented in the Appendix and summarized in Table 1. They show a Famennian echinoderm diversity consisting of 117 genera including 16 genera of blastoids and 101 genera of crinoids. These figures include one new blastoid genus included in Thomas & Haude (2006) and 11 new genera of cladids from the Saalfeld fauna that await formal description.

Sepkoski (2002) recognized 39 genera of Famennian blastoids and crinoids. His analysis included data from Lane *et al.* (1997) on echinoderms from the Hongguleleng Formation but not the later works published by Waters *et al.* (2003). To accurately portray the differences in known Famennian echinoderm diversity prior to and as a result of our studies, we have removed the taxa described by Lane *et al.* (1997) from the Sepkoski compilation. The revised Sepkoski database now contains 24 Famennian blastoid and crinoid genera including 13 Famennian crinoid genera not recorded in any of the known Famennian echinoderm faunas. This means that only 11 genera of crinoids and blastoids (*Petaloblastus*, *Doryblastus*, *Taxocrinus*, *Paraclidocrinus*, *Gilbertsocrinus*, *Platycrinites*, *Triacrinus*, *Allagecrinus*, *Pentececrinus*, *Scytalocrinus* and *Decadocrinus*) had actually been collected from Famennian-aged rocks in the collecting history captured by Sepkoski. The other genera were range-through taxa. These 13 range-through crinoid genera include four genera of flexibles, *Clidochirus*, *Dactylocrinus*, *Euryocrinus* and *Wachsmuthicrinus*, the camerate genera *Aorocrinus*, *Megistocrinus* and *Oenochoacrinus*, the disparid genera *Desmacriocrinus*, *Synbathocrinus* and *Halysiocrinus* and the cladid genera *Hallocrinus*, *Costalocrinus* and *Lasiocrinus*. The bibliographic compilation of Palaeozoic crinoids by Webster (2003) allows an evaluation of the pre- and post-Famennian stratigraphic ranges for the taxa reported by Sepkoski to ensure that Famennian 'range-throughs' are based on valid occurrences. Based on the information in the Webster bibliography, we invalidate the Famennian occurrences of *Clidochirus* (the Mississippian occurrence is based on a questionable identification), *Hallocrinus* (the Iranian report by Webster *et al.* (2001) was later shown to be Mississippian and not *Hallocrinus* by Webster *et al.* 2003) and *Desmacriocrinus* (no pre-Mississippian occurrences reported). We consider *Dactylocrinus*, *Euryocrinus*, *Wachsmuthicrinus*, *Aorocrinus*, *Megistocrinus*, *Oenochoacrinus*, *Synbathocrinus*, *Halysiocrinus*, *Costalocrinus* and *Lasiocrinus* to be range-through taxa with no recorded Famennian occurrences.

Table 1 and the Appendix combine the generic diversity reported in Famennian blastoid and crinoid faunas with the validated range-through data from the Sepkoski database to produce our best estimate of Famennian blastoid and crinoid diversity. Recognizing that the actual Famennian

Table 1. *Generic diversity of Famennian echinoderms*

Genera	Total	Sepkoski 2002	NW China	SW China	Australia	US	Morocco	England	Saalfeld	Etroeungtian Germany	Iran
Blastoidea	15 (1)	2	12	0	0	0	1	1	0	2 (1)	0
Crinoidea											
Flexibles	9	5	5	0	1	2	0	1	0	2	1
Camerates	22	5	8	2	1	2	0	4	1	6	1
Disparids	12	4	2	1	3	1	2	0	1	3	0
Cladids	47 (11)	5	11	2	0	8	2	7	3 (11)	12	1
Total	105 (12)	21	38	5	5	13	5	13	16	26	3

New genera not yet formally described are indicated in parentheses. The Sepkoski column excludes partial data from NW China, and includes validated range-through occurences. The US column includes faunas from New York state, Missouri, colorado and Idaho. The Germany column includes historic Etroeugtian faunas plus data from Thomas & Haude (2006). A new fauna being described from Saalfeld, Germany, is listed separately because it comes from a different geological and sedimentological setting. Because some genera have been found in more than one geographic location, the Total column can be different from the sum of the individual columns. See Appendix for raw data.

dataset has been validated for taxonomy and stratigraphic occurrence but that the taxa in the Sepkoski database have only been validated for potential stratigraphic occurrence, approximately 117 genera of echinoderms occur in the Famennian, including 16 genera of blastoids and 101 genera of crinoids.

The increase in familial level diversity in Famennian echinoderms is equally impressive. Sepkoski (2002) recognized two families of blastoids and five families of crinoids when we remove the taxa reported in Lane *et al.* (1997) as discussed above. Our current compilation recognizes nine families of blastoids and 44 families of crinoids or a sevenfold increase in the familial diversity recorded in Sepkoski (2002) for Famennian echinoderms. Because we have identified taxa from all the families of Famennian echinoderms recorded by Sepkoski, there are no range-through families.

Significance of Famennian echinoderm faunas

Echinoderms were significant elements in Palaeozoic benthic marine communities, but they suffered significant reduction in diversity in extinction events that occurred at the Eifelian–Givetian and the Givetian–Frasnian boundaries (Webster *et al.* 1998). Fourteen families of crinoids and two families of blastoids went extinct during those extinction events. Primitive genera of the camerates, flexibles, disparids and cladids were especially hard hit. Harper & Rong (2001) recently noted that Devonian brachiopod diversity peaked in the Emsian. Similarly, Copper (1998) illustrated the stepwise extinction of the Atrypida during the Devonian with, in terms of numbers of genera, a maximum peak in the Emsian and highest extinction rates in the Givetian. In contrast, the Frasnian–Famennian boundary did not mark a major extinction event at the generic level for stemmed echinoderms as it apparently did for corals, brachiopods, cephalopods, tentaculids and conodonts.

Famennian marine faunas in general and echinoderm faunas in particular are rare in eastern North America and Europe, providing meagre data for a mass extinction event at the Frasnian–Famennian boundary. We believe the paucity and composition of Famennian marine faunas in Europe and North America reflect a tectonically derived sedimentological megabias similar to the megabias invoked by Smith *et al.* (2001) as an alternative explanation to the Cenomanian extinction event. Various authors such as Girard & Lécuyer (2002), Racki (1998) and Copper (2002) all argue that sedimentological and climatic changes in the Late Devonian are at least in part tectonically derived.

Based on patterns of extinction and origination using data from the Sepkoski database, Bambach *et al.* (2003) concluded that the Late Devonian was an extended time of elevated extinction. Considering both origination and extinction, these authors concluded that the marked loss of diversity in Late Devonian was a result of low origination rates rather than accelerated extinction rates. This conclusion is consistent with tectonically driven (in part at least) changes in sedimentological regime and climate focused in the tropical oceans of what is now eastern North America and western Europe.

Existing evidence suggests that rebound from the Devonian extinction events was slow. Poty (1999) concluded that after the Late Frasnian coral extinction, the long interval during which rugose corals were almost totally missing from the shallow marine platforms was probably caused mainly by cool climatic conditions and associated argillaceous-dominated environments and by the development of restricted marine environments. The re-establishment of corals can be correlated with the development of more open marine environments (still mainly siliciclastic, but including limestones) in the mid- to late-Famennian. Historically, data from North America and Europe supported the concept of echinoderm extinction at the Fransnian–Famenian boundary, with the Famennian and lowermost Mississippian characterized by rare low-diversity faunas. In this model, abundant, high-diversity echinoderm communities were not re-established until the middle Tournaisian (Kammer & Ausich 2006).

The taxonomic composition of newly discovered or redescribed Famennian echinoderm faunas clearly indicates that diversification and reradiation in the aftermath of the Devonian extinction event(s) were well underway before the close of the Famennian and were geographically concentrated in Asia outside the tectonic influence of the Appalachians. The dramatic increase in diversity at the generic and familial level suggests that rates of origination among Famennian echinoderms were very high, but were localized. As the number of identified Famennian echinoderm genera continues to grow, so does the number of important Permo-Carboniferous families that have their origins extended into the Famennian, a conclusion predicted by Lane *et al.* (1997) and Waters *et al.* (2003). The origins of the 'Age of Crinoids' discussed in Kammer & Ausich (2006) lie in the Famennian and not in the Early Mississippian.

Iterative echinoderm community reorganization

Camerate crinoids generally dominated Late Silurian through Middle Mississippian pelmatozoan echinoderm assemblages forming the Middle Palaeozoic crinoid macroevolutionary fauna

(Ausich *et al.* 1994; Kammer & Ausich 2006). Early Mississippian assemblages, particularly those located on carbonate platforms, were dominated by monobathrid camerates (Lane 1972). During the Early Mississippian cladid crinoids were typically more diverse in clastic-dominated areas, suggesting cladids and camerates differed in habitat preference. Cladids dominated virtually all echinoderm assemblages after the Osagean–Meramecian boundary (early Viséan) and formed the Late Palaeozoic crinoid macroevolutionary fauna. Waters & Maples (1991) concluded that the community reorganization resulted from a combination of changing sedimentological regime and increasing predation pressure. During the transition, many echinoderm assemblages were numerically dominated by platycrinid camerates and blastoids even though cladids were the most generically diverse elements of the faunas. Webster *et al.* (2003) noted a similar numerical abundance of camerates when describing a cladid-dominated Early Mississippian fauna from southeastern Iran.

In terms of generic diversity global Famennian echinoderm communities are dominated by cladid crinoids and are positioned between older Frasnian and younger Tournaisian echinoderm communities in the Middle Palaeozoic crinoid macroevolutionary fauna. Both the Frasnian and Tournaisian are periods of high carbonate production globally, although the details of carbonate depositional environments were very different during the two time intervals. The Famennian was a time of significantly reduced carbonate production (Copper 2002). The final transition from the Middle to Late Palaeozoic crinoid macroevolutionary fauna took place during a time of changing sedimentological regime as discussed above. All of our Famennian echinoderm communities either lived in mixed carbonate/siliciclastic or siliciclastic-dominated environments even though detailed depositional environments varied widely.

The Famennian echinoderm faunas can be divided into camerate/blastoid-dominated assemblages and cladid crinoid-dominated assemblages (Table 2) although the relationships with sedimentological regime and predation are less clear. A blastoid/camerate crinoid assemblage (>50% total diversity) can be delineated in the faunas from NW China, and from the Etroeungtian in Germany (Lane *et al.* 2001*b*). The most abundant and diverse of the Famennian echinoderm communities is from the Hongguleleng Formation, which is 200 m of mixed carbonates and siliciclastics deposited as part of an accretionary wedge off the eastern margin of Kazakhstania (Waters *et al.* 2003). Blastoid diversity and abundance in this fauna is significantly higher than in other Famennian echinoderm assemblages. This fauna is numerically dominated

Table 2. *Composition (as a percentage) of Famennian echinoderm faunas from each location*

Genera	Total	NW China	SW China	Australia	US	Morocco	England	Saalfeld	Etroeungtian Germany	Iran
Blastoidea	14	32	0	0	0	20	8	0	12	0
Crinoidea										
Flexibles	8	13	0	20	15	0	8	0	8	33
Camerates	19	21	40	20	15	0	31	6	23	33
Disparids	10	5	20	60	8	40	0	6	12	0
Cladids	50	29	40	0	62	40	54	88	46	33
Total	100	100	100	100	100	100	100	100	100	100

Percentages are based on the total fauna including known, but undescribed genera with a total diversity of 112 genera.

by one blastoid genus, *Sinopetaloblastus*, and one platycrinid crinoid genus, *Chinacrinus*. The two taxa combine for more than 50% of the total specimens collected from the Hongguleleng Formation, a situation remarkably similar to collections from the Monteagle Limestone (Genevievian; Mississippian) from Alabama in which one blastoid genus, *Pentremites*, and one platycrinid, *Platycrinites*, combine for more than 60% of total specimens collected (Horowitz & Waters 1972).

Faunas from the United States and Saalfeld, Germany, are dominated by cladids (>50%), have fewer camerates and no blastoids. The fauna from the type Devonian in England is split in terms of diversity between these assemblages. Flexible crinoids are ubiquitous members of Famennian echinoderm communities, with their absence in faunas from SW China and Saalfeld, Germany, a notable exception.

Palaeogeography

Palaeogeographically Famennian crinoids can be grouped into two broad realms, here referred to as Palaeotethys and Rheic. The Palaeotethys realm includes localities in NW China, SW China, Australia and Iran, each on different tectonic plates which bordered the Palaeotethys. The Rheic realm includes those within Laurussia (Europe and North America) and northern Africa (Morocco), all of which bordered the Rheic Ocean (Golonka 2002). At the species level all Famennian taxa are endemic at each of the localities in both realms. At the generic level range-through genera and genera with their first occurrence in the Famennian may occur in the other or both realms at different or approximately coeval times later.

Table 3 shows that the blastoids are much more diverse in the Palaeotethys realm (12 genera) than in the Rheic realm (two identified genera and one new genus). Flexibles are almost twice as diverse in the Palaeotethys realm than in the Rheic realm. Camerates are only slightly more diverse in the Palaeotethys realm and disparids are divided evenly between the two realms. Cladids are more than twice as diverse in the Rheic realm (26 identified genera and 11 new genera from Saalfeld) as in the Palaeotethys realm (16 genera). The reasons for these differences are uncertain, but may reflect differences in environments, including increased tolerance for clastic sedimentation. However, they could also be the result of literature bias, in that European and North American echinoderm reports dating well back into the 1800s vastly outnumber those from the Palaeotethys which are mostly within the past two decades.

Table 3. *Number of identified Famennian genera reported from the Palaeotethys and Rheic Realms*

Taxonomic group	Paleotethys realm	Rheic realm
Blastoids	12	3 (1)
Crinoids		
Flexibilia	5	3
Camerata	11	9
Disparida	6	6
Cladida	16	26 (11)
Total	50	47 (12)

Numbers in parentheses represent new genera under study from Germany. Table does not include range-through genera that are not known in the Famennian.

The numbers of genera in Table 3 suggest that the Palaeotethys was a refugium for the blastoids in the Famennian and the Rheic realm was a greater refugium for the cladids with few genera common to both areas, yet both areas have approximately the same total number of genera. This suggests that there was minimal seaway connections between the two realms.

It should also be noted that there are no Famennian blastoids reported from SW China, Australia, the United States or Iran; no flexibles are recognized in SW China or Morocco, no camerates in Morocco, no disparids in England or Iran, and no cladids from Australia. The lack of these different groups at the various localities may be artifacts of collection or may be the result of environmental controls.

Comparing the number of genera in each of the localities (Table 1) the decreasing diversity is: NW China (38 genera), the Etroeungtian of Germany (26), Saalfeld (16), England (13), North America (13), SW China (5), Australia (5), Morocco (5) and Iran (3). We believe that these numbers reflect environmental differences; future collection at these known localities will probably not significantly add to the relative numbers with the exception of SW China, which we consider to be inadequately investigated. However, the discovery of new localities could grossly alter the relative numbers within the two realms. We consider that the Palaeotethys realm, part of Russia and South America are vastly underinvestigated. The discovery of the Hongguleleng locality in the 1990s is a classic example of one locality adding immensely to the known Famennian diversity.

Conclusions

Recent advances in the understanding of Famennian echinoderm communities on a global basis allow us to reinterpret patterns of Late Devonian extinction and rebound. Echinoderms suffered decreases in diversity at the Eifelian–Givetian and the Givetian–Frasnian boundaries but were less impacted by diversity depletions at the Frasnian–Famennian boundary. With a generic diversity of approximately 117 genera, Famennian echinoderm communities are less diverse than Middle Devonian or Early Mississippian communities, but the decline is not as precipitous as previously believed. The rebound from depletions in echinoderm diversity in the Late Devonian was well underway in the Famennian. Famennian echinoderm communities can be divided into camerate/blastoid-dominated assemblages and cladid crinoid-dominated assemblages, presaging the transition of crinoid macroevolutionary faunas. We also recognize that all Famennian species are endemic and that Famennian genera can be divided into two palaeogeographic realms, Palaeotethys and Rheic, with very few genera common to both realms. The absence of some of the major echinoderm groups in the various localities may be artifacts of collection or environmental control that future studies may resolve. Vast areas in Russia, Australia and Asia that have been inadequately mapped and searched should add significantly to the known Famennian blastoid and crinoid faunas and diversity.

The authors acknowledge many fruitful discussions on echinoderms over the years with C. G. Maples and N. G. Lane (deceased). The manuscript was improved significantly by two anonymous reviews. J. A. W.'s field research in China was supported by a grant from the National Geographic Society, and NSF research grants EAR-9117673, EAR-9404729, and the Appalachian Faculty Fellows programme.

Appendix

Table A1. *Crinoid and blastoid species described from various Famennian aged localities*

	Sepkoski*	NW China	SW China	Australia	Colorado	New York	Idaho	Missouri	Morocco	England	Saalfeld	Germany	Iran
Blastoidea													
Fissiculata													
Phaenoschismatidae													
Junggaroblastus hoxtolgayensis	×	×											
Phaenoschismatid, gen. and sp. indet.										×			
Orophocrinidae													
Orophocrinus devonicus	×	×											
Codasteridae													
Emuhablastus planus		×											
Astrocrinidae													
Tripoblastus plicatus		×											
Neoschismatidae													
Hadroblastus n. sp.		×											
Troosticrinida													
Troosticrinidae													
Uyguroblastus conicus	×	×											
Pentremitida													
Hyperoblastidae													
Petaloblastus ovalis												×	

(*Continued*)

Table A1. *Continued*

	Sepkoski*	NW China	SW China	Australia	Colorado	New York	Idaho	Missouri	Morocco	England	Saalfeld	Germany	Iran
Petaloblastus boletus	×											×	
?*Petaloblastus ovalis*										×			
Sinopetaloblastus jinxingae	×	×											
Sinopetaloblastus grabaui		×											
Breimeriblastus pyramidalis		×											
Breimeriblastus gracilis		×											
Conoblastus invaginatus		×											
Hyperoblastus emuhaensis		×											
Orbitremitidae													
Doryblastus melonianus	×											×	
Houiblastus devonicus	×	×											
Granatocrinida													
Granatocrinidae													
Xinjiangoblastus ornatus	×	×											
n.g., n. sp. (Thomas & Haude 2006)												×	
Order, family, and genus indeterminate										×			
Crinoidea													
Flexibilia													
Taxocrinida													
Taxocrinidae													
Eutaxocrinus chinaensis	×	×											
Eutaxocrinus boulongourensis		×											
Eutaxocrinus basellus		×											
Eutaxocrinus risehensis													×
Eutaxocrinus sp.												×	×
Taxocrinus anomalus		×											
Taxocrinus macrodactylus	×									×		×	
Taxocrinus stultus										×			
Taxocrinus sp.					×								
Taxocrinid indet. 1					×							×	
Taxocrinid indet. 2					×								
Synerocrinidae													
Euonychocrinus websteri		×											
Sagenocrinida													
Sagenocrinidae													
Forbesiocrinus inexpectans	×	×											
Forbesiocrinus sp.				×									
Clidocrinus	I												
Dactylocrinus	×												
Euryocrinus	×												
Paraclidochirus	×						×						
Wachsmuthicrinus	×												

(*Continued*)

Table A1. *Continued*

	Sepkoski*	NW China	SW China	Australia	Colorado	New York	Idaho	Missouri	Morocco	England	Saalfeld	Germany	Iran
Dactylocrinidae													
Labrocrinus granulatus		×											
Family, genus, and species unknown												×	
Camerata													
Diplobathrida													
Rhodocrinitidae													
Rhipidocrinus schmidti												×	
Gilbertsocrinus sp.	×											×	
Dimerocrinitidae													
Gnarycrinus lanei		×											
Monobathrida													
Periechocrinidae													
Athabascocrinus orientale		×											
Megistocrinus	×												
Periechocrinacea uncertain												×	
Amphoracrinidae													
Amphoracrinus sp.		×											
Hexacrinitidae													
Hexacrinites pinnulata		×											
Hexacrinites persiaensis													×
Adelocrinus hystrix										×		×	
Parahexacrinidae													
Agathocrinus junggarensis	×	×											
Actinocrinitidae													
Abactinocrinus devonicus		×											
Actinocrinites zhaoae		×											
Actinocrinites batheri										×			
Eumorphocrinus porteri										×			
Carpocrinidae													
Cylicocrinus												×	
Dichocrinidae													
Dichocrinus cf. *D. expansus*												×	
Dichocrinidae uncertain												×	
Melocrinitidae													
Melocrinites bainbridgensis						×							
Melocrinites conicus			×										
Melocrinites stellatus			×										
Hapalocrinidae													
Changninocrinus sphaeroidea			×										
Platycrinitidae													
Chinacrinus xinjiangensis	×	×											
Chinacrinus nodosus		×											
Chinacrinus species A		×											

(*Continued*)

Table A1. *Continued*

	Sepkoski*	NW China	SW China	Australia	Colorado	New York	Idaho	Missouri	Morocco	England	Saalfeld	Germany	Iran
Chinacrinus species B		×											
Oenochoacrinus	×												
Platycrinites sp.										×			
?*Platycrinites wunstorfi*												×	
Platycrinites guttifer	×											×	
Platycrinitid indeterminate 1											×		
Wacrinus millardensis				×									
Wacrinus caseyensis				×									
Aorocrinus	×												
Strimplecrinus dyerensis					×								
Disparida													
Calceocrinidae													
Deltacrinus asiaticus	×	×											
Desmacriocrinus	I												
Halysiocrinus	×												
Eohalysiocrinus												×	
Playfordicrinus kellyensis				×									
Jaekelicrinus murrayi				×									
Pisocrinidae													
Parapisocrinus sp.									×				
Triacrinus granulatus	×										×	×	
Triacrinus pyriformis											×	×	
Allagecrinidae													
Allagecrinus	×								×				
Catillocrinidae													
?Catillocrinid sp.				×									
Catillocrinid undesignated									×				
Anamesocrinidae													
Anamesocrinus tieni		×										×	
Haplocrinitidae													
Haplocrinites bipyramidatus			×										
Synbathocrinidae													
Synbathocrinus	×												
Cladida													
Codiacrinidae													
Codiacrinus sp.											×		
Ovalocrinus pyriformis			×										
Quasicydonocrinus typicus			×										
"Poteriocrinus" mespiliformis											×		
?*Elicrinus weyeri*									×				
Pentececrinus	×							×					
Cyathocrinidae													
Cyathocrinus													

(Continued)

Table A1. *Continued*

	Sepkoski*	NW China	SW China	Australia	Colorado	New York	Idaho	Missouri	Morocco	England	Saalfeld	Germany	Iran
Lecythocrinidae													
Cestocrinus												×	
Barycrinidae													
Pellecrinus												×	
Euspirocrinidae													
Parisocrinus	×												
Genus indet.												×	
Botryocrinidae													
Costalocrinus	×												
? Parisocrinus nodosus	×	×											
? Parisocrinus conicus		×											
Plicodendrocrinidae													
N. gen., sp. 1											×		
?N. gen., sp.											×		
Mastigocrinidae													
Quantoxocrinus clarkei						×							
? Quantoxocrinus singulocirrus												×	
Lasiocrinus sp.	×												
N. gen., n. sp.											×		
Glossocrinidae													
Glossocrinusn. sp.													×
?*Glossocrinus salebrosus*										×			
?*Glossocrinus* sp.													
Catactocrinus singulocirrus												×	
Propoteriocrinus												×	
N. gen., n. sp.											×		
Amabilicrinidae													
Bufalocrinus torus					×								
Hallocrinus	I												
?*Hallocrinus sp.*													×
Amabilicrinus whidbornei										×			
Mrakibocrinus brockwinkeli									×				
Moroccocrinus ebbighauseni									×				
Glossocrinacea, family uncertain													
N. gen., sp. 1											×		
Genus indet.		×											
Poteriocrinitidae													
Poteriocrinites transcisus										×			
Indet. primitive poteriocrinid													
N. gen. sp. 1											×		
N. gen. sp. 2											×		

(*Continued*)

Table A1. *Continued*

	Sepkoski*	NW China	SW China	Australia	Colorado	New York	Idaho	Missouri	Morocco	England	Saalfeld	Germany	Iran
N. gen. sp. 3											×		
N. gen.? sp. 4											×		
Rhenocrinidae													
Cydrocrinus sp.													×
Dendrocrinina new family													
N. gen., n. sp.											×		
N. gen., n. sp.											×		
Scytalocrinidae													
Scytalocrinus barumensis	×									×			
Scytalocrinus sp. indet.												×	
Julieticrinus romeo		×											
?*Hypselocrinus bisonensis*					×								
Histocrinus ? *cheni*		×											
Histocrinus sp.		×											
Cercidocrinidae													
Cercidocrinus												×	
Eumhacrinus tribrachiatus		×											
Genus and species indet.										×			
Bridgerocrinidae													
Bridgerocrinus discus	×	×											
Bridgerocrinus stadiodactylus										×			
Bridgerocrinus delicatulus		×											
?*Bridgerocrinus arachnoideus*										×			
Bridgerocrinus sp.												×	
Blothrocrinidae													
Blothrocrinus sp.					×								
Culmicrinus cylindricus													×
Aphelecrinidae													
Aphelecrinus tensus										×			
?*Aphelecrinus* sp.										×			
Cosmetocrinus sp.												×	
Cosmetocrinus parvus	×	×											
Paracosmetocrinus sp.												×	

(*Continued*)

Table A1. *Continued*

	Sepkoski*	NW China	SW China	Australia	Colorado	New York	Idaho	Missouri	Morocco	England	Saalfeld	Germany	Iran
Sostronocrinidae													
Sostronocrinus mundus										×			
Sostronocrinus quadribrachiatus		×											
Sostronocrinus minutus		×											
Sostronocrinus ratingensis												×	
Sostronocrinus paprothae												×	
Sostronocrinus pauli												×	
Sostronocrinus sp. indet.												×	
Amadeusicrinus subpentagonalis		×											
Decadocrinidae													
Grabauicrinus xinjiangensis		×											
Grabauicrinus constrictus		×											
Grabauicrinus elongatus		×											
Grabauicrinus rugosus		×											
"Decadocrinus" usitatus	×	×											
? *Eireocrinus coloradoensis*					×								
Pachylocrinidae													
? *Pachylocrinus plumifer*										×			
Graphiocrinidae													
? *Graphiocrinus* sp.		×											
Holcocrinus asiaticus	×	×											
"Gilmocrinus" albus					×								
Cladid family indet.													
Thuringocrinus saalfeldianus											×		
Cladid indeterminate 1											×		
Tarassocrinus synchlydus					×								
Cladid indeterminate 2		×											
Cladid indeterminate 3		×											
Cladida indet. 1					×								
Cladida indet. 2					×								
Cladida indet. 3					×								
Cladida indet. 4					×								
Cladida indet. 5					×								

*Famennian occurrences for taxa marked 'I' have been invalidated.

References

AUSICH, W. I., KAMMER, T. W. & BAUMILLER, T. K. 1994. Demise of the middle Paleozoic crinoid fauna: A single extinction event or rapid faunal turnover? *Paleobiology*, **20**, 345–361.

BAMBACH, R. K., KNOLL, A. H. & WANG, S. G. 2003. Origination, extinction and mass depletions of marine diversity. *Paleobiology*, **30**, 533–542.

CHEN, Z. T. & YAO, J. H. 1993. *Palaeozoic Echinoderm Fossils of Western Yunnan, China*. Geological Publishing House, Beijing.

COPPER, P. 1998. Evaluating the Frasnian-Famennian mass extinction: Comparing brachiopod faunas. *Acta Palaeontologica Polonica*, **43**, 137–154.

COPPER, P. 2002. Reef development at the Frasnian/Famennian mass extinction boundary. *Palaeogeography, Palaeoclimatology, Palaeoecology*, **181**, 27–65.

GIRARD, C. & LÉCUYER, C. 2002. Variations in Ce anomalies of conodonts through the Frasnian/Famennian boundary of Poland (Kowala – Holy Cross Mountains): Implications for the redox state of seawater and biodiversity. *Palaeogeography, Palaeoclimatology, Palaeoecology*, **181**, 299–311.

GOLONKA, J. 2002. Plate-tectonic maps of the Phanerozoic. *In*: KIESSLING, W., FLUEGEL, E. & GOLONKA, J. (eds) *Phanerozoic Reef Patterns*. Society for Sedimentary Geology, Special Publications, **72**, 21–75.

HARPER, D. A. T. & RONG, J.-Y. 2001. Paleozoic brachiopod extinctions, survival and recovery: Patterns within the rhynchonelliformeans. *Geological Journal*, **36**, 317–328.

HOROWITZ, A. S. & WATERS, J. A. 1972. A Mississippian echinoderm site in Alabama. *Journal of Paleontology*, **46**, 660–665.

JELL, P. A. & JELL, J. S. 1999. Crinoids, a blastoid and a cyclocystoid from the Upper Devonian Reef Complex of the Canning Basin, Western Australia. *Memoirs of the Queensland Museum*, **43**, 201–236.

KAMMER, T. W. & AUSICH, W. I. 2006. The 'Age of Crinoids': A Mississippian biodiversity spike coincident with widespread carbonate ramps. *Palaois*, **21**, 238–248.

LANE, N. G. 1972. Synecology of Middle Mississippian (Carboniferous) crinoid communities in Indiana. *Paleontology, Section 7*. International Geological Congress, **24**, 89–94.

LANE, N. G., WATERS, J. A. & MAPLES, C. G. 1997. *Echinoderm faunas of the Hongguleleng Formation, Late Devonian (Famennian), Xinjiang-uygur Autonomous Region, People's Republic of China*. The Paleontological Society, Memoir, **47**, 1–43.

LANE, N. G., MAPLES, C. G. & WATERS, J. A. 2001*a*. Revision of Late Devonian (Famennian) and some Early Carboniferous (Tournaisian) crinoids and blastoids from the type Devonian area of North Devon. *Palaeontology*, **44**, 1043–1080.

LANE, N. G., MAPLES, C. G. & WATERS, J. A. 2001*b*. Revision of Etroeungtian crinoids and blastoids from Germany. *Paläontologische Zeitschrift*, **75**, 233–252.

POTY, E. 1999. Famennian and Tournaisian recoveries of shallow water Rugosa following late Frasnian and late Strunian major crises, southern Belgium and surrounding areas, Hunan (South China) and the Omolon region (NE Siberia). *Palaeogeography, Palaeoclimatology, Palaeoecology*, **154**, 11–26.

RACKI, G. 1998. Frasnian–Famennian biotic crisis: Undervalued tectonic control? *Palaeogeography, Palaeoclimatology, Palaeoecology*, **141**, 177–198.

SEPKOSKI, J. J. JR. 2002. *A Compendium of Fossil Marine Animal Genera*. (edited by JABLONSKI, D. & FOOTE, M.). Bulletins of American Paleontology, **363**.

SMITH, A. B., GALE, A. S. & MONKS, N. E. A. 2001. Sea-level change and rock-record bias in the Cretaceous; a problem for extinction and biodiversity studies. *Paleobiology*, **27**, 241–253.

THOMAS, E. & HAUDE, R. 2006. Echinodermen. *In*: Deutsche Stratigraphische Kommission (eds) *Stratigraphie von Deutschland VI. Unterkarbon (Mississippium)*. Schriftenreihe der Deutschen Gesellschaft für Geowissenschaften, **41**, 183–197.

WATERS, J. A. & MAPLES, C. G. 1991. Mississippian pelmatozoan community reorganization; a predation-mediated faunal change. *Paleobiology*, **17**, 400–410.

WATERS, J. A., MAPLES, C. G. *ET AL*. 2003. A quadrupling of Famennian pelmatozoan diversity: New Late Devonian blastoids and crinoids from the Late Devonian (Famennian) of Northwest China. *Journal of Paleontology*, **77**, 922–948.

WEBSTER, G. D. 2003. *Bibliography and Index of Paleozoic Crinoids, Coronates and Hemistreptocrinoids, 1758–1999*. Geological Society of America, Special Paper, **363**.

WEBSTER, G. D. & WATERS, J. 2009. Late Devonian echinoderms from the Hongguleleng Formation of northwestern China. *In*: KÖNIGSHOF, P. (ed.) *Devonian Change: Case Studies in Palaeogeography and Palaeoecology*. Geological Society, London, Special Publications, **314**, 263–287.

WEBSTER, G. D., LANE, N. G., MAPLES, C. G., WATERS, J. A. & HOROWITZ, A. S. 1998. Frasnian–Famennian extinction was a non-event for crinoids, blastoids, and bryozoans. Geological Society of America Annual Meeting, *Program with Abstracts*, **30**, 30–31.

WEBSTER, G. D., HAFLEY, D. J., BLAKE, D. B. & GLASS, A. 1999. Crinoids and stelleroids (Echinodermata) from the Broken Rib Member, Dyer Formation (Late Devonian, Famennian) of the White River Plateau, Colorado. *Journal of Paleontology*, **73**, 461–486.

WEBSTER, G. D., YAZDI, M., DASTANPOUR, M. & MAPLES, C. 2001. Preliminary analysis of Devonian and Carboniferous crinoids and blastoids from Iran. *Travaux de l'Institut Scientifique, Rabat, Série Géologie et Géographie Physique (2000)*, **20**, 108–115.

WEBSTER, G. D., MAPLES, C. G., MAWSON, R. & DASTANPOUR, M. 2003. *A cladid-dominated Early Mississippian crinoid and conodont fauna from Kerman Province, Iran and revision of the glossocrinids and rhenocrinids*. The Paleontological Society, Memoir, **60**.

WEBSTER, G. D., BECKER, R. T. & MAPLES, C. G. 2005. Biostratigraphy, paleoecology and taxonomy of Devonian (Emsian and Famennian) crinoids from southeastern Morocco. *Journal of Paleontology*, **79**, 1052–1071.

WEBSTER, G. D., MAPLES, C. G. & YAZDI, M. 2007. Late Devonian and Early Mississippian echinoderms from central and northern Iran. *Journal of Paleontology*, **81**, 1101–1113.

Upper Devonian miospore and conodont zone correlation in western Europe

M. STREEL

Paleobotany, Paleopalynology, Micropaleontology Unit, Department of Geology, University of Liège, Belgium
(e-mail: Maurice.Streel@ulg.ac.be)

Abstract: The stratigraphical occurrence of 38 Upper Devonian miospore taxa is compared to some miospore and conodont zones in 28 intercalibrated levels. The accurate position of 15 miospore First Occurrence Biohorizons and one Last Occurrence Biohorizon in terms of the conodonts available in a few regions of western Europe is discussed in detail and their correlation evaluated.

Publications on the stratigraphical distribution of Devonian miospores effectively started with the classic work of Naumova (1953) on the Russian Platform. Since then, the Devonian has been zoned by various palynologists, mainly in Europe and North America, but also elsewhere around the world. Major works provide reviews and references containing detailed biostratigraphic synthesis, while illustrating most of the characteristic miospores (McGregor 1979*a*, *b*; Richardson & McGregor 1986; Richardson & Ahmed 1989; McGregor & Playford 1992; Avkhimovitch *et al.* 1993). The zonal concept used in these papers is most often the Assemblage Zone, defined, at best, by a combination of different criteria such as the appearance and disappearance of selected miospore structural or sculptural features and the first and last occurrence of selected taxa, two of them giving their name to the zone. Sometimes, the assemblages are Acme Zones (Avkhimovitch *et al.* 1993).

Streel *et al.* (1987) separated Oppel Zones and Interval Zones for the miospores of the whole Devonian in the Ardenne–Rhine regions. The Oppel Zone is difficult to define empirically because judgment may vary as to how many and which of the selected diagnostic taxa need to be present to identify the zone (Hedberg 1976). The Interval Zone is an interval between two distinctive Biohorizons. The advantage of the Interval Zone and Biohorizon concepts is that they allow unequivocal correlations with Interval Zones based on other fossils. First (or rarely last) occurrence(s) of single species (First Occurrence Biohorizon (FOB) or Last Occurrence Biohorizon (LOB) were located in continuous miospore-bearing marine sequences, preferably in uniform lithologies. The marine sequences do not normally contain the best preserved and diversified miospores because transport may have altered their morphology. But their study provides the unique possibility to obtain reliable correlation with marine faunas. The quality of these correlations depends on the distance in time and space between the miospore data on one hand and the nearest reliable faunal data (here conodonts) on the other. It depends also on the kind of stratigraphy (litho- versus bio-) used for indirect correlation.

Therefore a Correlation Quality Index (CQI: 1 to 6, best to worst) was proposed (Streel & Loboziak 1994, 1996; Streel *et al.* 2000*a*). In the present paper a simplified version of the CQI is used (CQI: A,B,C, best to worst) which characterizes the correlations between conodonts and miospores occurring in one and the same section (A) or in different but correlated sections (B), or not correlated (C) at all, by independent faunas or floras. The distance between the correlated sections may be indicated by a plus sign (+) at short distance (hundreds of metres to a few kilometres), a minus sign (−) at long distance (tens of kilometres) or two minus signs at (− −) very long distance (hundreds or more kilometres). The CQI for the biohorizons described in this paper are also indicated on Figures 1 to 5. Of course the correlation value depends also on the reliability of the reference conodont fauna itself, a subject which is not discussed here (see Johnson 1992; Bultynck 2007).

Chronostratigraphic notations refer to the *Subcommission on Devonian Stratigraphy Newsletters* 19 to 22 (2003 to 2007). As substages are not yet formally accepted, initial capital letters are avoided. Intercalibrated levels (first line on Figs 2, 4, 6, 7, 8) are intended only to help in recognizing stratigraphical steps in the western European Upper Devonian, but not to be used as a new zonation. They should replace the intercalibrated levels published by Streel *et al.* (1987). Each level is characterized by miospore-based biohorizons (<u>*X*></u>) or/and nominal taxa of Oppel Zones (<u>X</u>) or/and changes in Conodont Zones. The occurrences of a few selected miospore taxa (X), helpful for Oppel Zones identification, are also given.

From: KÖNIGSHOF, P. (ed.) *Devonian Change: Case Studies in Palaeogeography and Palaeoecology*.
The Geological Society, London, Special Publications, **314**, 163–176.
DOI: 10.1144/SP314.9 0305-8719/09/$15.00

BOULONNAIS REGION											
Chron.	Lithostratigraphy				Biostratigraphy						
	Formations	Members			Conodonts					Miospores	
			(B)	(A)	Faunas	Old Zones	New Zones	MN Z.	CQI	Biohorizons	Zone bases
l. Fam	Ste Godeleine										
				205			*triangularis ?*		B – –	*K. dedaleus* FOB	DV
FFB											
				1221							BA plic
u. Frasn.	Hydrequent			305			*linguiformis-* U. *rhenana ?*	13	B – –	*G. gracilis* FOB	BA grac
		"dolomitic bed"									
				24			*rhenana ?*	12?	B – –	*R. bricei FOB*	BA pre-grac
		Gris	a–b		*'Ag.coeni'*			? 8-11			
	Ferques	Parisienne	a–f				*-jamieae*	or			
m. Frasn.		Bois	a–e	vw 5			L. *hassi-*	? 7-10	C+	*L. media* FOB	BM
		Fiennes	a–d								
		Pâtures	a–d			M. *asymmetricus*	*punctata*	? 5			
			c	P	*A.gigas*						
	Beaulieu	Noces		O 5		L. *asymmetricus*	*transitans-*	? 4	A	*V.bulliferus* FOB	BJ
l. Frasn.			a–b	N			*falsovialis*				
		Cambresèque	b	32	IX *A.alata*			2? or 3			
GFB?			a								
		Bastien	a–b								
Giv. pars	Blacourt	Couderousse		H 26	V		M/U. *varcus*		A	*C. concinna* FOB	TCo
		Griset	g		IV	*ansatus*	L-M. *varcus*				

Fig. 1. Late Givetian to early Famennian in the Boulonnais region. Column headings: (A), number of samples in text; (B), subdivisions in Brice (2003). Abbreviations: FFB, Frasnian–Famennian boundary; GFB, Givetian–Frasnian boundary.

A list of the 38 miospore species is given in Table 1.

The Givetian–Frasnian boundary (384 Ma, Kaufmann 2006) and the lower and middle Frasnian (Figs 1 and 2)

Miospores occur with conodonts of the Givetian–Frasnian transitional beds in the Ferques railroad section in the Boulonnais region, northern France. They allow rather good correlation (CQI A) with the conodont zonation which was demonstrated there by Bultynck (in Brice *et al.* 1979, table III; emended by observations made by Coen in Brice *et al.* 1981). In the Couderousse Member of the Blacourt Formation, fauna V corresponds to the Mid or/and Upper *varcus* Zones (Bultynck in Brice 1988; Brice *et al.* 2002; P. Bultynck pers. comm. 2008). The basal beds of the succeeding Beaulieu Formation are not present in this section. A few metres above the first shales of the Cambresèque Member occurs fauna IX with *Ancyrodella rotundiloba alata* (Brice et *al.* 1979) which has a rather low occurrence within the 'old' *asymmetricus* Zone, now (Ziegler & Sandberg 1990) the *falsiovalis*

Intercalibrated levels:	1	2	3	4
Chronostratigraphy:	u.Givet.	l.Frasn.	m.Frasn.	m.Frasn.
New conodont zones:	dispar.-fals.	falsiovalis.	trans. to punct.	hassi to jam.
Old or alternative conodont zones:		MN 1 to 3	MN 4 to 6	MN 7 to 11
Miospore ass. zones (Rich. & McGr. 1986):	opt.-triang.	opt.-triang.	oval.-bull.	oval.-bull.
Miospore Oppel / interval zones (Streel et al. 1987):	TCo	TCo	BJ	BM
CQI for miospore biohorizon (X>)			*A*	*C+*
Diagnostic taxa of miospore zones *(nominal taxa)*				
Chelinospora concinna	X	X	X	X
Samarisporites triangulatus	X	X	X	X
Cirratriradites jekhowskyi			**X**	X
Verrucosisporites bulliferus			**X>**	**X**
Hystricosporites multifurcatus			X	X
Lophozonotriletes media				**X>**
Pustulatisporites rugulatus				X

Fig. 2. Intercalibrated levels and diagnostic miospore taxa from upper Givetian to middle Frasnian.

BOULONNAIS REGION					EIFEL REGION			
Lithostratigraphy			Biostratigraphy		Lithostrat.		Biostratigraphy	
Formation	Member	(A)	Conodonts	Miospores	Formations	(A)	Conodonts	Miospores Biohorizons
Blacourt	**Couderousse**	H 27	U. *varcus* M. *varcus* L. *varcus*	*S. triangulatus*			U. *varcus* M. *varcus* L. *varcus*	
					Kerpen (upper) **Kerpen (lower)** **Cûrten**	44	*hemiansatus CQI A*	*S. triangulatus* FOB

Fig. 3. *Samarisporites triangulatus* occurrence in the Givetian of the Boulonnais and Eifel regions. (A), Number of samples cited in text.

(partim) and the *transitans* Zones or MN zones 1 to 4 of Klapper (1989). The fauna IX probably belongs to MN zone 2 or even 3 on a new finding of *A. rugosa* in the same sample 32 (P. Bultynck pers. comm. 2008). The first occurrence of *Ancyrodella gigas* noted by Coen (in Brice *et al.* 1981) in the Noces Member of the Beaulieu Formation is immediately below the base of the 'old' Middle *asymmetricus* Zone, now (Ziegler & Sandberg 1990), the mid Frasnian *punctata* Zone or MN zones 5 and 6 of Klapper (1989). Becker (2002, p. 135) suggested that the base of next higher, shaly Pâture Member correlates with the conodont MN 5.

Miospores of the late mid-Frasnian beds occur in a few quarries in the Ferques Formation of the Boulonnais region (Brice *et al.* 1981, fig. 1). In the 'La Parisienne' Quarry, *Ancyrognathus triangularis euglypheus* (now *Ag. coeni*) is present, indicating the Lower *hassi* or *jamieae* conodont Zone or MN zones 8 to 11 of Klapper (1989) (P. Bultynck pers. comm. 2008).

Four successive miospore Oppel Zones, *S. triangulatus–Ancyrospora ancyrea* var. *ancyrea* (TA), *S. triangulatus–Chelinospora concinna* (TCo), *Verrucosisporites bulliferus–Cirratriradites jekhowskyi* (BJ) and *Verrucosisporites bulliferus–Lophozonotriletes media* (BM), are present in the range of these conodont zones in the Boulonnais region. The last three are described as zones I, II and III by Loboziak & Streel (1981), being partly renamed in the same paper. The three zones are completely renamed in Streel *et al.* (1987, fig. 7).

Intercalibrated levels:	5	6	7	8
Chronostratigraphy:	u.Frasn.	u.Frasn.	u.Frasn.-l.Fam.	l.Fam.
New conodont zones:	rhenana?	U.rhen.-linguiformis?	U.rhen.-triangularis?	triang.? to u.crep.
Old or alternative conodont zones:	MN 12?	MN 13		
Miospore ass. zones (Rich. & McGr. 1986):		torq.-grac.	torq.-grac.	torq.-grac.
Miospore phases (Streel *et al.* 1987):	'IV A to B'	'IV C to D2'	'IV D3-E'	'V' - GH
New miospore Oppel/interval zones (this paper)	BA pre-grac.	BA grac.	BA plic.	DV
CQI for miospore biohorizon (X>):	*B – –*	*B – –*		*B – –*
Diagnostic taxa of miospore zones (nominal taxa)				
Samarisporites triangulatus	X	X	X	
Cirratriradites jekhowskyi	X	X	X	
Verrucosisporites bulliferus	X	X	X	X
Hystricosporites multifurcatus	X	X	X	X
Lophozonotriletes media	X	X	X	
Pustulatisporites rugulatus	X	X	X	
Cymbosporites acanthaceus	**X**	X	X	
Rugospora bricei	***X>***	X	X	X
Cristatisporites deliquescens	X	X	X	
Diducites poljessicus	X	X	X	X
Diducites mucronatus	X	X	X	X
Teichertospora torquata		X	X	X
Grandispora gracilis		***X>***	X	X
Retusotriletes planus		X	X	X
Diducites plicabilis			**X**	X
Retusotriletes incohatus				X
Lophozonotriletes lebedianensis				X
Diducites versabilis				**X**
Knoxisporites dedaleus				***X>***

Fig. 4. Intercalibrated levels and diagnostic miospore taxa from upper Frasnian to lower Famennian.

Chron.	ARDENNE REGION							
	Lithostratigraphy			Biostratigraphy				
	Formations	Members	(A)	Conodonts Old Zones	Conodonts New Zones	CQI	Miospores Biohorizons	Miospores Zone bases
u* Fam.								LE
	Comblain-au-Pont		111	L.-M. *costatus*	*U. expansa*	A		
			22			*B* – –	*R. lep.lepidophyta* FOB	LL
	Evieux		20'd	L. *costatus*	M. *expansa*		*V. hystricosus* FOB.	VH
		Fontin						
u. Fam.	Comblain-la-Tour		54	*styriacus* U. *velifer*	*postera* ? U. *trachytera*	*C* +	*G. cornuta* FOB	VCo
	Montfort	Bon-Mariage	19	M. *velifer*	? L. *trachytera*	C +	*R. macroreticulata* FOB	GF mac
			13	L. *velifer*	U* *marginifera*	A,C +	*G. microseta* FOB	GF mic
	Souverain-Pré							
m. Fam. Part	Esneux		13	L. *marginifera*	L. *marginifera*	A,C –	*G. fam.famenensis* FOB	
			4				*G. fam.minuta* FOB	GF pre-mic

Fig. 5. Middle to uppermost Famennian in the Ourthe Valley. (A), number of samples cited in text.

Richardson & McGregor (1986) described two Assemblage Zones (*Contagisporites optivus* var. *optivus–Cristatisporites* (now *Samarisporites*) *triangulatus Zone* and *Archaeoperisaccus ovalis–Verrucosisporites bulliferus* Zone) in about the same timespan. The limit between these Assemblage Zones corresponds to the lower limit of the BJ Zone (Streel *et al.* 1987, fig. 13). For that reason, the *C. optivus* var. *optivus–C. triangulatus* Zone, the base of which is characterized by the first appearance of *Samarisporites triangulatus*, includes the TA and TCo Oppel Zones. Richardson & McGregor (1986) also use the Ferques railroad section of Loboziak & Streel (1980) as a reference

Intercalibrated levels:	9	10	11	12
Chronostratigraphy:	m.Fam.	m.Fam.	u.Fam.	u.Fam.
New conodont zones:	u*.crep.to rhomb.	l. to u.marg..	u*.marg.	l.trach.?
Old or alternative conodont zones:			l.vel.	m.vel.?
Miospore ass. zones (Rich. & McGr. 1986):	torq.-grac.	(immensus ?)	?	?
Miospore Oppel / interval zones (Streel *et al.* 1987):	'V'-GH	GFl	GFm	GFu
New miospore Oppel/interval zones (this paper)	DV	GF pre-mic	GF mic	GF mac
CQI for miospore biohorizon (X>):		*A and C–*	*A and C+*	*C+*
Diagnostic taxa of miospore zones (nominal taxa)				
Verrucosisporites bulliferus	X	X	X	X
Hystricosporites multifurcatus	X	X	X	X
Rugospora bricei	X			
Diducites poljessicus	X	X	X	X
Diducites mucronatus	X	X	X	X
Teichertospora torquata	X	X	X	X
Grandispora gracilis	X	X	X	X
Retusotriletes planus	X	X	X	X
Diducites plicabilis	X	X	X	X
Retusotriletes incohatus	X	X	X	X
Lophozonotriletes lebedianensis	X	X	X	X
Diducites versabilis	X	X	X	X
Knoxisporites dedaleus	X	X	X	X
Grandispora famenensis var. minuta		**X>**	X	X
Grandispora famenensis var. famenensis		**X>**	X	X
Grandispora microseta			**X>**	X
Retispora macroreticulata				**X>**

Fig. 6. Intercalibrated levels and diagnostic miospore taxa from middle to upper Famennian.

Intercalibrated levels:	13	14	15	16	17	18	19
Chronostratigraphy:	u.Fam.	u.Fam.	u.Fam.	u.Fam.	u.Fam.	u.Fam.	u.Fam.
New conodont zones:	u.trach.	l.post.	u.post.	l.exp.	l.exp.	m.exp.	m.exp.
Old or alternative conodont zones:	u.vel.	l.styr.	m.styr	u.styr.	l.cost.	l.cost.	l.cost.
Miospore ass. zones (Rich. & McGr. 1986):	?	?	flex-corn	flex-corn	flex-corn	flex-corn	?
Miospore Oppel / interval zones (Streel *et al.* 1987):	GFu	GFu?	VCo	VCo	VCo	VCo	VH
New miospore Oppel/interval zones (this paper)	GF mac	GF mac?					
CQI for miospore biohorizon (X>):			*C+*				*B+*
Diagnostic taxa of miospore zones <u>(nominal taxa)</u>							
Verrucosisporites bulliferus	X	X	X	X	X	X	
Hystricosporites multifurcatus	X	X	X	X	X	X	
Diducites poljessicus	X	X	X	X	X	X	X
Diducites mucronatus	X	X	X	X	X	X	X
Teichertospora torquata	X	X	X	X	X	X	X
Grandispora gracilis	X	X	X	X	X	X	X
Retusotriletes planus	X	X	X	X	X	X	X
Diducites plicabilis	X	X	X	X	X	X	X
Retusotriletes incohatus	X	X	X	X	X	X	X
Lophozonotriletes lebedianensis	X	X	X	X	X	X	
<u>Diducites versabilis</u>	X	X	**<u>X</u>**	X	X	X	X
Knoxisporites dedaleus	X	X	X	X	X	X	X
Grandispora famenensis var. minuta	X	X	X	X	X	X	X
Grandispora famenensis var. famenensis	X	X	X	X	X	X	X
Grandispora microseta	X	X	X	X	X	X	X
Retispora macroreticulata	X	X	X	X	X	X	X
Retusotriletes phillipsii			X	X	X	X	
<u>Grandispora cornuta</u>			**<u>X></u>**	X	X	X	X
Rugospora radiata			X	X	X	X	X
<u>Vallatisporites hystricosus</u>							**<u>X></u>**
<u>Apiculiretusispora verrucosa</u>							**<u>X</u>**
Grandispora echinata							X

Fig. 7. Intercalibrated levels and diagnostic miospore taxa in the upper Famennian.

section for the base of their *Archaeoperisaccus ovalis–Verrucosisporites bulliferus* Zone which obviously includes both BJ and BM Oppel Zones.

Three biohorizons were selected by Streel & Loboziak (1996) and Streel *et al.* (2000*a*, fig. 3) in the TA–BJ timespan.

The *Samarisporites triangulatus* FOB (CQI A) is not present in the Boulonnais but in the Eifel region. It occurs in sample 44 in the lower part of the Kerpen Formation (Fm) at Kerpen, Hillesheim Syncline, Eifel region (Loboziak *et al.* 1991, fig. 5). Six samples from the underlying Cürten Formation, 108 to 162 m, did not contain *S. triangulatus*. The Kerpen Formation at Kerpen and the Cürten Formation are correlated with the conodont *ensensis–bipennatus* Subzone (Weddige 1984, 1988, p. 150). The base of the Kerpen Fm is now (K. Weddige, pers. comm. 2004) considered to belong to the upper part of the *hemiansatus* Zone (Fig. 3). The typical radial extensions of the zona of *S. triangulatus* in the Kerpen Fm are very short, possibly marking an early stage in the development of this species, but known also in beds occurring higher than the *C. concinna* FOB in Poland (Turnau 1996, pl. IV, fig. 6) . In the Blacourt Fm (Boulonnais region), containing the conodont Mid- and/or Upper *varcus* Zone (Fig. 3), the typical radial extensions are better developed and more conspicuous (see Loboziak & Streel 1980, pl. 2, fig. 12). Such occurrence of an 'early form' of *S. triangulatus* in the upper part of the *hemiansatus* Zone is not found by Turnau & Racki (1999) in Poland and Marshall *et al.* (2007, fig. 8) in Scotland, who place, therefore, the *S. triangulatus* FOB in the Middle–Upper *varcus* Zone, i.e. higher than their *C. concinna* FOB (see below). This discrepancy might result from these regions belonging to another phytogeographic realm (J. Marshall pers. comm. 2007).

The *Chelinospora concinna* FOB (CQI A) occurs in sample 26, unit H in the Couderousse Member of the Blacourt Fm in the Ferques railroad section (Boulonnais region, northern France; Loboziak & Streel 1988, fig. 1). Two samples from a 3 m interval below sample 26 did not contain *C. concinna*, but these samples immediately overlie a reef, 190 m thick, that lacks palynomorphs. The *C. concinna* FOB might belong to the conodont Mid *varcus* or/and Upper *varcus* Zones. However Turnau (1996) and Turnau & Racki (1999) in Poland and Marshall *et al.* (2007) in Scotland recommend a slightly lower position, i.e. the Lower or Mid *varcus*, now the transition between the *rhenanus* and *ansatus* Givetian conodont Zones.

The *V. bulliferus* FOB (CQI A) occurs in sample 05 in unit O which belongs to the Noces Member of

Intercalibrated levels:	20	21	22	23	24	25	26	27	28
Chronostratigraphy:	u.Fam.	u*.Fam	u*.Fam	u*.Fam	u*.Fam	u*.Fam	u*.Fam	u*.Fam	Carbon.
New conodont zones:	m.exp.	u.exp	u.exp	u.exp-l.praes ?	l.praes.	m.praes.	u.praes.	u.praes.	sulcata
Old or alternative conodont zones:	l.cost.	l.cost.	m.cost.	m.cost	m.cost.	u.cost	kockeli	kockeli	kuehni
Miosp. ass. zones (Rich. & McGr. 1986):	pus.-lep.	pus.-lep.	pus.-lep.	pus.-lep.	pus.-lep.	nit.-verr.	nit.-verr.	nit.-verr.	nit.-verr.
Miosp. interv./ass. zones (Streel *et al.* 1987):	LL	LL	LL	LE	LE	LN	LN	VI	VI
CQI for miospore biohorizon (X>):	*B - -*					*A*	*A*		
Diagn. taxa of miosp. Z. *(nominal taxa)*									
Diducites poljessicus	X	X	X	X	X	X	X	X	X
Diducites mucronatus	X	X	X	X	X	X	X	X	X
Teichertospora torquata	X	X	X	X	X				
Grandispora gracilis	X	X	X	X	X	X	X		
Retusotriletes planus	X	X	X	X	X	X	X	X	X
Diducites plicabilis	X	X	X	X	X	X	X		
Retusotriletes incohatus	X	X	X	X	X	X	X	**X**	X
Diducites versabilis	X	X	X	X	X	X	X	X	
Knoxisporites dedaleus	X	X	X	X	X	X	X		
Grandispora famenensis var. minuta	X	X	X						
Grandispora famenensis var. famenensis	X	X	X						
Grandispora microseta	X	X	X						
Retispora macroreticulata	X	X	X						
Grandispora cornuta	X	X	X	X	X	X	X	X	X
Rugospora radiata	X	X	X	X	X	X	X		
Vallatisporites hystricosus	X	X	X	X	X	X	X		
Apiculiretusispora verrucosa	X	X	X	X	X				
Grandispora echinata	X	X	X	X	X	X	X	X	X
Retispora lepidophyta	***X>***	X	X	**X**	X	**X**	*X>*		
Knoxisporites literatus	**X**	X	X	X	X	X	X	X	X
Tumulispora rarituberculatus	X	X	X	X	X	X	X	X	X
Indotriradites explanatus				***X>***	X	X	X	X	X
Vallatisporites verrucosus				X	X	X	X	X	X
Vallatisporites vallatus						X	X	**X**	X
Verrucosisporites nitidus						***X>***	X	X	X

Fig. 8. Intercalibrated levels and diagnostic miospore taxa from upper Famennian to lowermost Carboniferous.

the Beaulieu Fm in the Ferques railroad section (Loboziak & Streel 1981, fig. 1). Unit O is a shale underlying a limestone (unit P) containing the conodont *punctata* Zone and overlying a dolomitic bed (unit N). *V. bulliferus* was absent in the five samples which have been studied below, in a 45 m interval above the base of the Beaulieu Fm. The *V. bulliferus* FOB probably belongs to the upper part of the lower Frasnian conodont *transitans* Zone (MN Zone 4), i.e. to a conodont zone above the *falsiovalis* Zone where the base of the Frasnian Stage has been defined at the first occurrence of *Ancyrodella rotundiloba* (but see Bultynck 1986; Sandberg *et al.* 1989).

The TCo Oppel Zone ranges into the lower Frasnian *falsiovalis* (partim) and *transitans* Zones or MN Zones 1 to 4 of Klapper (1989). The BJ Oppel Zone and the next BM Oppel Zone are present in the middle Frasnian *punctata* Zone until the Lower *hassi* or *jamieae* Zone or MN zones 5 to 11 of Klapper (1989). Unfortunately the upper part of the Beaulieu Fm and the lower part of the Ferques Fm contain only very poor conodont faunas (Brice *et al.* 1981, fig. 2).

The *L. media* FOB (CQI C+) occurs in sample vw5 in the lower part of the Ferques Fm (Bois Member) in the 'Bois' Quarry, i.e. above the conodont *punctata* Zone known from the underlying Beaulieu Fm but below *Ancyrognathus coeni* indicating the conodont Lower *hassi* to *jamieae* Zone, near the top of the Ferques Fm in the 'La Parisienne' Quarry (Brice *et al.* 1981). R. T. Becker (pers. comm. 2007) suggests that the Bois Member fall in the middle of the interval from MN Zones 7 to 10. One sample in the Ferques Fm below vw5 and four samples in the Beaulieu Fm do not have *L. media*. The Ferques railroad section, the 'Bois' Quarry and the 'La Parisienne' Quarry are localities that are only a few kilometres apart.

The upper Frasnian to the lower Famennian and the Frasnian–Famennian boundary (376 Ma, Kaufmann 2006) (Fig. 4)

Miospores and conodonts are poorly correlated during this timespan. Upper Frasnian and Frasnian–Famennian boundary (FFB) miospore zones are best displayed in the upper part of the Hydrequent Fm in the 'Briqueterie de Beaulieu' (Boulonnais area, northern France) which does not contain any conodonts. *Ancyrognathus coeni* occurring near the top of the preceding Ferques Fm in the 'La Parisienne' Quarry (500 m north of the 'Briqueterie de Beaulieu'), and indicating the conodont Lower *hassi* or *jamieae* Zone, is about 85 m lithostratigraphically below the miospore productive part of the Hydrequent Fm. The stratigraphic

Table 1. *List of species*

Species	References
Apiculiretusispora verrucosa	(Caro-Moniez) Streel in Becker *et al.* (1974)
Chelinospora concinna	Allen (1965)*
Cirratriradites jekhowskyi	Taugourdeau-Lantz (1967)*
Corbulispora sp.	*in* Loboziak & Streel (1981, pl.1 fig. 5)
Cristatisporites deliquescens	(Naumova) Archangelskaya (1972)*
Cymbosporites acanthaceus	(Kedo) Obukhovskaya (2000)*
Diducites mucronatus	(Kedo) Van Veen (1981)*
Diducites plicabilis	Van Veen (1981)
Diducites poljessicus	(Kedo) Van Veen (1981)
Diducites versabilis	(Kedo) Van Veen (1981)
Grandispora cornuta	Higgs (1975)*
Grandispora echinata	Hacquebard (1957)*
Grandispora famenensis var. famenensis	(Naumova) Streel in Higgs *et al.* (2000)
Grandispora famenensis var. minuta	(Naumova)/(Nekriata) Streel in Higgs *et al.* (2000)
Grandispora gracilis	(Kedo) Streel in Becker *et al.* (1974)
Grandispora microseta	(Kedo) Streel in Higgs *et al.* (2000)
Hystricosporites multifurcatus	(Winslow) Mortimer & Chaloner (1967)*
Indotriradites explanatus	(Luber) Playford (1990)*
Knoxisporites dedaleus	(Naumova) Streel in Becker *et al.* (1974)
Knoxisporites literatus	(Waltz) Playford (1963)*
Lophozonotriletes lebedianensis	Naumova (1953)
Lophozonotriletes media	Taugourdeau-Lantz (1967)
Pustulatisporites rugulatus	(Taugourdeau-Lantz) Loboziak & Streel (1981)
Retispora lepidophyta	(Kedo) Playford (1976)*
Retispora macroreticulata	(Kedo) Byvsheva (1985)*
Retusotriletes incohatus	Sullivan (1964)*
Retusotriletes phillipsii	Clendening *et al.* (1980)
Retusotriletes planus	Dolby & Neves (1970)*
Rugospora bricei	Loboziak & Streel (1989)
Rugospora radiata	(Jushko) Byvsheva (1985)
Samarisporites triangulatus	Allen (1965)
Teichertospora torquata	(Higgs) McGregor & Playford (1990)*
Tumulispora rarituberculata	(Luber) Playford (1991)*
Vallatisporites hystricosus	(Winslow) Byvsheva (1985)
Vallatisporites verrucosus	Hacquebard (1957)
Vallatispotites vallatus	Hacquebard (1957)
Verrucosisporites bulliferus	Richardson & McGregor (1986)
Verrucosisporites nitidus	Playford (1964)*

*The references of these taxa can be found in the following citations in the reference list: Avkhimovitch *et al.* 1993; Loboziak & Streel 1981; Loboziak *et al.* 1983; Streel & Loboziak 1996; Streel *et al.* 1987.

position of the late *rhenana* to *triangularis* conodont zones and of the FFB in the Hydrequent Fm is known by acritarch correlation (CQI B−) with the Hony section in the Ardenne area (Streel *et al.* 2000*a*, figs 13 and 16, 2000*b*). Magnetostratigraphy susceptibility correlation of the FFB between the 'La Serre' section in southern France and the section of Hony in east Belgium is given by Crick *et al.* (2002). The obvious conclusion of this correlation is that the BA Oppel Zone *plicabilis* Interval Zone (see below) extends across the FFB.

Establishing a miospore zonation is essentially a step by step process and it is important that the zones erected at any given time reflect these steps. Loboziak & Streel (1981) used the 'phase concept' of Van der Zwan (1980) as an Oppel Zone that was not yet controlled by application in other localities and which therefore might just have local significance. Phase zones IV and V were so far the last zones unnamed (in Streel *et al.* 1987) although they have been used for correlation in other regions and even other phytogeographic areas (Avkhimovitch *et al.* 1988, 1993; Melo & Loboziak 2003).

Phase zone IV is renamed here as *Rugospora bricei–Cymbosporites acanthaceus* (BA) Oppel Zone and phase zone V as *Knoxisporites dedaleus–Diducites versabilis* (DV) Oppel Zone.

The BA Zone is subdivided in three parts by new interval zones, characterized respectively by the absence (pre-grac.) or entry of *Grandispora gracilis* (grac.) at the base of IV C and the entry of *Diducites plicabilis* (plic.) at the base of IV D3 (see Loboziak *et al.* 1983). It should be noted that the base of the *Grandispora gracilis–Samarisporites* sp. cf *Acanthotriletes hirtus* (GH) Oppel Zone was poorly defined (Streel *et al.* 1987, fig. 9, p. 221). The Oppel Zone GH is therefore withdrawn here.

Although it has a poorly defined base in the USA (Streel *et al.* 2000*a*, p. 131), the *Auroraspora torquata–Grandispora gracilis* assemblage Zone clearly starts in the upper Frasnian conodont MN zone 13 of Klapper (1989) (likely within the range of *Palmatolepis juntianensis*; J. Over pers. comm. 1998) and continues more or less at least to the middle Famennian *rhomboidea* Zone (Richardson & McGregor 1986, fig. 7).

Three biohorizons were selected by Streel & Loboziak (1996) in the BA–DV timespan of the Hydrequent Fm. Conodonts have not been found in this section.

The *Rugospora bricei* FOB (CQI B––) occurs in sample 24, immediately below the 'banc dolomitique' in the upper Hydrequent Fm of the 'Briqueterie de Beaulieu' (Loboziak *et al.* 1983, fig. 1). Two poor samples, without *R. bricei*, occur within the 2 m below the biohorizon, and another one 25 m below. According to the correlations discussed above, the R. *bricei* FOB is in the Upper *hassi*, *jamieae or rhenana* Zone. However, correlation with eastern Europe (the Pripyat Depression, Belarus and the Timan-Pechora Province, Russia indicates (Obukhovskaya *et al.* 2000) that this biohorizon should fit into the Upper *rhenana* Zone or the MN Zone 12. However, as these regions belong to another phytogeographic realm, this correlation is questionned.

The *Grandispora gracilis* FOB (CQI B––) occurs in sample 305 in the upper part of the Hydrequent Fm of the 'Briqueterie de Beaulieu' (Loboziak *et al.* 1983, fig. 1). Six samples in a 4.5 m interval below sample 305 above the 'banc dolomitique' do not contain *G. gracilis*. According to the correlations discussed above, the *G. gracilis* FOB should fit into the Upper *rhenana* or *linguiformis* Zone or the MN13 Zone.

The *Knoxisporites dedaleus* FOB (CQI B––) occurs in sample 205 in the uppermost Hydrequent Fm of the 'Briqueterie de Beaulieu' (Loboziak *et al.* 1983, fig. 1). Sixteen samples, rich in miospores but lacking *K. dedaleus*, have been studied in the 6 m of sediments immediately below sample 205. According to the correlations discussed for the BA Oppel Zone, the *K. dedaleus* FOB should fit into the *triangularis* zones or even higher in the Famennian.

The middle to upper Famennian transition (Figs 5 and 6)

The four Famennian substages have not yet been formally defined, therefore we refer to Streel (2005) where we have adopted the base of the Uppermost *crepida*, the base of the Uppermost *marginifera* and the base of the Upper *expansa* conodont Zones to mark respectively the bases of the middle, upper and uppermost Famennian.

Miospores are poorly represented in the lower–middle Famennian transition of western Europe and eastern North America (see the 'Lower–middle Famennian vegetation crisis' in Streel *et al.* 2000*a* and Streel 2007). This is matched by the occurrence in most of the lower–middle Famennian (from the conodont *triangularis* to the *rhomboidea* Zones at least) of the single Oppel Zone DV in western Europe and the single *Auroraspora torquata–Grandispora gracilis* assemblage Zone in eastern North America. A tentative attempt to subdivide the *torquata–gracilis* Zone into a lower part without, and an upper part with ?*Lagenicula* cf. *Hymenozonotriletes immensus*, reallocated by Turnau (2002) to the genus *Tergobulasporites*, reaching the *marginifera* conodont zone, has been published by Richardson & Ahmed (1989). However, this species might well be a megaspore, having a limited distribution range, and therefore might be unsuitable for correlation.

It is not until the *marginifera* Zone that the miospore Oppel Zone GF, *Grandispora gracilis–Grandispora famenensis* (named the GM Zone in papers older than Streel *et al.* 1987) occurs in western Europe, initiating a succession of miospore zones characterized by species of the genus *Grandispora* (Higgs *et al.* 2000). The GF Zone is subdivided in three parts by interval zones, characterized respectively by the absence (pre-mic.) or entry (mic.) of *Grandispora microseta* and by the entry of *Retispora macroreticulata* (mac.).

The *Grandispora famenensis* FOB (CQI A and C−) occurs with G. *famenensis* var. *minuta*, a variety with reduced ornamentation, in sample 4 in the Esneux Fm at the Esneux locality, Dinant Synclinorium, Ardenne region (Loboziak *et al.* 1995). G. *famenensis* var. *famenensis* occurs in sample 13, 11 m higher. Conodonts of the Lower *marginifera* Zone are known from another lithological unit (Souverain-Pré Fm) in the same section about 23 m above sample 4 (CQI A). Lithological correlation (CQI C−) suggests that the equivalent of the conodont Lower *marginifera* Zone occurs also in the upper part of the Esneux Fm, in a section studied, 25 km to the west, at Modave-Pont de Bonne (Dreesen & Thorez 1994, p. 170).

The *Grandispora microseta* FOB (CQI A and C+) and the *Retispora macroreticulata* FOB (CQI C+) occur successively and in close proximity in the lower part of the Montfort Fm of the Comblain-au-Pont 'Bon-Mariage' section, Dinant synclinorium, Ardenne region (Bouckaert *et al.* 1968, hors-texte I). Sample 13, 9 m below sample 19, is known to lack *R. macroreticulata* but have *G. microseta.* The best documented conodont faunas is given by Dreesen *et al.* (1986, fig. 1) and their correlation (CQI A) with the miospore Zones is indicated by Streel (1986, fig. 2). *G. microseta* first occurs in the Uppermost *marginifera* conodont Zone. *R. macroreticulata* probably corresponds to the next Lower *trachytera* conodont Zone. Conodonts have been restudied in a parallel section (Comblain-la-Tour) 4 km distant, by Dreesen & Thorez (1994, p. 175) demonstrating the Lower *trachytera* Zone at about the same level (CQI C+).

The upper Famennian in the Condroz Sandstones of Belgium (Fig. 7)

Dreesen *et al.* (1986, fig. 1) indicate that, due to unfavourable facies, the stratigraphic interval Upper *trachytera* to Lower *expansa* Zones in the Condroz Sandstones in Belgium lacks the characteristic conodonts of the 'standard zonation' of Ziegler & Sandberg (1990). The late GF mac Interval Zone, defined above and the Oppel Zone VCo (*Diducites versabilis–Grandispora cornuta*) cover the stratigraphic interval where characteristic conodonts are poorly present.

The base of the VCo Oppel Zone, marked by the first occurrence of *G. cornuta, Retusotriletes phillipsii* and *Rugospora radiata*, matches the base of the *Rugospora flexuosa* (now *radiata*)*–Grandispora cornuta* Assemblage Zone of Richardson & McGregor (1986). These authors (p. 21 and fig. 7) and also Streel & Loboziak (1994, fig. 2) have tried to evaluate the respective control by faunas of both zones. The VCo Oppel Zone base is obviously not older (CQI C+) than the Upper *trachytera* conodont Zone (See *G. cornuta* FOB). The '*flexuosa–cornuta*' Assemblage Zone, however, might well start in the middle Famennian part of the *marginifera* conodont Zone if the poor lithological correlation between a few faunas and the rich miospore assemblages in the USA is confirmed.

The *Grandispora cornuta* FOB separates the GF and VCo miospore zones.

The *Grandispora cornuta* FOB (CQI C+) was found in the now almost inaccessible locality of the lower part (sample 36) of the Evieux Fm, in the La Gombe/Montfort section, Dinant Synclinorium, Ardenne region (Bouckaert *et al.* 1971, fig. 6; Bouckaert & Streel 1974; Thorez *et al.* 1977, p. 18). Ten samples (from an interval between 50 and 180 m below sample 36) contained miospores lacking *G. cornuta. Scaphignathus velifer velifer* (first occurrence in the Latest *marginifera* conodont Zone) is known 162 m below sample 36 in the same section. *G. cornuta* first occurrence is also present in sample 54, above the base of the Evieux Fm in the Comblain-au-Pont 'Bon-Mariage' section at a level believed to correspond (see CQI C+ in Comblain-la-Tour section) to the *postera* conodont Zone (Streel 1986; Streel *et al.* 2003, fig. 2). These conodonts have been restudied by Dreesen & Thorez (1994, p. 175) in the parallel section of Comblain-la-Tour. They have proposed an Upper *trachytera* Zone at a lithostratigraphic level slightly below sample 54 of the Comblain-au-Pont section.

The *Vallatisporites hystricosus* FOB marks the top of the VCo Zone and the base of the *Apiculiretusispora verrucosa–Vallatisporites hystricosus* or VH Zone.

The *Vallatisporites hystricosus* FOB (CQI A) is found in sample 20'd, 10 m below the top of the Evieux Fm (Maziane *et al.* 1999, fig. 3; Streel *et al.* 2007, fig. 1). Eleven samples without *V. hystricosus* are known below this sample in the same section and the same formation. The first occurrence of the Late *expansa* conodont Zone is known (Dreesen *et al.* 1993; Streel & Hartkopf-Fröder 2005; Streel *et al.* 2007) from the Comblain-au-Pont Fm in the same section, about 28 m higher than Bed 20'd. Conodonts of the Middle *expansa* Zone occur (Dreesen *et al.* 1993, p. 23) in the underlying Evieux Fm of the Esneux railway section, 3 km from Chanxhe (CQI B+) where the Fontin Event has been traced in the VCo Oppel Zone (Streel 1999, p. 203–205). Consequently, the *V. hystricosus* FOB is in the Middle *expansa* Zone but higher than the Fontin Event (Streel & Marshall 2006, table 2, grid E6).

The uppermost Famennian and the Devonian–Carboniferous boundary (361 Ma, Kaufmann 2006) (Figs 8 and 9)

Almost all published papers recently on the Strunian as a chronostratigraphic unit refer to the old (now disused) 'Fa2d', a base being defined by the *Retispora lepidophyta* FOB at about the level of the Epinette Event. The *R. lepidophyta* FOB is an excellent marker, the species reaching, higher in the sequences, sometimes 50% of the miospore assemblages, and having a worldwide distribution in continental and neritic facies. However, it is unfortunately not matched by any well defined conodont limit. Therefore the uppermost Famennian

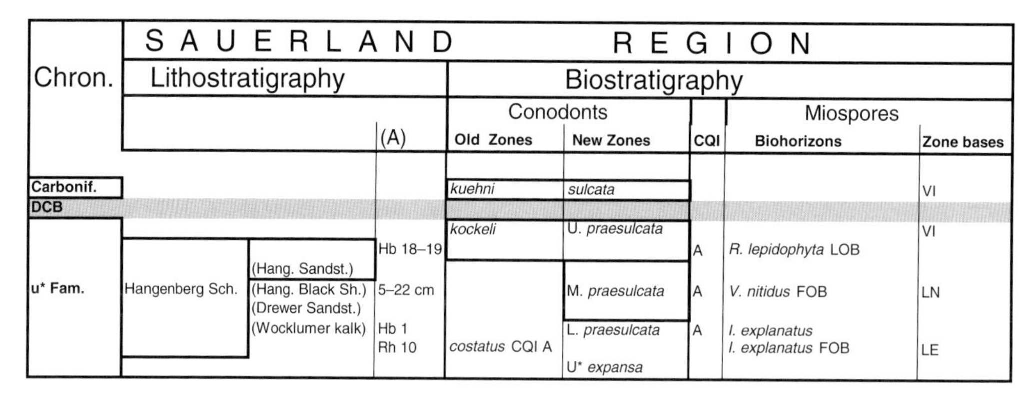

Fig. 9. Uppermost Famennian to the lowermost Carboniferous in the Sauerland region. (A), Number of samples cited in text. DCB, Devonian–Carboniferous boundary.

Substage base, at the base of the Upper *expansa* conodont Zone, and a reference section for neritic facies (Strunian) were proposed by Streel (2002, 2005) and Streel *et al.* (2003, 2005, 2006, 2007).

Richardson & Ahmed (1989, fig. 5) and Avkhimovitch & Richardson (1996) have proposed respectively to separate the lower part of the *Vallatisporites pusillites (sensu lato)–Retispora lepidophyta* Zone of Richardson & McGregor (1986) as an *Apiculiretusispora fructicosa* (now *verrucosa*)*–V. pusillites* Subzone (1988) or as a *V. pusillites–Knoxisporites literatus* PLi Zone (1996). They correlate the base of these (sub)zones with the base of the old (now disused) 'Fa2d' in Belgium (starting in the Middle *expansa* conodont Zone) but also with the base of the Cattaraugus Fm, equivalent to the Uppermost *marginifera* conodont Zone in marine sediment, after Kirchgasser & Oliver (1993, fig. 1) and Kirchgasser (2000). Such contradictions might depend on the diachronous character of the Catskill facies. As long as this situation is not clarified, these miospore zone subdivisions cannot be taken into consideration here.

The transition from the upper Famennian to the Carboniferous is covered by six conodont zones (from Middle *expansa* to *sulcata*), by three miospore Interval Zones, i.e. the *Retispora lepidophyta–Knoxisporites literatus, R. lepidophyta–Indotriradites explanatus*, and *R. lepidophyta–Verrucosisporites nitidus* (respectively LL, LE, LN), and by one Assemblage Zone, i.e. the *Vallatisporites vallatus–Retusotriletes incohatus* Zone (VI) which crosses the Devonian–Carboniferous boundary (DCB).

The LL Interval Zone now includes (Maziane *et al.* 1999) the former LV Zone (Streel *et al.* 1987) and could be further subdivided by the first occurrence of *Tumulispora rarituberculatus* and the sudden change in abundance from *R. lepidophyta lepidophyta* to *R. lepidophyta minor* almost at the base of the Upper *expansa* Zone (Maziane-Serraj *et al.* 2007). In the Sauerland, Germany, the extinction of *R. lepidophyta* is announced by the disappearance of peat swamps which produced *Diducites plicabilis*, followed by a strong reduction of the proportion of *R. lepidophyta* (from 30% to 1 or 2%; Higgs *et al.* 1993) suggesting also a progressive reduction of the related swamp-margin environment, which seems to disappear temporarily soon after, together with other swamp-margin environments characterized by species such as *Vallatisporites hystricosus* and *Auroraspora asperella* (Streel 1999). These miospore events partly correspond to, and immediately succeeded, the Hangenberg Event, a sedimentary cycle constituted of a transgression (the Hangenberg Black Shale) and a deep regression (the Hangenberg Sandstone and Shale) (Bless *et al.* 1993). The regression can be correlated by miospores with the glacial episode known in Western Gondwana (Streel *et al.* 2000*a*, Melo & Loboziak 2003).

The complete extinction of *Retispora lepidophyta* immediately below the base of the Carboniferous System as defined by the first occurrence of the conodont *sulcata* Zone, is well known around the world (Streel 1986; Higgs *et al.* 1993; Loboziak *et al.* 1993; Streel & Loboziak 1996). It corresponds to the change from the LN Zone to the VI Zone. The VI Assemblage Zone is poorly defined, the two nominal species being present below the top of the LN Zone. Its base corresponds to the *Retispora lepidophyta* LOB.

The *Retispora lepidophyta* FOB (CQI A and B+) is found in sample 22, 2 m below the top of the Evieux Fm in the Chanxhe section, Dinant Synclinorium, Ardenne region (Maziane *et al.* 1999, fig. 3; Streel *et al.* 2007, fig. 1). Fourteen samples without this species are known below these samples in the same section and the same formation. The first occurrence of the Late *expansa* conodont Zone is known (Dreesen *et al.* 1993;

Streel *et al.* 2007) from the Comblain-au-Pont Fm in Bed 111 of the same section (CQI A), about 20 m higher than samples 22. Conodonts of the Middle *expansa* Zone occur (Dreesen *et al.* 1993, p.23) in the underlying Evieux Fm of the Esneux railway section, 3 km from Chanxhe (CQI B+), where the Fontin Event has been traced in the VCo Oppel Zone (Streel 1999, p. 203–205). The lower part of the Comblain-au-Pont Fm contains abundant large specimens (var. *lepidophyta*) of *R. lepidophyta* (Streel 1966; Maziane *et al.* 2002) as in the Refrath 1 Borehole (Bergisch Gladbach–Paffrath Syncline, Germany) which contains a Middle *expansa* Zone (Streel & Hartkopf-Fröder 2005). Therefore, the *R. lepidophyta* FOB is in the Middle *expansa* Zone.

The *Indotriradites explanatus* FOB (CQI B−) is found in sample Rh10 in the greenish silty shales (Hangenberg Schiefer equivalent) of the Riescheid section, Remscheid Altena Anticline, Sauerland, Germany (Higgs & Streel 1984, fig. 3). Three samples in the underlying 2.5 m interval lacked *I. explanatus* (Higgs & Streel 1994). The undivided *costatus* conodont Zone (Lane & Ziegler in Paproth & Streel 1982) was found in almost the same bed (equivalent to the conodont Lower or Middle *praesulcata* Zone ?). Another, better dated sample is from 50 cm below the top of the Wocklum Kalk at Hasselbachtal (28 km east of Riescheid), in the same anticline (Higgs & Streel 1994). It is a single sample (Hb1) in the latest part of the Lower *praesulcata* conodont Zone (CQI B−), which occurs 15 cm above this sample. The Middle *praesulcata* Zone occurs 30 cm above this sample, i.e. 20 cm below the top of the Wocklum Kalk (Becker *et al.* 1984, p. 189). However, no samples with miospores are known below this single sample. Consequently, the *I. explanatus* FOB is in the late part of the Upper *expansa* Zone or in the Lower *praesulcata* Zone.

The *Verrucosisporites nitidus* FOB (CQI A) is found in a sample collected from the 5 to 22 cm interval above the base of the Hangenberg Black Shale, i.e. on top of the Wocklum Kalk, at Hasselbachtal section, Remscheid Altena Anticline, Sauerland, Germany. Two specimens of *V. nitidus* have small (3 μm) verrucate ornaments, which fall within the lower part of the morphological range of the species. The presence of the Middle *praesulcata* Zone (see *I. explanatus* FOB) 20 cm below the top of the Wocklum Kalk in the same section (CQI A) allows us to assign the *V. nitidus* FOB to the Middle *praesulcata* Zone.

The *Retispora lepidophyta* LOB (CQI A) can be observed in sample Hb 18-19 in Bed 85 of the Hangenberg Schiefer of the Hasselbachtal section, Remscheid Altena Anticline, Sauerland (Higgs & Streel 1984, figs. 5 and 6). Six samples in the overlying 14 cm did not yield *R. lepidophyta* but were dominated by simple laevigate taxa. The *sulcata* conodont zone occurs 14 cm higher than the *R. lepidophyta* LOB.

The author is indebted to J. E. A. Marshall (Southampton) and R. T. Becker (Münster) for reviewing this manuscript and for having made suggestions which led to a much improved work. P. Bultynck (Brussels) is also gratefully acknowledged for his contribution to the Givetian and Frasnian part of the conodont stratigraphy.

References

AVKHIMOVITCH, V. I. & RICHARDSON, J. B. 1996. Correlation of Late Devonian marginal marine and non-marine sediments from eastern North America, western Europe and Belorussia. *International Union of Geological Sciences, Subcommission on Devonian Stratigraphy Newsletter*, **12**, 58–61.

AVKHIMOVITCH, V. I., BYVSHEVA, T. V., HIGGS, K., STREEL, M. & UMNOVA, V. T. 1988. Miospore systematics and stratigraphic correlation of Devonian-Carboniferous boundary deposits in the European part of the USSR and Western Europe. *Courier Forschungsinstitut Senckenberg*, **100**, 169–191.

AVKHIMOVITCH, V. I., TCHIBRIKOVA, E. V. *ET AL.* 1993. Middle and Upper Devonian miospore zonation of Eastern Europe. *Bulletin Centre de Recherche Elf Exploration Production*, **17**, 79–147.

BECKER, G., BLESS, M. J. M., STREEL, M. & THOREZ, J. 1974. Palynology and ostracode distribution in the Upper Devonian and basal Dinantian of Belgium and their dependence on sedimentary facies. *Mededelingen Rijks Geologische Dienst, new serie*, **25**(2), 9–99.

BECKER, R. T. 2002. Frasnian Goniatites from the Boulonnais (France) as indicators of regional sealevel changes. *Annales de la Société géologique du Nord (Lille) 2e série*, **9**, 129–140.

BECKER, R. T., BLESS, M. J. M. *ET AL.* 1984. Hasselbachtal, the section best displaying the Devonian–Carboniferous boundary beds in the Rhenish Massif (Rheinisches Schiefergebirge). *Courier Forschungsinstitut Senckenberg*, **67**, 181–191.

BLESS, M. J. M., BECKER, R. T., HIGGS, K., PAPROTH, E. & STREEL, M. 1993. Eustatic cycles around the Devonian–Carboniferous Boundary and the sedimentary and fossil record in Sauerland (Federal Republic of Germany). *Annales de la Société géologique de Belgique*, **115**, 689–702.

BOUCKAERT, J. & STREEL, M. 1974. General information. *In*: BOUCKAERT, J. & STREEL, M. (eds) *International Symposium on Belgian Micropaleontological Limits, Namur 1974, Guidebook*. Geological Survey of Belgium, Brussels.

BOUCKAERT, J., STREEL, M. & THOREZ, J. 1968. Schéma biostratigraphique et coupes de référence du Famennien belge. Note préliminaire. *Annales de la Société géologique de Belgique*, **91**, 317–336.

BOUCKAERT, J., STREEL, M. & THOREZ, J. 1971. Le Famennien et les couches de transition Dévonien-Carbonifère dans la vallée de l'Ourthe (sud de Liège, synclinorium de Dinant). *Congrès et Colloques de l'Université de Liège*, **55**, 25–46.

BRICE, D. (ed.) 1988. *Le Dévonien de Ferques, Bas-Boulonnais (N. France)*. Université de Bretagne occidentale, Rennes, Biostratigraphie du Paléozoïque, **7**.

BRICE, D. 2003. Brachiopod assemblages in the Devonian of Ferques (Boulonnais, France). Relations to palaeoenvironments and global eustatic curves. *Bulletin of Geosciences*, **78**(4), 405–417.

BRICE, D., BULTYNCK, P., DEUNFF, J., LOBOZIAK, S. & STREEL, M. 1979. Données biostratigraphiques nouvelles sur le Givétien et le Frasnien de Ferques (Boulonnais). *Annales de la Société géologique du Nord (Lille)*, **98**, 325–344.

BRICE, D., COEN, M., LOBOZIAK, S. & STREEL, M. 1981. Précisions biostratigraphiques relatives au Dévonien supérieur de Ferques (Boulonnais). *Annales de la Société géologique du Nord (Lille), 2e série*, **9**, 61–74.

BRICE, D., MISTIAEN, B. & ROHART, J.-C. 2002. Progrès dans la connaissance des flores et faunes dévoniennes du Boulonnais (1971–2001). *Annales de la Société géologique du Nord (Lille)*, **100**, 159–166.

BULTYNCK, P. 1986. Accuracy and reliability of conodont zones, the *Polygnathus asymmetricus* Zone and the Givetian–Frasnian boundary. *Bulletin Institut Royal des Sciences Naturelles de Belgique, Sciences de la Terre*, **56**, 269–280.

BULTYNCK, P. 2007. Limitations on the application of the Devonian standard conodont zonation. *Geological Quarterly*, **51**, 339–344.

CLENDENING, J. A., EAMES, L. E. & WOOD, G. D. 1980. *Retusotriletes phillipsii* n. sp., a potential Upper Devonian guide palynomorph. *Palynology*, **4**, 15–22.

CRICK, R. E., ELLWOOD, B. B. *ET AL.* 2002. Magnetostratigraphy susceptibility of the Frasnian/Famennian boundary. *Palaeogeography, Palaeoclimatology, Palaeoecology*, **181**, 67–90.

DREESEN, R. & THOREZ, J. 1994. Parautochtonous – autochtonous carbonates and conodont in the Late Famennian (Uppermost Devonian) Condroz Sandstones of Belgium. *Courier Forschungsinstitut Senckenberg*, **168**, 159–182.

DREESEN, R., SANDBERG, C. A. & ZIEGLER, W. 1986. Review of Late Devonian and Early Carboniferous conodont biostratigraphy and biofacies models as applied to the Ardenne Shelf. *Annales de la Société géologique de Belgique*, **109**, 27–42.

DREESEN, R., POTY, E., STREEL, M. & THOREZ, J. 1993. *Late Famennian to Namurian in the Eastern Ardenne, Belgium*. IUGS Subcommission on Carboniferous Stratigraphy, Liège.

HEDBERG, H. H. (ed.) 1976. *International Stratigraphical Guide. A Guide to Stratigraphic Classification, Terminology, and Procedure*. International Subcommission on Stratigraphic Classification (ISSC), Wiley, Chichester.

HIGGS, K. & STREEL, M. 1984. Spore stratigraphy at the Devonian–Carboniferous boundary in the northern "Rheinisches Schiefergebirge", Germany. *Courier Forschungsinstitut Senckenberg*, **67**, 157–179.

HIGGS, K. & STREEL, M. 1994. Palynological age for the lower part of the Hangenberg Shales in Sauerland, Germany. *Annales de la Société géologique de Belgique*, **116**, 243–247.

HIGGS, K., STREEL, M., KORN, D. & PAPROTH, E. 1993. Palynological data from the Devonian–Carboniferous boundary beds in the new Stockum Trench II and the Hasselbachtal boreholes, northern Rhenish Massif, Germany. *Annales de la Société géologique de Belgique*, **115**, 551–557.

HIGGS, K., AVKHIMOVITCH, V. I., LOBOZIAK, S., MAZIANE-SERRAJ, N., STEMPIEN-SALEK, M. & STREEL, M. 2000. Systematic study and stratigraphic correlation of the *Grandispora* complex in the Famennian of northwest and eastern Europe. *Review of Palaeobotany and Palynology*, **112**, 207–228.

JOHNSON, J. G. 1992. Belief and reality in Biostratigraphic Zonation. *Newsletter on Stratigraphy*, **26**(1), 41–48.

KAUFMANN, B. 2006. Calibrating the Devonian Time Scale: A synthesis of U-Pb ID-TIMS ages and conodont stratigraphy. *Earth-Science Reviews*, **76**, 175–190.

KIRCHGASSER, W. T. 2000. Correlation of stage boundaries in the Appalachian Devonian eastern United States. *Courier Forschungsinstitut Senckenberg*, **225**, 271–284.

KIRCHGASSER, W. T. & OLIVER, W. A. JR. 1993. Correlation of stage boundaries in the Appalachian Devonian, Eastern United States. *IUGS, Subcommission on Devonian Stratigraphy Newsletter*, **10**, 5–8.

KLAPPER, G. 1989. The Montagne Noire Frasnian (Upper Devonian) conodont succession. *In*: MCMILLAN, N.-J., EMBRY, A. F. & GLASS, D. J. (eds) *Devonian of the World, Vol. 3, Paleontology, Paleoecology and Biostratigraphy*. Canadian Society of Petroleum Geologists, Memoir, **14**, 70–92.

LOBOZIAK, S. & STREEL, M. 1980. Miospores in Givetian to lower Frasnian sediments dated by conodonts from the Boulonnais, France. *Review of Palaeobotany and Palynology*, **29**, 285–299,

LOBOZIAK, S. & STREEL, M. 1981. Miospores in middle-upper Frasnian to Famennian sediments partly dated by conodonts (Boulonnais), France. *Review of Palaeobotany and Palynology*, **34**, 49–66.

LOBOZIAK, S. & STREEL, M. 1988. Synthèse palynostratigraphique de l'intervalle Givetien-Famennien du Boulonnais (France). *In*: BRICE, D. (ed.) *Le Dévonien de Ferques, Bas-Boulonnais (N. France)*. Université de Bretagne occidentale, Rennes, Biostratigraphie du Paléozoïque, **7**.

LOBOZIAK, S., STREEL, M. & VANGUESTAINE, M. 1983. Miospores et acritarches de la Formation d'Hydrequent (Frasnien supérieur à Famennien inférieur, Boulonnais, France). *Annales de la Société géologique de Belgique*, **106**, 173–183.

LOBOZIAK, S., STREEL, M. & WEDDIGE, K. 1991. Miospores, the *lemurata* and *triangulatus* levels and their faunal indices near the Eifelian/Givetian boundary in the Eifel (F.R.G). *Annales de la Société géologique de Belgique*, **113**, 299–313.

LOBOZIAK, S., STREEL, M., CAPUTO, M. V. & MELO, J. H. G. 1993. Middle Devonian to Lower Carboniferous miospores from selected boreholes in Amazonas and Parnaiba Basins (Brazil): additional data, synthesis, and correlation. *Documents du Laboratoire de Géologie de Lyon*, **125**, 277–289.

LOBOZIAK, S., AVKIMOVITCH, V. I. & STREEL, M. 1995. Miospores from the type locality cf the Esneux Formation, Middle Famennian of the Ourthe valley, eastern Belgium; *Annales de la Société géologique de Belgique*, **117**, 95–102.

McGregor, D. C. 1979*a*. Spores in Devonian stratigraphical correlation. *In*: House, M. R., Scrutton, C. T. & Bassett, M. G. (eds) *The Devonian System*. Special Papers in Palaeontology, **23**, The Palaeontological Association, London, 163–184.

McGregor, D. C. 1979*b*. Devonian miospores of North America. *Palynology*, **3**, 31–52.

McGregor, D. C. & Playford, G. 1992. Canadian and Australian Devonian spores, zonation and correlation. *Geological Survey of Canada, Bulletin*, **438**, 1–125.

Marshall, J. E. A., Astin, T. R., Brown, J. F., Mark-Kurik, E. & Lazauskiene, J. 2007. Recognizing the Kačák Event in the Devonian terrestrial environment and its implications for understanding land-sea interactions. *In*: Becker, R. & Kirchgasser, W. T. (eds) *Devonian Events and Correlations*. Geological Society, London, Special Publications, **278**, 133–155.

Maziane, N., Higgs, K. & Streel, M. 1999. Revision of the late Famennian miospore zonation scheme in eastern Belgium. *Journal of Micropaleontology*, **18**, 17–25.

Maziane, N., Higgs, K. & Streel, M. 2002. Biometry and paleoenvironment of *Retispora lepidophyta* (Kedo) Playford 1976 and associated miospores in the latest Famennian nearshore marine facies, eastern Ardenne (Belgium). *Review of Palaeobotany and Palynology*, **118**, 211–226.

Maziane-Serraj, N., Hartkopf-Fröder, C., Streel, M. & Thorez, J. 2007. Palynomorph distribution and bathymetry in the Chanxhe section (eastern Belgium), reference for the neritic late to latest Famennian transition (Late Devonian). *Geologica Belgica*, **10**(3–4), 170–175.

Melo, J. H. G. & Loboziak, S. 2003. Devonian–Early Carboniferous miospore biostratigraphy of the Amazon Basin, Northern Brazil. *Review of Palaeobotany and Palynology*, **124**, 131–202.

Naumova, S. N. 1953. *Spore-pollen Assemblages of the Russian Platform and their Stratigraphic Significance*. Trudy Instituta Geologicheskikh Nauk, Akademiia Nauk SSSR, 143, Geologicheskikh Seriia **60**, [in Russian].

Obukhovskaya, T. G., Avkhimovitch, V. I., Streel, M. & Loboziak, S. 2000. Miospores from the Frasnian–Famennian Boundary deposits in Eastern Europe (the Pripyat Depression, Belarus and the Timan Pechora Province, Russia) and comparison with Western Europe (Northern France). *Review of Palaeobotany and Palynology*, **112**, 229–246.

Paproth, E. & Streel, M. (eds) 1982. *Devonian–Carboniferous transitional beds of the Northern "Rheinisches Schiefergebirge"*. IUGS Working Group on the Devonian–Carboniferous boundary, Liège.

Richardson, J. B. & Ahmed, S. 1989. Miospores, zonation and correlation of Upper Devonian sequences from western New York State and Pennsylvania. *In*: McMillan, N- J., Embry, A. F. & Glass, D. J. (eds) *Devonian of the World, Vol. 3, Paleontology, Paleoecology and Biostratigraphy*. Canadian Society of Petroleum Geologists, Memoirs, **14**, 541–558.

Richardson, J. B. & McGregor, D. C. 1986. Silurian and Devonian spore zones of the Old Red Sandstone Continent and adjacent regions. *Geological Survey of Canada, Bulletin*, **364**, 1–79.

Sandberg, C. A., Poole, F. G. & Johnson, J. G. 1989. Upper Devonian of Western United States. *In*: McMillan, N.-J., Embry, A. F. & Glass, D. J. (eds), *Devonian of the World, Vol. 3, Paleontology, Paleoecology and Biostratigraphy*. Canadian Society Petroleum Geology, Memoirs, **14**, 183–220.

Streel, M. 1966. Critères palynologiques pour une stratigraphie détaillée du Tnla dans les bassins ardenno-rhénans. *Annales de la Société géologique de Belgique*, **89**, 65–96.

Streel, M. 1986. Miospore contribution to the upper Famennian–Strunian event stratigraphy. *Annales de la Société géologique de Belgique*, **109**, 75–92.

Streel, M. 1999. Quantitative palynology of the Famennian events in the Ardenne–Rhine Regions. *Abhandlungen der Geologischen Bundesanstalt, Wien*, **54**, 201–212.

Streel, M. 2002. The Uppermost Famennian around the World (definition, biostratigraphical and sedimentological context). *IUGS Subcommission on Devonian Stratigraphy Newsletter*, **18**, 55–61.

Streel, M. 2005. Subdivision of the Famennian Stage into four substages and correlation with the neritic and continental miospore zonation. *IUGS Subcommission on Devonian Stratigraphy Newsletter*, **21**, 14, 16–17; **22**, 16.

Streel, M. 2007. West Gondwanan and Euramerican climate impact on early Famennian to latest Viséan miospore assemblages. *IUGS Subcommission on Devonian Stratigraphy Newsletter*, **22**, 53–57.

Streel, M. & Hartkopf-Fröder, C. 2005. Late Famennian correlation by miospores between the Refrath 1 Borehole (Bergisch Gladbach-Paffrath Syncline, Germany) and the reference section of Chanxhe (Dinant Syncline, Belgium). *In*: Steemans, P. & Javaux, E. (eds) *Pre-Cambrian to Palaeozoic Palaeopalynology and Palaeobotany*. Carnets de Géologie/Notebooks on Geology, Brest, Memoir, **2005/02**, Abstract 10.

Streel, M. & Loboziak, S. 1994. Observations on the establishment of a Devonian and Lower Carboniferous high-resolution miospore biostratigraphy. *Review of Palaeobotany and Palynology*, **83**, 261–273.

Streel, M. & Loboziak, S. 1996. Middle and Upper Devonian miospores. *In*: Jansonius, J. & McGregor, D. C. (eds) *Palynology: Principles and Applications*. American Association of Stratigraphic Palynologists Foundation, **2**, 575–587.

Streel, M. & Marshall, J. E. A. 2006. Devonian–Carboniferous boundary global correlations and their paleogeographic implications for the assembly of Pangaea. *In*: Wong, Th. E. (ed.) *Proceedings of the XVth International Congress on Carboniferous and Permian Stratigraphy*, Utrecht, the Netherlands, 10–16 August 2003. Royal Netherlands Academy of Arts and Sciences, 481–496.

Streel, M., Higgs, K., Loboziak, S., Riegel, W. & Steemans, P. 1987. Spore stratigraphy and correlation with faunas and floras in the type marine Devonian of the Ardenne-Rhenish regions. *Review of Palaeobotany and Palynology*, **50**, 211–229.

STREEL, M., CAPUTO, M. V., LOBOZIAK, S. & MELO, J. H. G. 2000*a*. Late Frasnian–Famennian climates based on palynomorph analyses and the question of the Late Devonian glaciations. *Earth Science Reviews*, **52**, 121–173.

STREEL, M., VANGUESTAINE, M., PARDO-TRUJILLO, A. & THOMALLA, E. 2000*b*. The Frasnian–Famennian boundary sections at Hony and Sinsin (Ardenne, Belgium): new interpretation based on quantitative analysis of palynomorphs, sequence stratigraphy and climatic interpretation. *Geologica Belgica*, **3**(3–4), 271–283.

STREEL, M., AVKHIMOVITCH, V. I. *ET AL.* 2003. Biostratigraphic correlation at the late or/and latest Famennian from Western, Central and Eastern European sections. State of the art. *IUGS Subcommission on Devonian Stratigraphy Newsletter*, **19**, 50–56.

STREEL, M., BELKA, Z. *ET AL.* 2005. Relation of the neritic microfaunas and continental microfloras with the conodont and other pelagic faunas within the latest part of the Famennian (with a few, new additional data and a synthetic correlation chart). *IUGS Subcommission on Devonian Stratigraphy Newsletter*, **21**, 17–20; **22**, 13–15.

STREEL, M., BRICE, D. & MISTIAEN, B. 2006. Strunian. *Geologica Belgica*, **9**(1–2), 105–109.

STREEL, M., MAZIANE-SERRAJ, N., MARSHALL, J. E. A. & THOREZ, J. 2007. A reference section for neritic facies at the transition Late to Latest Famennian. *IUGS Subcommission on Devonian Stratigraphy Newsletter*, **22**, 34–40.

THOREZ, J., STREEL, M., BOUCKAERT, J. & BLESS, M. J. M. 1977. Stratigraphie et paléogéographie de la partie orientale du Synclinorium de Dinant (Belgique) au Famennien supérieur: un modèle de bassin sédimentaire reconstitué par analyse pluridisciplinaire sédimentologique et micropaléontologique. *Mededelingen Rijks Geologische Dienst, new serie*, **28**, 17–32.

TURNAU, E. 1996. Miospore stratigraphy of Middle Devonian deposits from Western Pomerania. *Review of Palaeobotany and Palynology*, **93**, 107–125.

TURNAU, E. 2002. Two new spore genera from Euramerica and their stratigraphic and geographic distribution. *Review of Palaeobotany and Palynology*, **118**, 261–268.

TURNAU, E. & RACKI, G. 1999. Givetian palynostratigraphy and palynofacies: new data from the Bodzentyn Syncline (Holy Cross Mountains, central Poland). *Review of Palaeobotany and Palynology*, **106**, 237–271.

VAN DER ZWAN, C. 1980. Aspects of Late Devonian and Early Carboniferous palynology of southern Ireland. III. Palynology of Devonian–Carboniferous transition sequences with special reference to the Bantry Bay area, Co. Cork. *Review of Palaeobotany and Palynology*, **30**, 165–286.

WEDDIGE, K. 1984. Zur Stratigraphie und Paläogeographie des Devons und Karbons von NE-Iran. *Senckenbergiana lethaea*, **65**, 179–223.

WEDDIGE, K. 1988. Eifel conodonts. *Courier Forschungsinstitut Senckenberg*, **102**, 103–110, 116–118, 132–133, 140–142, 150.

ZIEGLER, W. & SANDBERG, C. A. 1990. The Late Devonian standard conodont zonation. *Courier Forschungsinstitut Senckenberg*, **121**, 1–115.

Middle Devonian microfloras from the Chigua Formation, Precordillera region, northwestern Argentina

C. R. AMENÁBAR

CONICET, National Research Council of Argentina, UBA, Department of Geological Sciences, Faculty of Exact and Natural Sciences, University of Buenos Aires, Ciudad Universitaria, Pabellón II (1428), Capital Federal, Argentina

(e-mail: amenabar@gl.fcen.uba.ar)

Abstract: Rich palynological assemblages have been obtained from the Middle Devonian Chigua Formation (Chinguillos Group, San Juan Province), western Precordillera, Argentina, at two new localities (Del Chaco and Don Agustín creeks). From the palyniferous levels at Del Chaco Creek, there are two assemblages: one is very rich in microplankton (assemblage 1) whereas the other is dominated by spores (assemblage 2). The level obtained from Don Agustín Creek, poorest in overall taxa, is also rich in microplankton (assemblage 3). Compared with coeval microfloras elsewhere, a late Emsian–early Eifelian, and an early Givetian age are proposed for the first and the second assemblages respectively. A third assemblage (assemblage 3) tentatively represents a time-span around the Givetian–Frasnian boundary. The proposed age based on the palynomorphs reinforces previous palaeontological records. Variation in the microplankton diversity, fluctuations in the microplankton/spore ratio and differences in acritarch morphology are used here to interpret changing proximity to palaeoshoreline and so to recognize fluctuations in relative sea level that occurred during the deposition of the Chigua Formation.

In the Precordillera area of northwestern Argentina, thick sequences of Devonian rocks corresponding to the Cuyo or Precordillera Basin crop out (Padula *et al.* 1967). Although the strata are well exposed and are widely distributed, stratigraphic studies and correlations between these sequences are particularly difficult to carry out, due to the scarcity of fossiliferous levels and biostratigraphic data. The age of some Devonian formations in the region has been deduced from lithological similarities among them or from their stratigraphic position in relation to other dated units (e.g. Punilla, Codo and Del Planchón formations; Baldis & Peralta 2000).

Previous palynological work on the Devonian of Argentina has focused mainly on the Los Monos Formation in the subsurface of the Tarija Basin (Barreda 1986; Volkheimer *et al.* 1986; Ottone 1996; Grahn & Gutiérrez 2001; Grahn 2003; di Pasquo 2007). However, there is still scant information about palynofloras from the Argentinian Precordillera. Earlier palynological papers include only a few, and these are purely biostratigraphic data with illustrated lists of species of the Talacasto and Punta Negra formations (Le Hérissé *et al.* 1997*a*; Rubinstein 1997, 2000) located in the Precordillera of San Juan Province. Recently, Rubinstein & Steemans (2007) described a palynological assemblage from the Villavicencio Formation that crops out in the Precordillera of the Mendoza Province, Argentina.

A detailed systematic palynological work of the Chigua Formation in the Precordillera of San Juan Province was part of the PhD dissertation of the author (Amenábar 2007). The first work on the Chigua Formation was at La Cortadera Creek in the western flank of the Del Volcán Range, Precordillera (Amenábar *et al.* 2006, 2007). The main objective of this study is to present the first palynological records for the Chigua Formation, at two new localities, Del Chaco and Don Agustín creeks.

Many levels have been sampled during different field trips in several localities in the region, but few of them are palynologically fertile. The strong tectonic activity that occurred not only during the Middle to Late Palaeozoic but also during the Andean Orogeny presumably affected the area, and it could be the cause of the destruction of the organic material in the sediments, thus it is not easy to find good palynological specimens in the Precordillera region. Moreover, the tectonic processes make it difficult to find complete sections in the area. Although in this work only three palyniferous samples are analysed, they have yielded rich microfloras with abundant and well-preserved palynomorphs, they represent an important tool to date the Chigua Formation and provide useful data for correlation with palynofloras of similar age from elsewhere.

Geological setting

The Precordillera is a fold-and-thrust belt, *c.* 400 km long and 80 km wide, located in northwestern Argentina, which includes Palaeozoic and Tertiary

From: KÖNIGSHOF, P. (ed.) *Devonian Change: Case Studies in Palaeogeography and Palaeoecology*. The Geological Society, London, Special Publications, **314**, 177–192.
DOI: 10.1144/SP314.10 0305-8719/09/$15.00

sediments (Bracaccini 1946; Heim 1952). It extends from the southern part of the La Rioja Province, through the San Juan Province, up to the northern part of Mendoza Province, striking in a north–south direction between the Frontal Cordillera to the west and Pampean Range to the east (Fig. 1a). As a geological unit, and based on stratigraphic and structural features (Baldis & Peralta 2000), it has traditionally been divided into western, central and eastern domains (Fig. 1a).

The Precordillera has been interpreted as a typical thin-skinned structure developed during the Andean Orogeny (e.g. von Gosen 1992; Cristallini & Ramos 2000; Alonso *et al.* 2005). The age of deformation is considered to be Miocene to Recent, with west to east progression of thrusting (Jordan *et al.* 1993; Zapata & Allmendinger 1996; Alvarez-Marron *et al.* 2006). However, the existence of pre-Andean deformation within the Palaeozoic sedimentary strata in the Precordillera has been widely recognized by several authors (e.g. Baldis & Sarudiansky 1975; Furque 1979). Evidence of pre-Carboniferous deformation is well known in the western and eastern Precordillera, where Mississippian deposits, such as the Malimán and El Ratón formations, overlie older Palaeozoic rocks (Chigua and Codo formations, respectively; Azcuy *et al.* 2000; Baldis & Peralta 2000) with a strong angular unconformity between them. In contrast, in the central Precordillera no pre-Tertiary deformation has been observed where Upper Carboniferous deposits (Paganzo Group) overlie Devonian sediments (Punta Negra Formation) with an erosional unconformity (paraconcordance) (e.g. Astini 1996).

Various hypotheses about the origin and the nature of this unconformity are still in discussion. Some authors attributed this type of contact to the Chanic Orogeny, but occurring at different times during the Devonian (e.g. Ramos 1984, 2004; Astini 1996). Others, however, related it to sea-level

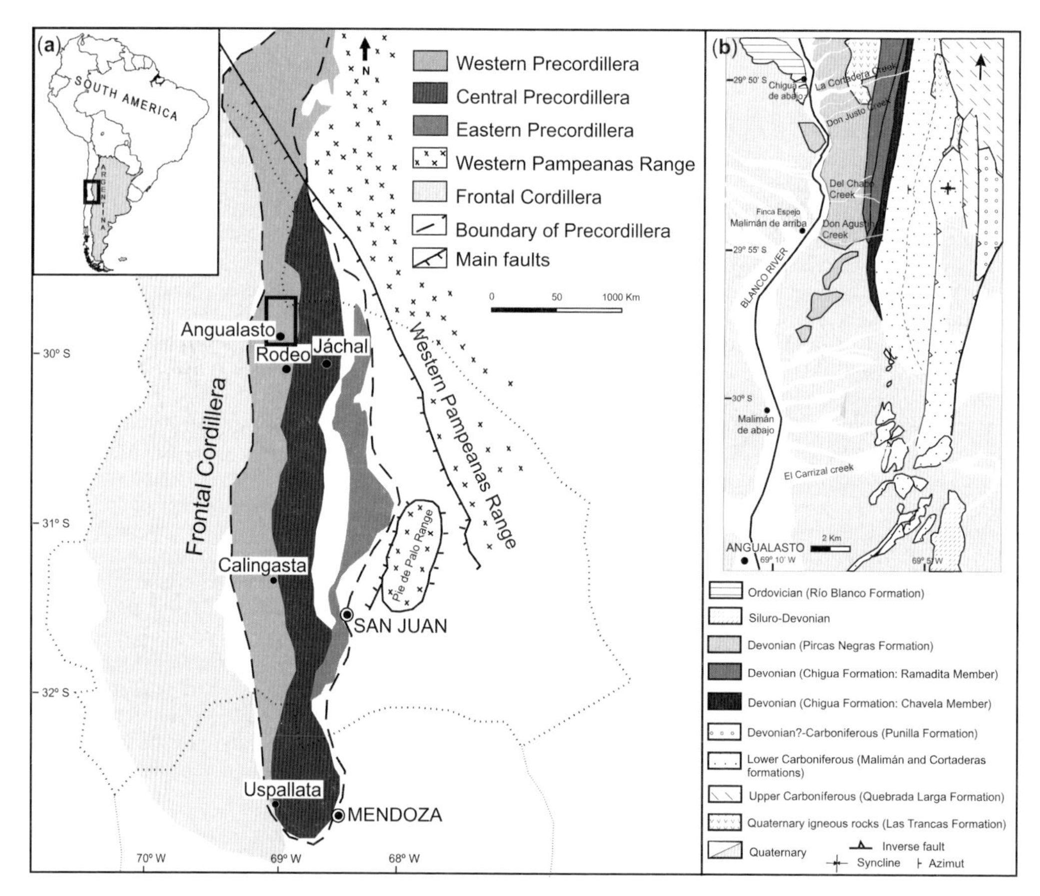

Fig. 1. (**a**) Map of South America showing the location of the Precordillera region and a detailed map of the Western Precordillera of Argentina (modified from Baldis & Peralta 2000). The box represents the study area. (**b**) Geological map and location of the Del Chaco and Don Agustín creeks in the San Juan Province, Argentina.

changes linked with the glacial episodes that occurred during the Late Devonian in southern South America (e.g. González Bonorino 1990).

Devonian rocks are well represented throughout the Precordillera; the biostratigraphic record is well constrained in the central Precordillera, where a varied faunal association composed of brachiopods, trilobites, ostracodes, bivalves, pelmatozoans and plant remains occurs in the Talacasto and Punta Negra formations (Baldis & Peralta 2000; Herrera & Bustos 2001).

This study focuses on sediments belonging to the Chinguillos Group, which crops out in the western Precordillera, in San Juan Province, *c.* 30°S and 69°W, near the town of Angualasto (Fig. 1a, b). The Chinguillos Group contains the Pircas Negras and Chigua formations, and forms a belt that extends to the east of the Blanco River, on the western slope of the Punilla and Del Volcán ranges (Fig. 1b). The Chigua Formation (700 m thick) is in tectonic contact with the underlying Pircas Negras Formation and is separated by an angular unconformity from the overlying Malimán Formation (Late Tournaisian?–Early Viséan; Amenábar 2007; Fig. 2c). The Chigua Formation is divided into two members, the lower Chavela (marine) Member and the upper Ramadita (paralic–continental) Member. The lithology of the Chavela Member, showing green-brown colour variation, consists mainly of shales with concretions and calcareous lenses and subordinate sandstone layers. The Chavela Member is distinguishable from the Ramadita Member by its lithology and coloration, having a clearer greenish colour and minor sandstone layers in the former unit. Some Quaternary igneous rocks corresponding to the Las Trancas Formation (Furque 1963) make the recognition of the geological structure of the Chigua Formation difficult (Baldis & Sarudiansky 1975; Fig. 2c).

The Chinguillos Group was considered to be deposited under marine conditions (Baldis & Sarudiansky 1975). Some fossils have been found in the unit, including marine invertebrates such as

Fig. 2. (**a**) Field photograph of the highly deformed rocks of the Chavela Member of the Chigua Formation with palynological samples BAFC-Pl 1797, 1798. (**b**) Field photograph of the carbonaceous shale of the Chavela Member, Chigua Formation, where sample BAFC-Pl 1780 was obtained. It is intercalated with sandstone beds bearing lycophyte remains. (**c**) Field photograph showing the contact between the Chigua and Malimán formations, where weakly folded Devonian rocks underlie Carboniferous homoclinal strata represented by a thick conglomerates and sandstones beds. The sample BAFC-Pl 1776 was recovered from shales located *c.* 150 m below the Devonian–Carboniferous boundary. Note the Las Trancas Formation (Quaternary igneous rocks) intruded in the Chigua Formation. The unconformity in Don Agustín Creek is just distinguishable.

the trilobites *Punillaspis argentina* Baldis, *Phacops chavelai* Baldis and Longobucco, and *Acanthopyge balliviani* Kozlowski, among others; the cephalopods *Tornoceras baldisii* Leanza and *Orthoceras* sp.; the cnidarian *Conularia* sp. (Baldis & Sarudiansky 1975); and a fossil flora that is represented by the herbaceous lycophytes *'Haplostigma' furquei* Frenguelli, *'H.' baldisii* Gutiérrez and ?*Cyclostigma* sp. (Gutiérrez 1996). All of these fossils were collected by colleagues in the area between La Cortadera and Del Chaco creeks (Fig. 1b; Baldis & Sarudiansky 1975). *Acanthopyge balliviani* is also recorded in Bolivian Devonian deposits indicating an Eifelian or Emsian age (depending on the author: Ahlfeld & Branisa 1960; Wolfart & Voges 1968), and the genus *Tornoceras* is recognized in Frasnian and Givetian deposits of Euramerica (Kullmann 1993). Thus, the co-occurrence of both taxa in the Chigua Formation allowed Baldis & Sarudiansky (1975) to consider that the stratigraphic range of the Chigua Formation spanned from the Emsian–Eifelian boundary through the Givetian–Frasnian boundary, meaning that the whole Middle Devonian is present. Based on these data, these authors tentatively proposed a Givetian age for the Chavela Member of the Chigua Formation.

The first detailed palynological studies from the Chavela Member of the Chigua Formation were carried out by Amenábar *et al.* (2006, 2007) at La Cortadera Creek. Palynological data from Del Chaco and Don Agustín creeks (Fig. 1a), situated to the south of the first studied locality, are herein presented.

Material and methods

Five palynological samples were collected from the Chavela Member of the Chigua Formation at Del Chaco Creek and eight samples from Don Agustín Creek (Fig. 1b). Only three samples (BAFC-Pl 1797, 1798, 1780) from Del Chaco Creek and one sample (BAFC-Pl 1776) from Don Agustín Creek were productive. All the samples were collected from fine-grained lithologies; BAFC-Pl 1797, 1798 and 1798 were recovered from shales of the highly deformed (folded) rocks of the unit (Fig. 2a), whereas BAFC-Pl 1780 was obtained from a carbonaceous shale. It is interbedded with sandstone beds (Fig. 2b) bearing megaflora remains belonging to a lycophyte; their specific identification is still under study. The only sample (BAFC-Pl 1776) recovered from Don Agustín Creek came from shales recognized for their green colour, which were situated stratigraphically beneath the Devonian–Carboniferous boundary (Fig. 2c).

The samples were processed using standard palynological techniques. The palynological localities have produced rich palynofloras and the preservation of the palynomorphs is generally good, showing yellow to light brown colour. Material from some samples, however, is altered by pyrite crystals. The microflora obtained from Del Chaco Creek contains microplankton (acritarchs, prasinophytes and green algae) and spores (acavate and cavate/pseudosaccate) in variable percentages, depending on the sample. The Don Agustín Creek material includes mainly poorly preserved microplankton taxa (acritarchs and prasinophytes), with most specimens showing dark coloration, as well as pyrite voids. Preservation of some palynomorph species (mainly specimens of the *Maranhites*) precludes a proper systematic assignment and they were left in open nomenclature.

Stratigraphic ranges of some index microplankton and spore species are shown in Figure 3, and some selected specimens are illustrated in Figure 4 and 5. Quantitative analysis of the palynological levels of the Chavela Member of the Chigua Formation and a list of the species of each level recovered from different localities are presented in Tables 1 and 2. The Ramadita Member of the Chigua Formation has also been sampled, but it was palynologically barren.

The palynological slides (BAFC-Pl) and residues are housed at the Laboratory of Palynology, Department of Geological Sciences, Faculty of Exact and Natural Sciences, University of Buenos Aires, Argentina.

Palynological data and age of the assemblages

From the three palyniferous samples (BAFC-Pl 1797, 1798, 1780; Fig. 2a) obtained from the Chavela Member of the Chigua Formation at Del Chaco Creek (Fig. 1b), two assemblages are recognized: **assemblage 1** (BAFC-Pl 1797) is characterized by a high diversity of microplankton (acritarchs, prasinophytes and green algae), consisting of 26 identified species (Tables 1 and 2). It is mainly composed of acanthomorphic acritarchs, some of which are large (Fig. 4). The spores are scarce (ten species identified) and they are present as subordinate elements in the assemblage. In contrast, **assemblage 2** (BAFC-Pl 1780; Fig. 2b) is composed of predominantly continental elements with *Grandispora* being especially abundant (Tables 1 and 2, Fig. 5(1)). Sample BAFC-Pl 1798 yielded a very scarce and poorly preserved palynomorph assemblage that is not age-diagnostic, and is not considered as part of the aforementioned

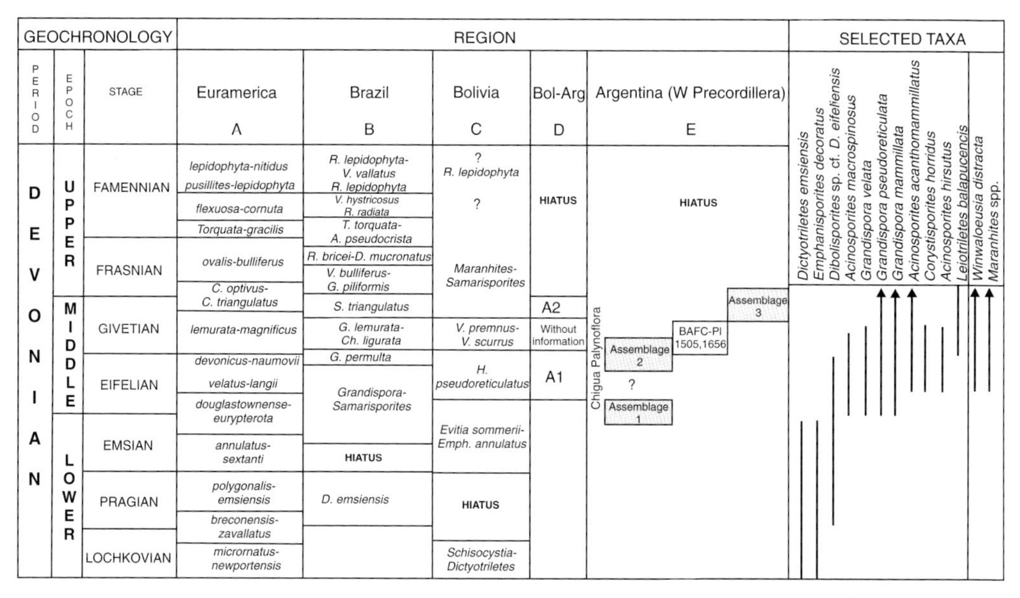

Fig. 3. Summary correlation chart of the assemblages studied here (E) with other Devonian biozones or palynofloras of Euramerica (A), Brazil (B), Bolivia (C), Argentina–Bolivia (D) and the stratigraphic ranges of selected spore and microplankton species from the Chavela Member of the Chigua Formation at the Del Chaco and Don Agustín creeks. References: A, Richardson & McGregor (1986); B, Melo & Loboziak (2003); C, Suárez Soruco & Lobo Boneta (1983), Limachi *et al.* (1996); D, di Pasquo (2007). In column E, the samples BAFC-Pl 1505 and 1656 recorded at the La Cortadera Creek (Amenábar *et al.* 2006, 2007) are also included.

assemblage. **Assemblage 3** corresponds to sample BAFC-Pl 1776 from Don Agustín Creek and yielded a low-diversity assemblage of microplankton (acritarchs and prasinophytes) which includes taxa with stratigraphic value.

Assemblage 1 includes some spore and microplankton species with biostratigraphic importance such as *Dictyotriletes emsiensis* Morphon, *Emphanisporites decoratus*, *Acinosporites hirsutus* and specimens of the genus *Grandispora* Fig. 5(1). There are numerous well-known South American palynofloras where specimens of the *D. emsiensis* Morphon have been registered; these include the assemblages of the Talacasto Formation, attributed to late Lochkovian–Emsian (Le Hérissé *et al.* 1997*a*) and the Villavicencio Formation, dated as late Pragian–early Emsian (Rubinstein & Steemans 2007), both in the Argentinean Precordillera, and also those of Solimões (Rubinstein *et al.* 2005), Paraná (Dino 1999) and Amazonas basins (Melo & Loboziak 2003) in Brazil. According to Melo & Loboziak (2003) the *Grandispora/Samarisporites* spp. (GS) Interval Zone, defined in the Amazon Basin and of late Emsian–early Eifelian age, is characterized by the appearance of large spinose pseudosaccate/zonate spores of the genera *Grandispora*, *Samarisporites* and *Craspedispora*; it can also contain some species from the Early Devonian, such as *Dictyotriletes emsiensis* and other related ones characterized by small size, but these do not persist into younger biozones. Thus, the presence of the *D. emsiensis* Morphon together with species of *Grandispora* allows attribution of a late Emsian–early Eifelian age (Early late to Middle early Devonian) of this assemblage (Fig. 3).

On the other hand, **assemblage 2** includes the spores *Acinosporites hirsutus*, *A. macrospinosus*, *A. acanthommammillatus*, *Corystisporites horridus*, *Dibolisporites* sp. cf. *D. eifeliensis*, *Grandispora mammillata* and *G. pseudoreticulata*, with *Leiotriletes balapucensis*, which are diagnostic species (Fig. 3). This assemblage is comparable to samples BAFC-Pl 1505 and BAFC-Pl 1656 from the Chavela Member of the Chigua Formation in the La Cortadera Creek (Fig. 3; Amenábar *et al.* 2006, 2007), as they share the spore species *Grandispora pseudoreticulata*, *Leiotriletes balapucensis* and *Dibolisporites* sp. cf. *D. eifeliensis, Acinosporites acanthomammillatus*. It can be attributed to the Middle Devonian (early Givetian). The assemblage of the Chavela Member of the Chigua Formation at La Cortadera Creek can be placed between the A1 and A2 assemblages from the Los Monos Formation (Tarija Basin), based on the presence of *Geminospora lemurata* and the absence of *Samarisporites triangulatus* in the microflora (Fig. 3).

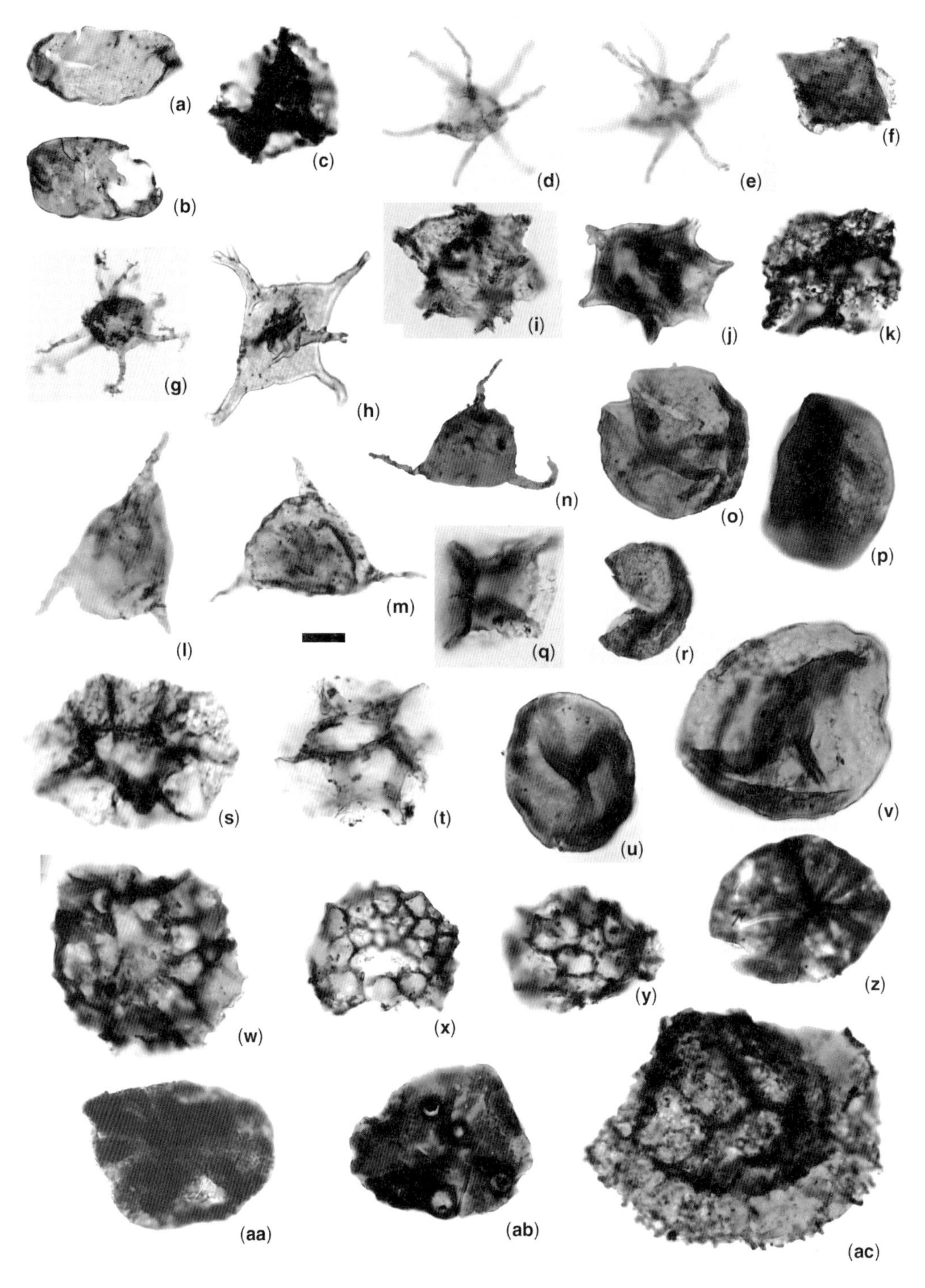

Fig. 4. Assemblage 1 recognized in the Chavela Member of the Chigua Formation at Del Chaco Creek. Scale bar: 20 μm in a, b, d, e, g, h, l–n, s, ac; 15 μm in c, k, o–r, t–z, aa, ab; 10 μm in f, i, j. Co-ordinates after England Finder graticule. (**a**, **b**) *Navifusa bacilla* BAFC-Pl 1797 (2), A28/2; BAFC-Pl 1797 (2), P21/4; (**c**) *Arkonites bilixus* BAFC-Pl 1797 (2), A50/2; (**d**, **e**) *Diexallophasis simplex* BAFC-Pl 1797 (2), S26; (**f**) *Duvernaysphaera angelae* BAFC-Pl 1797 (2), A33/1; (**g**) *Multiplicisphaeridium ramispinosum* BAFC-Pl 1797 (2), L25/1; (**h**) *Exochoderma arca* BAFC-Pl 1797

According to di Pasquo (2007), assemblages from the Los Monos Formation are comparable to the Euramerican biozones *D. devonicus–G. naumovii* and *G. lemurata–C. magnificus* Richardson & McGregor (1986) and to the Interval zones GS, Per and LLi defined by Melo & Loboziak (2003) in the Amazonas Basin, respectively (Fig. 3). Other species in assemblage 2, such as *Acinosporites hirsutus*, *A. macrospinosus*, *Acanthotriletes horridus* and *Grandispora mammillata*, are present in Givetian palynofloras of southern Euramerica and western Gondwana (e.g. Richardson & McGregor 1986), and confirm the age attributed to this assemblage (Fig. 3).

Assemblage 3 has yielded only three identifiable taxa: the acritarchs *Winwaloeusia distracta*, *Polyedryxium embudum* and unattributed species of the prasinophyte genus *Maranhites*; the latter were grouped under the term *Maranhites* spp. and they were left in open nomenclature due to their poor preservation (strong pyritization and dark colour; Fig. 5(2)). Several indeterminable spore specimens of the genus *Apiculiretusispora* were also recovered, but in very low numbers (Table 1). *Winwaloeusia distracta* is known from the Lower Devonian (late Pragian–early Emsian) of the Villavicencio Formation, Argentina (Rubinstein & Steemans 2007), Tunisia and Algeria (Deunff 1966, 1977; Jardiné 1972), in the Lochkovian of France (Deunff 1980) and may extend into the Middle and Upper Devonian (early Famennian) in North Africa, France (Deunff 1980) and Belgium (Stockmans & Williere 1969). In Brazil, it has also been reported from upper Lochkovian rocks of the Solimões Basin (Rubinstein *et al.* 2005), and from the Givetian–Frasnian strata of the Paraná and Amazonas basins (Quadros 1999; Le Hérissé 2001). In addition, *Maranhites* is a typical component of the Givetian–Frasnian palynofloras of South America (Díaz Martínez *et al.* 1999; Dino 1999; Quadros 1999), and throughout the world (e.g. Le Hérissé & Deunff 1988; Colbath 1990), with its greatest species diversity in Frasnian and Famennian assemblages (Quadros 1999).

Dino (1999) described 'Spore-Assemblage 6' (late Givetian to late Frasnian) in the Paraná Basin, and characterized it by a proliferation of *Maranhites* spp. and *Pseudolunulidia* spp. Moreover, Le Hérissé (2001) defined 'Zone B2' in the Amazonas Basin, from the early Frasnian. It is divided into two subzones, B2A and B2B. The latter contains *Winwaloeusia distracta*, among other acritarchs, and is characterized by mainly large leiosphaerids, *Maranhites* and *Tasmanites*. A decline of overall phytoplankton diversity occurs in this subzone, in comparison with the older biozones. Thus, taking into account the presence of *Winwaloeusia distracta* together with *Maranhites* spp. in the assemblage Fig. 5(2), the age of assemblage 3 is tentatively placed around the Givetian–Frasnian boundary (Fig. 3). Moreover, the presence of abundant specimens of the genus *Maranhites*, some of which are large (100 μm on average; Fig. 5(2), has been documented by Díaz Martínez *et al.* (1999) as characteristic of the Frasnian assemblages of Bolivia, where the unusual size of the specimens suggests favourable conditions for development of the phytoplankton. More palynological information will be necessary to facilitate a reliable age determination of assemblage 3 and thus to correlate it with the rest of the assemblages obtained from other localities in the Precordillera area, and also with other microfloras elsewhere in the world.

As a result, the age of the Chavela Member of the Chigua Formation is based on the palynological data from the Del Chaco and Don Agustín creeks. It supports the previous age proposed by Baldis & Sarudiansky (1975), who attributed the unit to the Middle Devonian, based on invertebrate fossils. The common spores and microplankton species confirm the affinity of the palynofloras of the Precordillera to those recognized in northern Argentina (Tarija Basin). They show a palaeogeographic connection between both areas during the Devonian, and it parallels the trilobite faunas of both regions (Baldis 1967).

Moreover, a similarity between the microfloras from southern South America and Euramerica during the Middle Devonian is sustained by the presence of numerous co-occurring spore species. The similarity between the two areas is due to narrowing of the Rheic Ocean that began during the Early Devonian and continued in the Middle Devonian, bringing continental masses closer and thus facilitating the exchange of taxa (cf. Le Hérissé *et al.* 1997*b*). There is a second-order marine transgression in the Middle Devonian allowing more communications between the intracratonic basins (Isaacson & Sablock 1989). As is well-documented

Fig. 4. (*Continued*) (2), U55/0-4; (**i**) *Estiastra improcera* BAFC-Pl 1797 (2), R23; (**j**) *Evittia* sp. cf. *E. cymosa* BAFC-Pl 1797 (2), A50/1; (**k**) *Quadrisporites variabilis* BAFC-Pl 1797 (2), J56; (**l–n**) *Veryhachium trispinosum* BAFC-Pl 1797 (2), A30/4; A48/1; H25; (**o, p, r, u, v**) *Dictyotidium* spp. BAFC-Pl 1797 (2), J58/2; N22/4; R55/4; S23/2; 1797 (2), O28/2; (**q**) *Polyedryxium embudum* BAFC-Pl 1797 (2), H22/3; (**s**) *Dictyotidium munificum* BAFC-Pl 1797 (2), U40; (**t**) *Cymatiosphaera canadensis* BAFC-Pl 1797 (2), O20/0-1; (**w–y**) *Dictyotriletes emsiensis* BAFC-Pl 1797 (2), T21; A51/1; O20; (**z**) *Emphanisporites rotatus* BAFC-Pl 1797 (2), B24/4; (**aa**) *Emphanisporites decoratus* BAFC-Pl 1797 (2), F47; (**ab**) Spore with pyrite and voids leaving BAFC-Pl (2), A33/1; (**ac**) *Grandispora velata* BAFC-Pl 1797 (2), A53/1.

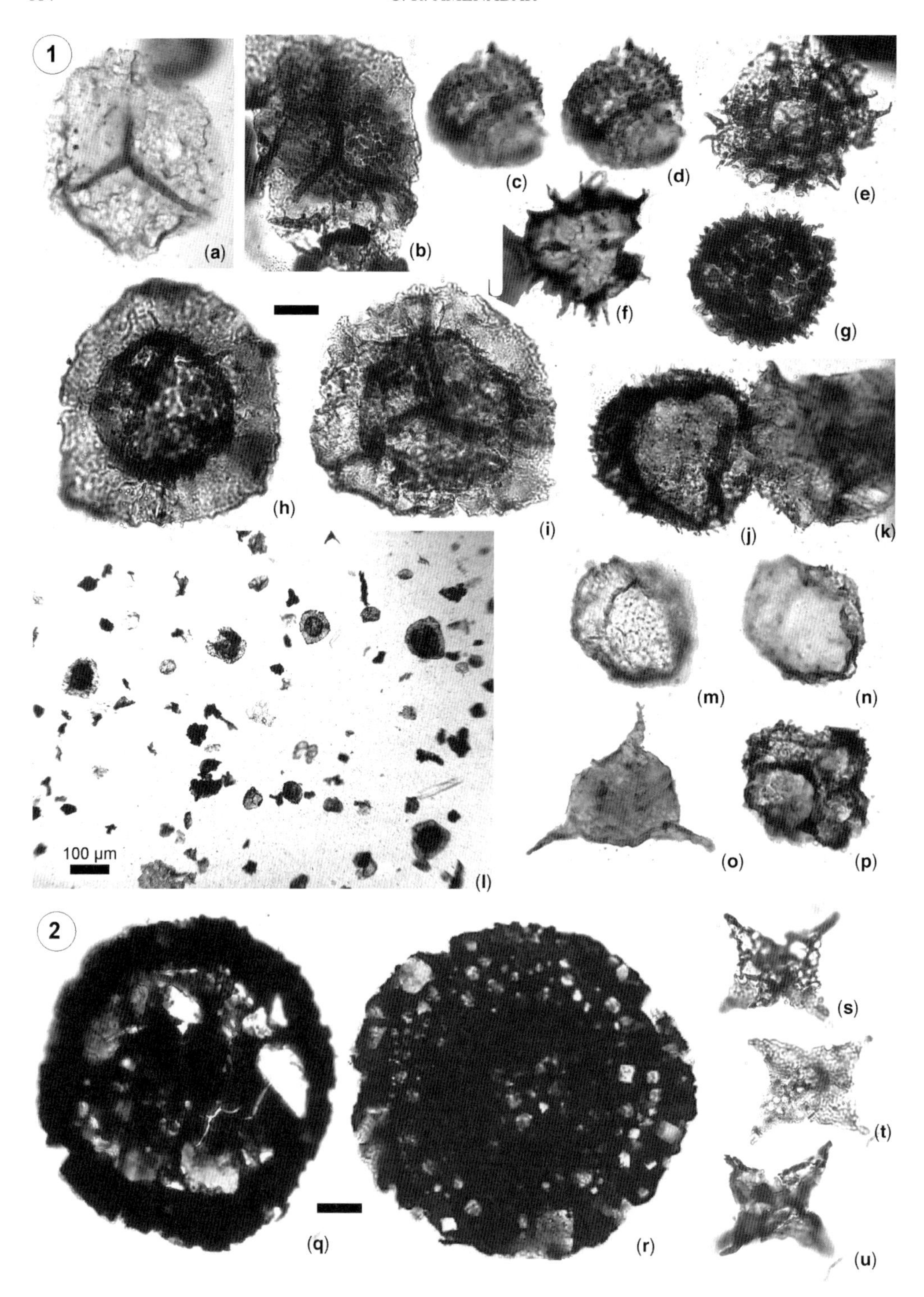

Fig. 5. (**1**) Assemblage 2 recognized in the Chavela Member of the Chigua Formation at Del Chaco Creek. Scale bar: 20 µm in a–k, p; 15 µm in m, n, o; except l. Co-ordinates after England Finder graticule. (**a**) *Leiotriletes balapucensis* BAFC-Pl 1780 (1), A44/3-4; (**b**) *Grandispora mammillata* BAFC-Pl 1780 (1), X3372-4; (**c**, **d**) *Acinosporites acanthomammillatus* BAFC-Pl 1780 (1), R41/1; (**e**) *Acinosporites macrospinosus* BAFC-Pl 1780 (1), J31;

by some authors (e.g. Le Hérissé *et al.* 1997*b*), the more or less marked phytoplankton provincialism during the Late Silurian–Early Devonian, includes a cold Malvinokaffric Realm in the southern high latitudes, which was determined by means of brachiopod distribution (Boucot 1975). However, this realm largely disappeared in the Middle Devonian, as evidenced by the records of brachiopods such as *Tropidoleptus carinatus* (e.g. Isaacson & Perry 1977; Fonseca & Melo 1987) and several microplankton species (Quadros 1999). This indicates that transgression brought an apparent migration of warm water forms into the cold water environment of high latitudes during this time interval.

Palaeoenvironmental considerations

Microplankton distribution is influenced by various local ecological conditions such as water temperature, depth, salinity, nutrient availability, light, turbidity, rate of reproduction and distance from shoreline (Le Hérissé *et al.* 1997*b*; Playford 2003). In addition, oceanic circulation, the presence of physical barriers and dispersal of the continental masses may have influenced the distribution of the microplankton assemblages (Raevskaya *et al.* 2004; Vecoli & Le Hérissé 2004).

As discussed by several authors (e.g. Staplin 1961; Smith & Saunders 1970; Gray & Boucot 1972; Riegel 1974; Dorning 1981; Wicander & Wright 1983; Wicander & Playford 1985; Wicander & Wood 1997), a relationship between microplankton diversity and morphotypes, increasing distance from a shore and/or increasing water depth have been demonstrated. Therefore, the quantitative distribution of microplankton as well as fluctuations in the microplankton/spore ratio is used here to interpret the depositional conditions of the samples in the Chavela Member of the Chigua Formation.

Assemblage 1 is characterized by high microplankton diversity, reaching 72.3% of the total assemblage (Tables 1 and 2). According to Dorning (1981), high diversity is represented by between ten and 90 species per sample (usually 25 to 60 species) and is indicative of an offshore marine environment, so the assemblage analysed herein indicates that the deposition occurred in a normal-marine near-to offshore environment. In addition, free communication with the open sea is apparent in this sample, not only because of the abundance of the microplankton, but also because of the presence of delicate process morphotypes, many of which are elongate and considered to indicate open marine forms (Smith & Saunders 1970; Vavdrová *et al.* 1993; Stricanne *et al.* 2004). Also, noteworthy is the unusual size of some specimens of *Verhyachium trispinosum* (approximately 100 μm; Fig. 4) suggesting that in the early Late Devonian (Frasnian) in the region of the Precordillera, there were favourable conditions for the development of the acritarchs (cf. Díaz Martínez *et al.* 1999) where a temperate-warm climate prevailed during this time (e.g. Streel *et al.* 2000). The scarce spores found in this assemblage were most likely transported from coastal swamps toward the marine depocentre. The high planktonic diversification in the Chavela Member of the Chigua Formation supports the interpretation of Padula *et al.* (1967) who proposed a transgressive phase that occurred during the Early Devonian and advanced from north to NW and continued though the Middle Devonian.

On the other hand, the increase in the relative abundance of spores seen in **assemblage 2** and the subsequent decrease of the microplankton/spore ratio indicates a closer proximity to the palaeoshoreline (Wicander & Wood 1997). This is because nearshore assemblages are characterized by a low diversity of microplankton (low diversity represents five to 15 species per sample; after Dorning 1981), and in **assemblage 2** the plankton reaches only 20.9% of the total of the assemblage (Tables 1 and 2).

The proximity to the coast or the existence of a marginal marine palaeoenvironment revealed in this sample is supported by the composition of the subordinate microplankton taxa in the assemblage, which include mainly the genera *Verhyachium* and *Quadrisporites* (Table 1, Fig. 5(1)). *Verhyachium* is considered an opportunistic genus if it dominates the assemblage or if it is the only taxon present (Lé Hérissé 2001, 2002). Three-spined forms of veryhachids were able to tolerate the unstable conditions of nearshore environments (Lé Hérissé 2002). Low planktonic diversity suggests physical–chemical conditions of a particular marine palaeoenvironment (i.e. environmental stress due to scarce nutrients,

Fig. 5. (*Continued*) (**f**) *Corystisporites horridus* BAFC-Pl 1780 (1), R43/3; (**g**) *Acinosporites hirsutus* BAFC-Pl 1780 (1), J39/0-2; (**h**) *Grandispora* sp. BAFC-Pl 1780 (1), R34; (**i**) *Grandispora pseudoreticulata* BAFC-Pl 1780 (1), R26; (**j**, **k**) *Dibolisporites* sp. cf. *D. eifeliensis* BAFC-Pl 1780 (1), W41/4; (**l**) assemblage rich in pseudosaccate spores of the genus *Grandispora*; (**m**, **n**) *Apiculiretusispora laxa* BAFC-Pl 1780 (1), A23/2; (**o**) *Veryhachium trispinosum* BAFC-Pl 1780 (1), N23/4; (**p**) *Quadrisporites variabilis* BAFC-Pl 1780 (1), S20. (**2**) Assemblage 3 recognized in the Chavela Member of the Chigua Formation at Don Agustín Creek. Scale bar: 20 μm in q–u. Co-ordinates after England Finder graticule. (**q**, **r**) *Maranhites* spp. BAFC-Pl 1776 (1), M35; BAFC-Pl 1776 (2), M20/4; (**s**–**u**) *Winwaloeusia distracta* BAFC-Pl 1776 (**1**), F36/4; A43/3; O26/2.

Table 1. *Number of specimens counted in the Chavela Member of the Chigua Formation, at Del Chaco and Don Agustin creeks, indicating the assemblages which correspond to each other*

Registered species	Del Chaco creek			Don Agustin creek
	1*		2	3
	1797 (1, 2)†	1798 (1)	1780 (1)	1776 (1, 2)
Microplankton				
Acritarchs				
Arkonites bilixus Legault, 1973	1			
Diexalophasis simplex Wicander and Wood, 1981	1			
Estiastra improcera Loeblich, 1969	1			
Evittia* sp. cf. *E. cymosa Loeblich, 1970	1			
Exochoderma arca Wicander and Wood, 1981	5			
?*Estiastra* sp.	4			5
***Gorgonisphaeridium* sp.**	1			
***Micrhystridium* spp.**	1		1	
***Multiplicisphaeridium ramispinosum*,** Staplin, 1961	1			
***Multiplicisphaeridium* sp.**	1			
Navifusa bacilla (Deunff) Playford, 1977	3			
Polygonium barredae Ottone, 1996	1			
Veryhachium polyaster Staplin, 1961	4		2	
***Veryhachium* spp.**	13			7
Veryhachium trispinosum (Eisenack) Deunff, 1954	9		2	
Winwaloeusia distracta (Deunff) Deunff, 1977				13
Prasinophytes				
Cymatiosphaera canadensis Deunff, 1954; ex Deunff, 1961	3			
Cymatiosphaera perimembrana Staplin, 1961	5		1	
Dictyotidium munificum (Wicander and Wood) Amenabar, di Pasquo, Carrizo and Azcuy, 2006	5			
***Dictyotidium* spp.**	12			
Duvernaysphaera angelae Deunff, 1964	1			
***Duvernaysphaera* sp.**	5			
***Maranhites* spp.**				17
***Leiosphaeridia* spp.**	4			
Polyedryxium embudum Cramer, 1964	1			1
Polyedryxium pharaonis Deunff, 1954; ex Deunff, 1961	4			
***Polyedryxium* sp.**		1		
***Pterospermella* sp.**	2			
Quadrisporites variabilis (Cramer) Ottone and Rossello, 1996	11		2	
Spores				
Acavate				
Acinosporites acanthomammillatus Richardson, 1965			1	
Acinosporites hirsutus (Brideaux and Radforth) Richardson and McGregor, 1982			5	
Acinosporites lindlarensis Riegel, 1968			1	
Acinosporites macrospinosus Richardson, 1965			2	
***Anapiculatisporites* spp.**	3			
***Ancyrospora* sp.**			1	
Apiculatasporites microconus (Richardson) McGregor and Camfield, 1982	1		1	
Apiculiretusispora laxa Amenabar, di Pasquo, Carrizo and Azcuy, 2006			1	
***Apiculiretusispora* sp.**			1	1
Corystosporites horridus Hacquebard, 1957			1	
Cyclogranisporites plicatus Allen, 1965			1	
Dibolisporites* sp. cf. *D. eifeliensis (Lanninger) McGregor, 1973			9	
***Dibolisporites* spp.**			6	
Dictyotriletes emsiensis (Allen) McGregor, 1973	6			

(*Continued*)

Table 1. *Continued*

Registered species	Del Chaco creek			Don Agustin creek
	1*		2	3
	1797 (1, 2)[†]	1798 (1)	1780 (1)	1776 (1, 2)
Emphanisporites decoratus Allen, 1965	3			
Emphanisporites* sp. cf. *E. decoratus Allen, 1965	39			
Emphanisporites rotatus McGregor, 1961; emend. McGregor, 1973	1			
***Granulatisporites* sp.**		1	3	
Leiotriletes balapucencis di Pasquo, 2007			21	
Leiotriletes trivialis Naumova, 1953	1			
***Leiotriletes* sp.**			1	
***Punctatisporites* sp.**		2	1	
***Retusotriletes* sp.**	1			
Cavate Pseudosaccate				
Grandispora velata (Richardson) McGregor, 1973	2			
Grandispora mammillata Owens, 1971	51		1	
Grandispora pseudoreticulata (Menendez and Pothe de Baldis) Ottone, 1996			135	
***Grandispora* spp.**			2	
Total	208	4	202	44

*Assemblage number (1, 2, 3)
[†]Sample number (BAFC-PL 1797 etc.)

variable salinity, etc.; Molyneux *et al.* 1996), possibly related to marginal marine palaeoenvironments. Additionally, *Quadrisporites* is considered a fossil green alga that can be present not only in open marine sediments, but also in marginal marfine or even in lacustrine deposits (Lé Hérissé 2002). The presence of well-preserved lycophyte remains recovered just below sample BAFC-Pl 1780 (Fig. 2b) suggests minimal transport, and also supports the interpretation of a continental to coastal palaeoenvironment for this assemblage.

Lastly, **assemblage 3** is characterized by a moderate abundance and low diversity of microplankton and spores, with the microplankton constituting 83.4% of the total assemblage (Tables 1 and 2). Despite the low abundance and diversity of the palynomorphs displayed in this sample, an increase in the microplankton/spore ratio can be observed, indicating once more that deposition occurred in a normal-marine, near- to offshore environment. On the other hand, the prasinophytes and acritarchs in general are correlated with platform marine environments, although the first ones could also be found in marginal environments where the salinity is variable (e.g. Dorning 1981; Molyneux *et al.* 1996). The exclusive occurrence of prasinophytes can indicate restricted marginal marine environments (Wicander & Playford 1985; Huysken *et al.* 1992). However, in assemblage 3 the presence of *Maranhites* with acritarchs (with processes), such as *Winwaloeusia distracta* (Fig. 5s–u), limits the kind of palaeoenvironment,

Table 2. *Number of species and percentage of palynomorphs identified in the Chavela Member of the Chigua Formation*

	Assemblage 1 1797* (1, 2)		Assemblage 2 1780 (1)		Assemblage 3 1776 (1, 2)	
	No. species	Per cent	No. species	Per cent	No. species	Per cent
Microplankon	26	72.3	5	20.9	5	83.4
Spores	10	27.7	19	79.1	1	16.6

*Sample number (BAFC-PL 1797 etc.)

being of marine platform instead of a marginal marine environment. Favourable conditions for phytoplankton productivity are indicated by the large size of specimens of *Maranhites* (about 120 μm; Fig. 5q, r), which have also been observed in Late Devonian assemblages from Bolivia (Vavdrová *et al.* 1993; Díaz Martínez *et al.* 1999), where specimens of *Maranhites* not only reached a large size but also were very abundant.

Conclusions

Devonian palynological data from the Chavela Member of the Chigua Formation are documented in detail for the first time at Del Chaco and Don Agustín creeks. This increases the knowledge about the Devonian palynofloras of the Precordillera of Argentina. These new data provide valuable stratigraphic and palaeoenvironmental information for these poorly known Devonian sections of the western Precordillera, and they open new perspectives for correlations with other assemblages of Argentina, and with the well-established miospore zonation schemes of the Amazonas Basin and western Europe.

The age of the Chavela Member of the Chigua Formation, given by the palynomorphs studied herein, ranges from the late Emsian–early Eifelian through the Givetian–Frasnian boundary. There is, however, an interval without information between the assemblages (Fig. 3). Assemblages 1 and 2 have several elements that are recognized in other global palynofloras, and they can be dated and correlated with more precision. In assemblage 3, palynological data locate it around the Givetian–Frasnian boundary, but further work is required to define its precise age. Moreover, palynological data from the new localities confirm previous preliminary age determinations based on invertebrate fossils.

Analysis of the microplankton/spore ratio combined with the morphotype composition and diversity of the microplankton assemblages allowed recognition of the changing proximity to the palaeoshoreline. It is therefore possible to propose a palaeoenvironmental assessment that reflects relative sea-level fluctuations in the Precordillera region.

In summary, the highest microplankton diversity occurs in assemblage 1, and suggests deposition in a near- to offshore marine environment during the Emsian–Eifelian transition; also, the moderate diversity registered in the assemblage and the worldwide distribution of the taxa indicates an open platform marine palaeoenvironment (cf. Pöthe de Baldis 2001). The scarce spores found in this assemblage would have been transported towards the marine depocentre from coastal swamps. The high plankton diversification observed in assemblage 1 resulted from the transgression occurring during the Early Devonian in the Precordillera area (Padula *et al.* 1967).

During the early Givetian, an increase in relative abundance of spores in assemblage 2 indicates a closer proximity to the palaeoshoreline, representing a marginal marine environment. This suggests the development of vegetated swamps near to the coast line. This interpretation is also sustained by the presence of carbonaceous shales with well-preserved plant remains. Thus, a continental to coastal palaeoenvironment as a result of a marine regression is deduced, at least locally, for the early Givetian.

The uppermost sample represented by assemblage 3 reflects a return to the previous transgressive trend, such that marine conditions around the Givetian–Frasnian boundary have been inferred. Evidence for this palaeoenvironment is the occurrence of acritarchs and prasinophytes in the assemblage, and a dramatic decline in the number of spores. During deposition of the Chavela Member of the Chigua Formation favourable conditions for the development of the microplankton prevailed, and the microplanktonic forms reached large sizes.

Finally, although it is still not possible to propose a detailed depositional history for the Chavela Member of the Chigua Formation due to the scarce and dispersed palyniferous samples studied herein, a transgressive phase interrupted by an episode of shoaling is postulated during the time span between the late Emsian–early Eifelian through the Givetian–Frasnian boundary. In consequence, these data allow emergent areas (e.g. continental or at least transitional areas) in the northwestern portion of southern South America to be delineated with more precision than previous proposals (e.g. Harrington 1967; Cuerda & Baldis 1971) for the Cuyo or Precordillera Basin. Additional sedimentological studies and new productive palynological samples will define second- and third-order transgressive–regressive cycles in the Devonian of the Precordillera.

I am indebted to Dr Peter Isaacson (University of Idaho) and Dr Ken Higgs (National University of Ireland) for their detailed language revision of the manuscript and their disinterested help. Drs R. Wicander and M. Streel are greatly thanked for their valuable suggestions and comments as reviewers. I thank Drs M. di Pasquo and P. Pazos for their assistance during the field work and Dr P. Koenigshof for the financial support during the stay of DEVEC field trip in San Juan Province, Argentina. I also thank Lic. Gustavo Holfeltz (FCEN, UBA) for processing the palynological samples. This research was funded by grants from the National Agency for the Promotion of Science and Technology of Argentina (PICTR 00313/03), the University of Buenos Aires

(UBACYT X428) and the National Research Council of Argentina (PIP 5518 CONICET). It is a contribution to IGCP 499.

References

AHLFELD, F. & BRANISA, L. 1960. *Geología de Bolivia*. Don Bosco, La Paz.

ALONSO, J. L., RODRÍGUEZ-FERNÁNDEZ, L. R., GARCÍA-SANSEGUNDO, J., HEREDIA, N., FARIAS, P. & GALLASTEGUI, J. 2005. Gondwanic and Andean structure in the Argentine central Precordillera: The Río San Juan section revisited. *6° International Symposium on Andean Geodynamics* (ISAG), *Barcelona, Extended Abstracts*, 36–39.

ALLEN, K. C. 1965. Lower and Middle Devonian spores of North and Central Vestspitsbergen. *Palaeontology*, **8**, 687–748.

ALVAREZ-MARRON, J., RODRÍGUEZ-FERNÁNDEZ, R., HEREDIA, N., BUSQUETS, P., COLOMBO, F. & BROWN, D. 2006. Neogene structures overprinting Palaeozoic thrust systems in the Andean Precordillera at 30°S latitude. *Journal of the Geological Society, London*, **163**, 949–964.

AMENÁBAR, C. R. 2007. *Palinoestratigrafía y paleoambiente de las Formaciones Chigua (Grupo Chinguillos, Devónico), Malimán y El Ratón (Grupo Angualasto, Carbonífero Inferior), Cuenca Uspallata-Iglesia. Comparación y correlación con otras palinofloras y caracterización del límite Devónico-Carbonífero en la región*. PhD thesis, University of Buenos Aires.

AMENÁBAR, C. R., DI PASQUO, M. M., CARRIZO, H. A. & AZCUY, C. L. 2006. Palynology of the Chigua and Malimán Formations in the Sierra del Volcán, San Juan province, Argentina. Part 1. Palaeomicroplankton and acavate smooth and ornamented spores. *Ameghiniana*, **43**, 339–375.

AMENÁBAR, C. R., DI PASQUO, M. M., CARRIZO, H. A. & AZCUY, C. L. 2007. Palynology of the Chigua and Malimán Formations in the Sierra del Volcán, San Juan province, Argentina. Part 2. Cavate, pseudosaccate and cingulizonate spores. *Ameghiniana*, **44**, 547–564.

ASTINI, R. A. 1996. Las fases diastróficas del Paleozoico medio en la Precordillera del Oeste Argentino-Evidencias estratigráficas. *13° Congreso Geológico Argentino y 3° Congreso de Exploración de Hidrocarburos*, **5**, 509–526.

AZCUY, C. L., CARRIZO, H. A. & CAMINOS, R. 2000. Carbonífero y Pérmico de las Sierras Pampeanas, Famatina, Precordillera, Cordillera Frontal y Bloque San Rafael. *In*: CAMINOS, R. (ed.) *Geología Argentina*. Instituto de Geología y Recursos Minerales, Anales, **29**(12), 261–318.

BALDIS, B. A. 1967. Some Devonian Trilobites of the Argentina Precordillera. *International Symposium on the Devonian System, Proceedings*, **2**, 789–796.

BALDIS, B. A. & PERALTA, S. H. 2000. Capítulo 10. Silúrico y Devónico de la Precordillera de Cuyo y Bloque de San Rafael. *In*: CAMINOS, R. (ed.) *Geología Argentina*, Instituto de Geología y Recursos Minerales, Anales, **29**(10), 215–238.

BALDIS, B. A. & SARUDIANSKY, R. 1975. El Devónico en el noroeste de la Precordillera. *Revista de la Asociación Geológica Argentina*, **30**, 301–329.

BARREDA, V. D. 1986. Acritarcos Givetiano-Frasnianos de la Cuenca del noroeste, provincia de Salta. Argentina. *Revista Española de Micropaleontología*, **18**, 229–245.

BOUCOT, A. J. 1975. *Evolution and Extinction Rate Controls*. Elsevier, Amsterdam.

BRACACCINI, I. O. 1946. Contribución al conocimiento geológico de la Precordillera sanjuaninomendocina. *Boletín de Informaciones Petroleras*, **258**, 259–274.

COLBATH, G. K. 1990. Devonian (Givetian–Frasnian) organic-walled phytoplankton from the limestone Billy Hills Reef Complex, Canning Basin, Western Australia. *Palaeontographica*, Abt B, **217**, 87–145.

CRAMER, F. H. 1964. Microplankton from three Paleozoic formations in the Province of Leon, NW Spain. *Leidse Geologische Mededelingen*, **30**, 253–361.

CRISTALLINI, E. O. & RAMOS, V. A. 2000. Thick-skinned and thin-skinned thrusting in the La Ramada fold and thrust belt: Crustal evolution of the High Andes of San Juan, Argentina (32° SL). *Tectonophysics*, **317**, 205–235.

CUERDA, A. J. & BALDIS, B. A. 1971. Silúrico y Devónico en Argentina. *Ameghiniana*, **8**, 128–164.

DEUNFF, J. 1954. Sur un microplancton du Dévonien du Canada recélant des types nouveaux d'Hystrichosphaeridés. *Compte Rendu de l'Académie des Sciences de Paris*, **239**, 1064–1066.

DEUNFF, J. 1961. Quelques précisions concernant les Hystrichosphaeridés du Dévonien du Canada. *Société Géologique de France, Comptes Rendus*, **8**, 216–218.

DEUNFF, J. 1964. Le genre *Duvernaysphaera* Staplin. *Grana Palynologica*, **5**, 210–215.

DEUNFF, J. 1966. Acritarches du Dévonien de Tunisie. *Bulletin de la Société Géologique de France*, **1**, 22–24.

DEUNFF, J. 1977. *Winwaloeusia*, genre nouveau d'acritarche du Dévonien. *Géobios*, **10**, 465–469.

DEUNFF, J. 1980 Le paléoplancton des Grès de Landévennec (Gédinnien de la Rade de Brest-Finistère): étude biostratigraphique. *Géobios*, **13**, 483–539.

DI PASQUO, M. M. 2007. Asociaciones palinológicas presentes en las Formaciones Los Monos (Devónico) e Itacua (Carbonífero Inferior) en el perfil de Balapuca, sur de Bolivia. Parte 1. Formación Los Monos. *Revista Geológica de Chile*, **34**, 98–137.

DÍAZ MARTÍNEZ, E., VAVRDOVÁ, M., BEK, J. & ISAACSON, P. E. 1999. Late Devonian (Famennian) Glaciation in Western Gondwana: Evidence from the Central Andes. *Abhandlungen der Geologischen Bundesanstalt*, **54**, 213–237.

DINO, R. 1999. Palynostratigraphy of the Silurian and Devonian sequence of the Paraná Basin, Brazil. *In*: RODRIGUES, M. A. C. & PEREIRA, E. (eds) *Ordovician–Devonian Palynostratigraphy in Western Gondwana: Update, Problems And Perspectives*. Faculdade de Geologia da Universidade Estatal do Rio de Janeiro, Resumos expandidos, 27–61.

DORNING, K. J. 1981. Silurian acritarch distribution in the Ludlovian shelf sea of South Wales and the

Welsh Borderland. *In*: NEAL, R. G. & BRASIER, M. D. (eds) *Microfossils from Recent and Fossil Shelf Seas*. Ellis Horwood, Chichester, 31–36.

FONSECA, V. M. M. & MELO, J. H. G. 1987. Occôrrencia de *Tropidoleptus carinatus* (Conrad) brachiopoda (Orthida) na Formaçâo Pimenteira, e sua importância paleobiogeografica. *10° Congress Brasileiro de Paleontologia, Rio de Janeiro, Anais*, **2**, 505–537.

FURQUE, G. 1963. Descripción geológica de la Hoja 17 b, Guandacol, provincias de La Rioja y San Juan. *Dirección Nacional de Geología y Minería* (Buenos Aires), *Boletín*, **92**, 1–104.

FURQUE, G. 1979. *Descripción geológica de la Hoja 18c, Jachal, Provincia de San Juan. Carta Geológico-Económica de la República Argentina, Escala 1:200.000*. Boletín del Servicio Geológico Nacional, Buenos Aires, **164**.

GONZÁLEZ BONORINO, G. 1990. El relieve de la Precordillera de Cuyo en el Paleozoico Tardío: el caso de la protocordillera. *11° Congreso Geológico Argentino*, **2**, 89–92.

GRAHN, Y. 2003. Silurian and Devonian chitinozoan assemblages from the Chaco-Paraná Basin, northeastern Argentina and Central Uruguay. *Revista Española de Micropaleontología*, **35**, 1–8.

GRAHN, Y. & GUTIÉRREZ, P. R. 2001. Silurian and Middle Devonian Chitinozoa from the Zapla and Santa Bárbara Ranges, Tarija Basin, northwestern Argentina. *Ameghiniana*, **38**, 35–50.

GRAY, J. & BOUCOT, A. J. 1972. Palynological evidence bearing on the Ordovician–Silurian paraconformity in Ohio. *Geological Society of American Bulletin*, **83**, 1299–1314.

GUTIÉRREZ, P. R. 1996. Revisión de las Licópsidas de la Argentina. 2. *Malanzania* Archangelsky, Azcuy *et* Wagner y *Haplostigma* Seward; con notas sobre *Cyclostigma* Haughton. *Ameghiniana*, **33**, 127–144.

HACQUEBARD, P. A. 1957. Plant spores in coal from the Horton Group (Mississippsan) of Nova Scotia. *Micropaleontology* **3**, 301–324.

HARRINGTON, H. J. 1967. Devonian of South America. *International Symposium of the Devonian System, Proceedings* **1**, 651–671.

HEIM, A. 1952. Estudios tectónicos de la Precordillera de San Juan: los ríos San Juan, Jáchal y Huaco. *Revista de la Asociación Geológica Argentina*, **7**, 11–70.

HERRERA, Z. A. & BUSTOS, U. D. 2001. Braquiópodos devónicos de la Formación Punta Negra, en el perfil del río de las Chacritas, Precordillera argentina. *Ameghiniana*, **38**, 367–374.

HUYSKEN, K. T., WICANDER, R. & ETTENSON, F. R. 1992. Palynology and biostratigraphy of selected Middle and Upper Devonian blackshale section in Kentucky. *Michigan Academician*, **24**, 355–368.

ISAACSON, P. E. & PERRY, D. G. 1977. Biogeography and morphological conservatism of *Tropidoleptus* (Brachiopoda, Orthida) during the Devonian. *Journal of Paleontology*, **51**, 1108–1122.

ISAACSON, P. E. & SABLOCK, P. E. 1989. Devonian System in Bolivia, Peru, and Northern Chile. *In*: MCMILLAN, N. J., EMBRY, A. F. & GLASS, D. J. (eds) *Devonian of the World*. Canadian Society of Petroleum Geologists Memoir, **14**, 719–728.

JARDINÉ, S. 1972. Microplancton (Acritarches) et limites stratigraphiques du Silurien terminal au Dévonien supérieur. *7° Congrès International de Stratigraphie et de Géologie du Carbonifère*, **1**, 313–323.

JORDAN, T. E., ALLMENDINGER, R. W., DAMANTI, J. F. & DRAKE, R. E. 1993. Chronology of motion in a complete thrust belt: The Precordillera, 30–31°S, Andes Mountains. *Journal of Geology*, **101**, 135–156.

KULLMANN, J. 1993. Paleozoic ammonoids of Mexico and South America. *20° International Congress Carboniferous and Permian Geology and Stratigraphy*, **1**, 557–562.

LEGAULT, J. A. 1973. Chitinozoa and Acritarcha of the Hamilton Formation (Middle Devonian), Southwestern Ontario. *Geological Survey of Canada, Bulletin*, **221**, 1–110.

LE HÉRISSÉ, A. 2001. Evolution of Devonian phytoplanktonic assemblages in the upper Erere Formation and Curua Group (Barreirinha and lower Curiri Formations) Tapajos River area, Amazon Basin, northern Brazil. *Ciencia Tecnica Petroleo. Secao: Exploracao de Petroleo*, **20**, 117–124.

LE HERISSÉ, A. 2002. Paleoecology, biostratigraphy and biogeography of late Silurian to early Devonian acritarchs and prasinophycean phycomata in well A161, western Libya, North Africa. *Review of Palaeobotany and Palynology*, **118**, 359–395.

LE HÉRISSÉ, A. & DEUNFF, J. 1988. Acritarches *et* Prasinophycés (Givétien supérieur-Frasnien moyen) de Ferques (Boulonnais-France). *In*: BRICE, D. (ed.) *Le Dévonien de Ferques, Bas-Boulonnais (N. France): Biostratigraphie du Paléozoïque*, **7**, Université de Bretagne occidentale, Brest, 103–152.

LE HÉRISSÉ, A., RUBINSTEIN, C. & STEEMANS, P. 1997*a*. Lower Devonian palynomorphs from the Talacasto Formation, Cerro del Fuerte Section, San Juan Precordillera, Argentina. *In*: FATKA, O. & SERVAIS, T. (eds) *Acritarcha in Praha 1996*. Acta Universitatis Carolinae Geologica, Lille, **40**, 497–515.

LE HÉRISSÉ, A., GOURVENNEC, R. & WICANDER, R. 1997*b*. Biogeography of Late Silurian and Devonian acritarchs and prasinophytes. *Review of Palaeobotany and Palynology*, **98**, 105–124.

LIMACHI, R., GOITIA, V. H., SARMIENTO, D. *ET AL*. 1996. Estratigrafía, geoquímica, correlaciones, ambientes sedimentarios y bioestratigrafía del Silúrico-Devónico de Bolivia. *12° Congreso Geológico de Bolivia, Tarija, Memorias*, **12**, 183–197.

LOEBLICH, A. R. 1970. Morphology, Ultrastructure and Distribution of Paleozoic Acritarchs. *Proocedings of the North American Paleontological Convention*, Part **G**, 705–788.

MCGREGOR, D. C. 1973. Lower and Middle Devonian spores of eastern Gaspé, Canada. I. Systematics. *Palaeontographica*, Abt B, **142**, 1–77.

MCGREGOR, D. C. & CAMFIELD, M. 1982. Middle Devonian miospores from the Cape de Bray, Weatherall, and Hecla Bay Formations of northeastern Melville Island, Canadian Arctic. *Geological Survey of Canada, Bulletin*, **348**, 1–105.

MELO, J. H. G. & LOBOZIAK, S. 2003. Devonian–Early Carboniferous miospore biostratigraphy of the

Amazon Basin, Northern Brazil. *Review of Palaeobotany and Palynology*, **124**, 131–202.

MOLYNEUX, S. G., LEHERISSÉ, A. & WICANDER, R. 1996. Paleozoic phytoplankton. *In*: JANSONIUS, J. & MCGREGOR, D. C. (eds) *Palynology: Principles and Applications*. American Association of Stratigraphical Palynologist Foundation, **2**, 493–529.

NAUMOVA, S. N. 1953. Spore-pollen complexes of the Upper Devonian of the Russian Platform and their stratigraphic significance. *Transactions of the Institute of Geological Sciences, Academy of Science*, SSSR, (Geol. Ser. 60), **143**, 1–200. [in Russian].

OTTONE, E. G. 1996. Devonian palynomorphs from the Los Monos Formation, Tarija Basin, Argentina. *Palynology*, **20**, 101–151.

OTTONE, E. G. & ROSSELLO, E. A. 1996. Palinomorfos devónicos de la Formación Tequeje, Angosto del Beni, Bolivia. *Ameghiniana*, **33**, 443–452.

OWENS, B. 1971. Miospores from the Middle and Early Upper Devonian rocks of the Western Queen Elizabeth Islands, Arctic Archipiélago. *Geological Survey of Canada, Paper*, **70-38**, 1–157.

PADULA, E., ROLLERI, E., MINGRAMM, A. R., CRIADO ROQUE, P., FLORES, M. A. & BALDIS, B. 1967. Devonian of Argentina. *International Symposium on the Devonian System, Proceedings*, **2**, 165–199.

PLAYFORD, G. 1977. Lower to Middle Devonian Acritarchs of the Moose River Basin, Ontario. *Geological Survey of Canada, Bulletin*, **279**, 1–87.

PLAYFORD, G. 2003. *Acritarchs and Prasinophyte Phycomata: A Short Course*. American Association of Stratigraphic Palynologists Foundation, **41**, 1–46.

PÖTHE DE BALDIS, E. D. 2001. Nuevos aportes sobre acritarcas y prasinofíeceas en la Formación Don Braulio (Llandoveriano temprano-medio), Precordillera Oriental, San Juan, Argentina. *Ameghiniana*, **38**, 419–428.

QUADROS, L. P. 1999. Silurian–Devonian acritarch assemblages from Paraná Basin: An update and correlation with Northern Brazilian basins. *In*: RODRIGUES, M. A. C. & PEREIRA, E. (eds) *Ordovician–Devonian Palynostratigraphy in Western Gondwana: Update, Problems and Perspectives*. Faculdade de Geologia da Universidade Estatal do Rio de Janeiro, Resumos expandidos, 105–145.

RAEVSKAYA, E., VECOLI, M., BEDNARCZYK, W. & TONNGIORGI, M. 2004. Billingen (Lower Arenig/Lower Ordovician) acritarchs from the East European Platform and their palaeogeographic significance. *Lethaia*, **37**, 97–111.

RAMOS, V. A. 1984. Patagonia: un continente a la deriva? *9° Congreso Geológico Argentino*, **2**, 311–328.

RAMOS, V. A. 2004. Cuyania, an exotic block to Gondwana: review of a historical success and the present problems. *Gondwana Research*, **7**, 1009–1026.

RICHARDSON, J. B. 1965. Middle Old Red Sandstone spore assemblages from the Orcadian Basin north-east Scotland. *Palaeontology*, **7**, 559–605.

RICHARDSON, J. B. & MCGREGOR, D. C. 1986. Silurian and Devonian spore zones of the Old Red Sandstone continent and adjacent regions. *Geological Survey of Canada, Bulletin*, **364**, 1–79.

RIEGEL, W. 1968. Die Mitteldevon-Flora von Lindlar (Rheinland). 2. Sporae dispersae. *Palaeontographica*, Abt B, **123**, 76–96.

RIEGEL, W. 1974. Phytoplankton from the Upper Emsian and Eifelian of the Rhineland, Germany. A preliminary report. *Review of Palaeobotany and Palynology*, **18**, 29–39.

RUBINSTEIN, C. V. 1997. Silurian Acritarchs from South America: A Review. *In*: FATKA, O. & SERVAIS, T. (eds) *Acritarcha in Praha 1996*. Acta Universitatis Carolinae Geologica, Lille, **40**, 603–629.

RUBINSTEIN, C. V. 2000. Middle Devonian palynomorphs from the San Juan Precordillera, Argentina: biostratigraphy and paleobiogeography. *1° Congreso Ibérico de Paleontologia, 14° Jornadas de la Sociedad Española de Paleontología y 8° International Meeting of IGCP 421*, Resúmenes, 274–275.

RUBINSTEIN, C. V. & STEEMANS, P. 2007. New palynological data from the Devonian Villavicencio Formation, Precordillera of Mendoza, Argentina. *Ameghiniana*, **44**, 3–9.

RUBINSTEIN, C. V., MELO, J. H. G. & STEEMANS, P. 2005. Lochkovian (earliest Devonian) miospores from the Solimões Basin, northwestern Brazil. *Review of Palaeobotany and Palynology*, **133**, 91–113.

SMITH, N. & SAUNDERS, R. S. 1970. Paleoenvironemnts and their control on acritarc distribution: Silurian of East Central Pennsylvania. *Journal of Sedimentary Petrology*, **40**, 324–333.

STAPLIN, F. L. 1961. Reef-controlled distribution of Devonian microplankton in Alberta. *Palaeontology*, **4**, 392–424.

STOCKMANS, F. & WILLIERE, Y. 1969. Acritarches du Famennien inférieur. *Mémories de la Classe des Sciences, Académie Royale de Belgique, Collection in 8°, série 2*, **38**, 1–63.

STREEL, M., CAPUTO, M. V., LOBOZIAK, S. & MELO, J. H. G. 2000. Late Frasnian–Famennian climates based on palynomorph analyses and the question of the Late Devonian glaciations. *Earth-Science Reviews*, **52**, 121–173.

STRICANNE, L., MUNNECKEC, A., PROSS, J. & SERVAIS, T. 2004. Acritarch distribution along an inshore–offshore transect in the Gorstian (lower Ludlow) of Gotland, Sweden. *Review of Palaeobotany and Palynology*, **130**, 195–216.

SUAREZ SORUCO, R. & LOBO BONETA, J. 1983. La fase compresiva Eohercínica en el sector oriental de la Cuenca Cordillerana de Bolivia. *Revista Técnica de Yacimientos Petrolíferos Fiscales Bolivianos*, **9**, 189–202.

VAVRDOVÁ, M., ISAACSON, P. E., DÍAZ MARTÍNEZ, E. & BEK, J. 1993. Devonian – Carboniferous boundary at Lake Titikaka, Bolivia: Preliminary palynological results. *12° Congrès International de la Stratigraphie et Géologie du Carbonifère et Permien, Buenos Aires, 1991, Compte Rendu*, **1**, 187–200.

VECOLI, M. & LÉ HÉRISSÉ, A. 2004. Biostratigraphy, taxonomic diversity and patterns of morphological evolution of Ordovician acritarchs (organic-walled microphytoplankton) from the northern Gondwana margin in relation to palaeoclimatic and palaeogeographic changes. *Earth-Science Reviews*, **67**, 267–311.

VOLKHEIMER, W., MELENDI, D. L. & SALAS, A. A. 1986. Devonian chitinozoans from northwestern

Argentina. *Neues Jahrbuch für Geologie und Paläontologie, Abhandlungen*, **173**, 229–251.

von Gosen, W. 1992. Structural evolution of the Precordillera (Argentina): The Rio San Juan section. *Journal of Structural Geology*, **14**, 643–667.

Wicander, E. R. & Wood, G. D. 1981. Systematics and Biostratigraphy of the organic-walled microphytoplankton from the Middle Devonian (Givetian) Silica Formation, Ohio, USA. *American Association of Stratigraphic Palynologists, Contribution Series*, **8**, 1–137.

Wicander, R. & Wright, R. P. 1983. Organic-walled microphytoplankton abundance and stratigraphic distribution from the Middle Devonian Columbus and Delaware Limestones of the Hamilton Quarry, Marion County, Ohio. *Ohio Journal of Science*, **83**, 2–13.

Wicander, R. & Playford, G. 1985. Acritarchs and spores from the Upper Devonian Lime Creek Formation, Iowa, U.S.A. *Micropaleontology*, **31**, 97–138.

Wicander, R. & Wood, G. D. 1997. The use of microphytoplankton and chitinozoans for interpreting transgressive/regressive cycles in the Rapid Member of the Cedar Valley Formation (Middle Devonian), Iowa. *Review of Palaeobotany and Palynology*, **98**, 125–152.

Wolfart, T. & Voges, A. 1968. Beitrage zur Kenntnis ides. Devons von Bolivien. *Beihefte zum Geologischen Jahrbuch, Beiheft*, **74**, 122, 214–215.

Zapata, T. R. & Allmendinger, R. W. 1996. Thrust front zone of the Precordillera, Argentina: A thick skinned triangle zone. *Bulletin of the American Association of Petroleum Geologists*, **80**, 350–381.

Middle Devonian microfloras and megafloras from western Argentina and southern Bolivia: their importance in the palaeobiogeographical and palaeoclimatic evolution of western Gondwana

M. DI PASQUO*, C. R. AMENÁBAR & S. NOETINGER

CONICET, National Research Council of Argentina, UBA, Department of Geological Sciences, Faculty of Pure and Natural Sciences University of Buenos Aires Ciudad Universitaria Pabellón II (1428) Capital Federal, Argentina

**Corresponding author (e-mail: medipa@gl.fcen.uba.ar)*

Abstract: The study of microfloras and megafloras from western Argentina and southern Bolivia presented here extends our knowledge of the biodiversity and succession of floristic events during the Middle Devonian, and hence improves the current biostratigraphy. Among floral remains, species attributable to '*Haplostigma*' are mostly recorded from the same *Grandispora pseudoreticulata* and other palynomorph-bearing sections at Balapuca (southern Bolivia) and Del Chaco and La Cortadera creeks in the Precordillera of Argentina. The northernmost record of *Grandispora pseudoreticulata* is recorded from a third palynoassemblage from the Pando x-1 corehole of northern Bolivia. A great similarity among all these assemblages (notably abundant *Grandispora pseudoreticulata*) suggests the identification of terrestrial connections, or at least proximity, of these local areas and other regions in South America during the Givetian. Comparison with other contemporary Gondwanan palynofloras shows cosmopolitan species (e.g. *Geminospora lemurata*, *Samarisporites triangulatus*, *Archaeozonotriletes variabilis*, *Chelinospora concinna*) along with some others with more restricted distribution (e.g. *Grandispora pseudoreticulata*, *Leiotriletes balapucensis*, *Acinosporites ledundae*). This pattern defines an Afro-South American Subrealm, which most likely results from the effects of palaeolatitude and, in a lesser way, local palaeoenvironmental conditions. On the other hand, such a level of cosmopolitanism supports previous palaeogeographical reconstructions where a narrow Rheic Ocean was developed between Euramerica and the northern parts of Africa and South America.

Supplementary material: Stratigraphical and geographical distributions of selected species recorded from Middle and Late Devonian Argentina, Bolivia and Brazil microfloras are available at: http://www.geolsoc.org.uk/SUP18335.

The microfloras and megafloras from southern Bolivia and western Argentina have recently provided important evidence to extend knowledge on the biodiversity and succession of floristic events during the Middle Devonian, and therefore contribute to the improvement of our current biostratigraphy. This contribution aims to present for the first time a record of both plant and palynological assemblages from Balapuca located on the Bolivian side of the Bermejo River (San Telmo–de las Pavas Range), and Del Chaco Creek, in the Volcán Range of the Precordillera of Argentina. Both outcrops contain plants attributed to '*Haplostigma*' Seward 1932 and mostly continental palynomorphs with many species of *Grandispora*. The Pando x-1 Borehole in northwestern Bolivia has provided scarce specimens of *Grandispora pseudoreticulata* from one core sample. These new records allow an update of the palaeogeographical and stratigraphical distribution of selected *Grandispora* and '*Haplostigma*' species based on comparison with coeval or similar floras from Gondwanan South America and beyond (Figs 1–3; see Supplementary Material). In addition, it enables an evaluation of the degree of endemism of these palaeofloras and hence the identification of when terrestrial connections between these areas existed. Also analysed is the interaction of the main factors that controlled the development of these floras such as palaeoclimate, palaeoenvironment and palaeogeography. In this sense, some taphonomical features are analysed and discussed considering the previous palaeoenvironmental interpretations. Finally, a palaeobiogeographical region is proposed based on the correlation of the assemblages involved and the stratigraphical and geographical distribution of *Grandispora pseudoreticulata* and '*Haplostigma*' for the Middle and Late Devonian of Western Gondwana.

From: KÖNIGSHOF, P. (ed.) *Devonian Change: Case Studies in Palaeogeography and Palaeoecology*.
The Geological Society, London, Special Publications, **314**, 193–213.
DOI: 10.1144/SP314.11 0305-8719/09/$15.00

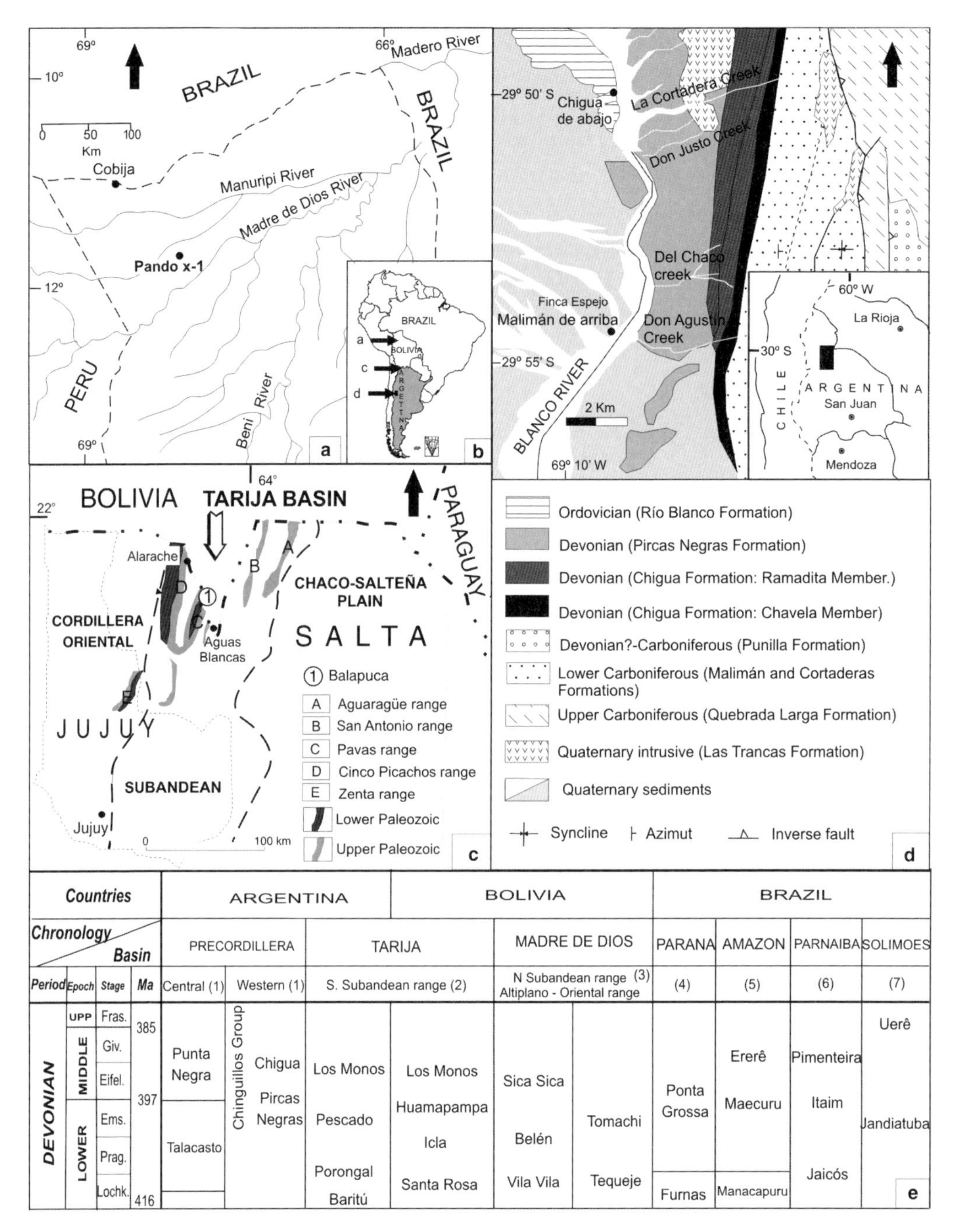

Fig. 1. (**a–d**) Location maps of the studied localities: (a) Pando x-1 borehole; (b) the three localities in South America; (c) Balapuca; and (d) La Cortadera and Del Chaco creeks. (**e**) Correlation of Devonian units of Argentina, Bolivia and Brazil. References: (1) Rubinstein (1999, 2000); (2) Starck (1996, 1999), Suárez Soruco (2000); (3) Díaz Martínez (1999), Suárez Soruco (2000); (4) Dino (1999); (5) Melo & Loboziak (2003), Grahn (2003); (6, 7) Grahn *et al.* (2003). Absolute time dates after Gradstein *et al.* (2004). Note that there are relatively huge differences between this scale and the one proposed more recently by Menning *et al.* (2006), where the Emsian–Eifelian boundary is dated at 392 Ma and the Givetian–Frasnian at 381 Ma.

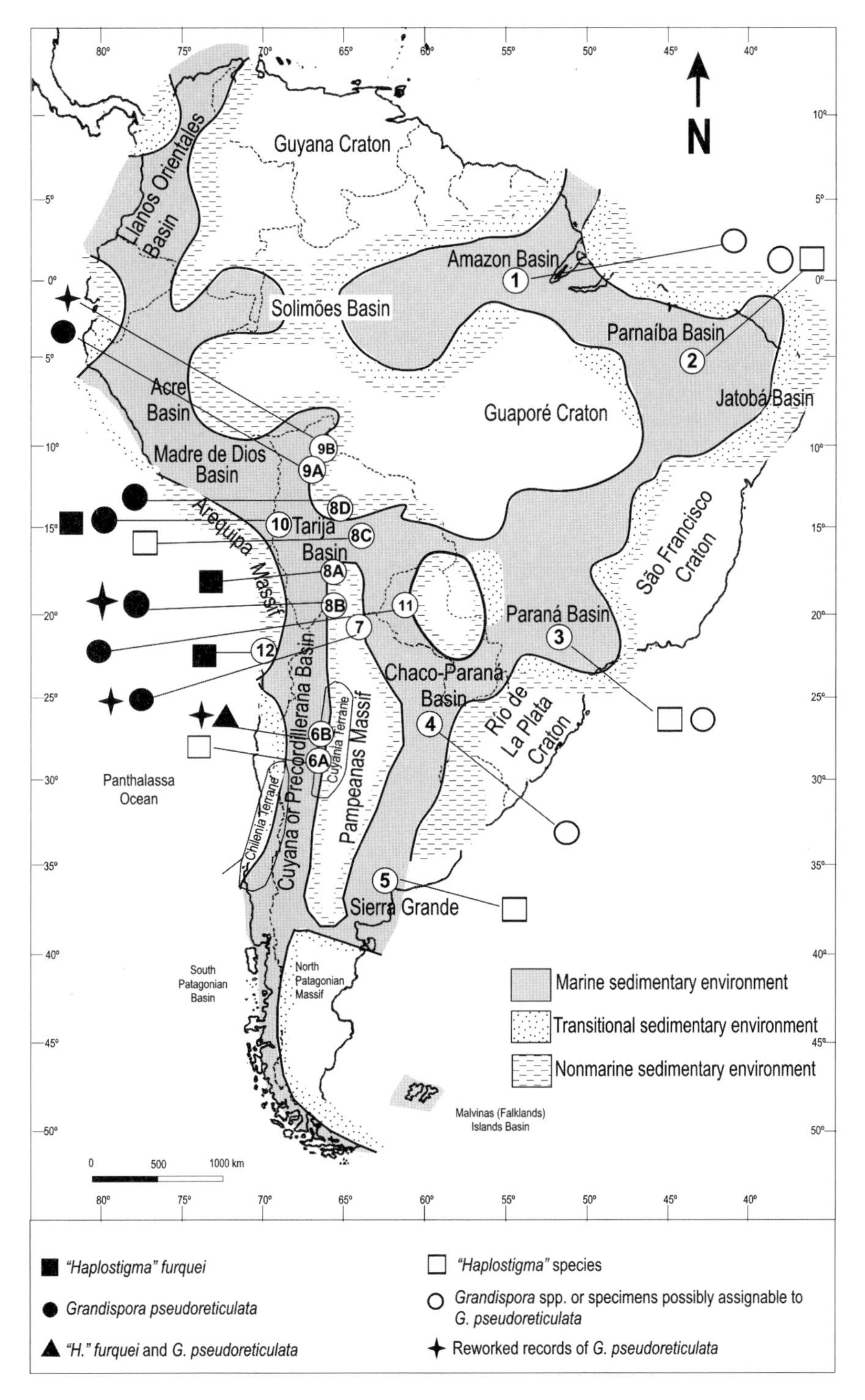

Fig. 2. Location of known occurrences of *Grandispora pseudoreticulata* and species of '*Haplostigma*' recorded in South America on a palaeogeographical reconstruction modified from Cuerda & Baldis (1971) and Melo (1989). Key: 1, Amazon Basin (Melo & Loboziak 2003). 2, Parnaíba Basin, Pimenteira Formation (Suárez Riglos 1975; Rodriguez *et al.* 1995). 3, Paraná Basin, Ponta Grossa Formation (Kräusel 1960; Daemon *et al.* 1967). 4, Chaco-Paraná Basin (Antonelli &

Summary of stratigraphy and palaeontology

The stratigraphical development, palaeontological content, facies and palaeoenvironments of the Devonian of southern Bolivia, northern Argentina and the western Precordillera of Argentina are of such an extent that numerous studies are still required to understand the geological history of each of the regions and the group in general. The Devonian rocks in northern Argentina are exposed in the Sub-Andean and Santa Barbara mountains plus in the subsurface of greater part of the Chaco-Salteña and the Chaco-Paranense Plains (Fig. 1c, e). They comprise essentially marine sedimentary rocks of the Tarija Basin that extends to central Bolivia and northwestern Paraguay. These marine deposits would have been interconnected with other neighbouring marine rocks registered in the Arizaro (northern Chile and the Puna of Argentina) and the Madre de Dios (northern Bolivia and southern Peru) basins together with other basins in eastern Argentina and Brazil based on common palaeontological content (Fig. 2; see Grahn 2005). In general terms, the very extensive lateral distribution of the carbonaceous mudstones has sourced the deposits of petroleum and gas that mainly occur in the Sub-Andean Bolivian region (e.g. Starck 1999). For this reason this particular basin has been extensively studied for petroleum exploration. Many multidisciplinary works based on subsurface and surface information have been carried out by different oil companies, although published accounts are less numerous (e.g. Suárez Soruco 2000; Dalenz Farjat *et al.* 2002; Albariño *et al.* 2002; Alvarez *et al.* 2003).

Inevitably the number of Middle and Late Devonian palaeobotanical and palynological records from different areas around the world is very variable. For South America these are still few in comparison to those from North America and Europe. For northern Argentina, there are a few publications on Devonian microfloras that contain systematic descriptions and illustrations of spores, acritarchs and chitinozoans, notably Volkheimer *et al.* (1986), Barreda (1986), Ottone (1996), Grahn & Gutierrez (2001) and Grahn (2002). Menéndez & Pöthe de Baldis (1967) and Pöthe de Baldis (1974, 1979) described and illustrated palynomorphs found in the Picuiba Borehole from northwestern Paraguay. For southern Bolivia, there are several stratigraphical, biostratigraphical and palaeobiogeographical contributions, some of which were illustrated; these include Lobo Boneta (1975), Suárez Soruco & Lobo Boneta (1983), Kimyai (1983), McGregor (1984), Wood (1984, 1994, 1995), Pérez Leyton (1990, 1991), Racheboeuf *et al.* (1993), Blieck *et al.* (1996), Grahn (2002, 2005) and di Pasquo (2005, 2007*a*, *b*). For northern Bolivia there are publications by Vavrdová *et al.* (1993, 1996), Ottone & Rossello (1996), Vavrdová & Isaacson (1999) and Díaz Martínez *et al.* (1999). Both Suárez Soruco & Lobo Boneta (1983) and Limachi *et al.* (1996) proposed biostratigraphical schemes for the Devonian to Permian based on lists of species but without illustration. Melo (2005) presented new palynostratigraphical results from some Devonian units in Bolivia but included neither lists nor illustrations of palynomorphs. Some of these works need to be revised; none of them are sufficiently detailed to understand the evolution of the microfloras in this region of Gondwana and cannot be applied with certainty to new localities or compared with other microfloras of more distant regions.

Plant occurrences in the Devonian of northern Argentina and Bolivia are rarely documented: Suárez Soruco (1988) contains illustrations, while other reviews lack them (Suárez Riglos 1975; Archangelsky 1993). The Bolivian material of the genus '*Haplostigma*' has not been revised previously and very few illustrations of the specimens were included in Branisa (1965) and Suárez Soruco (1988).

In the Precordillera Range in San Juan Province, northwestern Argentina, Devonian deposits belonging to the Chinguillos Group (Pircas Negras and Chigua formations, see Figs 1, 2), are present to the east of the Blanco River, in the western slope of the Punilla and Volcán ranges of the Cuyana or Precordillera Basin (Padula *et al.* 1967). The Chigua Formation (*c.* 700 m) overlies the Pircas Negras Formation with a thrust contact and is separated by an angular unconformity from the overlying Malimán Formation (Early Viséan; Amenábar 2006, 2007*b*). The Chigua Formation is divided

Fig. 2. (*Continued*) Ottone 2006). 5, Sierras Australes, Lolén Formation (Cingolani *et al.* 2002). 6, Cuyana or Precordillerana Basin; 6A, Punta Negra Formation (Baldis & Peralta 2000); 6B, Chigua Formation (Gutiérrez 1996; Amenábar *et al.* 2006) and Malimán Formation (Early Viséan; Amenábar 2006) as a reworked form. 7, Tarija Basin, Los Monos (Middle Devonian) (Ottone 1996) and Macharetí and Mandiyutí Goups (Late Carboniferous; see di Pasquo & Azcuy 1997; di Pasquo 2003) as a reworked form. 8A, Huamampampa Formation (Finca Carlazo; Branisa 1965). 8B, Los Monos Formation, Balapuca (di Pasquo 2007*a*) and Itacua formations (Early Viséan; di Pasquo 2007*b*). 8C, Huamampampa, Los Monos, Iquirí formations (Campo Redondo (Suárez Riglos 1975), Lajas (Wood 1995)). 8D, Los Monos, Iquirí formations (Bermejo River, Santa Cruz; Pérez Leyton 1991). 9, Madre de Dios Basin: 9A, Pando x-1 Borehole (this contribution); 9B, Kaka Formation (Serpukhovian; Azcuy & Ottone 1987) as a reworked form. 10, Vila Molino, Titicaca lake (Vavrdová & Isaacson 2000) and Sica Sica (Branisa 1965). 11, Picuiba Borehole (Menéndez & Pöthe de Baldis 1967). 12, El Toco Formation (Antofagasta region; Moisan & Niemeyer 2005).

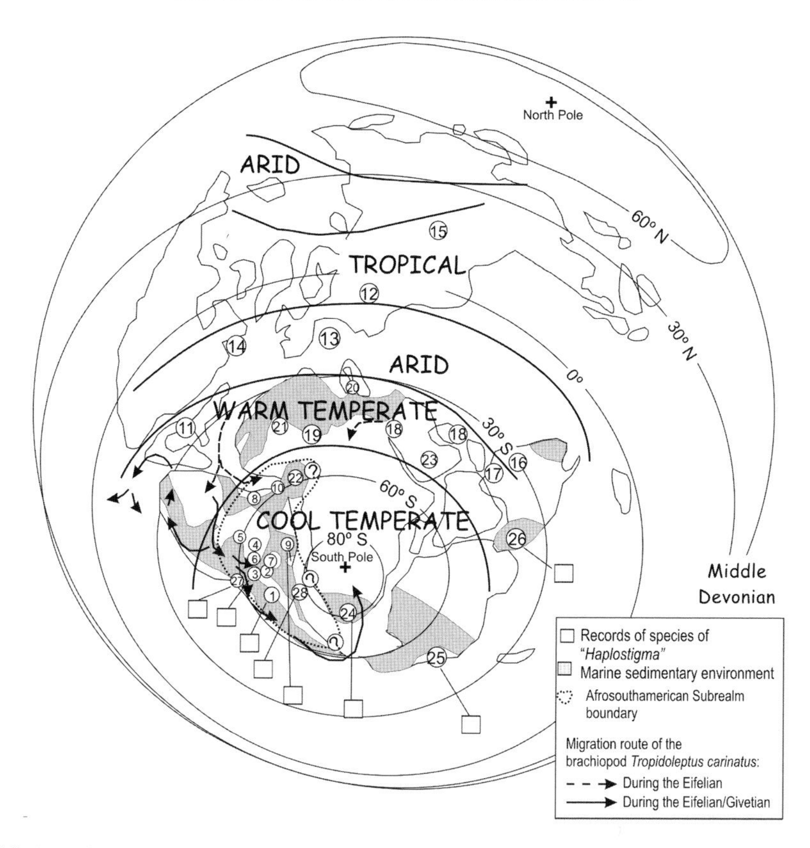

Fig. 3. Middle Devonian palaeogeographical map (modified from Isaacson & Sablock in Wood 1995) with palaeoclimatical zones (based on Scotese *et al.* 1999) showing the marine deposits with the migration route of the brachiopod *Tropidoleptus carinatus* (after Fonseca & Melo 1987) and the location of known Middle Devonian palynological assemblages, some of which contain *Grandispora pseudoreticulata* and other endemic species of the Afro-South American Subrealm. Key. Argentina: 1A, Chigua Formation (Amenabar 2007*a*, *b*, 2009); 1B, Villavicencio Formation (Rubinstein & Steemans 2007); 1C, Talacasto Formation (Le Hérissé *et al.* 1997); 1D, Punta Negra Formation (Rubinstein 1999, 2000). 2, Los Monos Formation (Barreda 1986; Ottone 1996). Bolivia: 3, Los Monos Formation (di Pasquo 2007*a*). 4, Limachi *et al.* (1996), Pérez Leyton (1991). 5A, Tomachi Formation (Vavrdová *et al.* 1996); 5B, Ottone & Rossello (1996). 6, Kimyai (1983), McGregor (1984), Blieck *et al.* (1996). Paraguay: 7, Menéndez & Pöthe de Baldis (1967), Pöthe de Baldis (1974, 1979). Brazil: 8, Maecurú, Ererê and Barreirinha formations (Melo & Loboziak 2003; Loboziak & Streel 1995); Jandiatuba and Uerê formations (Quadros 1988; Grahn *et al.* 2003). 9A, Ponta Grossa Formation (Loboziak & Streel 1995; Dino 1999; Quadros 1999); 9B, Daemon *et al.* (1967; biozones correlated to Dino 1999); 9C, Oliveira (1997); 10, Pimenteira Formation (Rodriguez *et al.* 1995). Old Red Sandstone Continent: 11, McGregor (1979), McGregor & Camfield (1976, 1982), Richardson & McGregor (1986), Ravn & Benson (1988). Pomerania: 12, Turnau (1996). France: 13, Le Hérissé & Deunff (1988). Canada: 14, Braman & Hills (1992), Cloutier *et al.* (1996). Russia: 15, Avchimovitch *et al.* (1993). Australia: 16, Hashemi & Playford (2005). 17, Colbath (1990), Playford & Dring (1981), Balme (1988). Iran: 18, Ghavidel-Syooki (1994, 2003), Hashemi & Playford (1998). Libya: 19A, Moreau-Benoit (1979, 1980), Loboziak & Streel (1989), Steemans *et al.* (2007*a*); 19B, Paris *et al.* (1985), Streel *et al.* (1988). Tunisia and Libya: 20, Loboziak & Streel (1995). Algeria: 21, Moreau-Benoit *et al.* (1993). Ghana: 22, Bär & Riegel (1974). Saudi Arabia: 23A, Loboziak (2000); 23B, Breuer *et al.* (2007*b*); 23C, Marshall *et al.* (2007). Records of species of '*Haplostigma*' in the rest of Gondwana: 24, Karoo Basin, South Africa (Anderson & Anderson 1985). 25, Antarctica (Edwards 1990; McLoughlin & Long 1994). 26, Australia, New South Wales (McLoughlin & Long 1994). 27, El Toco Formation, Antofagasta, Chile (Moisan & Niemeyer 2005). 28, Lolén Formation, Sierras Australes, Argentina (Cingolani *et al.* 2002).

into two members, the lower Chavela (marine) and the upper Ramadita (marine–continental) (see Baldis & Peralta 2000). The lithology mainly consists of shales with concretions and calcareous lenses with subordinate sandstone layers, with rock colours varying from green to brown. The succession is rich in fossiliferous levels, including flora and marine invertebrates. The latter include the trilobites *Punillaspis argentina* Baldis and *Phacops chavelai* Baldis & Longobucco, the cephalopods *Tornoceras baldisi* Leanza and *Orthoceras* sp., and the cnidarian *Conularia* sp. (e.g. Furque 1956, 1963; Baldis & Sarudiansky 1975; Baldis & Longobucco 1977), whilst the palaeoflora is composed of the herbaceous lycophytes '*Haplostigma*' *furquei* Frenguelli and ?*Cyclostigma* sp. (Frenguelli 1952, 1954; Gutierrez 1996; Gutierrez & Archangelsky 1997). New palynological data from Devonian deposits of the Malimán area are presented by Amenábar (2006, 2007*a, b*, 2009) and Amenábar *et al.* (2006, 2007) (see also Supplementary Material). There are other few recent palynological contributions from the Precordillera Range of western Argentina but those with illustrated lists of palynomorphs include Le Hérissé *et al.* (1997), Rubinstein (1993, 1999, 2000) and Rubinstein & Steemans (2007). It is emphasized that the Precordillera has suffered strong tectonic activity not only during the Middle to Late Palaeozoic but also during the Andean Orogeny (cf. Alonso *et al.* 2005). This would be the main reason for poor preservation (or absence) of organic matter in most of the Devonian sediments and the difficulty of finding more complete sections.

Most of the palynological references mentioned above are part of the database (see Supplementary Material), along with others from Brazil, Africa, Antarctica, Australia and the rest of the world. It is also interesting to note that, in the Gondwana region, faunas and floras contain significant numbers of cosmopolitan taxa which allow correlation with palaeoequatorial assemblages (e.g. Streel & Loboziak 1996). But a quite variable degree of regional microflora and some disparities in the vertical range of species may have some palaeogeographical and palaeoclimatic meaning, which is discussed later.

Description of studied material

The characteristics, comparisons and age of the palynoassemblages collected from three localities are summarized as follows (see Figs 1–6).

Balapuca section (22°31′00″S, 64°26′00″W)

Recently, di Pasquo (2005, 2007*a, b*) published detailed stratigraphical and palynological information from the Middle Devonian Los Monos and the Lower Carboniferous Itacua Formations at Balapuca (Fig. 1). The upper section of the Los Monos Formation, *c.* 50 m thick, appears to be unconformable below the Itacua Formation (early Viséan) and is mainly composed of mudstones interbedded with sandstones representing tempestite and normal shelf deposits. According to di Pasquo (2005), the stratigraphical distribution of the species found in the Los Monos Formation at Balapuca enabled recognition of two assemblages composed of abundant plant debris (cuticles, tracheids, other brown to black phytoclasts) and amorphous organic matter; the proportion of major palynological groups (spores, acritarchs, prasinophytes, chitinozoans) is variable in different samples but spores are always dominant over the microplankton. In assemblage 1 (A1) the dominant elements are of continental origin, with many specimens of diverse species of *Grandispora* associated with scarce marine palynomorphs, represented by acritarchs (*Leiosphaeridia*, *Navifusa*) and chitinozoans. Additionally, fragmented moulds of hyolithes, crinoid stems (see Fig. 4d) and trilobites plus the lycophytes referred to '*Haplostigma*' *furquei* (BAFC-Pb 16999 to 17003; BAFC-Pl 1281) by di Pasquo (2007*a*) were recorded. The plant remains are composed of fragmented carbonaceous compressions and impressions of the herbaceous stems (Fig. 4). This section is composed of dark siltstones interbedded with bioturbated rippled sandstones and is interpreted as deposited on a shallow marine shelf.

The overlying assemblage 2 (A2) (BAFC-Pl 1268 and 1269) is slightly more diverse, with abundant phytoclasts (tracheids, cuticles, charcoal) and many specimens of *Grandispora* and megaspores attributed to *Biharisporites* along with rare prasinophytes and acritarchs (generally spheroidal forms without processes) and chitinozoans. Only plant remains appeared with the palynomorphs in a black diamictitic deposit with a mudstone matrix that includes clasts of the underlying unit (see di Pasquo 2007*a*). This section was probably deposited in the same marginal palaeoenvironment as A1 but invertebrate fossils were not recovered. Di Pasquo (2007*a*) attributed A1 to the late Eifelian mainly based on the absence of *Geminospora lemurata* and other related species that first appear in the Givetian, and A2 to the early late Givetian to the base of the Frasnian, due to the presence chiefly of *Samarisporites triangulatus* and its correlation to the *optivus–triangulatus* Biozone Richardson & McGregor (1986), and the *Trg* Zone of Melo & Loboziak (2003) (Fig. 6).

Pando x-1 Borehole (11°36′07″S, 67°56′45″W)

Vavrdová *et al.* (1996) provided the first palynological study of this completely cored drilling located in northern Bolivia (see Figs 1, 2). More than 60 samples from the Devonian to Permian section of this borehole were collected by M.d.P. during a field trip to Bolivia in 2000 with permission of Pluspetrol S.A. This yielded 59 productive samples from which 26 levels between 1929–1932 m and 1203–1206 m correspond with certainty to the Devonian. They include 14 samples that cover the section between 1559.30 and 1189.09 m that was not sampled by Vavrdová *et al.* (1996). Preliminary examination of all the samples enabled the recognition only at 1260/63 m deep

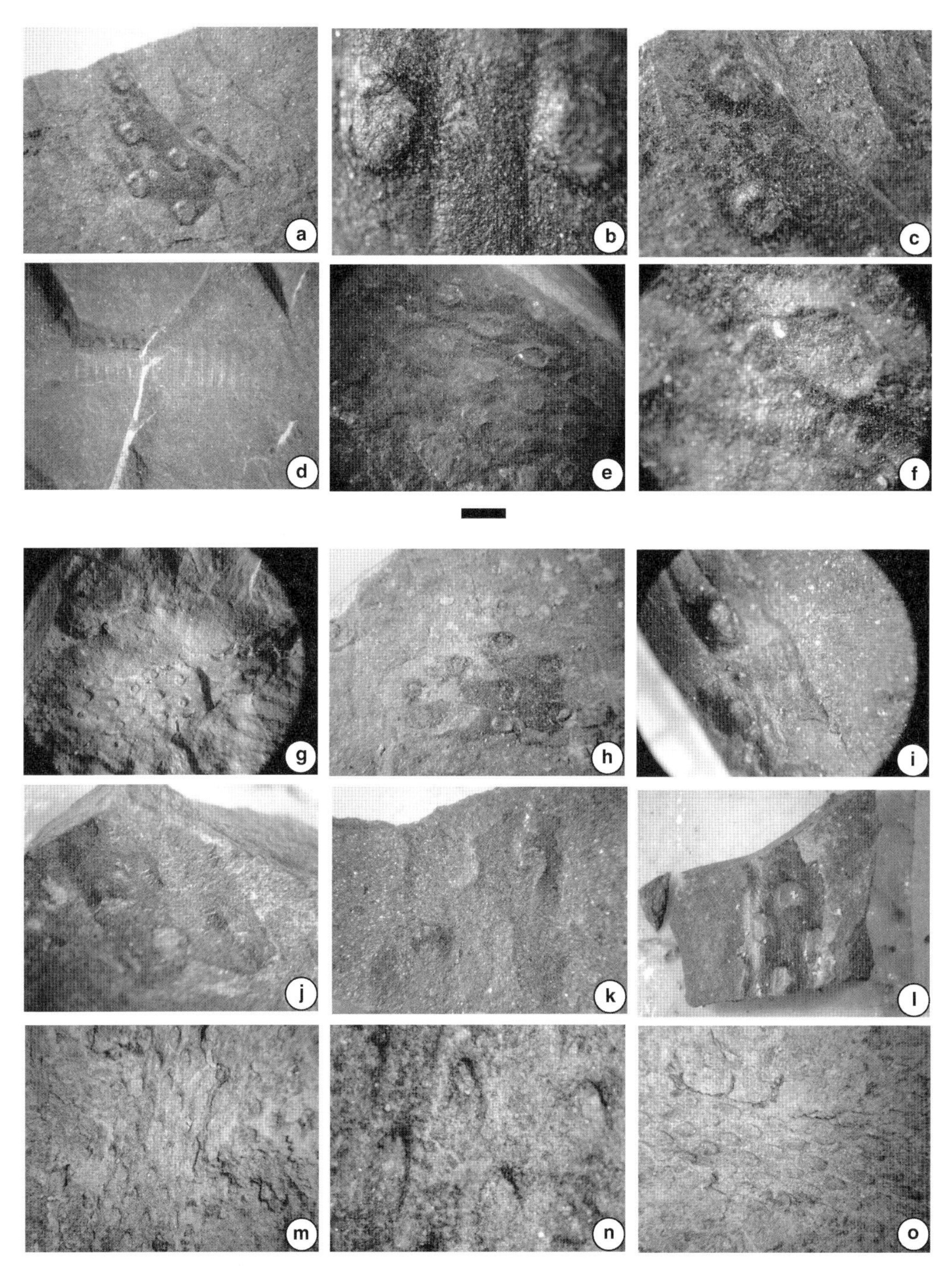

Fig. 4. '*Haplostigma*' species from Bolivia and Argentina. (**a**) '*Haplostigma*' sp. BAFC-Pb 17003. Scale bar 3.3 mm. (**b**) Detail of the scars rib and furrows of (a). Scale bar 0.6 mm. (**c**) Detail of the disposition of the scars of (a). Scale bar 1.3 mm. (**d**) Impression of transported crinoid stem. Scale bar 3 mm. (**e**) '*Haplostigma*' sp. BAFC-Pb 16999. Scale bar 3.8 mm. (**f**) Detail of the scar of (e). Scale bar 1.4 mm. (**g**) '*Haplostigma*' sp. BAFC-Pb 17000. Scale bar 3.5 mm. (**h**) Detail of the scars in the counterpart of (g). Scale bar 1.7 mm. (i) Detail of the scars, ribs and furrows from part of (g). Scale bar 1 mm. (j, k) Part and counterpart '*Haplostigma*' sp. BAFC-Pb 17002. Scale bar 3.1 mm. (**l**) '*Haplostigma*' sp. BAFC-Pb 17001. Scale bar 0.3 mm. (**m–o**) '*Haplostigma*' sp. from Precordillera, BAFC-Pb 16998. (m, o) Scale bar 3.1 mm. (n) Detail of the phyllotaxis and scars. Scale bar 7.5 mm.

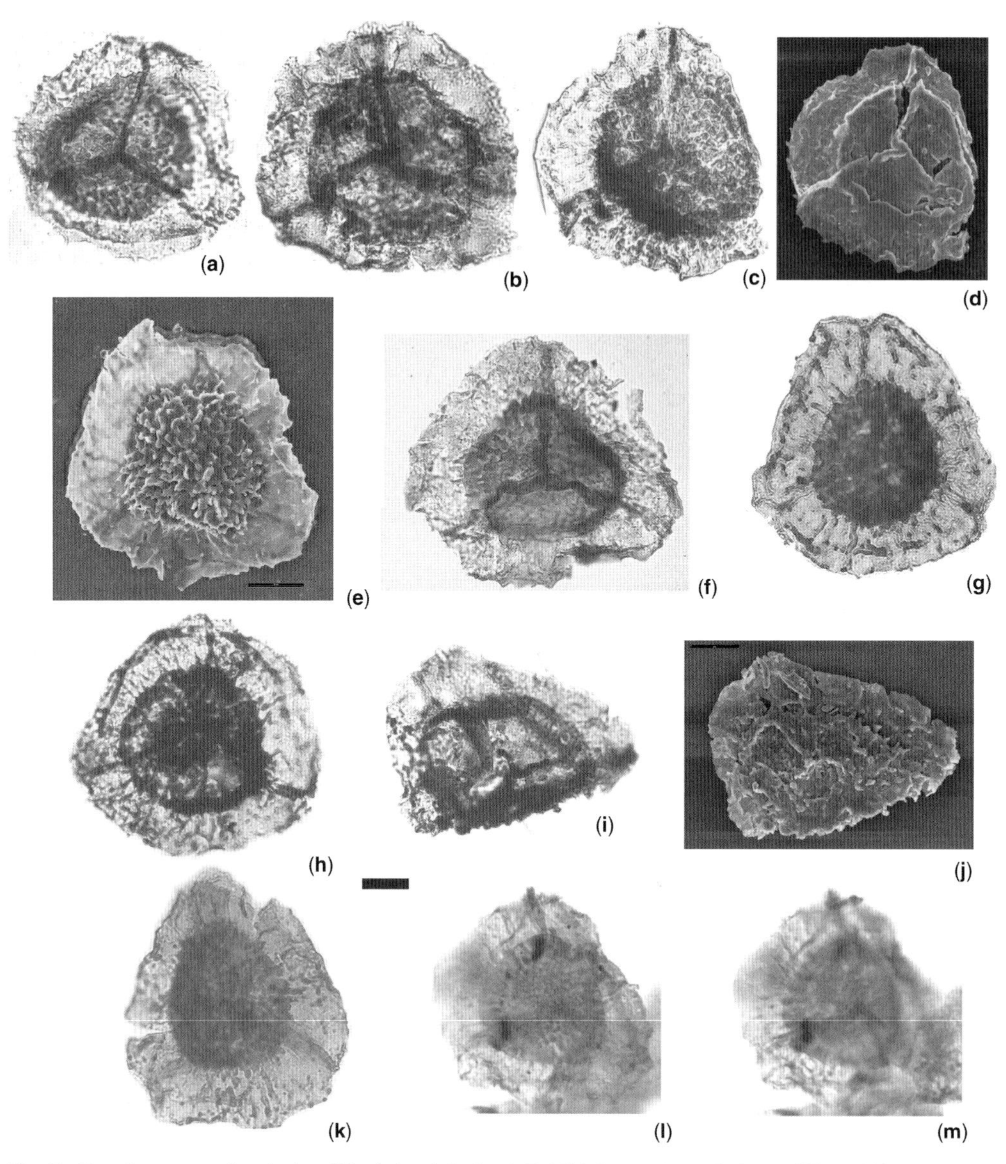

Fig. 5. *Grandispora pseudoreticulata* (Menéndez & Pöthe de Baldis) Ottone. (a, c–f). Precordillera (Chigua Formation, La Cortadera Creek): (**a**) BAFC-Pl 1656 (1): H23; (**c**, **d**) BAFC-Pl 1505; (**e**, **f**). BAFC-Pl 1655(SEM) Q44. (**b**) Precordillera (Chigua Formation, Del Chaco Creek), BAFC-Pl 1780 (1): J25/4. (**g**) Precordillera (Malimán Formation, La Cortadera Creek), reworked specimen. BAFC-Pl 1508 (4): P44. (h–j). Los Monos Formation (Balpuca, San Telmo Range, Tarija department): (**h**) BAFC-Pl 1269(1): Y35/2; (**i**, **j**). BAFC-Pl 1269 (SEM) J45/2. (**k**) Itacuamí Formation (Tuyunti Creek, Salta), reworked specimen. BAFC-Pl BAFC-Pl 1154 (1): Y39/4. (**l**, **m**). Pando x-1 Borehole, Bolivia. BAFC-Pl 1534 (5): M61/1. Scale bar 20 μm.

(BAFC-Pl 1534) of scarce specimens attributable to *Grandispora pseudoreticulata* (Fig. 5l, m). This assemblage would probably be akin to the Frasnian due to the presence of *Acinosporites eumammillatus* and several species of genera such as *Maranhites*, *Ancyrospora* and *Hystricosporites*. More detailed studies are in progress.

La Cortadera and Del Chaco creeks (29°53′S, 69°7′W)

Palynoassemblages from the Chigua Formation (Chavela Member), exposed in the eastern part of the Blanco River, Del Volcán Range, western Precordillera Argentina (Figs 1 and 2), were recently

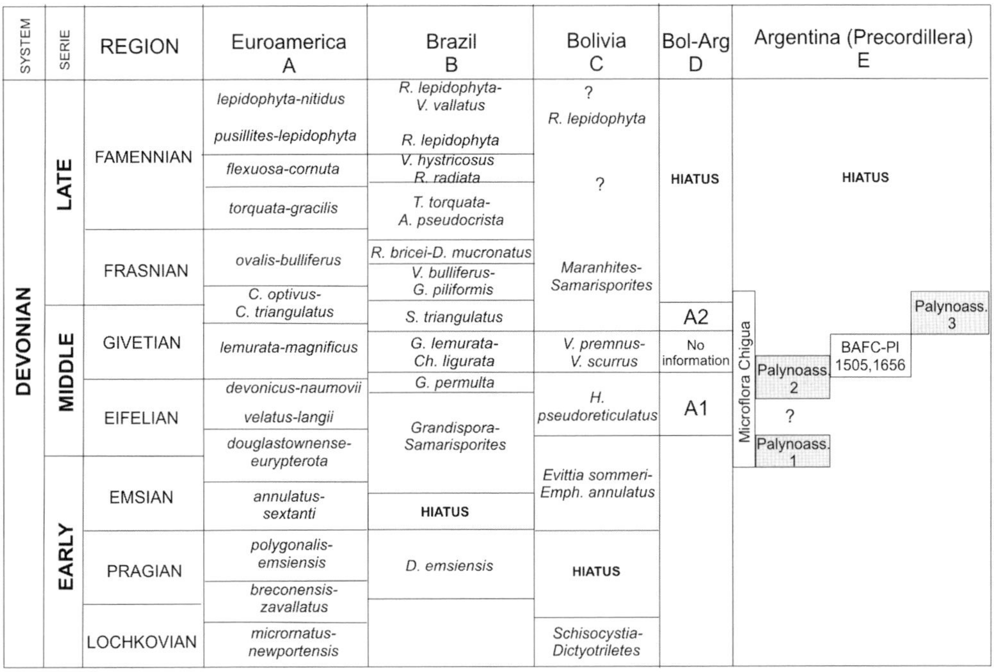

SYSTEM	SERIE	REGION	Euroamerica A	Brazil B	Bolivia C	Bol-Arg D	Argentina (Precordillera) E
DEVONIAN	LATE	FAMENNIAN	lepidophyta-nitidus pusillites-lepidophyta flexuosa-cornuta torquata-gracilis	R. lepidophyta-V. vallatus R. lepidophyta V. hystricosus R. radiata T. torquata-A. pseudocrista	? R. lepidophyta ?	HIATUS	HIATUS
		FRASNIAN	ovalis-bulliferus C. optivus-C. triangulatus	R. bricei-D. mucronatus V. bulliferus-G. piliformis S. triangulatus	Maranhites-Samarisporites	A2	Microflora Chigua Palynoass. 3
	MIDDLE	GIVETIAN	lemurata-magnificus	G. lemurata-Ch. ligurata G. permulta	V. premnus-V. scurrus	No information	BAFC-Pl 1505,1656 Palynoass. 2
		EIFELIAN	devonicus-naumovii velatus-langii douglastownense-eurypterota	Grandispora-Samarisporites	H. pseudoreticulatus	A1	? Palynoass. 1
	EARLY	EMSIAN	annulatus-sextanti	HIATUS	Evittia sommeri-Emph. annulatus		
		PRAGIAN	polygonalis-emsiensis breconensis-zavallatus	D. emsiensis	HIATUS		
		LOCHKOVIAN	micrornatus-newportensis		Schisocystia-Dictyotriletes		

Fig. 6. Summary correlation chart of the assemblages studied (D, E) from the Balapuca, La Cortadera and Del Chaco sections, with other Devonian biozones or assemblages of Euroamerica (A), Brazil (B) and Bolivia (C), and the stratigraphical ranges of selected spore species recovered. References: A, Richardson & McGregor (1986); B, Melo & Loboziak (2003); C, Suárez Soruco & Lobo Boneta (1983), Limachi *et al.* (1996); D, di Pasquo (2007*a*); E, Amenábar *et al.* (2006), Amenábar (2007*b*). Correlation chart modified from di Pasquo (2007*a*).

studied by Amenábar *et al.* (2006, 2007) and Amenábar (2007*a*, *b*, 2009). The Chigua Formation (Devonian) unconformably underlies the Carboniferous sediments of the Malimán Formation at La Cortadera Creek. But, in the Del Chaco Creek the top and base of the Devonian section are delimited by faults. The fossiliferous samples collected at La Cortadera (BAFC-Pl 1505, 1656) were obtained from carbonaceous shales which also contain trilobites (e.g. *Punillaspis argentina* Baldis). The microflora has yielded more or less the same ratio of *Grandispora pseudoreticulata* and microplankton specimens. Amenábar (2007*a*, *b*, 2009) attributed the assemblage from La Cortadera Creek to the Middle Devonian (early Givetian), based on the range of key species that are recognized in the *D. devonicus–G. naumovii* (*DN*) and *G. lemurata–C. magnificus* (*LM*) Richardson & McGregor (1986) and the *GS*, *Per* and *LLi* Interval zones Melo & Loboziak 2003 (Fig. 6). This assemblage would represent the time between the A1 and A2 assemblages from the Los Monos Formation at Balapuca.

Samples from the Del Chaco Creek (BAFC-Pl 1797, 1798, 1780) are characterized by miospores, acritarchs, prasinophytes and chitinozoans, with many specimens of *Grandispora pseudoreticulata* (Fig. 5). Two assemblages are distinguished (Fig. 6): one obtained from shales (BAFC-Pl 1797) is very rich in microplankton (A1) while the other (BAFC-Pl 1780) has spores as the dominant element (A2). Lycophyte remains are present in fine sandstones interbedded within the section bearing the latter sample; they comprise mainly impressions of stems parallel to bedding planes at several levels (Fig. 4). The third level (BAFC-Pl 1798) has very few palynomorphs so it is not considered further.

Assemblage 1 (A1) is characterized by high microplankton diversity (26 species = 72% of the total of the association) and spores as subordinate elements (10 species = 28%). Some biostratigraphically important spores are *Dictyotriletes emsiensis* Morphon, *Emphanisporites decoratus*, *Acinosporites hirsutus* and *Grandispora* spp. The *D. emsiensis* morphon has been recorded in the Talacasto Formation (Central Precordillera of Argentina), where the assemblage was attributed to the late Lochkovian–Emsian (Le Hérissé *et al.* 1997). In Brazil, it is recognized in the late Lochkovian–Emsian deposits from the Solimões (Rubinstein *et al.* 2005), Paraná (Dino 1999) and Amazon basins (Melo & Loboziak 2003). According to Melo & Loboziak (2003) the *Grandispora/Samarisporites* spp. (*GS*) Interval Zone (late Emsian–early Eifelian) is characterized by the appearance of large spinose pseudosaccates/zonates spores of the genera *Grandispora*, *Samarisporites* and *Craspedispora*. Some species such as *Dictyotriletes emsiensis* are more typical of the Early Devonian but they do not persist into younger biozones. Thus, palynoassemblage 1 was attributed to the late Emsian–early Eifelian due to the co-occurrence of *Dictyotriletes emsiensis* morphon and species of *Grandispora* (Fig. 6).

In contrast, assemblage 2 (A2) is dominated by continental elements with 19 species of spores and five species of microplankton representing 79% and 21% respectively. Pseudosaccate spores of the genus *Grandispora* are especially dominant (e.g. *Grandispora mammillata*, *G. pseudoreticulata*) and other diagnostic spore species are *Acinosporites hirsutus*, *A. macrospinosus*, *A. acanthomammillatus*, *Corystisporites horridus*, *Dibolisporites* sp. cf. *D. eifeliensis* and *Leiotriletes balapucensis*. This level is correlated to the levels BAFC-Pl 1505 and 1656 recorded from La Cortadera Creek on the basis of common species including some of those mentioned above. These taxa together with *Acinosporites hirsutus*, *A. macrospinosus*, *Acanthotriletes horridus* and *Grandispora mammillata* enable this assemblage to be assigned to the early Givetian and correlate to the *D. devonicus–G. naumovii* (*DN*) and *G. lemurata–C. magnificus* (*LM*) Zones Richardson & McGregor (1986) and the *GS*, *Per* and *LLi* Interval Zones Melo & Loboziak 2003 (see Fig. 6).

Considering the co-occurrence of trilobites comparable to those of the *Acanthopyge balliviani* zone (Eifelian or Emsian, according to different authors, see Baldis & Saurdiansky 1975) defined in Bolivia, with the cephalopod genus *Tornoceras* (Frasnian) in the Chigua Formation, Baldis & Sarudiansky (1975) suggested that the range of the unit would be from the Emsian–Eifelian boundary up to the Givetian–Frasnian boundary. Therefore, these authors indicated that the presence of *Tornoceras* could reduce the age range of *A. balliviani* and thus they propose a tentative Givetian age for the Chavela Member of the Chigua Formation.

Methods

The plant specimens were studied with a stereo microscope under magnifications between ×8 and ×50. The illustrations were taken with a Sony Cyber-shot DSC-P200 7.2 megapixel camera. Standard palynological isolation was carried out on mudstone samples to obtain the organic residues. They were extracted from crushed samples, treated first with hydrochloric acid and then hydrofluoric acid to remove carbonate, silica and silicates, respectively, and finally mounted on slides with glycerin jelly. Identification of the palynomorphs was undertaken using both Leitz Orthoplan and Nikon Eclipse 80i binocular transmitted light microscopes, with ×1000 maximum magnification. The photomicrographs were obtained with Motic (2.0 megapixels) and Pax-it (3.1 megapixels) video cameras. The illustrations are indicated by the BAFC-Pl acronym followed by the England Finder reference. The microfossils were processed and deposited at the Department of Geology of the Faculty of Natural and Pure Sciences (University of Buenos Aires) under the prefix BAFC-Pl (palynology) along with megafossils (BAFC-Pb for plants). Surface details of the spores were photographed under SEM with a Phillips series XL model 30 at the Natural Science Argentine Museum 'Bernardino Rivadavia' (MACN) of Buenos Aires (Argentina). The specimens were coated with a 200–300 Å gold-palladium film using a sputter-coater 'termo VG Scientific SC 7620'. Following SEM study, cover slips bearing the isolated spores were removed and mounted on slides with glycerin jelly to be re-illustrated under the light microscope.

Grandispora and 'Haplostigma' species: commented records

A detailed taxonomic study of both *Grandispora pseudoreticulata* and the '*Haplostigma*' species is beyond the scope of this paper and will be the subject of other contributions currently in preparation. This paper includes commented references about the presence of both taxa in South America and other related species illustrated from different Middle and Late Devonian assemblages of the rest of the world (see Figs 2 and 3). The Supplementary Material summarizes selected palynological references from the world based on selected species mainly from the Emsian to Frasnian of Argentina, Bolivia and Brazil, where the complete authority of species is included; if not, it is specified in the text.

'Haplostigma' spp. (Figs 1–4)

Several records of the lycopsid '*Haplostigma*' Seward 1932 are known from South America and some other places of Gondwana (Figs 2 and 3); however, because of problems of nomenclature that still exist with the generic status, it is considered here with inverted commas. Plumstead (1967) suggested that the specimens illustrated by Seward

(1932, pl. XXIII, figs 2, 6, 7; pl. XXIV, figs 11, 13) from the Lower Witteberg Group, Cape Fold Belt of South Africa, were truly assignable to the type species '*Haplostigma*' *irregularis* (Schwarz) Seward 1932. She also added that the specimens should show evidence of a vascular strand and foliar appendages longer than 10 mm. This feature changes the diagnosis of the genus '*Haplostigma*'. Subsequently, Anderson & Anderson (1985) emended the genus '*Haplostigma*' based on the description of two different species, '*Haplostigma*' *kowiensis* (Plumstead) Anderson & Anderson 1985 and '*Haplostigma*' *irregularis* (Schwarz) Seward 1932, and included in the latter species the records of '*Haplostigma*' *furquei* Frenguelli 1952 emend. Gutierrez 1996 from the Middle to Late Devonian of the Precordillera, Argentina. The description noted the occasional forking of the axes as well as a central vascular trace that was sometimes evident and included the presence of appendages ('leaves') that would be found, rarely, as squat conical with an acute tip and diverging at 50–60° from the axis. Because of this emendation of Anderson & Anderson (1985) is unclear, the diagnosis of the genus still remains inconclusive (see also Cingolani *et al.* 2002).

The resemblance of '*Haplostigma*' *furquei* to '*Haplostigma*' *irregularis* is remarkable, although a revision of the material is needed to confirm this in agreement with McLoughlin & Long (1994). In this context, the specimens illustrated here from the Los Monos Formation at Balapuca (Figs 1, 2 and 4; see also di Pasquo 2007*a*), resemble '*Haplostigma*' *furquei* and '*Haplostigma*' *irregularis* but they lack spiny lateral appendages due to fragmentation thus preventing a more precise assignment at this time. More material recently collected by M.d.P. from new localities in southern Bolivia is under study. Many of these specimens are very similar to the ones illustrated by Branisa (1965) as '*Haplostigma*' *furquei* from Finca Carlazo (east of Tarija city; pl. 51, figs 2, 3) and as '*Haplostigma*' cf. *furquei* from Sica Sica (pl. 50, figs 1–3). Finally, the revision of the records of '*Haplostigma*' *furquei* presented by Gutierrez (1996) is accepted here where the presence of this species is mentioned in the El Toco Formation, Angosturas range, Antofagasta, northern Chile. Moreover, recently Moisan & Niemeyer (2005) mentioned new findings of this taxon from the same place.

However, the specimens from the Precordillera of Argentina (Fig. 4) are somewhat similar to the 'Lepidodendroid fragments' from the Middle Devonian deposits of the Falkland Islands (Halle 1911) in the phyllotaxis and shape of the scars, but differ in the absence of any evidence of a leaf trace, and the axis is narrower. Seward & Walton's 'Lepidodendroid stems' (1922) from Halfway Cove (West Falkland) include the specimens found by Halle (1911), even remarking on the resemblance between the specimen (Halle 1911, pl. 6, fig. 3) and theirs (Seward & Walton 1992, pl. XIX, fig. 2). Subsequently Archangelsky (1983) reassigned these specimens to his new species *Malanzania antiqua* (see also Gutierrez & Archangelsky 1997).

Other illustrated records are referred to as '*Haplostigma*' sp. by Cingolani *et al.* (2002) from the Givetian Lolén Formation (Ventana Range, Buenos Aires Province), '*Haplostigma*' *baldisii* Gutierrez & Archangelsky 1997 from the Chigua Formation (Precordillera) of Argentina, and '*Haplostigma*' *lineare* (Walkom) McLoughlin & Long 1994 from Beacon Orthoquartzite (Middle Devonian), Southern Victoria Land, Antarctica. The latter has grooves like '*Haplostigma*' *furquei* but bears fusiform leaf scars without preserved appendages. In contrast, Retallack (1995) described new records of this lycophyte from the same locality, but he assigned the stems to '*Haplostigma*' *irregularis*. Even though this report did not include photographs, a reconstruction of the plant showing the oval scars and short mucronate, keeled 'leaves' is pictured. '*Haplostigma*' *irregularis* was also illustrated by Kräusel (1960) from Ponta Grossa Formation (Middle Devonian), Paraná Basin, Brazil (see Figs 1–3). Other records are cited in the literature but they cannot be confirmed since they were not illustrated and the materials were not available. These are (see Figs 2, 3): '*Haplostigma*' *irregularis* from Pimenteira Formation, Parnaíba Basin (Suárez Riglos 1975); '*Haplostigma*' *furquei*, from Huamampampa, Los Monos and Iquirí formations, at Campo Redondo (Suárez Riglos 1975) and Lajas (Wood 1995); and other localities of the central part of Bolivia (Suárez Soruco 1988).

The similarity between the herbaceous lycophyte '*Haplostigma*' *furquei* from Precordillera and Bolivia with '*Haplostigma*' *irregularis* from Brazil and South Africa and others from Antarctica and the Falklands Islands is remarkable (Anderson and Anderson 1985; Edwards 1990), although they still require detailed morphotaxonomical study. Due to the fact that none of the lycophyte-bearing Middle Devonian deposits cited above yielded palynomorphs and they may also continue into the early Frasnian, they are not trustworthly for detailed biostratigraphy. But, at least the records of both '*Haplostigma*' and *Grandispora pseudoreticulata* and other key taxa from the same stratigraphic levels at Balapuca or from interbedded ones in Del Chaco and La Cortadera creeks (see Figs 1, 2) enable a more accurate age for these plant assemblages. On the other hand, the mentioned similarity among those plant species or specimens should probably be enough to support a common origin for them during this time.

Palynology

Grandispora pseudoreticulata (Menéndez & Pöthe de Baldis) Ottone 1996 is a trilete camerate spore distinguishable from others of the genus by its characteristic distal-polar exoexine sculpture of cones, spines and biform elements mostly fused to form concentric irregular, anastomosing ridges and mainly discrete on the flange (see Fig. 5). It was described for the first time as '*Hymenozonotriletes*' *pseudoreticulatus* by Menéndez & Pöthe de Baldis (1967, pp. 168–169, pl. 1 figs C, D, E) and later reassigned by Ottone (1996) to *Grandispora*. Currently, the original specimens are lost, so it was not possible to directly compare our material. Nevertheless, we have revised slides from the '*Hymenozonotriletes*' *pseudoreticulatus* zone of Suárez Soruco & Lobo Boneta (1983) from Bolivia provided by Lobo Boneta, and plenty of specimens of this species and the subsurface Devonian material from Argentina described by Ottone (1996). Additionally, M.d.P. collected surface and subsurface samples during several field trips to Bolivia, which have provided comparison specimens of this and related species.

Grandispora pseudoreticulata was illustrated from the Middle Devonian of Paraguay (Menéndez & Pöthe de Baldis 1967), Middle Late Devonian of Argentina (Ottone 1996; Amenábar *et al.* 2007*b*) and Bolivia (Ottone & Rosello 1996; Vavrdová & Isaacson 2000; di Pasquo 2007*a*). Unfortunately, the only record for this species outside South America was in a list, but not illustrated, by Bär & Riegel (1974) from the Middle Devonian of Ghana. The assemblage comprises some other species also cited by Menéndez & Pöthe de Baldis (1967) from the Middle Devonian of Paraguay (see Figs 2, 3). This taxon is used as a zonal index species in biostratigraphical schemes proposed for the Tarija Basin (di Pasquo 2007*c*), mainly that proposed by Limachi *et al.* (1996), who described the '*Hymenozonotriletes*' *pseudoreticulatus* Zone, which in turn is correlated to the '*Haplostigma*' zone Limachi *et al.* 1996, attributed to the Eifelian to early Frasnian and partially to the *Tropidoleptus carinatus* brachiopod zone Limachi *et al.* 1996 (late Givetian–Frasnian) (see Figs 2, 3, 6). Therefore, its stratigraphical range is well established as late Eifelian to early Famennian on the basis of the presence of Euramerican index species.

This species is also recognized as reworked in the Early Carboniferous (early Viséan) Malimán Formation (Río Blanco Basin, Argentina; Amenábar 2006, 2007*b*), from the Itacua Formation at Balapuca (overlying the Los Monos Formation; di Pasquo 2007*b*), the Serpukhovian Kaka Formation (Madre de Dios Basin, Bolivia; Azcuy & Ottone 1987; Fasolo *et al.* 2006), and in the Late Carboniferous Macharetí and Mandiyutí Groups from the Tarija Basin (di Pasquo & Azcuy 1997; di Pasquo 2003; see Figs 2 and 5). *Grandispora pseudoreticulata*, as *Indotriradites variabilis* Pérez Loinaze (2005), from the Malimán and Cortaderas (Mississippian) formations, is here interpreted as a reworked form from the Devonian Chigua Formation (see Amenábar 2006). Evidence to support this interpretation is presented and discussed by di Pasquo and Azcuy (1997) and Amenábar (2006) and di Pasquo (2003, 2007*b*).

Several species present in our assemblages show clear morphological characters allowing unambiguous identification (e.g. *Leiotriletes balapucensis*, *Archaeozonotriletes variabilis*, several species of *Apiculiretusispora*, *Geminospora lemurata*, *Samarisporites triangulatus*). Ravn & Benson (1988) and Breuer *et al.* (2007*a*) have also commented on the highly variable and intergradational complex of large miospores assignable to *Grandispora*. Ravn & Benson (1988) noted that many of the specimens recovered from the Emsian–Eifelian of Georgia (USA) resemble known species but were difficult to assign, so they were illustrated and left in open nomenclature. Breuer *et al.* (2007*a*) found a continuous morphological intergradation within a single form-species *Grandispora libyensis* in the Middle Devonian of Libya. This is also the case of *Grandispora pseudoreticulata* that intergrades especially with *G. mammillata* in our assemblages. Other similar or very similar species assigned to the genus *Grandispora* or allied genera such as *Samarisporites*, '*Calyptosporites*' or '*Hymenozonotriletes*', defined and illustrated from global the Middle to Late Devonian spore literature, are compared as described below (a–k; see also Supplementary Material).

(a) *Camptozonotriletes caperatus* McGregor and *Grandispora* cf. *protea* (Naumova) Moreau-Benoit in Cloutier *et al.* (1996, pl. 1, figs 5, 14) from the middle Frasnian Escuminac Formation, eastern Québec, Canada. These spores, along with *Grandispora douglastounense*, were recorded in the same level as other characteristic spores from the Emsian–Eifelian such as *Dictyotriletes emsiensis*, so these authors have interpreted all of them as reworked.

(b) *Perotrilites meonacanthus* var. *rugosus* Kedo in Turnau (1996, pl. 2, fig. 5) from western Pomerania assigned to the late Eifelian–mid Givetian (*Rhabdosporites langii* and *Aneurospora extensa* zones Avchimovitch *et al.* 1993) resembles *Grandispora mammillata*.

(c) '*Hymenozonotriletes*' *domanicus* Naumova in Braman & Hills (1992, pl. 14, figs 14, 15, pl. 15, fig. 1) from the Givetian to Frasnian Imperial Formation of northwestern Canada. The sculpture on distal face forms a pseudoreticulum but not like *G. pseudoreticulata*.

(d) *Grandispora macrotuberculata* (Archangelskaya) McGregor 1973 in Melo & Loboziak (2003, pl. 4, fig. 9) and *Grandispora mammillata* Owens 1971 in Melo and Loboziak (2003, pl. 3, fig. 20) from the late Eifelian and *Auroraspora pseudocrista* Ahmed 1980 in Melo & Loboziak (2003, pl. 1, fig. 6) from the Frasnian–early Famennian of the Amazon Basin in Brazil.

(e) *Samarisporites eximius* (Allen) Loboziak & Streel (1989) in Rodrigues *et al.* (1995, pl. 1, fig. 2) and *Samarisporites praetervisus* (Naumova) Allen (1965) in Rodrigues *et al.* (1995, pl. 1, fig. 1) from Parnaíba Basin in Brazil, are considered here as likely records of *Grandispora pseudoreticulata* (see Fig. 2).

(f) *Samarisporites* sp. A and B in Daemon *et al.* (1967, pl. 1, figs 14, 15) from Paraná Basin in Brazil are also considered here as likely records of *Grandispora pseudoreticulata* (see Fig. 2).

(g) *Grandispora ?macrotuberculata* (Archangelskaya) McGregor 1973 in Boumendjel *et al.* (1988, pl. 2, fig. 3) from the Givetian Gazelle Formation, D'Illizi Basin, Sahara Algeria. In this assemblage the common species are *Geminospora lemurata*, *Grandispora inculta*, *Grandispora mammillata*, *Grandispora velata*, *Rhabdosporites parvulus*, *Samarisporites praetervisus*, *Emphanisporites annulatus*, *Emphanisporites rotatus* and some chitinozoans (e.g. *Ancyrochitina taouratinensis* Boumendjel, *Linochitina jardinei* Boumendjel, *Alpenachitina eisenacki* Dunn & Miller).

(h) *Camptozonotriletes leptohymenoides* Balme (1988, pl. 8, figs 1–4, size range 100–170 μm) from the early Frasnian Gneudna Formation, Carnarvon Basin, Western Australia. Balme (1988) noted that it is very close to *Grandispora pseudoreticulata*, being differentiated only by its apparently smaller size (85–114 μm). Scarce specimens of this species were recorded only from the lower half of this sequence, so Balme gave little significance to its apparent stratigraphical restriction. The close similarity of these species remains unresolved and it is important to compare material. This information would confirm the migration path of the parental plant eastwards during the latest Givetian to early Frasnian. Balme's assemblage has 44 species of which only the following are common to South America: *Ancyrospora langii*, *Emphanisporites annulatus*, *Emphanisporites rotatus*, *Geminospora lemurata*, *Cymbosporites hormiscoides*, *Gneudaspora divellomedia*, *Rhabdosporites langii*, *Samarisporites triangulatus*, *Verrucosisporites scurrus*. The Frasnian plant assemblages from Gneuda Formation are dominated by progymnosperms, but during the Givetian plant species are less known (see Balme 1988).

(i) *?Calyptosporites* sp. A in Paris *et al.* (1985, pl. 21, fig. 1) from the early Eifelian of Libya.

(j) Spores similar to *Grandispora pseudoreticulata* were recently studied by Wellman & Gensel (2004). These authors presented a detailed morphological study including SEM and TEM of the sporangia and spores of the enigmatic Lower Devonian (*douglastownense–eurypterota* Zone Richardson & McGregor 1986, late Emsian–earliest Eifelian) plant *Oocampsa catheta* Andrews *et al.*, which is considered intermediate between the trimerophytes and progymnosperms. In addition, they analysed dispersed spores assigned to *Grandispora douglastownense* and *Grandispora ?macrotuberculata* considered to possibly represent forms derived from *O. catheta*. They concluded that *G. douglastownense* and *G. ?macrotuberculata* are probably end members of the same spore complex and most likely are dispersed camerate spores produced by *Oocampsa catheta*.

(k) Another similar form (*Camptozonotriletes caperatus*) was illustrated with an SEM by Wellman (2006) but is different in possessing verrucate–vermiculate distal sculpture with many radial folds on the flange. *Camptozonotriletes caperatus* appears in the Pragian–Emsian *polygonalis–emsiensis* Zone Richardson & McGregor (1986).

As the inception of both *G. douglastownense* and *C. caperatus* occurred prior to *G. pseudoreticulata*, it would be interesting to investigate whether one of these taxa could be the ancestor of *G. pseudoreticulata* and related species of the late Eifelian South American microfloras. On the other hand, the abundant presence of *Grandispora pseudoreticulata* and '*Haplostigma*' in the same stratigraphic levels at Balapuca, Del Chaco and La Cortadera creeks, leads us to speculate on a possible relationship among them.

Approaches to Middle and Late Devonian palaeogeography and palaeobiogeography

Some previous palaeobiogeographical proposals for the Middle and Late Palaeozoic were summarized and updated by Wnuk (1996), who distinguished the Euramerican Realm and Gondwanan Realm, the latter including the South Gondwanan Temperate region. Wnuk provided a brief synthesis of previous palaeontological data to sustain the Middle and Late Devonian palaeobiogeographical scheme, but indicated that additional work on Middle and Late Devonian phytogeography is needed. The Gondwanan Realm encompasses the largest land area during the Palaeozoic and includes the continents of South America, Antarctica, Africa, India and Australia plus smaller regions such as Madagascar, Arabia, New Zealand and the peripheral Tibetan, Iranian, Turkish and China Plates (Li Xingxue 1986; see Fig. 3). Although the

Devonian floristic development of this realm is not well understood due to the uneven knowledge of its fossil record, Streel & Loboziak (1996) defined, for the Middle and Late Devonian, the Northern Euramerica and Southern Euramerica–Western Gondwana phytogeographical provinces. The former is based on the southern limit of *Archaeoperisaccus*, while the latter includes eastern Canada and western Europe based on the resemblance of its palynoassemblages.

Although most palynomorphs had a cosmopolitan distribution during the Devonian, Bär & Riegel (1974) suggested that provincialism would be recognized at the specific level for Gondwana. In this sense, Loboziak & Streel (1995) have shown that the miospore assemblages from western Gondwana are dominated, in the Middle Devonian, by some endemic forms such as *Camarozonotriletes? concavus*, *Craspedispora ghadamisensis*, *Grandispora libyensis* and *G. permulta*. The frequent association of *Grandispora pseudoreticulata* with other endemic but stratigraphically important spore species (e.g. *Grandispora daemonii*, *G. permulta*, *Leiotriletes balapucensis*, *Acinosporites ledundae*, *A. eumammillatus*, *Apiculiretusispora laxa*, *Apiculatisporis grandis*, *Retusotriletes paraguayensis*), as well as with the *'Haplostigma'* flora (see Figs 2, 3), shows a certain degree of endemism of the floras mainly in South America (embracing Bolivia, Brazil, Paraguay, Chile, Peru and Uruguay) and northwestern Africa (e.g. Ghana). All these taxa allow the establishment of the Afro-South American (ASA) Subrealm as the southwestern part of the Gondwana palaeophytogeographic Province during the Middle and early Late Devonian. Its northeastern boundary may be situated somewhere between Ghana and North Africa as not all the characteristic species from the ASA Subrealm occur in those assemblages (see Loboziak & Streel 1995). In the same way, the current evidence sustains the point of view that the vegetation was not uniform worldwide even though herbaceous lycophytes were widespread during the Middle Devonian (see McLoughlin 2001). The ASA northwestern boundary should be established somewhere between Venezuela and Colombia and the rest of South America based on different megafloras of the former region, which Edwards & Benedetto (1985; see also Berry 1996) integrated to the Old Red Sandstone Continent. It provided the most diverse assemblages of Devonian plant fossils including lycophytes, trimerophytes and progymnosperms, but unfortunately nothing is known about the palynomorphs accompanying these floras (see Berry *et al.* 1993; Berry & Edwards 1995 and references therein).

Knowledge about Devonian palynofloras related to floral assemblages from South Africa, the Falkland Islands, Antarctica and Australia (see Anderson & Anderson 1985; Playford 1990; Edwards 1990; McGregor & Playford 1992) is still scarce. McGregor & Playford (1992) presented a detailed comparison between the microfloras from Canada and Australia with other parts of the world including western Gondwana. They suggested that enough similarities between these countries and the rest of world exist to sustain long-distance biostratigraphical correlation during the Middle and Late Devonian. But there are many qualitative dissimilarities between palynofloras recorded from Australia and the rest of the world, including South America, during this lapse that probably reflect phytogeographical–palaeoclimatic differences (see also Streel & Loboziak 1996). Recent palynological studies from Middle Devonian subsurface deposits in Saudi Arabia (Breuer *et al.* 2007*b*; Marshall *et al.* 2007) have shown several common species with Euramerica and the ASA Subrealm (e.g. *Geminospora lemurata*, *Samarisporites triangulatus*, *Dibolisporites eifeliensis*, *Acinosporites acanthomammillatus*, *Verrucosisporites scurrus*). However, the endemic species of the ASA Subrealm were not recorded in Saudi Arabia.

During the Middle Devonian this floral subrealm appears to have developed between 55°S and 75°S palaeolatitude based on the reconstruction of palaeoclimatic zones from Scotese *et al.* (1999; see Fig. 3). The co-occurrence of some Euramerican miospore species in the assemblages of a cooler ASA Subrealm sustains some terrestrial connection between the palaeoequatorial continent Laurussia with Gondwana, with the east–west Rheic Ocean almost closed or at least not so extensive, and with some land masses connecting both continents (see Fig. 3; e.g. Streel *et al.* 1990; Edwards 1990; McGregor & Playford 1992; Cloutier *et al.* 1996; Steemans *et al.* 2007*a*; Marshall *et al.* 2007). Recently, Steemans *et al.* (2007*b*) have presented new palynological evidence from Saudi Arabia that the same palaeogeographical scenario, with a narrow Rheic Ocean between Gondwana and Euramerica, existed at least since the Lochkovian. Scotese *et al.* (1999) explained that during the 'Hot House' periods, the warm and cool temperate belts extended to the pole and the polar climate zone did not exist. The development of relatively gradual climatic changes is supported by palaeontological records during the Middle Devonian (e.g. Streel & Loboziak 1996), such as the records of brachiopods like *Tropidoleptus carinatus* (e.g. Fonseca & Melo 1987), fossil plants like *Haskinsia* (see Cingolani *et al.* 2002) and several miospores and microplankton species. This scenario is concordant with, for example, the palaeogeographical reconstruction presented by Heckel & Witzke (1979) used by Streel *et al.* (1990) and Wood (1995) to map Givetian–Frasnian phytogeography. A juxtaposition of Venezuela and eastern North America is also supported by Edwards & Benedetto (1985), due to the fact that both regions share the same flora (see also Berry *et al.* 1993). Later, Berry (1996) presented another explanation for this common origin of both floras: Venezuela (and Colombia) could have been a displaced terrane accreted to the north of South America during the collision of Laurussian and Gondwanan forming Pangaea. Even this interpretation could have been possible because, as was mentioned above, common palynomorphs are registered almost from the palaeoequator towards both poles. This fact reinforces that, at least during Middle Devonian time, Laurussia and Gondwana were connected and the main factor producing this subtle palaeobiogeographical distribution could have been the palaeolatitudinal climatic gradient (e.g. Streel & Loboziak 1996). In fact, as already noted, the latter authors used several endemic taxa such as *Archaeoperisaccus* (e.g. Braman & Hills 1985; Streel *et al.* 1990; Hashemi & Playford 2005) to delimit the Northern Euramerica phytogeographical province. The floral distribution of the northern hemisphere during the same time probably involved more than one endemic flora (see McGregor & Playford 1992; Berry 1996), and that appeared to be climatically influenced as well.

Other factors that influence the palaeobiogeographical distribution of plants relate to palaeogeography and the ability of plants to migrate short or long distances (homosporous *versus* heterosporous), to cross barriers (e.g. water bodies or mountains) and the time involved in this process. This subject was discussed extensively by Streel *et al.* (1990), McGregor & Playford (1992), Steemans *et al.* (2007*b*) and Marshall *et al.* (2007). Even though the long-distance dispersion of plants is not easy, cosmopolitan species are evidence of this type of migration. For example, *Archaeozonotriletes variabilis* shows a slightly diachronic pattern as it first appeared in the late Eifelian of the Old Red Sandstones (cf. Richardson & McGregor 1986) and later in the Givetian–Frasnian of South America, Saudi Arabia, Libya and Australia and in the Frasnian of Russia. This diachronism in its range reinforces a stronger connection between western Europe and eastern Canada with northern South America and Africa, at least during the Givetian. It is also probable that changes in sea level (regressions) have favoured such connections at least for short times. In contrast, the dispersion of endemic species such as *Grandispora pseudoreticulata* must be controlled by the nature of the parent plant, i.e. heterosporous *versus* homosporous plants, with some palaeoenvironmental and palaeoclimatic requirements, especially miospore dispersal (e.g. large size to be air-dispersed), that have prevented their migration over long distances (see also McGregor & Playford 1992).

Palaeoenvironmental considerations

A close affinity among the microfloras from the Precordillera and the north of Argentina and southern Bolivia (Tarija Basin) is supported by several common palynomorph species, invertebrates and plant fossils, which reinforces a palaeogeographic connection between both areas during the Middle Devonian (see Fig. 3; Baldis 1967). Although palyniferous levels are scarce in Del Chaco Creek, the microplankton/spore ratio was useful to interpret palaeoenvironmental change along the Chavela Member (Amenábar 2009). This shows a tendency to more terrestrial input through the sequence from the late Early Devonian (Emsian–Eifelian) to the Middle Devonian (early Givetian). These palaeoenvironmental considerations are coincident with previous palaeontological records (see Baldis & Peralta 2000). Additional sedimentological studies and new palynological samples will permit and improve recognition of transgressive–regressive cycles in the Devonian of the Precordillera.

Albariño *et al.* (2002) and Alvarez *et al.* (2003) established for the Tarija Basin a general model of distribution of facies in a sequence-stratigraphic framework, integrating unpublished palaeontological and sedimentological data to correlate the successions. They concluded that wave-dominated marine siliciclastic platforms were developed during the Ludlow to Frasnian interval. The deposition would have been controlled by eustasy marked by at least three intervals of forced regressions as shown by sand bodies deposited basinward. It is significant that the Balapuca outcrop yielded both palynomorphs and plants from the same stratigraphic levels. These recent studies show that *Grandispora* species frequently dominate in some levels where the microplankton is scarce or absent. Scarce and fragmented marine macrofossils are indicated in A1 from Balapuca associated with the plant remains, supporting a shallow-water marine platform environment. This interpretation is compatible with the general framework presented by Albariño *et al.* (2002) where the late Emsian to late Givetian Balapuca section is located on the border of the basin, thus mostly representing marginal palaeoenvironments. Thus, on the basis of these data Figure 2 gives a more accurate delimitation of the emergent areas (i.e. continental or at least transitional areas) of the studied regions.

Conclusions

Middle and Late Devonian (Eifelian to Frasnian) palynomorphs in Argentina, Bolivia and neighbouring areas include cosmopolitan species, such as *Geminospora lemurata*, *Samarisporites triangulatus*, *Archaeozonotriletes variabilis* and *Chelinospora concinna*. Others with more restricted distribution include *Grandispora pseudoreticulata*, *Leiotriletes balapucensis* and *Acinosporites ledundae*. Among the floral remains, species attributable to '*Haplostigma*' are found from the same stratigraphic levels at Balapuca, Del Chaco and La Cortadera creeks. The northernmost record of *Grandispora pseudoreticulata* is from a third palynoassemblage of the Pando x-1 corehole of northern Bolivia.

'*Haplostigma*'-bearing beds are dated accurately, based on associated palynoassemblages. These new records extend our knowledge of biodiversity and the succession of floristic events during the Middle Devonian, and hence give their biostratigraphic position.

A great similarity among these assemblages includes the co-occurrence of *Grandispora pseudoreticulata* and '*Haplostigma*' species. This suggests that there were terrestrial connections between (or at least proximity of) these local areas and other regions in South America mainly during the Givetian.

We offer an improved interpretation of the current palaeobiogeographic knowledge of *Grandispora pseudoreticulata* and other endemic palynomorphs and associated '*Haplostigma*' flora. The local distribution of these taxa allows the definition of the Afro-South American Subrealm; it most likely results from the effects of palaeolatitude and, to a lesser extent, to local palaeoenvironmental conditions (such as marine versus continental environments). On the other hand, the presence of cosmopolitan together with endemic species supports previous palaeogeographical reconstructions where a narrow Rheic Ocean was developed between Euramerica and the northern parts of Africa and South America.

More studies are needed in order to analyse the accuracy of the relatively restricted palaeogeographical distribution of all South American taxa. This will offer greater insight into the connection of this subrealm with its neighbouring regions. Additionally, it is important to define whether all '*Haplostigma*' species found in southwestern Gondwana belong to the same taxon ('*H.*' *irregularis*?). The recovery of its associated palynological assemblage (and its *in situ* spores) is also necessary.

Special acknowledgement is made to Lic. Gustavo Holfeltz (FCEN, UBA) for processing the samples and to Lic. Fabián Tricárico (Natural and Sciences Museum 'Bernardino Rivadavia' at Buenos Aires) for assistance in SEM studies. Drs John Marshall and Philippe Steemans are greatly thanked for their helpful suggestions, and John Marshall, Peter Isaacson and Scott Hayashi for detailed language revision of the manuscript. This research was supported by funds of the 'Agencia Nacional de Promoción Científica y Tecnológica' PICTR 00313/03, the University of Buenos Aires UBACYT X 428 and the 'Consejo Nacional de Investigaciones Científicas y Técnicas' PIP 5518 CONICET. It is a contribution to IGCP 499.

References

ALBARIÑO, L., DALENZ-FARJAT, A., ALVAREZ, L., HERNANDEZ, R. & PÉREZ LEYTON, M. 2002. Las secuencias sedimentarias del Devónico en el Subandino sur y el Chaco, Bolivia y Argentina. *5° Congreso de Exploración y Desarrollo de Hidrocarburos, Mar del Plata*, CD Trabajos Técnicos.

ALONSO, J. L., RODRÍGUEZ-FERNÁNDEZ, L. R., GARCÍA-SANSEGUNDO, J., HEREDIA, N., FARIAS, P. & GALLASTEGUI, J. 2005. Gondwanic and Andean structure in the Argentine central Precordillera: The Río San Juan section revisited. *6° International Symposium on Andean Geodynamics (ISAG), Barcelona, Extended Abstracts*, 36–39.

ALVAREZ, L. A., DALENZ-FARJAT, A., HERNÁNDEZ, R. M. & ALBARIÑO, L. 2003. Integración de facies y biofacies en un análisis secuencial en plataformas clásticas devónicas del sur de Bolivia y noroeste argentino. *Asociación Argentina de Sedimentología Revista*, **10**, 103–121.

AMENÁBAR, C. R. 2006. Significado estratigráfico de palinomorfos retrabajados en la Formación Malimán (Viséano) en la Sierra del Volcán, Provincia de San Juan, Argentina. Resultados preliminares. *Revista Brasileira de Paleontologia*, **9**, 21–32.

AMENÁBAR, C. R. 2007*a*. New palynological assemblage from the Chigua Formation (late Early–Middle Devonian), at Del Chaco Creek, Volcán Range, Precordillera Argentina. *In*: ACEÑOLAZA, G. F., VERGEL, M., PERALTA, S. & HERBST, R. (eds), *Field Meeting of the IGCP 499-UNESCO 'Devonian Land-Sea Interaction: Evolution of Ecosystems and Climate'* (DEVEC), *San Juan*, 2007, 92–96.

AMENÁBAR, C. R. 2007*b*. *Palinoestratigrafía y paleoambiente de las Formaciones Chigua (Grupo Chinguillos, Devónico), Malimán y El Ratón (Grupo Angualasto, Carbonífero Inferior), Cuenca Uspallata-Iglesia. Comparación y correlación con otras palinofloras y caracterización del límite Devónico-Carbonífero en la región*. PhD Thesis, University of Buenos Aires.

AMENÁBAR, C. R. 2009. Middle Devonian microfloras from the Chigua Formation, Precordillera region, northwestern Argentina. *In*: KÖNIGSHOF, P. (ed.) *Devonian Change: Case Studies in Palaeogeography and Palaeoecology*. Geological Society, London, Special Publications, **314**, 177–192.

AMENÁBAR, C. R, DI PASQUO, M. M., CARRIZO, H. A. & AZCUY, C. L. 2006. Palynology of the Chigua and Malimán Formations in the Sierra del Volcán, San Juan province, Argentina. Part 1. Palaeomicroplankton and acavate smooth and ornamented spores. *Ameghiniana*, **43**, 339–375.

AMENÁBAR, C R, DI PASQUO, M. M., CARRIZO, H. & AZCUY, C. L. 2007. Palynology of the Chigua (Devonian) and Malimán (Carboniferous) formations in the Volcán Range, San Juan Province, Argentina. Part II. Cavate, pseudosaccate and cingulizonate spores. *Ameghiniana*, **44**, 547–564.

ANDERSON, J. M. & ANDERSON, H. M. 1985. *Palaeoflora of Southern Africa. Prodromus of South African Megafloras, Devonian to Lower Cretaceous.* A. A. Balkema, Rotterdam.

ANTONELLI, J. & OTTONE, E. G. 2006. Palinología de coronas del Devónico y Carbonífero Superior del pozo YPF. SE.EC.X-1. (El Caburé) Provincia de Santiago del Estero, Argentina. *13° Simposio Argentino de Paleobotánica y Palinología, Bahía Blanca, Resúmenes*, 22.

ARCHANGELSKY, S. 1983. Una nueva licophyta herbácea del Devónico de las Islas Malvinas, Argentina. *Revista Técnica de Yacimientos Petrolíferos Fiscales Bolivianos*, **9**, 129–135.

ARCHANGELSKY, S. 1993. Consideraciones sobre las Floras Paleozoicas de Bolivia. *In*: SUAREZ-SORUCO, R. (ed.) *Fósiles y Facies de Bolivia. Vol. II Invertebrados y Paleobotánica.* Revista Técnica de Yacimientos Petrolíferos Fiscales Bolivianos, **13–14**, 167–172.

AVCHIMOVITCH, V. I., TCHIBRIKOVA, E. V. *ET AL.* 1993. Middle and Upper Devonian miospore zonation of Eastern Europe. *Bulletin des Centres de Recherches de Exploration-Production Elf-Aquitaine*, **17**, 79–147.

AZCUY, C. L. & OTTONE, G. 1987. Datos palinológicos de la Formación Retama en la Encañada de Beu, Río Alto Beni (Bolivia). *4° Congreso Latinoamericano de Paleontología, Santa Cruz de la Sierra, Actas*, **1**, 235–249.

BALDIS, B. A. 1967. Some Devonian trilobites of the Argentina Precordillera. *International Devonian System (Calgary), Proceedings*, **2**, 789–796.

BALDIS, B. A. & LONGOBUCCO, M. 1977. Trilobites devónicos de la Precordillera noroccidental (Argentina). *Ameghiniana*, **14**, 145–161.

BALDIS, B. A. & PERALTA, S. H. 2000. Capítulo 10. Silúrico y Devónico de la Precordillera de Cuyo y Bloque de San Rafael. *In*: CAMINOS, R. (ed.) *Geología Argentina*. Instituto de Geología y Recursos Minerales, Anales, **29**(10) 215–238.

BALDIS, B. A. & SARUDIANSKY, R. 1975. El Devónico en el noroeste de la Precordillera. *Revista de la Asociación Geológica Argentina*, **30**, 301–329.

BALME, B. E. 1988. Miospores from Late Devonian (early Frasnian) strata, Carnarvon Basin, Western Australia. *Palaeontographica*, Abt B, **209**, 109–166.

BÄR, P. & RIEGEL, W. 1974. Les microflores des séries Paléozoïques du Ghana (Afrique Occidentale) et leur relations paléofloristiques. *Sciences Géologiques Bulletin*, **27**, 39–58.

BARREDA, V. D. 1986. Acritarcos Givetiano-Frasnianos de la Cuenca del Noroeste, Provincia de Salta. Argentina. *Revista Española de Micropaleontología*, **18**, 229–245.

BERRY, C. M. 1996. Diversity and distribution of Devonian Protolepidodendrales (Lycopsida). *Palaeobotanist*, **45**, 209–216.

BERRY, C. M. & EDWARDS, D. 1995. New species of the lycophyte *Colpodexylon* Banks from the Devonian of Venezuela. *Palaeontographica*, Abt B, **237**, 59–74.

BERRY, C. M., CASAS, J. E. & MOODY, J. M. 1993. Diverse Devonian plant assemblages from Venezuela. *Document Laboratoire Géologie Lyon*, **125**, 29–42.

BLIECK, A., GAGNIER, P. Y. *ET AL.* 1996. New Devonian fossil localities in Bolivia. *Journal of South American Earth Sciences*, **9**, 295–308.

BOUMENDJEL, K., LOBOZIAK, S., PARIS, F., STEEMANS, P. & STREEL, M. 1988. Biostratigraphie des miospores et des chitinozoaires du Silurien supérieur et du Dévonien dans le bassin d'Illizi (S.E. du Sahara algérien). *Géobios*, **21**, 329–357.

BRAMAN, D. R. & HILLS, L. V. 1985. The spore genus *Archaeoperisaccus* and its occurrence within the Upper Devonian Imperial Formation, District of Mackenzie, Canada. *Canadian Journal of Earth Sciences*, **22**, 1118–1132.

BRAMAN, D. R. & HILLS, L. V. 1992. Upper Devonian and Lower Carboniferous miospores, western District of Mackenzie and Yukon Territory, Canada. *Palaeontographica Canadiana*, **8**, 1–97.

BRANISA, L. 1965. *Los fósiles guía de Bolivia*. Boletín del Servicio Geológico de Bolivia, **6**, La Paz.

BREUER, P., FILATOFF, J. & STEEMANS, P. 2007*a*. Some considerations on Devonian miospore taxonomy. *Carnets de Géologie, Memoir*, **2007/01**, 3–8.

BREUER, P., AL-GHAZI, A., AL-RUWAILI, M., HIGGS, K. T., STEEMANS, P. & WELLMAN, C. H., 2007*b*. Early to Middle Devonian miospores from northern Saudi Arabian. *Revue de Micropaléontologie*, **50**, 27–57.

CINGOLANI, C. A, BERRY, C. M., MOREL, E. & TOMEZZOLI, R. 2002. Middle Devonian lycopsids from high southern palaeolatitudes of Gondwana (Argentina). *Geological Magazine*, **139**, 641–649.

CLOUTIER, R., LOBOZIAK, S., CANDILIER, A. & BLIECK, A. 1996. Biostratigraphy of the Upper Devonian Escuminac Formation, eastern Québec, Canada: A comparative study based on miospores and fishes. *Review of Palaeobotany and Palynology*, **93**, 191–215.

COLBATH, G. K. 1990. Palaeobiogeography of Middle Palaeozoic organic-walled phytoplankton. *In*: MCKERROW, W. S. & SCOTESE, C. R. (eds) *Palaeozoic Palaeogeography and Biogeography*. Geological Society, London, Memoirs, **12**, 207–213.

CUERDA, A. J. & BALDIS, B. A. 1971. Silúrico y Devónico en Argentina. *Ameghiniana*, **8**, 128–164.

DAEMON, R. F., QUADROS, L. P. & SILVA, L. C. 1967. Devonian palynology and biostratigraphy of the Paraná Basin. *In*: BIGARELLA, J. J. (ed.) *Problems in Brazilian Devonian Geology*. Boletim Paranaense Geociências, **21–22**, 99–132.

DALENZ FARJAT, A., ALVAREZ, L. A., HERNÁNDEZ, R. M. & ALBARIÑO, L. M. 2002. Cuenca Siluro-Devónica del Sur de Bolivia y del Noroeste Argentino: algunas interpretaciones. *5° Congreso de Exploración y Desarrollo de Hidrocarburos, Mar del Plata*, CD Trabajos Técnicos.

DI PASQUO, M. M. 2003. Avances sobre palinología, bioestratigrafía y correlación de las asociaciones presentes en los Grupos Macharetí y Mandiyutí,

Neopaleozoico de la Cuenca Tarija, provincia de Salta, Argentina. *Ameghiniana*, **40**, 3–32.

DI PASQUO, M. M. 2005. Resultados palinológicos preliminares de estratos del Devónico y Carbonífero en el perfil de Balapuca, sur de Bolivia. *16° Congreso Geológico Argentino, La Plata, Actas*, **4**, 293–298.

DI PASQUO, M. M. 2007*a*. Asociaciones palinológicas presentes en las Formaciones Los Monos (Devónico) e Itacua (Carbonífero Inferior) en el perfil de Balapuca, sur de Bolivia. Parte 1. Formación Los Monos. *Revista Geológica de Chile*, **34**, 98–137.

DI PASQUO, M. M. 2007*b*. Asociaciones palinológicas presentes en las Formaciones Los Monos (Devónico) e Itacua (Carbonífero Inferior) en el perfil de Balapuca, sur de Bolivia. Parte 2. Formación Itacua e interpretación estratigráfica y cronología de las formaciones Los Monos e Itacua. *Revista Geológica de Chile*, **34**, 163–198.

DI PASQUO, M. M. 2007*c*. State of the art of the Devonian palynological records in the northern Argentina, southern Bolivia and northwestern Paraguay. *In*: ACEÑOLAZA, G. F., VERGEL, M., PERALTA, S. & HERBST, R. (eds) *Field Meeting of the IGCP 499–UNESCO 'Devonian Land–Sea Interaction: Evolution of Ecosystems and Climate' (DEVEC), San Juan*, 2007, 70–73.

DI PASQUO, M. M. & AZCUY, C. L. 1997. Palinomorfos retrabajados en el Carbonífero Tardío de la Cuenca Tarija (Argentina): Su aplicación a la datación de eventos diastróficos. *Revista Universidade Guarulhos, Série Geociências*, **2** (no. especial), 28–42.

DÍAZ MARTÍNEZ, E. 1999. Estratigrafía y paleogeografia del Paleozoico Superior del norte de los Andes Centrales (Bolivia y sur del Perú). *In*: MACHARÉ, J., BENAVIDES, V. & ROSAS, S. (eds) *Volumen Jubilar No. 5 '75 Aniversario Sociedad Geológica del Perú'*. Boletin de la Sociedad Geológica del Perú, **5**, 19–26.

DÍAZ MARTÍNEZ, E., VAVRDOVÁ, M., BEK, J. & ISAACSON, P. E. 1999. Late Devonian (Famennian) glaciation in western Gondwana: evidence from the Central Andes. *Abhandlungen der Geologischen Bundesanstalt*, **54**, 213–237.

DINO, R. 1999. Palynostratigraphy of the Silurian and Devonian sequence of the Paraná Basin, Brazil. *In*: RODRIGUES, M. A. C. & PEREIRA, E. (eds) *Ordovician–Devonian Palynostratigraphy in Western Gondwana: Update, Problems and Perspectives*. Faculdade de Geologia da Universidade Estatal do Rio de Janeiro, Resumos expandidos, 27–61.

EDWARDS, D. 1990. Silurian–Devonian paleobotany: problems, progress and potential. *In*: TAYLOR, T. N. & TAYLOR, E. T. (eds) *Antarctic Paleobiology. Its Role in the Reconstruction of Gondwana*. Spinger-Verlag, New York, **8**, 90–101.

EDWARDS, D. & BENEDETTO, J. L. 1985. Two new species of herbaceous lycopods from the Devonian of Venezuela with comments on their taphonomy. *Palaeontology*, **28**, 599–618.

FASOLO, Z., VERGEL, M. M., OLLER, J. & AZCUY, C. 2006. Nuevos datos palinológicos de la Formación Kaka (Viseano – Serpukhoviano) en la Encañada de Beu, Subandino Norte de Bolivia. *Revista Brasileira de Paleontologia*, **9**, 53–62.

FONSECA, V. M. M. & MELO, J. H. G. 1987. Occôrrencia de *Tropidoleptus carinatus* (Conrad) brachiopoda (Orthida) na Formaçâo Pimenteira, e sua importância paleobiogeografica. *10° Congress Brasileiro de Paleontologia, Rio de Janeiro, Anais*, **2**, 505–537.

FRENGUELLI, J. 1952. *'Haplostigma furquei'* n. sp., del Devónico de la Precordillera de San Juan. *Revista de la Asociación Geológica Argentina*, **7**, 5–10.

FRENGUELLI, J. 1954. Plantas devónicas de la quebrada de La Charnela en la Precordillera de San Juan. *Notas Museo de La Plata (n.s.), Paleontología*, **17**(102), 359–376.

FURQUE, G. 1956. Nuevos depósitos devónicos y carbónicos en la Precordillera sanjuanina. *Revista de la Asociación Geológica Argentina*, **11**, 46–71.

FURQUE, G. 1963. Descripción geológica de la Hoja 17 b, Guandacol, provincias de La Rioja y San Juan. *Dirección Nacional de Geología y Minería (Buenos Aires), Boletín*, **92**, 1–104.

GHAVIDEL-SYOOKI, M. 1994. Upper Devonian acritarchs and miospores from the Geirud Formation in central Alborz Range, Northern Iran. *Journal of Science Iran Review*, **5**, 103–122.

GHAVIDEL-SYOOKI, M. 2003. Palynostratigraphy of Devonian sediments in the Zagros Basin, southern Iran. *Review of Palaeobotany and Palynology*, **127**, 241–268.

GRADSTEIN, F. M., OGG, J. G. & SMITH, A. G. 2004. *A Geologic Time Scale*. Cambridge University Press, Cambridge.

GRAHN, Y. 2002. Upper Silurian and Devonian chitinozoa from central and southern Bolivia, central Andes. *Journal of South American Earth Sciences*, **15**, 315–326.

GRAHN, Y. 2003. Silurian and Devonian chitinozian assemblages from the Chaco-Paraná Basin, northeastern Argentina and central Uruguay. *Revista Española de Micropaleontología*, **35**, 1–8.

GRAHN, Y. 2005. Devonian chitinozoan biozones of Western Gondwana. *Acta Geologica Polonica*, **55**, 211–227.

GRAHN, Y. & GUTIERREZ, P. R. 2001. Silurian and Middle Devonian Chitinozoa from the Zapla and Santa Bárbara Ranges, Tarija Basin, northwestern Argentina. *Ameghiniana*, **38**, 35–50.

GRAHN, Y., LOBOZIAK, S. & MELO, J. H. G. 2003. Integrated correlation of Late Silurian (Přídolí *s.l.*) – Devonian chitinozoans and miospores in the Solimões Basin, northern Brazil. *Acta Geologica Polonica*, **53**, 283–300.

GUTIERREZ, P. R. 1996. Revisión de las Licópsidas de la Argentina. 2. *Malanzania* Archangelsky, Azcuy *et* Wagner y *Haplostigma* Seward; con notas sobre *Cyclostigma* Haughton. *Ameghiniana*, **33**, 127–144.

GUTIERREZ, P. R. & ARCHANGELSKY, S. 1997. *Haplostigma baldisii* sp. nov. (Lycophyta) del Devónico de la Precordillera de San Juan, Argentina. *Ameghiniana*, **34**, 275–282.

HALLE, T. G. 1911. *On the geological structure and history of the Falkland Islands*. Geological Institute, University Upsala, Bulletin, **11**.

HASHEMI, H. & PLAYFORD, G. 1998. Upper Devonian palynomorphs of the Shishtu Formation, Central Iran

Basin, east-central Iran. *Palaeontographica*, Abt B, **246**, 115–211.

HASHEMI, H. & PLAYFORD, G. 2005. Devonian spore assemblages of the Adavale Basin, Queensland (Australia): Descriptive systematics and stratigraphic significance. *Revista Española de Micropaleontología*, **37**, 317–417.

HECKEL, P. H. & WITZKE, B. J. 1979. Devonian world palaeogeography determined from the distribution of carbonates and related lithic palaeoclimatic indicators. *In*: HOUSE, M. R., SCRUTTON, C. T. & BASSETT, M. G. (eds) *The Devonian System*. Palaeontological Association, Special Paper, **23**, 99–123.

KIMYAI, A. 1983. Palaeozoic microphytoplankton from South America. *Revista Española de Micropaleontología*, **15**, 415–426.

KRÄUSEL, R. 1960. Spongiophyton *nov. gen. (Thallophyta) e* Haplostigma *Seward (Pteridophyta) no Devoniano inferior do Paraná*. Departamento Nacional Producción Minera, División Geología y Minería, Monografías, **15**.

LE HÉRISSÉ, A. & DEUNFF, J. 1988. Acritarches et Prasinophycés (Givétien supérieur-Frasnien moyen) de Ferques (Boulonnais-France). *In*: BRICE, D. (ed.) *Le Dévonien de Ferques, Bas-Boulonnais (N. France): Biostratigraphie du Paléozoïque*, **7**, Université de Bretagne Occidentale, Brest, 104–152.

LE HÉRISSÉ, A., RUBINSTEIN, C. & STEEMANS, P. 1997. Lower Devonian Palynomorphs from the Talacasto Formation, Cerro del Fuerte Section, San Juan Precordillera, Argentina. *In*: FATKA, O. & SERVAIS, T. (eds) *Acritarcha in Praha 1996*. Acta Universitatis Carolinae Geologica, Lille, **40**, 497–515.

LE HÉRISSÉ, A., SERVAIS, T. & WICANDER, R. 2000. Devonian acritarchs and related forms. *Courier Forschungsinstitut Senckenberg*, **220**, 195–205.

LI XINGXUE, 1986. The mixed Permian Cathaysia-Gondwana flora. *Palaeobotanist*, **35**, 211–222.

LIMACHI, R., GOITIA, V. H. *ET AL.* 1996. Estratigrafía, geoquímica, correlaciones, ambientes sedimentarios y bioestratigrafía del Silúrico-Devónico de Bolivia. *12° Congreso Geológico de Bolivia, Tarija, Memorias*, **12**, 183–197.

LOBO BONETA, J. 1975. Sobre algunos palinomorfos del Devónico Superior y Carbónico Inferior de la zona subandina sur de Bolivia. *Revista Técnica Yacimientos Petrolíferos Fiscales Bolivianos* (Anales de la IV Convención Nacional de Geología, tomo 1), **4**, 159–175.

LOBOZIAK, S. 2000. Middle to early Late Devonian miospore biostratigraphy of Saudi Arabia. *In*: SA'ID AL-HAJRI, S. & OWENS, B. (eds) *Stratigraphic Palynology of the Palaeozoic of Saudi Arabia*. Gulf Petrolink, Bahrain, 134–145.

LOBOZIAK, S. & STREEL, M. 1989. Middle-Upper Devonian miospores from the Ghadamis Basin (Tunisia-Libya): Systematics and stratigraphy. *Review of Palaeobotany and Palynology*, **58**, 173–196.

LOBOZIAK, S. & STREEL, M. 1995. West Gondwanan aspects of the Middle and Upper Devonian miospore zonation in North Africa and Brazil. *Review of Palaeobotany and Palynology*, **86**, 147–155.

MCGREGOR, D. C. 1979. Devonian miospores of North America. *Palynology*, **3**, 31–52.

MCGREGOR, D. C. 1984. Late Silurian and Devonian spores from Bolivia. *Academia Nacional de Ciencias de Córdoba, Miscelánea*, **69**, 1–43.

MCGREGOR, D. C. & CAMFIELD, M. 1976. Upper Silurian? to Middle Devonian spores of the Moose River Basin, Ontario. *Geological Survey of Canada*, Bulletin **263**, 1–63.

MCGREGOR, D. C. & CAMFIELD, M. 1982. Middle Devonian miospores from the Cape de Bray, Weatherall, and Hecla Bay Formations of northeastern Melville Island, Canadian Arctic. *Geological Survey of Canada*, Bulletin **348**, 1–105.

MCGREGOR, D. C. & PLAYFORD, G. 1992. Canadian and Australian Devonian spores: Zonation and correlation. *Geological Survey of Canada*, Bulletin, **438**, 1–125.

MCLOUGHLIN, S. 2001. The breakup history of Gondwana and its impact on pre-Cenozoic floristic provincialism. *Australian Journal of Botany*, **49**, 271–300.

MCLOUGHLIN, S. & LONG, J. A. 1994. New record of Devonian plants from southern Victoria Land, Antarctica. *Geological Magazine*, **131**, 81–90.

MARSHALL, J., MILLER, M. A., FILATOFF, J. & AL-SHAHAB, K. 2007. Two new Middle Devonian megaspores from Saudi Arabia. *Revue de Micropaléontologie*, **50**, 73–79.

MELO, J. H. G. 1989. The Malvinokaffric Realm in the Devonian of Brazil. *In*: MCMILLAN, N. J., EMBRY, A. F. & GLASS, D. J. (eds) *Devonian of the World*. Canadian Society of Petroleum Geologists, Memoir, **14**, 669–703.

MELO, J. H. G. 2005. Palynostratigraphy of some Paleozoic rock units of Bolivia: additional results. *6° Congreso de Exploración y Desarrollo de Hidrocarburos, Mar del Plata*, CD Trabajos Técnicos.

MELO, J. H. G. & LOBOZIAK, S. 2003. Devonian – Early Carboniferous miospore biostratigraphy of the Amazon Basin, Northern Brazil. *Review of Palaeobotany and Palynology*, **124**, 131–202.

MENÉNDEZ, C. A. & PÖTHE DE BALDIS, E. D. 1967. Devonian spores from Paraguay. *Review of Palaeobotany and Palynology*, **1**, 161–172.

MENNING, M., ALEKSEEV, A. S. *ET AL.* 2006. Global time scale and regional stratigraphic reference scales of Central and West Europe, East Europe, Tethys, South China and North America as used in the Devonian–Carboniferous–Permian Correlation Chart 2003 (DCP 2003). *Palaeogeography, Palaeoclimatology, Palaeoecology*, **240**, 318–372.

MOISAN, P. & NIEMEYER, H. 2005. *Haplostigma* Seward in the Devonian of Northern Chile and its paleogeographical distribution in Gondwana. *Gondwana 12, Mendoza*, Abstracts, Academia, Nacional de Ciencias, Córdoba, 255.

MOLYNEUX, S. G., LE HÉRISSÉ, A. & WICANDER, R. 1996. Paleozoic phytoplankton. *In*: JANSONIUS, J. & MCGREGOR, D. C. (eds) *Palynology: Principles and Applications*. American Association of Stratigraphic Palynologists Foundation, **2**, 493–529.

MOREAU-BENOIT, A. 1979. Les spores du Dévonien de Libye, 1ère partie. *Cahiers de Micropaléontologie*, **1979** (4), 3–58.

MOREAU-BENOIT, A. 1980. Les spores du Dévonien de Libye, 2ème partie. *Cahiers de Micropaléontologie*, **1980** (1), 3–53.

MOREAU-BENOIT, A., COQUEL, R. & LATRECHE, S. 1993. Etude palynologique du Dévonien du Bassin d'Illizi (Sahara oriental algérien). Approche biostratigraphique. *Géobios*, **26**, 3–31.

OLIVEIRA, S. F. 1997. *Palinologia da sequência devoniana da Bacia do Paraná no Brasil, Paraguai e Uruguai: Implicações biocronoestratigráficas, paleoambientais e paleogeográficas*. DSc thesis, Uniersity of São Paulo, Brazil.

OTTONE, E. G. 1996. Devonian palynomorphs from the Los Monos Formation, Tarija Basin, Argentina. *Palynology*, **20**, 101–151.

OTTONE, E. G. & ROSSELLO, E. A. 1996. Palinomorfos devónicos de la Formación Tequeje, Angosto del Beu, Bolivia. *Ameghiniana*, **33**, 443–452.

PADULA, E., ROLLERI, E., MINGRAMM, A. R., CRIADO ROQUE, P., FLORES, M. A. & BALDIS, B. 1967. Devonian of Argentina. *International Symposium on the Devonian System, Proceedings*, **2**, 165–199.

PARIS, F., RICHARDSON, J. B., RIEGEL, W., STREEL, M. & VANGUESTAINE, M. 1985. Devonian (Emsian-Famennian) palynomorphs. *Journal of Micropaleontology*, **4**, 49–82.

PÉREZ LEYTON, M. 1990. *Palynomorphes du Devonien Moyen et Superieur de la Coupe de Bermejo-La Angostura (Sud-Est de la Bolivie)*. Master dissertation, Faculté des Sciences, Université de Liège, Belgium.

PÉREZ LEYTON, M. 1991. Miospores du Devonien Moyen et Superieur de la coupe de Bermejo–La Angostura (Sud-Est de la Bolivie). *Annales de la Société Géologique de Belgique*, **113**, 373–389.

PÉREZ LOINAZE, V. 2005. Some trilete spores from the Lower Carboniferous strata of the Rio Blanco Basin, western Argentina. *Ameghiniana*, **42**, 481–488.

PLAYFORD, G. 1990. Proterozoic and Paleozoic Palynology of Antarctica: a review. *In*: TAYLOR, T. N. & TAYLOR, E. T. (eds) *Antarctic Paleobiology. Its Role in the Reconstruction of Gondwana*. Spinger-Verlag, New York, **6**, 51–70.

PLAYFORD, G. & DRING, R. S. 1981. *Late Devonian acritarchs from the Carnarvon Basin, Western Australia*. Palaeontological Association, Special Papers in Palaeontology, **27**, 1–78.

PLUMSTEAD, E. P. 1967. A general review of the Devonian fossil plants found in the Cape. *Palaeontologia Africana*, **10**, 1–83.

PÖTHE DE BALDIS, E. D. 1974. El microplancton del Devónico medio de Paraguay. *Revista Española de Micropaleontología*, **6**, 367–379.

PÖTHE DE BALDIS, E. D. 1979. Acritarcos y quitinozoos del Devónico Superior de Paraguay. *Palinología*, **1**, 161–177.

QUADROS, L. P. 1988. Zoneamento bioestratigráfico do Paleozóico inferior e medio (seçâo marinha) da Bacia do Solimôes. *Boletim Geociências Petrobras*, **2**, 95–109.

QUADROS, L. P. 1999. Silurian–Devonian acritarch assemblages from Paraná Basin: An update and correlation with Northern Brazilian basins. *In*: RODRIGUES, M. A. C. & PEREIRA, E. (eds) *Ordovician-Devonian Palynostratigraphy in Western Gondwana: Update, Problems and Perspectives*. Faculdade de Geologia da Universidade Estatal do Rio de Janeiro, Resumos expandidos, 105–145.

RACHEBOEUF, P., LE HÉRISSÉ, A., PARIS, F., BABIN, C., GUILLOCHEAU, F., TRUYOLS-MASSONI, M. & SUÁREZ-SORUCO, R. 1993. Le Dévonien de Bolivie: biostratigraphie et chronostratigraphie. *Comptes Rendus de ĺAcademie de Sciences de Paris*, **317**, 795–802.

RAVN, R. L. & BENSON, D. G. 1988. Devonian miospores and reworked acritarchs from southeastern Georgia, USA. *Palynology*, **12**, 179–200.

RETALLACK, G. J. 1995. Compaction of Devonian lycopsid stems from the Beacon Heights Orthoquartzite, southern Victoria Land. *Antarctic Journal Review*, **30**, 42–44.

RICHARDSON, J. B. & MCGREGOR, D. C. 1986. Silurian and Devonian spore zones of the Old Red Sandstone continent and adjacent regions. *Geological Survey of Canada*, Bulletin, **364**, 1–79.

RODRIGUES, R., LOBOZIAK, S., MELO, J. H. G. & ALVES, D. B. 1995. Geochemical characterization and miospore biochronostratigraphy of the Frasnian anoxic event in the Parnaíba Basin, northeast Brazil. *Bulletin des Centres de Recherches de Exploration–Production Elf-Aquitaine*, **19**, 319–327.

RUBINSTEIN, C. V. 1993. Primer registro de miosporas y acritarcos del Devónico Inferior, em el Grupo Villavicencio, Precordillera de Mendoza, Argentina. *Ameghiniana*, **30**, 219–220.

RUBINSTEIN, C. V. 1999. Primer registro palinológico de la Formación Punta Negra (Devónico Medio-Superior), de la Precordillera de San Juan, Argentina. *In*: *10° Simposio Argentino de Paleobotánica y Palinología*. Asociación Paleontológica Argentina, Buenos Aires, Publicación Especial, **6**, 13–18.

RUBINSTEIN, C. V. 2000. Middle Devonian palynomorphs from the San Juan Precordillera, Argentina: biostratigraphy and paleobiogeography. *1° Congreso Ibérico de Paleontologia, 14° Jornadas de la Sociedad Española de Paleontología y 8° International Meeting of IGCP 421*, Resúmenes, 274–275.

RUBINSTEIN, C. V. & STEEMANS, P. 2007. New palynological data from the Devonian Villavicencio Formation, Precordillera of Mendoza, Argentina. *Ameghiniana*, **44**, 3–9.

RUBINSTEIN, C. V., MELO, J. H. G. & STEEMANS, P. 2005. Lochkovian (earliest Devonian) miospores from the Solimões Basin, northewestern Brazil. *Review of Palaeobotany and Palynology*, **133**, 91–113.

SCOTESE, C. R., BOUCOT, A. J. & MCKERROW, W. S. 1999. Gondwanan palaeogeography and palaeoclimatology. *Journal of African Earth Sciences*, **28**, 99–114.

SEWARD, A. C. 1932. Fossil plants from the Bokkeveld and Witteberg Series of Cape Colony. *Quarterly Journal of the Geological Society of London*, **88**, 358–369.

SEWARD, A. C. & WALTON, J. 1922. On fossil plants from the Falkland Islands. *Quarterly Journal of the Geological Society of London*, **79**, 1–313.

STARCK, D. 1996. Facies continentales en el Siluro-Devónico de la cuenca del Noroeste. Provincia de Salta, Argentina. *12° Congreso Geológico de Bolivia, Tarija, Memorias*, 231–238.

STARCK, D. 1999. Los sistemas petroleros de la Cuenca de Tarija. *4° Congreso de Exploración y Desarrollo de Hidrocarburos, Mar del Plata, Actas*, **1**, 63–82.

STEEMANS, P., BREUER, P., GERRIENNE, P., PRESTIANNI, C., STREEL, M. & DE VILLE DE GOYET, F. 2007*a*. Givetian megaspores from Libya and Belgium. *In*: PEREIRA, Z., OLIVEIRA, J. T. & WICANDER, R. (eds) *CIMP Lisbon'07*. INETI, Portugal, Abstracts, 79–81.

STEEMANS, P., WELLMAN, C. H. & FILATOFF, J. 2007*b*. Palaeophytogeographical and palaeoecological implications of a spore assemblage of earliest Devonian (Lochkovian) age from Saudi Arabia. *Palaeogeography, Palaeoclimatology, Palaeoecology*, **250**, 237–254.

STREEL, M. & LOBOZIAK, S. 1996. Middle and Upper Devonian miospores. *In*: JANSONIUS, J. & MCGREGOR, D. C. (eds) *Palynology: Principles and Applications*. American Association of Stratigraphic Palynologists Foundation, **2**, 575–587.

STREEL, M., PARIS, F., RIEGEL, W. & VANGUESTAINE, M. 1988. Acritarch, chitinozoan and spore stratigraphy from the Middle and Late Devonian of northeast Libya. *In*: EL-ARNAUTI, A., OWENS, B. & THUSU, B. (eds) *Subsurface Palynostratigraphy of Northeast Libya*. Garyounis University Publications, Benghazi, Libya, 111–128.

STREEL, M., FAIRON-DEMARET, M. & LOBOZIAK, S. 1990. Givetian–Frasnian phytogeography of Euramerica and western Gondwana based on miospore distribution. *In*: MCKERROW, W. S. & SCOTESE, C. R. (eds) *Palaeozoic Palaeogeography and Biogeography*. Geological Society, London, Memoirs, **12**, 291–296.

SUÁREZ RIGLOS, M. 1975. Algunas consideraciones biocronoestratigráficas del Silúrico-Devónico en bolivia. *1° Congreso Argentino de Paleontología y Bioestratigrafía, Tucumán*, 1974, Actas, **1**, 293–317.

SUÁREZ SORUCO, R. 1988. *Estudio bioestratigráfico del ciclo Cordillerano de Bolivia*. Academia Nacional de Ciencias de Bolivia, Cochabamba.

SUÁREZ SORUCO, R. 2000. Compendio de geología de Bolivia. *Revista Técnica Yacimientos Petrolíferos Fiscales bolivianos*, **18**, 1–213.

SUÁREZ SORUCO, R. & LOBO BONETA, J. 1983. La fase compresiva Eohercínica en el sector oriental de la Cuenca Cordillerana de Bolivia. *Revista Técnica de Yacimientos Petrolíferos Fiscales Bolivianos*, **9**, 189–202.

TURNAU, E. 1996. Miospore stratigraphy of Middle Devonian deposits from Western Pomerania. *Review of Palaeobotany and Palynology*, **93**, 107–125.

VAVRDOVÁ, M. & ISAACSON, P. E. 1999. Late Famennian phytogeographyc provincialism: Evidence for a limited separation of Gondwana and Laurentia. *Abhandlungen der Geologischen Bundesanstalt*, **54**, 453–463.

VAVRDOVÁ, M. & ISAACSON, P. E. 2000. Palynology of selected Devonian strata, Western Gondwana. *Zentralblatt für Geologie und Paläontologie*, **1**, 799–821.

VAVRDOVÁ, M., ISAACSON, P. E., DÍAZ, E. & BEK, J. 1993. Devonian – Carboniferous boundary at Lake Titikaka, Bolivia: preliminary palynological results. *12° Congrès International de la Stratigraphie et Géologie du Carbonifère et Permien, Buenos Aires, 1991, Compte Rendu*, **1**, 187–200.

VAVRDOVÁ, M., BEK, J., DUFKA, P. & ISAACSON, P. E. 1996. Palynology of the Devonian (Lochkovian to Tournaisian) sequence, Madre de Dios Basin, northern Bolivia. *Bulletin of the Czech Geological Survey (Vestník Ceského geologického ústavu)*, **71**, 333–349.

VOLKHEIMER, W., MELENDI, D. L. & SALAS, A. 1986. Devonian chitinozoans from northwestern Argentina. *Neues Jahrbuch für Geologie und Paläontologie*, **173**, 229–251.

WELLMAN, C. H. 2006. Spore assemblages from the Lower Devonian 'Lower Old Red Sandstone' deposits of the Rhynie outlier, Scotland. *Transactions of the Royal Society of Edinburgh: Earth Sciences*, **97**, 167–211.

WELLMAN, C. H. & GENSEL, P. G. 2004. Morphology and wall ultrastructure of the spores of the Lower Devonian plant *Oocampsa catheta* Andrews *et al.*, 1975. *Review of Palaeobotany and Palynology*, **130**, 269–295.

WNUK, C. 1996. The development of floristic provinciality during the Middle and Late Paleozoic. *Review of Palaeobotany and Palynology*, **90**, 5–40.

WOOD, G. D. 1984. A stratigraphic, paleoecologic and paleobiogeographic review of the acritarchs *Umbellasphaeridium deflandrei* and *Umbellasphaeridium saharicum*. *9° International Congress on Carboniferous Stratigraphy and Geology, Illinois, Compte Rendu*, **2**, 191–211.

WOOD, G. D. 1994. *Togachitina*, a new bilayered chitinozoan genus from the Devonian of the Sierras Subandinas region, Bolivia. *Palynology*, **18**, 195–204.

WOOD, G. D. 1995. The Gondwanan Acritarch *Bimerga bensonii* gen. et sp. nov.: Paleogeographic and biostratigraphic importance in the Devonian Malvinokaffric Realm. *Palynology*, **19**, 221–231.

A Silurian–Devonian marine platform-deltaic system in the San Rafael Block, Argentine Precordillera–Cuyania terrane: lithofacies and provenance

M. J. MANASSERO[1]*, C. A. CINGOLANI[1,2] & P. ABRE[2]

[1]*Centro de Investigaciones Geológicas (UNLP-CONICET), Facultad de Ciencias Naturales y Museo, Universidad Nacional de La Plata, 1-644, (1900)-La Plata, Argentina*

[2]*División Científica de Geología-Facultad Ciencias Naturales y Museo, Universidad Nacional de La Plata, Paseo del Bosque s/n, (1900)-La Plata, Argentina*

**Corresponding author (e-mail: manasser@cig.museo.unlp.edu.ar)*

Abstract: The San Rafael Block is included as a part of the pre-Andean region, in the southern sector of the Argentine Precordillera–Cuyania terrane, within the western Gondwana margin. The Río Seco de los Castaños Formation (Upper Silurian–Lower Devonian) is one of the major marine-siliciclastic pre-Carboniferous units, and is interpreted as a distal to proximal silty platform-deltaic system. The dominant sedimentary processes were wave and storm action and the source areas were located to the east, close to the study area. The rocks are mainly of immature arkosic sandstones showing both recycled orogen and continental block provenances. Sedimentological characteristics of conglomerate-filled channels and an organic-matter-rich bed are described. X-ray diffraction analyses of the clay minerals from the sequences show that very low-grade metamorphic conditions acted during the Early Carboniferous. Geochemical analyses indicate moderate to strong weathering, and potassium metasomatism. Zr/Sc ratios lower than 22, no important enrichments of Zr, Th/Sc ratios, high Sc and Cr concentration and the Eu-anomalies indicate a provenance from a less evolved upper continental crust. T_{DM} ages and ε_{Nd} are within the range of the Mesoproterozoic basement and Palaeozoic supracrustal rocks from the Precordillera–Cuyania terrane. Probable sources, tectonic setting and land–sea interactions are discussed.

The southern Pacific South America Gondwana margin (Fig. 1a) is characterized during the Palaeozoic by the presence of orogenic belts orientated approximately north–south (Ramos *et al.* 1986). They were accreted to the cratonic areas during the Cambrian (Pampean), Mid-Ordovician (Famatinian) and Late Devonian (Gondwanian) tectonic cycles (Fig. 1b). The Argentine Precordillera or Cuyania composite terrane in the sense of Ramos *et al.* (1986) is related to the Famatinian cycle and lies eastward of the present-day Andes. Four sectors constitute this composite terrane: (a) the Precordillera thin-skinned fold and thrust belt that was generated by shallow east-dipping flat-slab subduction of the Nazca plate; (b) the Pie de Palo area, (c) San Rafael and (d) Las Matras blocks. This terrane had been considered stratigraphically and faunally unique to South America mainly for the Lower Palaeozoic carbonate and siliciclastic deposits overlying an igneous-metamorphic crust of 'Grenville age' (Ramos *et al.* 1998; Sato *et al.* 2004, and references therein). The Precordillera–Cuyania terrane has been the object of several lines of research during recent years, attempting to constrain its allochthonous or para-autochthonous origin with respect to Gondwana. One of the tectonic interpretations suggests that the Precordillera–Cuyania was detached from Laurentia in Cambrian times, was transferred to western Gondwana during the Early to Middle Ordovician, and was amalgamated to the early proto-Andean margin of Gondwana by the Mid-Late Ordovician (Thomas & Astini 2003 and references therein). Other studies have claimed a para-autochthonous-to-Gondwana origin based on strike-slip displacements from the South Africa–Antarctica regions (Aceñolaza *et al.* 2002; Finney *et al.* 2003). A late Middle to Late Ordovician time of docking for the Precordillera–Cuyania is constrained by a variety of geological and palaeontological evidence (Ramos 2004 and references therein). The terrane deformation linked to the collision started in the Ordovician, and continued until the time of approach of the Chilenia terrane during the Late Devonian, against the Pacific side (Fig. 1b).

In this tectonic scenario, Silurian–Devonian siliciclastic depocentres of foreland basins were developed in the Precordillera–Cuyania terrane. One of them is preserved in the San Rafael Block, within the southern sector of this terrane (Fig. 2).

From: KÖNIGSHOF, P. (ed.) *Devonian Change: Case Studies in Palaeogeography and Palaeoecology*. The Geological Society, London, Special Publications, **314**, 215–240.
DOI: 10.1144/SP314.12 0305-8719/09/$15.00

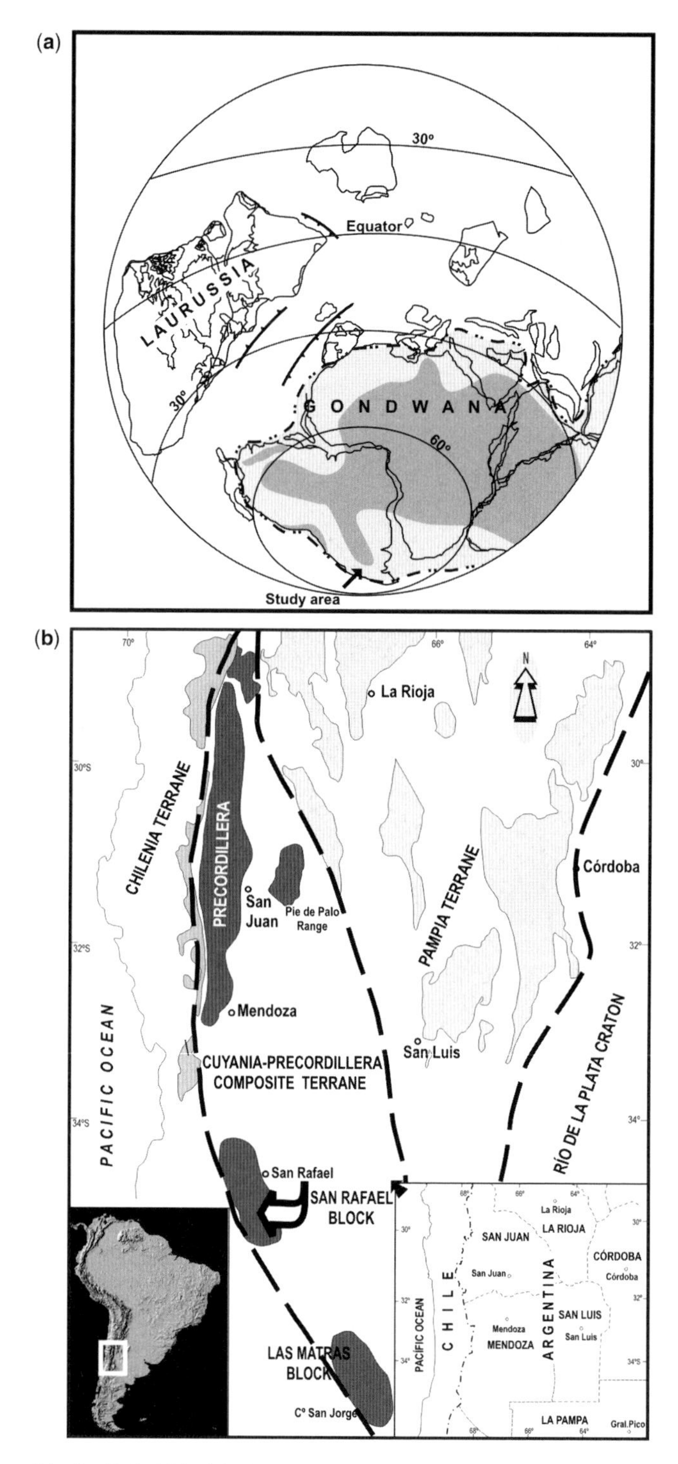

Fig. 1. (**a**) Location of the San Rafael Block in the Upper Silurian–Lower Devonian palaeogeographic reconstruction (after Torsvik & Cocks 2004). In Gondwana continent land masses are dark grey, marine transgressions are grey. (**b**) Regional location map showing the pre-Andean San Rafael Block in the Argentine Precordillera–Cuyania composite terrane and arrangement of the adjacent tectonic terranes.

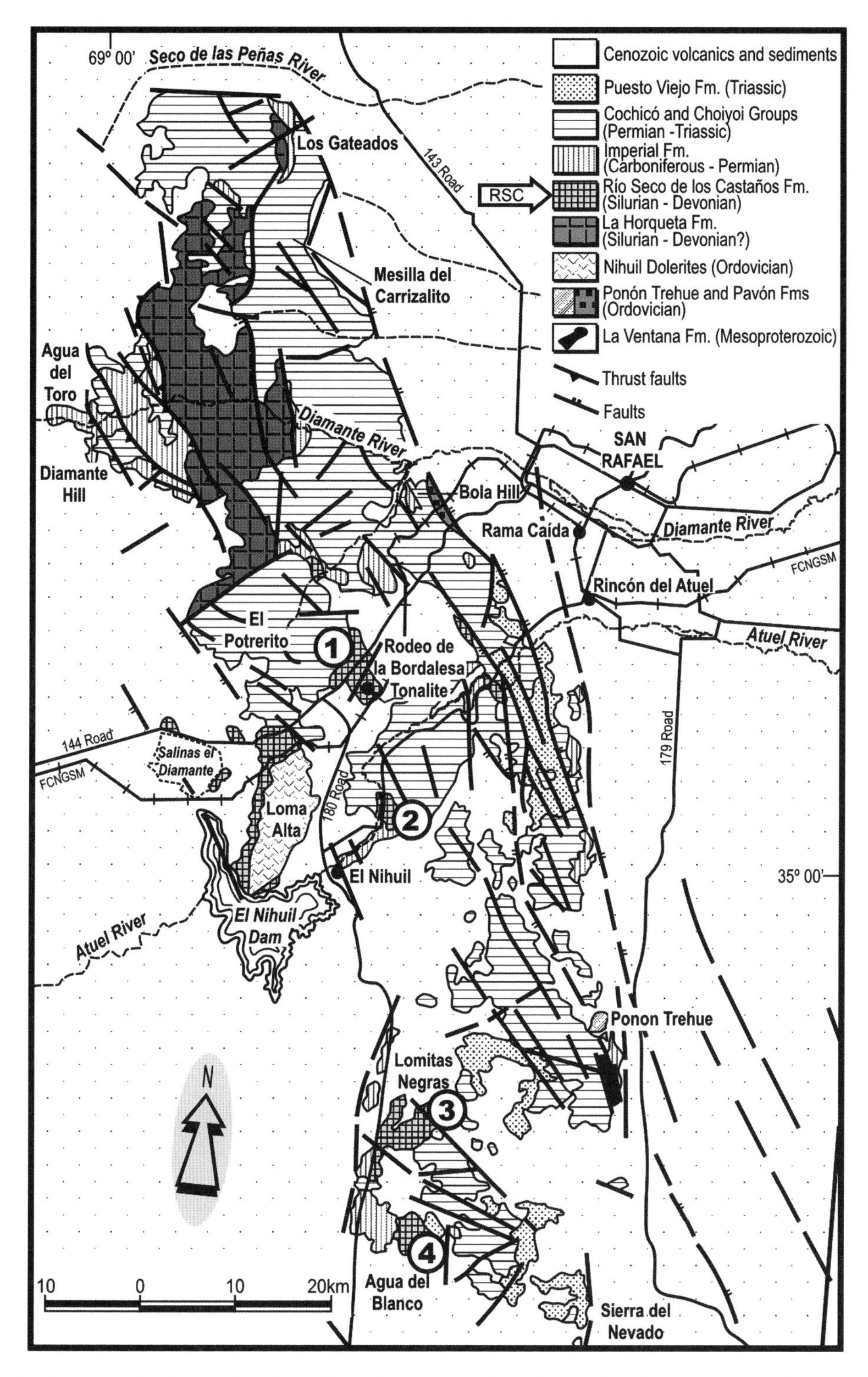

Fig. 2. Geological sketch map of the NW–SE trending San Rafael Block (simplified from Dessanti 1956; González Díaz 1972; Nuñez 1976). Main outcrops of the Río Seco de los Castaños Formation mentioned in text: 1, Road 144–Rodeo Bordalesa; 2, Atuel creek (type section); 3, Lomitas Negras; and 4, Agua del Blanco sections.

The San Rafael Block is a NW–SE trending morpho-structural entity located in the south-central part of the Mendoza Province, Argentina. It is mainly composed of isolated outcrops of 'pre-Carboniferous units' (Mesoproterozoic to Devonian in age), Upper Palaeozoic sedimentary and volcaniclastic rocks, Permian–Triassic volcaniclastic and magmatic complexes and an extended Cenozoic volcanism. In an integrated lithostratigraphic column (Fig. 3) the stratigraphic sequence, relationships, rock types and ages of the different units that composed the San Rafael Block are summarized. The Río Seco de los Castaños Formation (González Díaz 1972, 1981) is one of the above-mentioned 'pre-Carboniferous units'. Based on the stratigraphical and palaeontological evidence, the age of the unit is constrained to between the Late Silurian and Early Devonian.

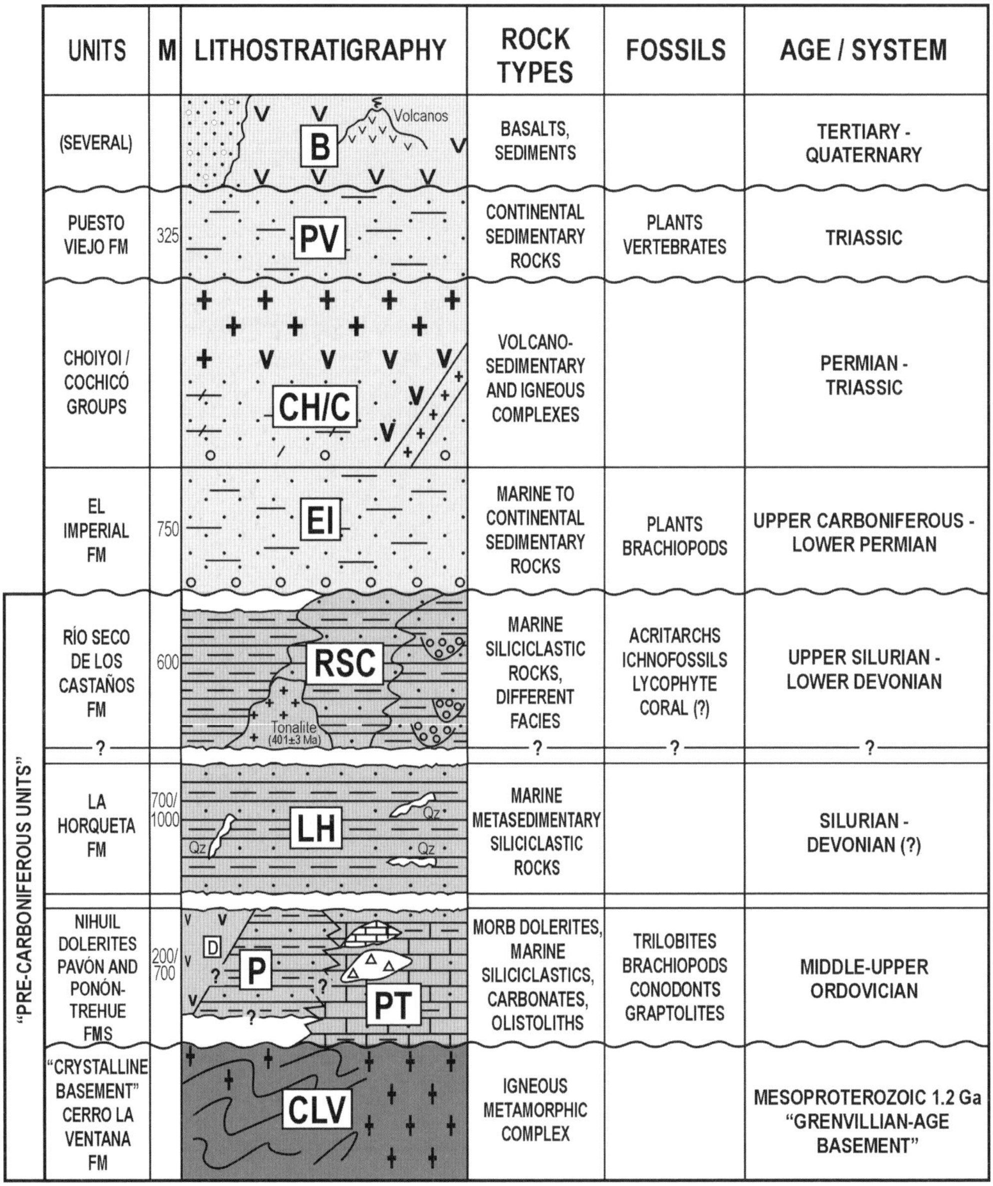

UNITS	M	LITHOSTRATIGRAPHY	ROCK TYPES	FOSSILS	AGE / SYSTEM
(SEVERAL)		B (Volcanos)	BASALTS, SEDIMENTS		TERTIARY - QUATERNARY
PUESTO VIEJO FM	325	PV	CONTINENTAL SEDIMENTARY ROCKS	PLANTS VERTEBRATES	TRIASSIC
CHOIYOI / COCHICÓ GROUPS		CH/C	VOLCANO-SEDIMENTARY AND IGNEOUS COMPLEXES		PERMIAN - TRIASSIC
EL IMPERIAL FM	750	EI	MARINE TO CONTINENTAL SEDIMENTARY ROCKS	PLANTS BRACHIOPODS	UPPER CARBONIFEROUS - LOWER PERMIAN
RÍO SECO DE LOS CASTAÑOS FM	600	RSC (Tonalite (401±3 Ma))	MARINE SILICICLASTIC ROCKS, DIFFERENT FACIES	ACRITARCHS ICHNOFOSSILS LYCOPHYTE CORAL (?)	UPPER SILURIAN - LOWER DEVONIAN
?			?	?	?
LA HORQUETA FM	700/1000	LH (Qz)	MARINE METASEDIMENTARY SILICICLASTIC ROCKS		SILURIAN - DEVONIAN (?)
NIHUIL DOLERITES PAVÓN AND PONÓN-TREHUE FMS	200/700	D, P, PT	MORB DOLERITES, MARINE SILICICLASTICS, CARBONATES, OLISTOLITHS	TRILOBITES BRACHIOPODS CONODONTS GRAPTOLITES	MIDDLE-UPPER ORDOVICIAN
"CRYSTALLINE BASEMENT" CERRO LA VENTANA FM		CLV	IGNEOUS METAMORPHIC COMPLEX		MESOPROTEROZOIC 1.2 Ga "GRENVILLIAN-AGE BASEMENT"

"PRE-CARBONIFEROUS UNITS"

Fig. 3. Integrated lithostratigraphic column-chart from the San Rafael Block. Note that the Río Seco de los Castaños Formation is the youngest of the 'pre-Carboniferous units'.

The focus of this paper is to describe and integrate the sedimentology, mineral composition, geochemistry and isotope data of the Río Seco de los Castaños unit. The combination of these different approaches can reveal the provenance and nature of the source areas and the tectonic setting of the sedimentary basin. At the same time, this study has yielded valuable insights into the crustal processes evolving land–sea interactions during Silurian–Devonian times at the San Rafael Block.

Fig. 4. (**a**) Panoramic view at the Atuel creek type section. The angular unconformity between the Río Seco de los Castaños Formation (RSC) and the Upper Palaeozoic sedimentary units (UPz) is shown. (**b**) Current and wave ripples at the top of sandstone beds (middle section). (**c**) Substratal sedimentary structures at the base of massive sandstones.

Geological setting

The Río Seco de los Castaños Formation was included in La Horqueta, a low grade metamorphic unit (Dessanti 1956), from which it was later differentiated based on its sedimentary characteristics by González Díaz (1981). The Río Seco de los Castaños Formation was assigned to the Devonian by Di Persia (1972), due to the presence of corals similar to *Pleurodyctium*. Contributions by Nuñez (1976) and Criado Roque & Ibañez (1979) described other sedimentary features of this foreland marine sequence. Poiré *et al.* (2002) recognized some trace fossil associations that helped to interpret different subenvironments of deposition within a wide siliciclastic marine platform. Neither the base nor the top of the Río Seco de los Castaños Formation are exposed. It is separated from the Carboniferous–Lower Permian El Imperial Formation (a fossiliferous marine/continental sedimentary unit) by an angular unconformity. During the Permian and Triassic, magmatic rocks and thick volcaniclastic complexes of the Cochicó-Choiyoy Groups were developed in the San Rafael Block (Fig. 3).

The main outcrops of the unit (which were dismembered by Mesozoic and Cenozoic tectonism) are rather isolated within the San Rafael Block and, as suggested by Cuerda & Cingolani (1998) and Cingolani *et al.* (2003*b*), they are located at the following sections (Fig. 2).

(a) *Road 144–Rodeo de la Bordalesa*. Outcrops where Rubinstein (1997) found acritarchs and other microfossils assigned to the Upper Silurian. Trace fossils such as the *Nereites–Mermia facies* were mentioned by some authors (Criado Roque & Ibañez 1979; Poiré *et al.* 2002). The tonalite body intruded into the Río Seco de los Castaños Formation, shows U–Pb (in zircons) and K–Ar (in biotite) crystallization ages of 401 ± 3 Ma (Lower Devonian; Cingolani *et al.* 2003*a*), which also constrain the sedimentation age of the Río Seco de los Castaños Formation.

(b) *Atuel Creek* (Figs 4 and 5). The type-section of the sequence (González Díaz 1981) is located in this creek. The two main outcrops are present about 12 km NE of El Nihuil town and near the Valle Grande dam. The Río Seco de los Castaños Formation comprises here about 600 m of tabular, green sandstones and mudstones with sharp contacts. This unit has regional folding and dips 50–72° to the SE or NE. Upper Palaeozoic horizontally bedded sedimentary rocks are found above the Río Seco de los Castaños Formation, displaying a notable angular unconformity with the Río Seco de los Castaños Formation (Fig. 4a). In the Atuel Creek area fragments

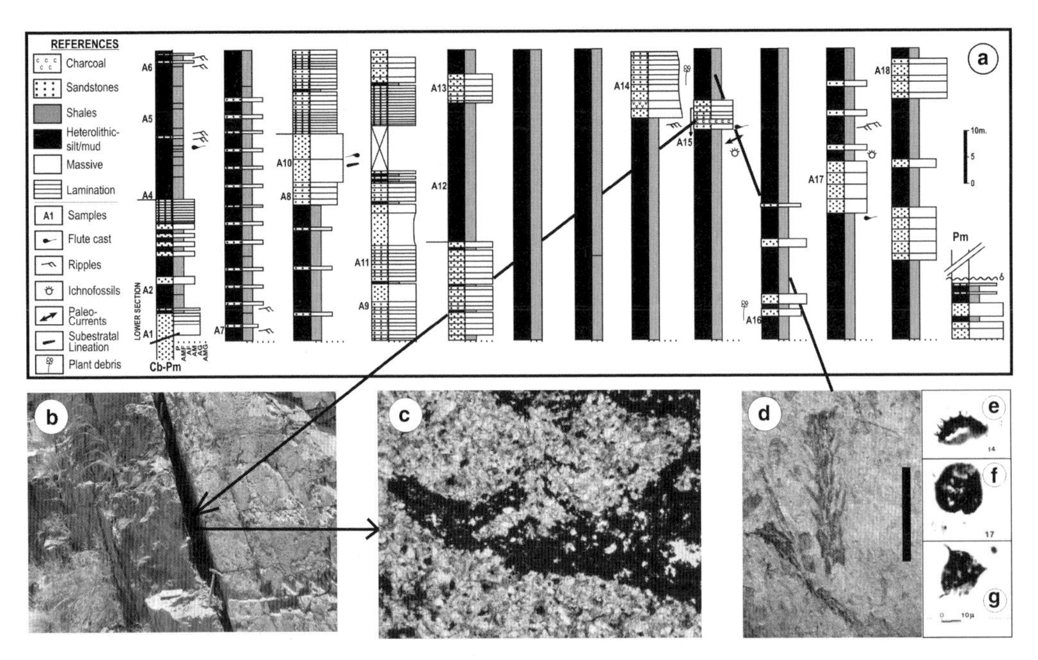

Fig. 5. (**a**) Sedimentary log of the Atuel creek section, with details of lithology and structures. (**b**) Charcoal bed field photograph and (**c**) photomicrograph detail showing the silty-quartz material (light colour) and the relicts of organic matter in black. (**d**) Detail of the fossil plants *Lycophytes* sp. (after Morel *et al.* 2006) black bar = 1 cm. (**e**) Microphotograph of spores and acritarchs (after Pöthe de Baldis, unpubl. data); (**f**) *Ammonidium* sp.; (**g**) *Lophosphaeridium* sp.

of fossil plants such as *Lycophytes* (Fig. 5d) are described and assigned to the Lower Devonian (Morel *et al.* 2006). Marine microfossils such as prasinophytes, spores and acritarchs were found by Pöthe de Baldis unpubl. data indicating shallow water conditions near the coastline and suggesting an Upper Silurian age (Fig. 5e, f and g).

(c) *Nihuil area*. The sequence assigned to the Río Seco de los Castaños Formation is overlying the Ordovician MORB-type dolerite rocks called 'El Nihuil mafic body' at the Loma Alta region (Cingolani *et al.* 2003*b*).

(d) *Lomitas Negras* (Fig. 6) *and Agua del Blanco* (Fig. 7) *areas*. The studied unit includes here the southernmost outcrops, where Di Persia (1972) mentioned a coral of Devonian age and conglomerates with limestone clasts bearing Ordovician fossils.

Methods

Sedimentology

During fieldwork, measurements and descriptions were made of several sedimentological sections and detailed sampling of all the lithological types was undertaken (Manassero *et al.* 2005). Thin sections were examined using optical microscopy to determine the textural and optical properties of minerals as well as paragenetic associations. Selected samples were tinted with alizarine red to determine feldspars under the optical microscope. Scanning electron microscopy (SEM) was used to examine textural features, mineral morphology and the diagenetic sequence of formation. The samples were collected along Road 144-Rodeo de la Bordalesa, Atuel creek and Lomitas Negras-Agua del Blanco sections (Fig. 2), and they will be referred to in separate sections. A total of 25 thin sections of medium-grained sandstones were studied under the microscope and quantitatively analysed with a Swift-type point counter. Four-hundred points were counted using the traditional method of Zuffa (1984), in which grains of plutonic rock fragments are counted as such rather than as mineral components. The results were plotted in Q–F–L ternary diagrams (Dickinson & Suczek 1979; Dickinson *et al.* 1983). The populations represented in each triangle include detrital grains, with the exception of micas, opaque minerals, chlorite, heavy minerals and carbonate grains. Chert was counted as a sedimentary rock fragment.

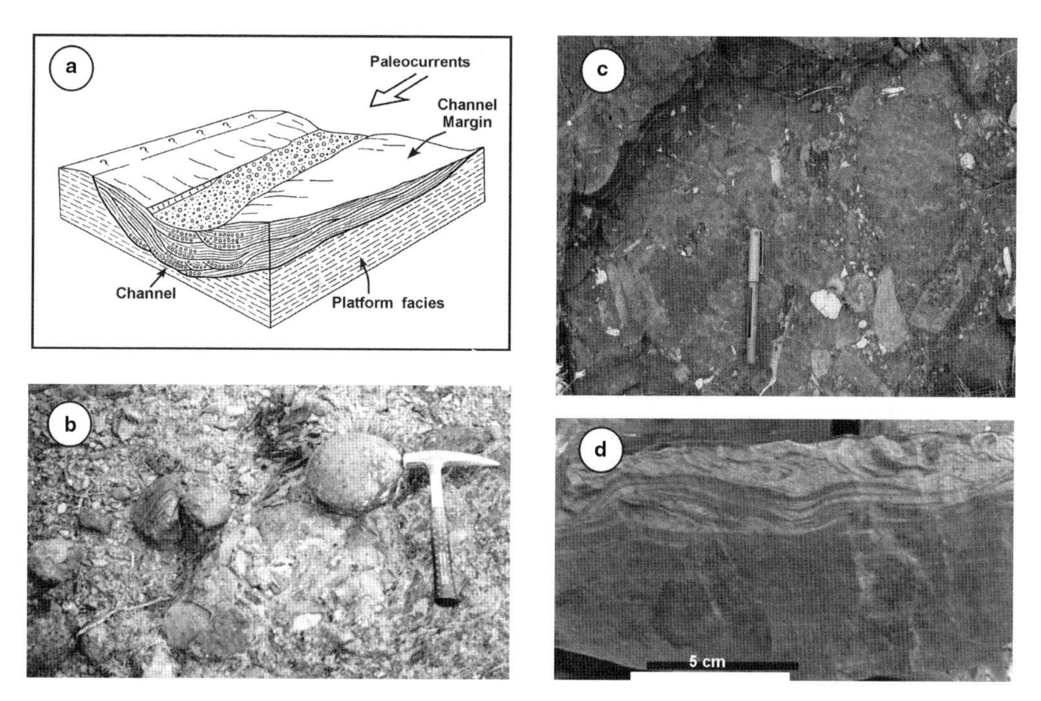

Fig. 6. Outcrop at Lomitas Negras. (**a**) Block diagram of the conglomerate channels. (**b**) Detail of rounded clasts of the conglomerates. (**c**) Detail of subangular clasts in the conglomerates. (**d**) Syndepositional sedimentary structures showing rapid deposition in slope areas due to high water saturation.

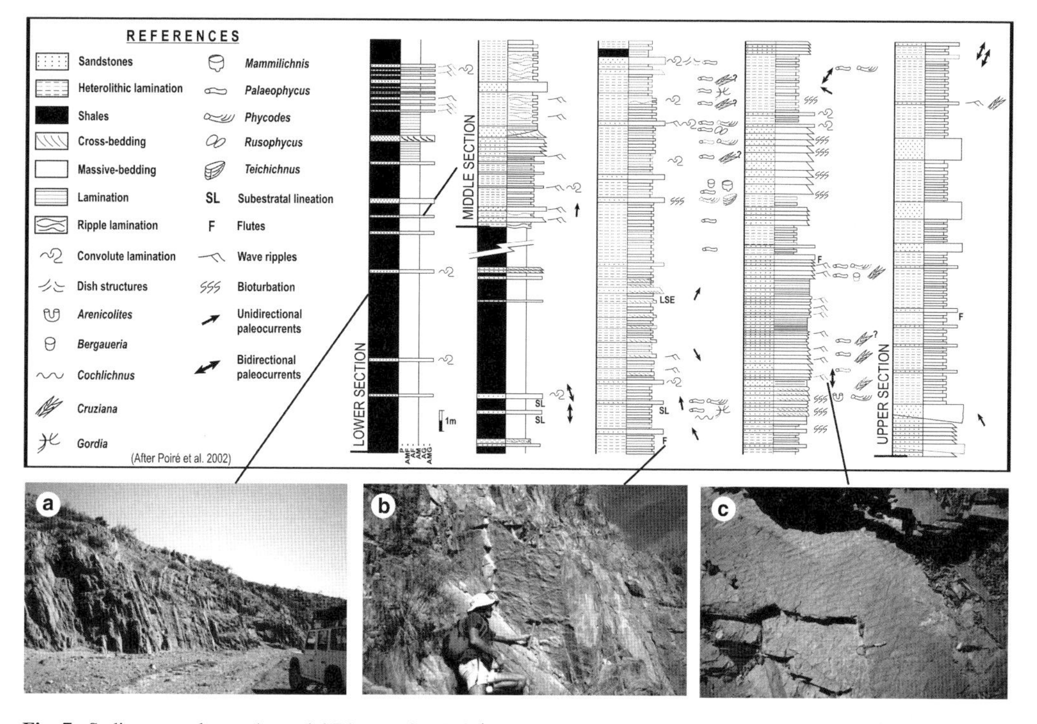

Fig. 7. Sedimentary log at Agua del Blanco after Poiré *et al.* (2002) showing lithology, grain size, structures, trace fossils and palaeocurrents. (**a**) General view of outcrops at the base of the section with dominant heterogeneous lithofacies. (**b**) Detail of substratal structure in beds bearing trace fossils. (**c**) Wave ripples at the top of sandy beds.

Diagenesis–metamorphism studies

Diagenesis consists of a dynamic suite of processes linked to the burial history of the sedimentary basin, and the conditions which favoured different diagenetic to very low-grade metamorphic reactions are recorded in both the diagenetic fabric and mineralogy of the resulting rocks. However, the use of diagenetic features to decipher the burial history of ancient basins implies the capacity to distinguish the products of the diagenetic reactions, which characterized the different regimes at various points in the basin's history. The objective of this study was to determine the degree of diagenesis or very low-grade metamorphism overprinted in the Río Seco de los Castaños Formation. The less than 2 μm fraction of 12 samples was analysed using standardized X-ray diffraction (XRD) methods. Sieving and settling velocity techniques were performed for grain size analysis following cement removal (Moore & Reynolds 1989). Sample preparation was done according to Kisch (1980, 1991). Organic matter was eliminated with H_2O_2; carbonates were removed with acetic acid. The XRD analyses were sequentially conducted on air-dried samples that were exposed to ethylene-glycol vapours for 24 hours, and heated to 550 °C. Semi-quantitative estimates of relative concentrations of clay minerals were based on the peak area method. Percentage evaluation was based on peak height and area, corrected by factors depending on the crystallinity of the mineral. Analyses were carried out at the Centro de Investigaciones Geológicas (La Plata, Argentina), using a Philips PW 2233/20 diffractometer, set at 36 kV and 18 mA, CuKα radiation, Ni-filter, wavelength 1.54 Å (vertical goniometer). Samples were studied in the range from 2° to 32° 2θ at a 2° 2θ/min scanning velocity and with a time constant of 1 second. At the same time, international standards were analysed under the same procedures already explained in order to determine the illite crystallinity index. The thickness of sample material on each glass slide was controlled by weighting 0.058 grams of sample prior to its deposition on the glass slide. The illite crystallinity values were determined by measuring the full-width at the high-medium of the peak (001) on the air-dried and ethylene-glycol treated diffractograms, using Winfit software (Krumm 1994; Warr & Rice 1994). These illite crystallinity values were standardized according to the regression curve obtained with the standards ($y = 1.3877x - 0.1959$, $R^2 = 0.9316$, where R^2 is the correlation coefficient).

Geochemistry

Whole-rock geochemistry of sedimentary rocks reflects the average composition of the crust that shed detritus into a certain basin (Taylor & McLennan 1985). Therefore, the characteristics and location of the source area(s) can be recognized in the ultimate composition of the sedimentary succession (McLennan & Taylor 1991). However, weathering, hydraulic sorting and diagenesis acting from initial erosion of a source rock(s) to the final burial of its detritus may modify the signatures of the source rock(s), and therefore these factors need to be evaluated in order to constrain the provenance of a sedimentary sequence (Nesbitt & Young 1982; Nesbitt *et al.* 1996). Fourteen pulp samples were prepared and analysed at ACME Labs, Canada. Major elements were obtained by inductively coupled plasma element spectroscopy (ICP-ES) on fusion beads (using $LiBO_2/Li_2B_4O_7$). The loss on ignition (LOI) was calculated by weight after ignition at 1000 °C. Mo, Cu, Pb, Zn, Ni, As, Cd, Sb, Bi, Ag, Au, Hg, Tl and Se were analysed by inductively coupled plasma mass spectroscopy (ICP-MS) after leaching each sample with 3 ml 2:2:2 $HCl–HNO_3–H_2O$ at 95 °C for one hour and later diluted to 10 ml. Rare earth elements (REE) and certain trace elements (Ba, Be, Co, Cs, Ga, Hf, Nb, Rb, Sn, Sr, Ta, Th, U, V, W, Zr, Y, La, Ce, Pr, Nd, Sm, Eu, Gd, Tb, Dy, Ho, Er, Tm, Yb, Lu) were analysed by ICP-MS following lithium metaborate/tetraborate fusion and nitric acid digestion.

Nd isotopes

Nd isotopes have been widely used as provenance indicators (e.g. McLennan *et al.* 1990; McLennan 1993). Nd isotopic signatures of terrigenous sedimentary rocks provide an average of the various sources from which the sediments were derived (McLennan 1989). Since the Sm/Nd ratio is modified during processes of mantle–crust differentiation it is possible to estimate the time at which the initial magma was separated from the upper mantle, also called the depleted mantle model age or T_{DM} (DePaolo 1981). When studying sedimentary rocks, the model age should be interpreted as the model ages of those rocks that have contributed in a higher degree to the Sm–Nd relationship of that sediment. The ε_{Nd} (t) indicates the deviation of the $^{143}Nd/^{144}Nd$ value of the sample from that of the standard CHUR (chondritic uniform reservoir; DePaolo 1981). To perform Sm–Nd analyses whole-rock samples were digested in acids (HF/HNO_3) after the addition of a combined spike of $^{149}Sm/^{150}Nd$. The Sm and Nd were separated in two stages of cation exchange columns; the first step used an AG-50x-X8 resin whereas the second step used teflon columns with a HDEHPLN-B50 anion resin. Samples were dried, dissolved in H_3PO_4 0.25 N and placed on simple (for samarium) or triple (for neodymium) Ta-Re filaments in order to quantify the elements using a thermal ionization mass spectrometer (TIMS). Sm–Nd analyses were performed using static mode on a VG sector 54 multicollector TIMS at the Laboratorio de Geología Isotópica da Universidade Federal do Rio Grande do Sul (LGI-CPGq/UFRGS), Porto Alegre, Brazil. Nd ratios were normalized to $^{146}Nd/^{144}Nd = 0.72190$ and calculated assuming $^{143}Nd/^{144}Nd_{(0)} = 0.512638$. Measurements for the La Jolla standard gave $^{143}Nd/^{144}Nd = 0.511859 \pm 0.000010$.

Results

Lithofacies analysis

The main components of shallow marine fine-grained siliciclastics are sandstones and mudstones. The conglomerates, in this case, are restricted to channel fills in certain areas of the basin. Five main lithotypes have been recognized in this platform (Table 1). In the following paragraphs, they will be described and compared with other facies schemes in order to attribute them to one or more processes of deposition (Aguirrezabala & García Mondéjar 1994; Martino & Curran 1990; Miller & Heller 1994; Melvin 1986).

Mudstones. These rocks constitute 50% to 90% of thin beds, and show greenish colours (HUE 5GY 3/2) usually with lamination to slight bioturbation commonly in repetitive sequences. Dark to light tonality changes are frequent but they are not related to textural grading. Some mudstones of this lithofacies are massive. The fine-grained sediments are the product of suspension and fallout from low-density turbidity currents (Stow & Piper 1984) deposited in low-energy conditions. The lack of tractive structures implies transport of bed load in distal areas of the platform. The dark tonality and the scarcity of organic activity suggest anoxic conditions in low-energy environments.

Heterolithics. This is a very common facies which is characterized by alternating beds of fine to very fine grey sandstones and laminated mudstones (Figs. 5 and 7). This sedimentary association comprises thin-bedded sandstones and intercalated green (HUE 5GY 5/2) mudstones, with good lateral continuity and tabular-planar beds a few centimetres thick. The sandstone/mudstone ratio is in the range of 1:2 to 1:4. The sandstones are massive but show wavy bedding structures in some cases. They exhibit sharp contacts and in many cases wave and current ripple structures (sharp rippled tops), as well as climbing ripples (Collinson & Thompson 1989). The current wave index is in the range of 13–16 (the wave ripples have an index of 3–4 and also the symmetry index is smaller than 2.2). The dominant internal structures are normal grading, and bioturbation. As shown in Figure 7, Poiré *et al.* (2002) have recognized *Arenicolites*, *Bergaueria*, *Cochlichnus*, *Cruziana*, *Gordia*, *Mammlichnis*, *Palaeophycus*, *Phycodes*, *Rusophycus* and *Teichichnus*. This facies represents a well-oxygenated environment and it is interpreted as a proximal or shallow marine platform, with dominance of a subtidal environment. The trace fossils are developed in soft substrates of moderate energy environment. The coarse sediments were carried into below-wave-base areas by storms. The finer sediments were deposited periodically due to diminished wave and storm actions.

Laminated siltstones. These rocks comprise bedded siltstones that range in thickness from several tens of centimetres to 1 m (Figs 5 and 7). They are intercalated with fine-grained sandstones with sharp contacts. Some coarser-grained beds show small-scale ripple cross-lamination. This facies represents fine-grained sediments deposited out of suspension in low-energy environments. They may also be associated to low-density gravity flows.

Sandstones. These rocks are medium-bedded grey and green sandstones (HUE 5GY 5/2). Grain sizes range from fine to medium sand. Sandstone beds are 10 to 150 cm thick, and sometimes separated by very thin (1–2 cm) dark grey mudstones. Trace fossils, such as *Cruziana*, are also found here. The sandstones are massive and have sharp contacts. Tops of beds show current and wave ripple marks (wave index 12–20) suggesting seawater depths of *c.* 20 m (Komar 1974; González Bonorino 1986). Deformational structures such as contorted beds, dish structures and scarce flute marks are present. Sedimentary structures such as hummocks and swales can be found in this facies (Walker *et al.* 1983; Cheel & Lecki 1993) suggesting rapid deposition and storm action on the platform. The presence of plant debris indicates that the continental source area was not far away (Fig. 5a, d). The massive beds are interpreted as deposited under wave and storm actions in a proximal platform, although they may also be a product of high-density gravity flows (Moulder & Alexander 2001). The erosive bases of some beds imply a high sedimentation rate. Cross-stratification, indicative of tractive currents, is very scarce within this facies. On the other hand the dominance of thin beds with fine sediments suggests the action of low-density gravity flows in the platform.

Within the last described facies a charcoal bed that might be a marker horizon was also found (Fig. 5a, b and c). It is composed of a mixture of silty-quartz, illite-kaolinite clays and amorphous

Table 1. *Schematic lithofacies description of the Río Seco de los Castaños Formation*

Facies	Contacts	Structures	Fossils/trace fossils	Stratomorphology	Interpretation
Mudstones	Sharp	Lamination-Massive	Acritarchs	Tabular planar lamination	Proximal to distal platform with wave and storm action–high density turbidite currents
Heterolithics	Sharp	Lamination Wave ripples Normal gradation	*Arenicolites Cruziana Palaeoplycus Rusophcus Rusophycus Telchichnus Gordia Phycodes*	Tabular planar	Shallow deposits in subtidal environments of low energy
Sandstones	Sharp	Massive Wave ripples Flute marks Convolute lamination Hummocks and swaleys	*Palaeophycus* plant debris charcoal bed acritarchs	Tabular planar	Suspension and fall out from low density turbidity currents in distal platform
Conglomerates	Erosive Base	Poor imbrication Two modal grain sizes	Limestone clasts with Ordovician fossils	Channels	Channels perpendicular to the coast with continuity along low angle platform
Laminated Siltstones	Sharp	Bedding	–	Tabular planar	Low energy environments

organic matter with total organic carbon of 1%. This bed is restricted to the Atuel creek section, where it is associated with beds bearing small plant debris (Morel *et al.* 2006). Transgressive sequences have been documented widely in the sedimentary record (Collinson 1968, 1978; Reading 1996). Wave-dominated deltas have facies sequences that coarsen upwards from shelf mud through silty-sand to wave- and storm-influenced sands, capped with lagoon or strand-plains where these peat beds could developed. This seems to be the case for the Atuel section (Fig. 4) where several prograding sequences with evidence for intense wave action have been described (Fig. 4b).

Conglomerates. This facies is usually restricted to channels (2 to 3 m wide and 1 m deep), which are filled with both clast- and matrix-supported conglomerates with erosive bases (Fig. 6a). They are present only in the Lomitas Negras section. The beds are usually lenticular and laterally discontinuous (Fig. 6b and c). They are poorly sorted and contain medium- to coarse-grained sandy matrix. Clasts can be rounded (Fig. 6b) or subangular (Fig. 6c) and they range from 2 to 10 cm long and show chaotic disposition without stratification. The clasts are mainly composed of mudstones, marls, limestones, siltstones, phyllites, quartz and feldspars. Some limestone clasts bear Ordovician fossils (Nuñez 1976; Criado Roque & Ibañez 1979). In some cases, the channels show normal grading, resulting from rapid settling of gravel in high-density gravity flows (Camacho *et al.* 2002) as shown in Figure 6d. At the base of each sequence, the channels tend to be more restricted and the conglomerates are better sorted. Channels are up to several metres wide at their tops and 2 or 3 m deep. The conglomerates tend to have a subvertical position due to the regional folding of the sequence. As they are harder than the associated fine- to medium-grained sedimentary rocks, they form small hills, which is probably the main reason for the name 'Lomitas' (small hills). This facies is interpreted as channels developed perpendicular to the coast. They not only transported a coarse bed load composed of allochthonous materials (removed from the coast by wave and storm action), but by-passed them to the west into deeper sectors of the platform. The thickness of sandstones and mudstones associated with this facies suggests a high-energy environment, combined with relative instability of the coastline and close continental source areas.

Sandstone petrography.

Petrographical analyses of sedimentary rocks have proved useful for determining provenance (Dickinson *et al.* 1983), but the resolution can be enhanced by the addition of geochemistry of minor and trace elements, considering that only the more stable minerals are preserved through weathering and diagenesis. Critical analysis of the results helps to reach new conclusions about the provenance variations. The minerals recorded by point counting are quartz (monocrystalline, polycrystalline and metamorphic), K-feldspar (microcline), plagioclase, opaque minerals, hematite and sedimentary or metamorphic rock fragments. The presence of detrital biotite and scarce muscovite suggests short transport and minimal reworking of sediments. Most of the medium-grained sandstones (2 to 1.5Φ) are wackes (more than 15% matrix) and are composed of subangular quartz, with normal and wavy extinction, feldspars and fragments of polycrystalline quartz (Fig. 8). Studies using scanning electronic microscopy reveal the presence of abundant mica flakes within the framework of these sandstones (Fig. 9).

The samples from the Lomitas Negras section show higher proportions of polycrystalline quartz. The rocks are classified as feldsphathic-wackes and quartz-wackes following Dott (1964). In the Q–F–L diagrams (Figs 10 and 11), the sandstones of the Río Seco de los Castaños Formation cluster in both the recycled orogen and continental block fields. It is important to underline that the feldspars and biotite are widely altered to chlorite, giving the typical greenish colours to the rocks. In the Lomitas Negras section the abundance of polycrystalline quartz places the samples in the recycled field. Although the data show some dispersion, we could assume an uplifted igneous–metamorphic basement or recycled orogen as source areas. The facies distribution and the palaeocurrent data suggest the first option, especially when considering that outcrops of basement rocks of Mesoproterozoic age (the Cerro La Ventana Formation; Cingolani & Varela

Fig. 8. Photomicrograph of medium-grained sand and angular quartz feldspathic-wackes (crossed nicols). Note the high matrix content and subangular character of framework minerals (low textural maturity). The sample was taken from the section at Atuel creek.

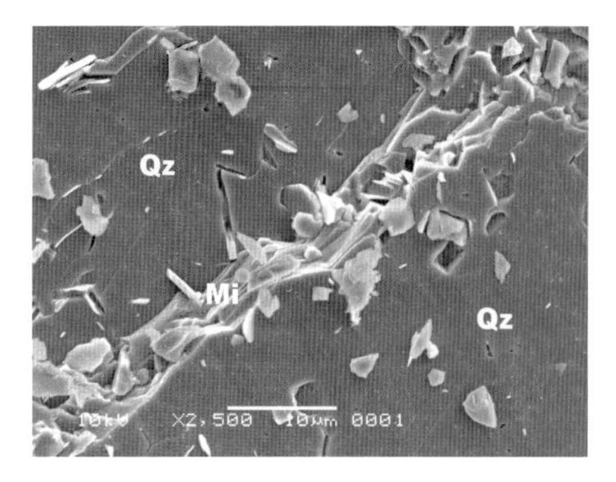

Fig. 9. Detail of Figure 8 using scanning electronic microscope, showing mica (Mi) with deformed cleavage between two quartz (Qz) crystals.

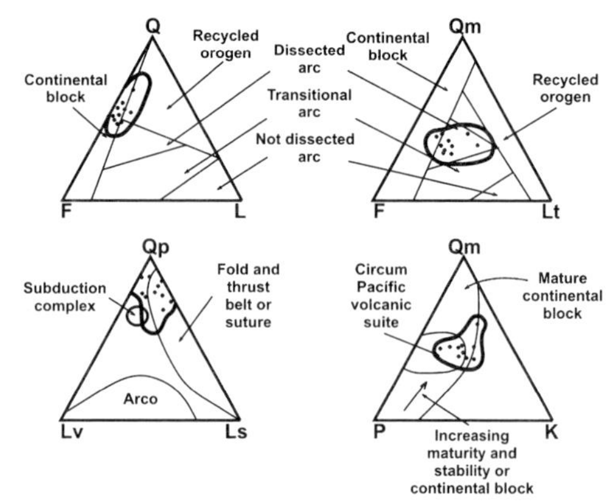

Fig. 11. Provenance ternary diagrams after Dickinson & Suczek (1979) and Dickinson *et al.* (1983) from sandstones at the Lomitas Negras section. Abbreviations: F, feldspars; K, K-feldspars, L, lithoclasts; Ls, sedimentary lithoclasts; Lt, total lithoclasts (including polycrystalline quartz); Lv, volcanic lithoclasts; P, plagioclases; Q, quartz (including polycrystalline quartz); Qm, monocrystalline quartz; Qp, polycrystalline quartz.

1999; Cingolani *et al.* 2005) are present to the east of the study area.

Clay composition and diagenesis

XRD analyses show that the clay mineral fraction (Fig. 12) is dominated by illite (range from 40% to 60%), followed by kaolinite (ranging from 25% to 40%), and chlorite which generally ranges from 10% to 20%, although it can rise to 35% when interlayered with smectite. Muscovite and interstratified chlorite/smectite are very scarce. Apart from the clay minerals, the fine fraction (less than 2 µm) contains very small amounts of quartz and plagioclases (see also petrographical description; chlorite may result as an alteration product of biotite). The illite crystallinity index (Fig. 13) obtained for the Río Seco de los Castaños Formation indicates that all samples belong to the low anchizone, thus confirming that the unit suffered only a very low-grade metamorphism (Criado Roque & Ibañez 1979;

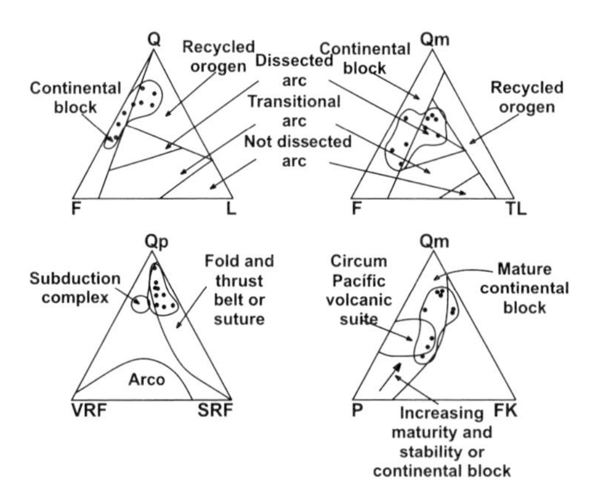

Fig. 10. Provenance ternary diagrams after Dickinson & Suczek (1979) and Dickinson *et al.* (1983) from sandstones from the Atuel creek section. Abbreviations: F, feldspars; FK, K-feldspars, L, lithoclasts; P, plagioclases; Q, quartz (including polycrystalline quartz); Qm, monocrystalline quartz; Qp, polycrystalline quartz; SRF, Sedimentary rock fragments; TL, total lithics; VRF, volcanic rock fragments.

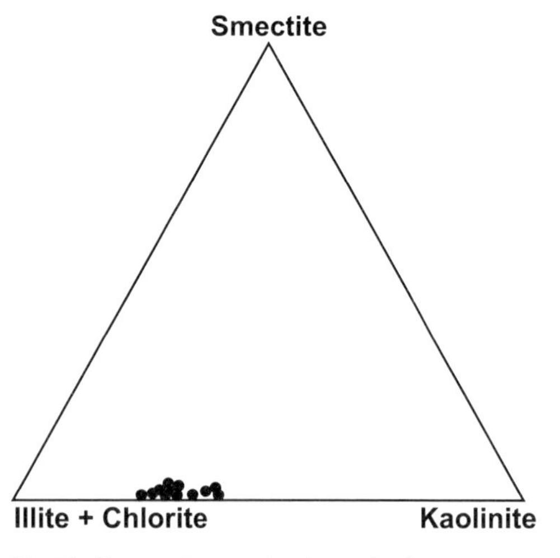

Fig. 12. Ternary diagram showing main clay composition of intercalation of claystones in the study sections. Illite and chlorite are dominant.

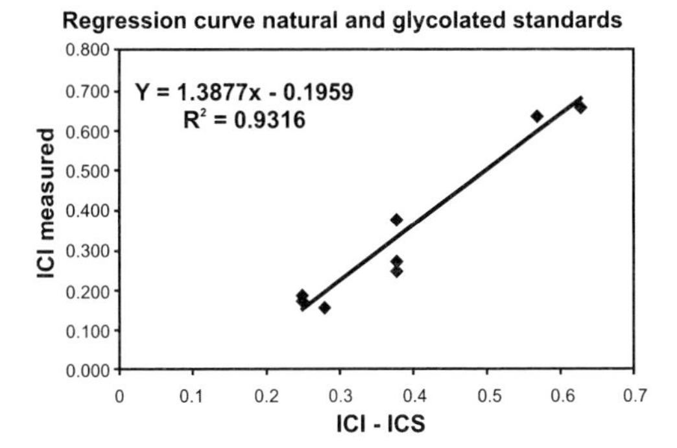

Fig. 13. Relationship between the illite crystallinity index (ICI) values of the standards as obtained at the laboratory (ICI measured) and the recommended ICI values of the standards (ICI-ICS). The regression curve obtained was used to recalculate the ICI values of all the samples in order to standardize the results and make them comparable to the worldwide established values used to separate diagenesis from very low-grade metamorphism.

González Díaz 1972). Recently, Cingolani & Varela (2008) presented the Rb–Sr isochronic whole-rock data on fine-grained samples (from the Río Seco de los Castaños Formation). The obtained age was 336 $\pm$ 23 Ma with an initial $^{87}Sr/^{86}Sr$ ratio of 0.7342 and MSWD (Mean Square Weighted Deviation) of 7.4.

Lithogeochemistry

Major elements. All the samples analysed from the Río Seco de los Castaños Formation are claystones, except for one siltstone (99S4) and one sandstone (HOR28). The differences in grain sizes explain the major element distribution (Table 2), which shows SiO_2 ranging from 52.9% to 59.9%, Al_2O_3 ranging from 18.57% to 21.4%, Fe_2O_3 concentrations between 7.68% and 9.65% and K_2O from 3.63% to 4.57% for the claystones. On the other hand, coarser fractions show SiO_2 concentrations between 70.44% and 76.82%, Al_2O_3 concentrations from 10.28% to 11.5%, Fe_2O_3 from 4.74% to 6.81%, while K_2O is between 1.41% and 1.72%. MgO and MnO concentrations as well as the LOI are lower for coarser grain-size samples. The effects of weathering on sedimentary rocks can be quantitatively assessed using the chemical index of alteration (CIA; Nesbitt & Young 1982). This index uses molecular proportions as follows: $CIA = \{Al_2O_3/(Al_2O_3 + CaO^* + Na_2O + K_2O)\} \times 100$. CaO* refers to the calcium associated with silicate minerals. Therefore, corrections for the measured CaO concentration regarding the presence of Ca in carbonates (calcite and dolomite) and phosphates (apatite) are needed. For this study, CaO was corrected for phosphate assuming that the P_2O_5 is entirely present in apatite. CIA values for the Río Seco de los Castaños Formation are between 61.1 and 78.7 (Table 2). In the A–CN–K diagram

Table 2. *Major elements (expressed in %) of the Río Seco de los Castaños Formation*

Sample	SiO_2	TiO_2	Al_2O_3	Fe_2O_3	MnO	MgO	CaO	Na_2O	K_2O	P_2O_5	LOI	SUM	CIA
Lomitas Negras and Agua del Blanco sections													
05AB1	58.47	1.03	18.62	7.68	0.08	2.92	0.48	1.49	4.00	0.18	4.80	99.75	72.05
05AB3	57.98	0.99	18.74	7.80	0.08	3.00	0.63	1.41	4.24	0.18	4.70	99.75	71.08
05AB5	55.72	1.01	19.66	8.17	0.09	3.07	1.61	1.65	4.17	0.19	4.40	99.74	66.96
05AB7	56.83	0.96	19.03	7.88	0.09	3.24	0.99	1.58	4.37	0.18	4.70	99.85	68.63
05LN2	56.65	1.01	19.88	7.97	0.08	3.17	0.46	1.57	4.34	0.19	4.40	99.72	72.18
05LN12	54.91	1.00	19.00	8.62	0.11	3.47	1.72	2.15	3.98	0.23	4.70	99.89	64.58
05LN17	56.34	0.95	19.10	8.30	0.08	3.16	0.88	1.43	4.41	0.23	5.00	99.88	70.03
05LN18	54.47	1.01	18.59	8.49	0.11	3.63	2.48	2.22	3.86	0.21	4.80	99.87	61.10
05LN22	55.45	0.96	18.57	8.88	0.10	3.83	1.21	2.16	3.72	0.23	4.80	99.89	66.79
Atuel Creek section													
05CA1	54.29	1.11	20.71	9.65	0.07	3.05	0.30	0.55	4.49	0.21	5.30	99.73	78.10
VG-2	52.94	1.14	21.41	9.17	0.07	3.14	0.25	0.88	4.57	0.14	5.00	98.71	76.67
Road 144-Rodeo de la Bordalesa section													
HOR22	59.93	1.14	19.17	8.59	0.07	1.92	0.28	0.71	3.63	0.18	3.96	99.58	78.74
HOR28	76.82	0.64	10.28	4.75	0.04	1.28	0.57	1.60	1.72	0.18	2.61	100.49	66.84
99S4	70.44	1.35	11.55	6.81	0.06	2.65	1.09	1.93	1.41	0.22	2.03	99.53	65.23
average	58.66	1.02	18.17	8.05	0.08	2.97	0.92	1.52	3.78	0.20	4.37	99.74	69.93
SD	6.47	0.15	3.07	1.14	0.02	0.64	0.63	0.50	0.95	0.02	0.90	0.36	5.07

LOI, loss on ignition (detection limit 0.1%). CIA, chemical index of alteration. SD, standard deviation. Detection limits are 0.01% for all elements, except for Fe_2O_3 which is 0.04%. Samples are claystones except for 99S4 which is a siltstone and HOR28 which is sandstone.

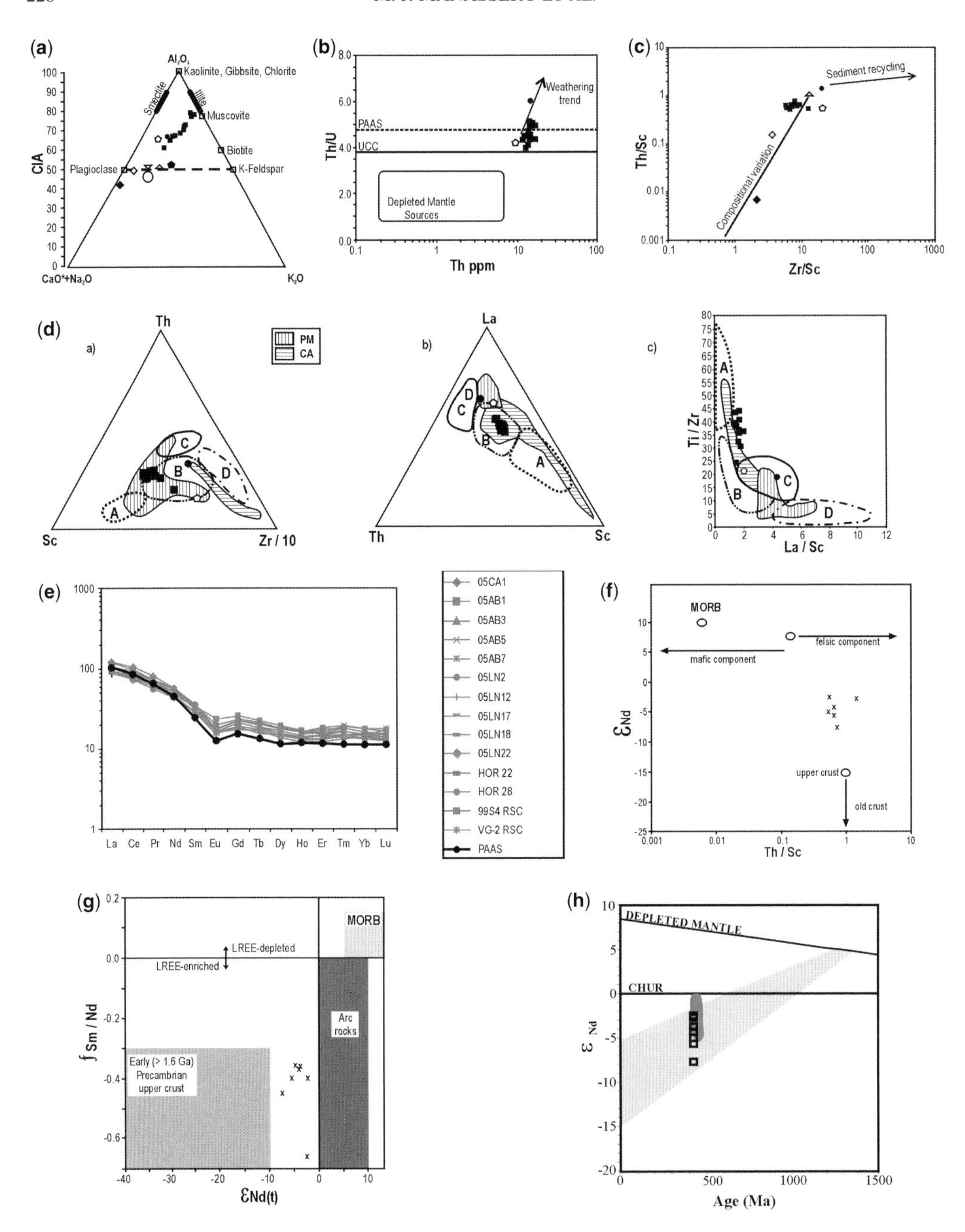

Fig. 14. Lithogeochemistry diagrams. (**a**) CIA: A–CN–K diagram constructed using molecular proportions of the oxides and with the CIA scale shown on the left. The average upper continental crust is plotted as an empty circle (Taylor & McLennan 1985), and idealized mineral compositions as empty squares. Solid pentagon, average granite; empty triangle, average adamellite; empty inverted triangle, average granodiorite; empty diamond, average tonalite; solid diamond, average gabbro (Nesbitt & Young 1989). Squares, claystones; empty pentagon, siltstone (99S4); circle, sandstone (HOR28). (**b**) Plot of Th/U versus Th (McLennan 1993). Squares, claystones; pentagon, siltstone (99S4); circle, sandstone (HOR28). PAAS, Post-Archaean Australian Shales pattern; UCC, upper continental crust. (**c**) Th/Sc versus Zr/Sc diagram after McLennan *et al.* (1993, 2006). Empty triangle, granodiorite (average upper crust); empty diamond, andesite; solid diamond, MORB. Squares, claystones; pentagon, siltstone (99S4); circle, sandstone (HOR28). (**d**) a, Th–Sc-Zr/10; b, La–Th–Sc; c, Ti/Zr versus La/Sc discriminatory plots after Bhatia & Crook (1986).

(A = Al_2O_3; CN = $CaO^* + Na_2O$; K = K_2O; Fig. 14a) the samples follow a general weathering trend which is parallel to subparallel to the A–CN join, regarding the average upper continental crust (UCC) composition (Fedo *et al.* 1995). However, deviations towards the A–K boundary are observed which could be a result of post-depositional metasomatic potassium enrichment (Nesbitt & Young 1989), as most of the samples are enriched in their K_2O concentration compared with the upper continental crust average value of 3.4% (McLennan *et al.* 2006). This metasomatism is responsible for the change of kaolinite to illite, and results in a CIA value lower than the pre-metasomatized one. It is therefore deduced that the Río Seco de los Castaños Formation is moderately to highly weathered.

Trace elements. Due to their immobile behaviour, trace elements (and in particular high field strength elements) are useful for provenance analysis because they preserve characteristics of the source rocks and therefore they reflect provenance compositions. Ratios such as Th/Sc, Th/U, Zr/Sc and Cr/V, along with the REE distribution provide some of the most useful data for provenance determination (Taylor & McLennan 1985). During weathering and/or recycling, there is a tendency for an elevation of the Th/U ratio above upper crustal igneous values of 3.8 to 4.0, because under oxidizing conditions U^{4+} oxidizes to the more soluble U^{6+} and is therefore more easily removed from the sediments than Th (McLennan 1993). Compared with Post-Archaean Australian Shale (PAAS) averages of Th (14.6 ppm) and U (3.1 ppm), most of the samples are depleted in Th and U (Table 3), although some samples are enriched in both. The Th/U ratios range from below to above the PAAS value of 4.7 but above the upper continental crust average, indicating weathering and/or recycling processes (Fig. 14b). As the CIA analysis indicates moderately to strongly weathered samples, it is deduced that weathering rather than recycling affected the samples. Another proxy to evaluate the presence or absence of recycling is the Zr/Sc ratio because Zr is strongly enriched in zircon which can be easily recycled, whereas Sc is present in labile phases (McLennan 1993). The Zr/Sc ratio for the Río Seco de los Castaños Formation (Table 3) is lower than the 13.13 value of the PAAS because Zr concentrations are lower than the PAAS average of 210 ppm and the Sc concentrations are higher than the PAAS average of 16 ppm. The exceptions are three samples which show Zr/Sc ratios between 12.75 and 21.26 due to their enrichment in Zr (HOR22 and 99S4) or depletion in Sc (HOR28) compared with the PAAS. The Th/Sc ratio indicates the degree of igneous differentiation of the source rocks since Th is an incompatible element whereas Sc is compatible in igneous systems (McLennan *et al.* 1990; McLennan 1993). The Th/Sc ratio for the Río Seco de los Castaños Formation (Table 3), except for the coarsest sample (HOR28), varies between 0.52 and 0.71, well below the PAAS value of 0.91 and the upper continental crust value of 0.79. The sandstone (HOR28) has a Th/Sc ratio of 1.4. The Zr/Sc and Th/Sc ratios indicate conclusively that for the Río Seco de los Castaños Formation recycling was not important and input was from a source geochemically less evolved than the average upper continental crust (Fig. 14c).

Cr, V, Ni and Sc are concentrated in mafic rocks, and therefore they are useful to evaluate the influence of a mafic source. The Cr/V ratio (Table 4) indicates the enrichment of Cr over other ferromagnesian trace elements. The main minerals which concentrate Cr over other ferromagnesians are chromites. The Y/Ni ratio indicates the concentration of ferromagnesian trace elements (such as Ni) compared with Y which represents a proxy for heavy REE. The Cr/V ratio is 0.79 $\pm$ 0.33 on average whereas the Y/Ni ratio is 0.86 $\pm$ 0.3 on average, plotting between the upper continental crust and the PAAS averages (diagram not shown) and indicating that although the source

Fig. 14. (*Continued*) A, Oceanic island arc; B, continental island arc; C, active continental margin; D, passive margin; PM, recent deep-sea turbidites derived from and deposited at a passive margin; CA, recent deep-sea turbidites derived from and deposited at a continental arc margin. The great dispersal of data showed by the PM and CA fields exemplify the difficulty of determining tectonic setting based only on geochemistry (data form McLennan *et al.* 1990). Squares, claystones; pentagon, siltstone (99S4); circle, sandstone (HOR28). (**e**) Chondrite-normalized REE patterns for the Río Seco de los Castaños Formation. PAAS pattern (Nance & Taylor, 1976) is drawn for comparison. Chondrite normalization factors are those listed by Taylor & McLennan (1985). $Eu/Eu^* = Eu_N/\{(Sm_N)(Gd_N)\}^{1/2}$. (**f**) $\varepsilon_{Nd}(t)$ versus Th/Sc ratio of samples from the Río Seco de los Castaños Formation, except sample 05LN13 for which geochemical analysis is not available. (**g**) Plots of $f_{Sm/Nd}$ versus $\varepsilon_{Nd}(t)$. $f_{Sm/Nd}$ values for the Río Seco de los Castaños Formation are in the range of variation of the basement (see text for discussion). (**h**) Plot of ε_{Nd} against age. Squares, Río Seco de los Castaños Formation; dark grey area, data from Lower Palaeozoic platform deposits from the San Rafael Block; light grey area, range of variation of ε_{Nd} for the basement rocks known as Cerro La Ventana Formation. The Río Seco de los Castaños Formation Nd system can be explained mainly by the basement rocks.

Table 3. *Trace elements (expressed in ppm) of the Río Seco de los Castaños Formation*

	Lomitas Negras and Agua del Blanco sections									Atuel Creek section		Road 144-Rodeo de la Bordalesa section				
	05AB1	05AB3	05AB5	05AB7	05LN2	05LN12	05LN17	05LN18	05LN22	05CA1	VG-2	HOR22	HOR28	99S4	Aver.	SD
Mo*	0.30	0.20	0.10	0.20	0.80	0.20	0.20	0.20	0.20	0.10	bd	bd	bd	2.35	0.44	0.63
Cu*	41.50	45.40	47.50	53.50	38.20	69.10	40.80	59.50	58.85	54.7	72.19	67.03	bd	13.54	50.91	15.17
Pb*	16.30	14.60	17.70	18.40	15.30	10.90	14.90	18.10	16.90	21.5	bd	18.50	bd	bd	16.65	2.62
Zn†	86.0	87.0	94.0	97.0	97.0	88.0	97.0	97.0	101.5	106.0	33.4	120.5	bd	bd	92.0	19.82
Ni¶	46.2	41.5	41.4	44.9	45.8	42.9	44.5	45.3	47.9	50.3	63.8	41.2	bd	22.7	44.49	8.47
As‡	9.40	8.00	6.50	4.90	26.50	3.60	19.70	7.90	8.30	17.6	47.3	17.54	bd	bd	14.77	11.87
Cd*	bd	bd	0.10	0.10	0.10	0.10	bd	0.10	0.10	bd	na	na	na	na	0.10	0.00
Sb*	0.20	bd	0.10	0.10	0.20	0.10	0.10	0.20	0.10	0.10	1.69	1.82	0.33	bd	0.37	0.60
Bi*	0.40	0.40	0.50	0.50	0.40	0.30	0.40	0.60	0.60	0.50	bd	0.99	bd	bd	0.38	0.29
Ag*	bd	bd	bd	bd	bd	bd	bd	bd	bd	bd	bd	bd	bd	bd		
Au‡	bd	bd	bd	bd	0.90	1.00	bd	3.00	0.80	bd	na	na	na	na	1.43	0.91
Hg§	bd	bd	0.01	0.02	0.01	0.01	0.01	0.02	0.04	bd	na	na	na	na	0.02	0.01
Tl*	0.10	0.10	0.10	0.10	0.10	0.10	0.10	0.10	0.10	0.10	bd	bd	bd	bd	0.10	0.00
Se‡	bd	bd	bd	bd	bd	bd	bd	bd	bd	bd	na	na	na	na		
Ba†	623.8	597.3	642.8	690.7	728.6	475.6	678.9	523.7	622.5	692.3	884.8	294.8	576.3	377.4	600.6	143.4
Be†	3.00	2.00	2.00	3.00	3.00	2.00	2.00	3.00	3.00	3.0	na	na	na	na	2.60	0.49
Co^	25.20	17.60	19.10	22.10	14.80	21.60	16.40	24.60	20.65	19.4	33.55	28.04	16.77	18.85	21.33	4.91
Cr~	102.6	102.6	109.5	109.5	123.2	88.9	109.5	116.3	112.9	116.3	149.5	100.3	103.7	247.5	150.2	59.4
Cs*	10.60	11.50	9.70	10.00	9.20	8.60	10.20	8.30	9.45	13.2	9.19	5.25	6.85	3.48	8.97	2.38
Ga‡	23.20	23.60	24.70	26.60	25.20	23.20	25.30	25.00	24.55	27.3	30.16	17.16	17.16	11.03	23.16	4.73
Hf*	6.00	5.50	5.10	5.20	4.70	4.80	4.50	4.20	4.15	5.6	5.05	7.37	5.83	9.78	5.56	1.42
Nb*	18.20	17.40	17.60	17.30	16.70	16.70	16.70	16.70	16.20	19.6	23.65	15.18	13.61	14.66	17.16	2.30
Rb*	176.4	174.7	179.1	186.8	186.9	183.4	197.0	182.0	170.6	189.8	219.0	105.1	174.7	57.5	170.2	39.1
Sn†	3.00	3.00	4.00	3.00	8.00	3.00	5.00	4.00	3.00	3.0	5.34	2.24	1.83	1.07	3.53	1.65
Sr‡	77.30	56.0	120.90	72.60	102.40	114.20	61.70	153.50	75.45	47.5	77.45	50.42	93.60	97.66	85.76	28.92
Ta*	1.40	1.40	1.40	1.40	1.50	1.40	1.40	1.50	1.35	1.5	1.84	2.28	1.80	2.76	1.64	0.40
Th^	14.00	14.00	14.70	13.80	13.20	12.10	12.80	14.00	12.20	16.4	16.13	11.63	14.10	9.29	13.45	1.76
U*	3.20	2.70	2.90	2.80	3.20	2.70	2.70	3.10	3.05	3.3	3.71	2.66	2.34	2.21	2.90	0.38
V$^{\#}$	154.0	148.0	151.0	156.0	171.0	160.0	168.0	161.0	160.5	182.0	227.7	108.3	133.7	127.6	157.8	26.5
W‡	33.20	29.00	27.60	31.40	23.40	37.90	18.00	23.10	20.20	15.9	8.87	267.67	28.05	104.18	47.75	64.72
Sc†	22.00	22.00	23.00	22.00	23.00	23.00	22.00	23.00	22.00	23.0	26.00	22.00	10.00	18.00	21.50	3.56
Zr*	201.9	180.0	163.5	161.5	157.7	151.9	140.5	138.0	131.4	183.2	179.1	280.5	202.9	382.6	189.6	64.6
Y*	36.90	36.20	38.50	37.40	28.60	32.80	32.10	36.20	33.45	39.0	40.62	34.77	39.47	41.84	36.28	3.5

na, Not analysed; bd, below detection limit. Detection limits: *0.1 ppm; †1 ppm; ‡0.5 ppm; §0.01 ppm; ^0.2 ppm; $^{\#}$8 ppm; ¶20 ppm; ~0.002 ppm. Aver., average; SD, standard deviation. Samples are claystones except for 99S4 which is a siltstone and HOR28 which is sandstone.

Table 4. *Element ratios of the Río Seco de los Castaños Formation*

Sample	Lomitas Negras and Agua del Blanco sections									Atuel Creek section		Road 144-Rodeo de la Bordalesa section				
	05AB1	05AB3	05AB5	05AB7	05LN2	05LN12	05LN17	05LN18	05LN22	05CA1	VG2	HOR22	HOR28	99S4	Aver.	SD
Eu/Eu*	0.68	0.65	0.66	0.58	0.71	0.67	0.71	0.61	0.66	0.63	0.81	0.69	0.69	0.69	0.68	0.05
Ce/Ce*	0.07	0.08	0.08	0.08	0.09	0.09	0.08	0.08	0.09	0.07	0.07	0.08	0.07	0.07	0.08	0.01
Th/Sc	0.64	0.64	0.64	0.63	0.57	0.53	0.58	0.61	0.55	0.71	0.52	0.53	0.62	1.41	0.66	0.22
Zr/Sc	9.18	8.18	7.11	7.34	6.86	6.60	6.39	6.00	5.98	7.97	21.26	12.75	6.89	20.29	9.48	4.90
Th/U	4.38	5.19	5.07	4.93	4.13	4.48	4.74	4.52	4.00	4.97	4.21	4.37	4.35	6.02	4.67	0.51
Nb/Y	0.49	0.48	0.46	0.46	0.58	0.51	0.52	0.46	0.48	0.50	0.35	0.44	0.58	0.34	0.48	0.07
Ti/Zr	30.58	32.97	37.03	35.64	38.40	39.47	40.54	43.88	43.55	36.32	21.09	24.43	38.00	18.88	34.34	7.62
La/Sc	1.75	1.65	1.67	1.72	1.50	1.39	1.59	1.59	1.50	1.97	2.04	1.55	1.53	4.37	1.85	0.72
La/Th	2.76	2.60	2.62	2.75	2.61	2.64	2.73	2.61	2.70	2.77	3.96	2.94	2.47	3.10	2.80	0.35
Cr/V	0.67	0.69	0.72	0.70	0.72	0.56	0.65	0.72	0.70	0.64	1.94	0.93	0.66	0.78	0.79	0.33
La_N/Yb_N	6.59	6.56	7.07	6.65	6.84	6.04	7.19	6.56	6.63	6.99	6.22	7.10	6.16	7.46	6.72	0.40
La_N/Sm_N	3.11	3.01	3.03	3.02	3.14	3.15	3.24	3.03	2.90	3.48	2.98	3.35	3.52	3.53	3.18	0.20
Tb_N/Yb_N	1.18	1.25	1.36	1.31	1.21	1.07	1.29	1.20	1.33	1.13	1.38	1.38	1.12	1.30	1.25	0.10
Cr/Th	7.33	7.33	7.45	7.93	9.33	7.35	8.55	8.31	9.25	7.09	26.64	8.62	9.27	7.36	9.41	4.84
Zr/Th	14.42	12.86	11.12	11.70	11.95	12.55	10.98	9.86	10.77	11.17	41.17	24.11	11.10	14.40	14.87	8.04
Zr/Nb	11.09	10.34	9.29	9.34	9.44	9.10	8.41	8.26	8.11	9.35	26.09	18.48	7.57	14.92	11.41	4.97
Zr/Y	5.47	4.97	4.25	4.32	5.51	4.63	4.38	3.81	3.93	4.70	9.14	8.07	4.41	5.14	5.19	1.49
Y/Ni	0.80	0.87	0.93	0.83	0.62	0.76	0.72	0.80	0.70	0.78	1.84	0.84	0.64		0.86	0.30
Sc/Th	1.57	1.57	1.56	1.59	1.74	1.90	1.72	1.64	1.80	1.40	1.94	1.89	1.61	0.71	1.62	0.29
Gd_N/Yb_N	1.30	1.32	1.48	1.37	1.24	1.29	1.39	1.39	1.36	1.25	1.58	1.62	1.26	1.48	1.38	0.11

See text for more details. Aver., average; SD, standard deviation. Subscript N denotes chondrite normalized values. $Eu/Eu^* = Eu_N/\{(Sm_N)(Gd_N)\}^{1/2}$. $Ce/Ce^* = Ce_N/\{(0.66La_N)(0.33Nd_N)\}$. Samples are claystones except for 99S4 which is a siltstone and HOR28 which is sandstone.

rock(s) was more mafic than the average upper continental crust composition, a major ophiolitic source can be ruled out. Nevertheless, the Cr/V ratio might be affected by V concentrations higher than the PAAS value (150 ppm) since V could have been fractionated from Cr during sedimentary processes such as diagenesis (Feng & Kerrich 1990). The high Cr concentration of sample 99S4 (247 ppm), which is well above the 110 ppm average value of the PAAS, suggests the presence of chromites, which could have been derived from a mafic source, or could have been reworked (chromites are resistant heavy minerals). Nevertheless, a Zr/Sc ratio of about 20 for sample 99S4 does not suggest significant reworking.

REE pattern. The shape of the REE pattern (including the presence or absence of an Eu anomaly) can provide information about both bulk compositions of the provenance and the nature of the dominant igneous process affecting the provenance (McLennan *et al.* 1990; McLennan & Taylor 1991). The chondrite-normalized REE diagram for the Río Seco de los Castaños Formation shows a moderately enriched light rare earth element (LREE) pattern, a negative Eu anomaly and a rather flat heavy rare earth elements (HREE) distribution (Fig. 14e and Table 5), being therefore essentially similar to the PAAS pattern. However, samples from the Río Seco de los Castaños Formation are enriched in the sum of REE compared with the PAAS, with concentrations of the elements between La and Nd varying from enriched to slightly depleted, but strongly enriched in elements between Sm and Lu. The Eu anomalies vary between 0.58 and 0.81 with an average value of 0.68, being in general higher than the average Eu anomaly for the PAAS (0.66) and for the upper continental crust (0.63). It is noteworthy that the sample with the highest Cr concentration (sample 99S4) displays a less negative Eu anomaly and the highest Eu concentration (Eu is almost double compared with the average value of the PAAS), supporting the influence of a depleted source. Sm/Nd ratios are in the range between 0.19 and 0.22, slightly higher than the average value for the PAAS (0.175).

Relationships between Th, Sc and Zr and La, Th and Sc can be useful to discriminate the tectonic setting of the depositional basin (Bhatia & Crook 1986). However, some dispersal of data is expected and caution on the interpretation is needed since detritus could have been transported across different tectonic settings (McLennan 1989). As shown in Figure 14d the samples plot within field B (continental arc settings); even those outside of any field show a trend towards the field of oceanic island arc setting (A).

Sm–Nd isotopic data

The Río Seco de los Castaños Formation samples (Table 6) shows ε_{Nd} (t) values (where $t = 420$ Ma,

Table 5. *Rare earth elements data (expressed in ppm) of the Río Seco de los Castaños Formation*

Sample	La*	Ce*	Pr†	Nd‡	Sm§	Eu†	Gd§	Tb¶	Dy§	Ho†	Er^	Tm¶	Yb§	Lu¶	ΣREE
Lomitas Negras and Agua del Blanco sections															
05AB1	38.60	89.70	10.06	38.90	7.80	1.56	6.36	1.09	6.61	1.29	3.89	0.62	3.96	0.57	211.01
05AB3	36.40	85.10	9.62	34.60	7.60	1.46	6.13	1.10	6.28	1.24	3.50	0.53	3.75	0.56	197.87
05AB5	38.50	86.80	9.76	36.30	8.00	1.59	6.72	1.17	6.81	1.38	3.78	0.59	3.68	0.63	205.71
05AB7	37.90	86.50	9.82	35.80	7.90	1.37	6.50	1.18	6.74	1.38	3.86	0.59	3.85	0.58	203.97
05LN2	34.40	78.70	8.89	33.50	6.90	1.40	5.21	0.96	5.28	1.03	3.04	0.47	3.40	0.48	183.66
05LN12	32.00	71.60	8.37	32.00	6.40	1.32	5.70	0.90	5.47	1.12	3.26	0.53	3.58	0.51	172.76
05LN17	35.00	74.40	8.85	32.10	6.80	1.43	5.65	0.99	5.44	1.10	3.32	0.50	3.29	0.48	179.35
05LN18	36.60	81.50	9.37	34.40	7.60	1.40	6.47	1.06	6.33	1.19	3.65	0.58	3.77	0.53	194.45
05LN22	32.95	74.65	8.54	33.30	7.15	1.38	5.65	1.05	5.98	1.13	3.16	0.52	3.36	0.49	179.30
Atuel Creek Section															
05CA1	45.40	102.50	11.23	40.70	8.20	1.53	6.78	1.16	6.99	1.37	4.01	0.67	4.39	0.62	235.55
VG-2	39.87	83.02	9.18	36.61	7.13	1.58	6.83	1.15	6.84	1.43	4.56	0.69	4.38	0.67	203.94
Road 144-Rodeo de la Bordalesa section															
HOR22	34.21	69.86	7.72	31.26	6.43	1.47	6.50	1.05	5.84	1.15	3.57	0.53	3.26	0.48	173.33
HOR28	43.70	93.35	10.13	40.05	7.80	1.70	7.22	1.21	6.71	1.35	4.34	0.64	3.96	0.60	222.77
99S4	36.80	77.70	8.90	37.44	7.77	2.06	7.78	1.29	7.24	1.41	4.39	0.63	4.00	0.58	198.00
average	37.31	82.53	9.32	35.50	7.39	1.52	6.39	1.10	6.33	1.26	3.74	0.58	3.76	0.56	197.26
SD	3.67	8.68	0.85	2.90	0.57	0.18	0.66	0.10	0.61	0.13	0.45	0.06	0.35	0.06	17.88

SD, standard deviation. Detection limits: *0.1 ppm; †0.02 ppm; ‡0.3 ppm; §0.05 ppm; ¶0.01 ppm; ^0.03 ppm. Samples are claystones except for 99S4 which is a siltstone and HOR28 which is sandstone.

Table 6. *Sm–Nd data of the Río Seco de los Castaños Formation*

Sample	Age (Ma)	Sm (ppm)	Nd (ppm)	$^{147}Sm/^{144}Nd$	$^{143}Nd/^{144}Nd$	Error (ppm)	ε_{Nd} (0)	ε_{Nd} (t)	$^{143}Nd/^{144}Nd$ (t)	T_{DM}^1 (Ma)	T_{DM}^2 (Ma)	$f_{Sm/Nd}$
Lomitas Negras and Agua del Blanco sections												
05LN13	420	6.34	30.63	0.12511	0.512240	9	−7.7	−3.8	0.511900	1366	1466	−0.36
05AB7	420	7.09	34.59	0.12393	0.512220	9	−8.1	−4.2	0.511883	1381	1490	−0.37
Atuel Creek section												
05CA1	420	3.96	21.97	0.10901	0.512000	9	−12.4	−7.7	0.511705	1494	1742	−0.45
VG-2	420	5.33	27.16	0.11860	0.512130	15	−9.9	−5.7	0.511804	1448	1604	−0.40
Road 144-Rodeo de la Bordalesa section												
HOR22	420	6.14	31.70	0.11710	0.512291	110	−6.8	−2.5	0.511969	1195	1363	−0.40
HOR28	420	4.31	39.02	0.06670	0.512144	70	−9.6	−2.7	0.511960	952	1376	−0.66
99S4	420	5.40	26.12	0.12500	0.512187	57	−8.8	−5.0	0.511843	1454	1549	−0.36

$f_{Sm/Nd} = (^{147}Sm/^{144}Nd)_{sample} / (^{147}Sm/^{144}Nd)_{CHUR} - 1$. $\varepsilon_{Nd}(0) = \{[(^{143}Nd/^{144}Nd)_{sample\ (t=0)} / 0.512638] - 1\} \times 10000$. $\varepsilon_{Nd}(t) = \{[(^{143}Nd/^{144}Nd)_{sample\ (t)} / (^{143}Nd/^{144}Nd)_{CHUR\ (t)}] - 1\} \times 10000$. $(^{147}Sm/^{144}Nd)_{CHUR} = 0.1967$. $(^{143}Nd/^{144}Nd)_{CHUR} = 0.512638$. t= 420 Ma. T_{DM}^1 (model ages) were calculated based on the depleted mantle model (DePaolo 1981) whereas T_{DM}^2 were calculated based on the three-stage model (DePaolo *et al.* 1991). Samples are claystones except for 99S4 which is a siltstone and HOR28 which is sandstone.

the proxy age of sedimentation) ranging from −2.5 to −7.7 (average −4.5 ± 1.7), $f_{Sm/Nd}$ (the fractional deviation of the sample $^{147}Sm/^{144}Nd$ from a chondritic reference) ranges from −0.36 to −0.66 (average −0.43 ± 0.10) whereas the $T_{DM}{}^{1}$ ages (calculated using the model of DePaolo 1981) range from 952 to 1494 Ma (average 1327 ± 178 Ma) and $T_{DM}{}^{2}$ ages (calculated using the model of DePaolo *et al.* 1991) range from 1363 to 1742 Ma (average 1513 ± 123 Ma). The ε_{Nd} values for the Río Seco de los Castaños Formation are between those typical for the upper continental crust or older crust and those typical for a juvenile component (Fig. 14 g). Regarding the relationship between ε_{Nd} (t) and Th/Sc ratio (Fig. 14f), the samples display a trend where those with the less negative ε_{Nd} values show the lowest Th/Sc ratios, indicating that the more juvenile the source the more depleted its geochemical signature. The exception is sample HOR28 which shows a low negative ε_{Nd} (t) but a high Th/Sc ratio (1.4). The plot of $f_{Sm/Nd}$ against ε_{Nd} (t) shows a data cluster between fields of arc-rocks and old crust. The $f_{Sm/Nd}$ values out of the range of variation of the upper crust (−0.4 to −0.5) could be indicating Sm–Nd fractionations due to secondary processes, and are therefore suspect (McDaniel *et al.* 1994).

Discussion and interpretation

Sedimentological studies and depositional environments

The relatively low diversity of subenvironments, dominance of fine to medium sedimentary grain sizes, lack of tractive sedimentary structures, and the significant thickness of the beds associated with gravity flow processes are typical of a distal (below wave base) to proximal marine platform-deltaic system (Fig. 15). In this case, the sedimentary input was continuous, as indicated by the absence of internal discontinuities. The basin was extended and the palaeoslope was very small (less than 1%). The dominant processes acting on this palaeoenvironment were wave and storm action, permitting the settling of fine material over the tractive processes during fair-weather times. The presence of plant debris such as *Lycophytes* in the Atuel and Lomitas Negras sections suggests proximity to vegetated areas. The hydraulic regimes were moderate and the sea-level changes in this sequence generated very few sedimentary unconformities, but widespread lateral bed continuity. The ichnofacies (Poiré *et al.* 2002) such as *Cruziana* increase towards the east and the *Nereites–Mermia* towards the west of the basin which consistent with the lithofacies interpretation of deeper sectors of the basin located to the west.

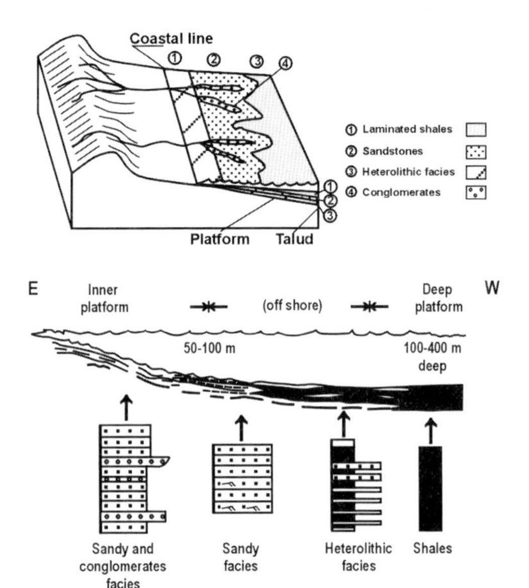

Fig. 15. Integrated block diagram showing general lithofacies distribution for the four study outcrops within the platform (adapted from Reading 1996).

Similar siliciclastic environments (probably equivalent stratigraphically), interpreted as over-filled sedimentary foreland systems with great thickness (high sedimentary rates) and low textural maturity, are found in the Villavicencio and Punta Negra formations both from the Precordillera terrane. They have been described by other authors (González Bonorino & Middleton 1976; Bustos 1996; Poiré & Morel 1996; Astini *et al.* 2005; Edwards *et al.* 2001; Peralta *et al.* 1995; Peralta 2003, 2005; Poiré *et al.* 2005). However, the channelled conglomerate and organic-matter-rich beds present in the Río Seco de los Castaños Formation, allow this unit to be distinguished from other similar environments found within the same terrane.

Diagenesis–metamorphism

The isotopic Rb–Sr ratios have been interpreted to provide the age of the low-grade metamorphism of the Río Seco de los Castaños Formation which occurred during the Late Devonian–Early Carboniferous. This metamorphic event could be linked to the final 'Chanic' tectonic phase that affected the Precordillera–Cuyania terrane (Ramos *et al.* 1986). The high Sr initial ratio suggests isotopic homogenization of the detritus which was derived from upper continental crust rocks. These Rb–Sr data also help to constrain the depositional age of the Río Seco de los Castaños Formation and the source areas.

Geochemical analyses

These data indicate moderate to strong weathering (CIA between 61 and 78 and Th/U ratios above 3.8–4), and potassium metasomatism. Zr/Sc ratios lower than 22 and no important enrichment of Zr (with some exceptions) indicate no recycling. Th/Sc ratios well below the averages for PAAS and upper continental crust, along with high Sc concentration, certain Cr enrichments and Eu anomalies less negative than PAAS and UCC, suggest a provenance from an unrecycled crust with an average composition similar to or slightly depleted compared with average upper continental crust composition. The unit seems to be related to an active margin.

Nd isotopes

T_{DM} ages are within the range of the Mesoproterozoic basement and Palaeozoic supracrustal rocks of the Precordillera terrane (Kay *et al.* 1996; Rapela *et al.* 1998; Cingolani *et al.* 2003*b*, 2005; Gleason *et al.* 2007) and the Western Pampeanas Ranges (Vujovich *et al.* 2005; Naipauer *et al.* 2005). The ε_{Nd} values of the Río Seco de los Castaños Formation are similar to those from sedimentary rocks from the Lower Palaeozoic carbonate–siliciclastic platform of the San Rafael Block, which show ε_{Nd} (t) between -0.4 and -4.9 (Cingolani *et al.* 2003*b*); they are also in the range of variation of ε_{Nd} values of the Mesoproterozoic basement of the San Rafael Block (the Cerro La Ventana Formation; Cingolani *et al.* 2005) recalculated at 420 Ma (Fig. 14 h). Although some $f_{Sm/Nd}$ values are below or above average values for the upper crust, all samples but one have $f_{Sm/Nd}$ values in the range of variation of the Cerro La Ventana Formation (Cingolani *et al.* 2005). Sample HOR28 shows a low $^{147}Sm/^{144}Nd$ ratio and a slightly negative $f_{Sm/Nd}$ value (-0.66) compared with the basement rocks, indicating that secondary processes might have fractionated Sm and Nd. Various studies have addressed processes that might alter the Sm–Nd isotopic signatures in detrital sediments. These include the alteration of Sm/Nd ratios and Nd isotopic signatures during weathering, diagenesis or sorting (McDaniel *et al.* 1994; Bock *et al.* 1994). Taking into account that sample HOR28 is a sandstone, has high Th/Sc ratios, low CIA values, and is one of the more recycled samples from this unit, it is deduced that most probably processes of sorting of LREE-enriched mineral phases might have altered its Sm–Nd isotopic signature.

In summary, the Cerro La Ventana Formation and the Ordovician carbonate–siliciclastic platform of the San Rafael Block provide a good fit to the Sm–Nd signature of the Río Seco de los Castaños Formation. Such a provenance for the Upper Silurian–Lower Devonian unit is in agreement with east to west palaeocurrents, because both sources are located to the east of the depositional basin of the Río Seco de los Castaños Formation (Fig. 16). This fact is also supported by the rather short transport deduced from the petrographical analyses (textural and compositional immaturity), as well as by the geochemical signature evidencing a non-recycled crust with an average composition similar to or depleted compared with average upper continental crust.

Provenance

A close spatial relationship between the depositional basin of the Río Seco de los Castaños Formation and the source rocks is supported by textural (e.g. subangular grains and high matrix content) and compositional (e.g. detrital mica flakes) immaturity of sandstones, the presence of plants debris and charcoal beds, and low Th/Sc ratios indicating no recycling. Petrographical analyses suggest source rocks from an igneous–metamorphic complex as well as a sedimentary source input, which included limestones bearing Ordovician fossils. Geochemical analyses and particularly the Th/Sc ratios, REE patterns and Eu anomalies further indicate that the source rocks have an average composition slightly less evolved than the average upper continental crust. The location and geochemical composition of the Mesoproterozoic basement rocks of the Cerro La Ventana Formation (igneous–metamorphic complex composed mainly of mafic to intermediate gneisses, micaschists, foliated quartz-diorites and tonalites, partially grading to amphibolites and migmatites, as well as pegmatitic and aplitic veins) and sedimentary rocks from the Ordovician carbonate–siliciclastic platform (Pavón and Ponón Trehué formations) fit the above-mentioned provenance constraints. Such a provenance location is further supported by palaeocurrents (Fig. 16). The Sm–Nd signature of the Río Seco de los Castaños Formation agrees well with the signature of both the Mesoproterozoic basement and the carbonate–siliciclastic platform (same range of variation of the ε_{Nd} (t) and T_{DM} ages), supporting such provenances (Fig. 14h).

Land–sea interactions

It is well known that the Devonian was a time of great changes not only of ecosystems but of climates as well, caused probably by complex interactions between the fast developing terrestrial biosphere, marine ecosystems and the atmosphere. The Río Seco de los Castaños Formation was deposited

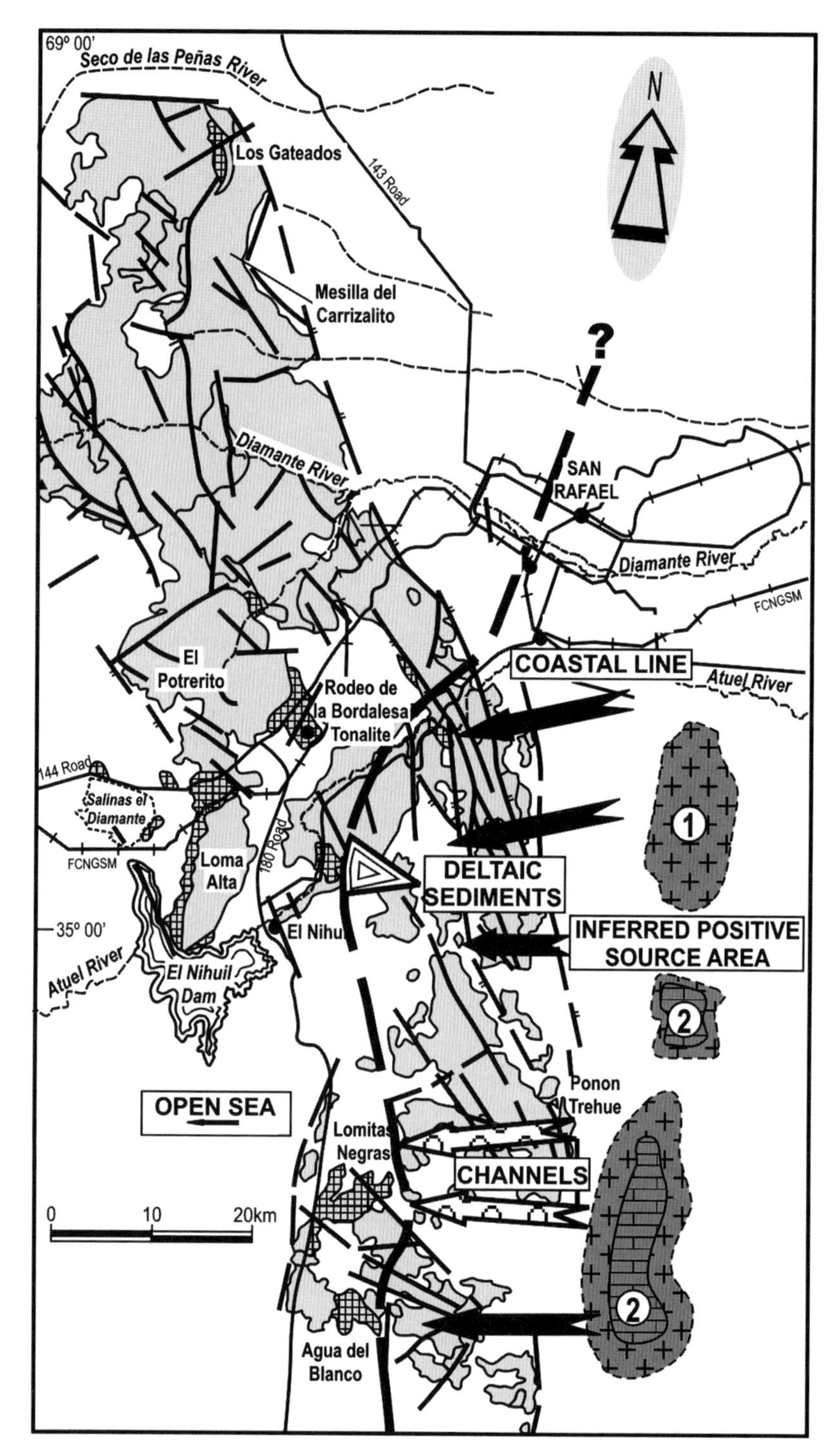

Fig. 16. Palaeogeographic interpretation. Suggested positive source areas towards the eastern side of the San Rafael Block during Upper Silurian–Lower Devonian times. key: 1, inferred location of the Mesoproterozoic crustal rocks; 2, Ordovician carbonate–siliciclastic platform (now exposed only at Ponón Trehué, see Figs 2 and 3). Interpretation of land–sea interactions: open sea towards the west of the coastal line; deltaic system and conglomerate channels on the eastern sector. For general references see Figure 2.

within a basin influenced by both terrestrial and marine environments. The continental source areas (Cerro La Ventana Formation and the Ordovician sedimentary units) were located not far away towards the east (Fig. 16). The detrital material was funnelled westwards (conglomerate channels) from these positive areas into the outer platform areas also laterally associated with a prograding deltaic system along coastal sectors. The basin deepened towards the west (open sea).

Conclusions

Considering the sedimentological and petrographical data, we conclude that the dominance of fine to medium grain sizes of the sedimentary rocks, lack of tractive sedimentary structures, and the important bed-thickness as well as associated gravity-flow deposition, are typical of a distal (below wave base) to proximal silty-siliciclastic marine platform-deltaic system.

The sedimentary input was continuous, as evidenced by the lateral bed continuity and absence of internal discontinuities; at the same time, the platform was extended with moderate hydraulic regimes and the palaeoslope seemed to be reduced. The dominant processes acting on the environment were wave and storm action.

The source areas were located to the east, close to the study area. The sandstone petrography shows both recycled orogen and continental block provenances. On the other hand the clay mineralogy shows that the fraction is dominated by illite (40–60%), kaolinite (25–40%) and chlorite (10–20%). The illite crystallinity index indicates very low-grade metamorphism for the sequence that occurred during the Early Carboniferous.

The presence of plant debris in the Atuel and Lomitas Negras sections suggests vegetated areas close to this Upper Silurian–Lower Devonian depocentre. Trace fossil distribution with *Cruziana* ichnofacies (indicative of a shallow environment in Agua del Blanco) to the east, and *Nereites* to the west (Road 144 outcrops) is consistent with a basin deepening towards the west.

Major element geochemistry suggests moderate to strong weathering and potassium metasomatism, in agreement with the clay mineral composition indicating an abundance of illite and kaolinite. Trace element (including REE) concentrations and ratios suggest a provenance from an unrecycled source with a composition similar to or depleted with respect to average upper continental crust.

Short transport of sediments is deduced from petrographical and sedimentological features, such as the presence of biotite and muscovite detrital flakes within sandstones, the high matrix content and subangular character of framework minerals (low textural maturity) as well as by the presence of plant debris. Facies distribution and Q–F–L diagrams indicating an uplifted igneous–metamorphic basement as source area are also in accordance with the geochemical signature. Furthermore, palaeocurrents towards the west and Sm–Nd signatures similar to those described for the Mesoproterozoic basement of the San Rafael Block (Cingolani *et al.* 2005) and the Ordovician platform, imply that the most probable sources are the Cerro La Ventana Formation and the carbonate–siliciclastic platform. The limestone conglomerate clasts also support a provenance from rocks that belong to a Lower Palaeozoic carbonate–siliciclastic platform, which is also located to the east.

A similar siliciclastic environment has been described for the Upper Silurian or Lower Devonian Villavicencio unit of the Precordillera (or Cuyania) terrane. However, the Río Seco de los Castaños Formation has two distinctive sedimentological characteristics: conglomerate channels and organic-matter-rich beds.

Continental source areas located to the east played an important role in the land–sea interactions during Upper Silurian–Lower Devonian times. The immature and poorly sorted detrital material was funnelled westwards (conglomerate channels) from these positive areas into the platform and deltaic systems. The ocean basin was open towards the west.

This contribution was supported by Argentine institution grants provided by PICT 07-10829 (ANPCYT) and PIP 5027-CONICET research projects. We are grateful to Prof. F. Chemale Jr, Prof. K. Kawashita and his technical group from UFRGS, Porto Alegre, Brazil, for the isotopic data. We thank Dr P. Königshof (Senckenberg Museum, Germany) and Prof. M. Namik Yalçin (Istanbul University, Turkey) for the invitation to participate in this volume, for discussions and constructive comments during the field excursion and for helping us to improve this work. We sincerely acknowledge N. Uriz and E. Morel (University of La Plata, Argentina) for participation during some fieldwork. Mario Campaña is thanked for technical assistance on the paper illustrations. We are very grateful to reviewers U. Linnemann (Germany) and S. Peralta (Argentina) for their valuable comments and suggestions that deeply enhanced our manuscript in many geological and editorial aspects. Finally, we are very grateful to C. Brett (USA) for linguistic improvement of the final version. This is a contribution to IGCP 499-UNESCO-IUGS 'Devonian land–sea interaction: evolution of ecosystems and climate'.

References

ACEÑOLAZA, F. G., MILLER, H. & TOSELLI, A. 2002. Proterozoic – Early Paleozoic evolution in Western South America – a discussion. *Tectonophysics*, **354**, 121–137.

AGUIRREZABALA, L. & GARCÍA MONDÉJAR, J. 1994. A coarse turbidite system with morphotectonic control (Middle Albian, Ondarroa, northern Iberia). *Sedimentology*, **41**, 383–407.

ASTINI, R., DÁVILA, F. *ET AL.* 2005. Cuencas de la Región Precordillerana. *In: Simposio Frontera Exploratoria de la Argentina. Congreso Exploración y Desarrollo de Hidrocarburos*, Mar del Plata, **6**, 115–145.

BHATIA, M. R. & CROOK, K. A. W. 1986. Trace element characteristics of graywackes and tectonic setting discrimination of sedimentary basins. *Contributions to Mineralogy and Petrology*, **92**, 181–193.

BOCK, B., MCLENNAN, S. M. & HANSON, G. N. 1994. Rare earth element redistribution and its effects on the neodymium isotope system in the Austin Glen Member of the Normanskill Formation, New York, USA. *Geochimica et Cosmochimica Acta*, **58**, 5245–5253.

BUSTOS, U. 1996. Modelo sedimentario alternativo para el Devónico de la Precordillera central sanjuanina: Formación Punta Negra. *Revista de la Asociación Argentina de Sedimentología*, **3**, 17–30.

CAMACHO, H., BUSBY, C. & KNELLER, B. 2002. A new depositional model for the classic turbidite locality at San Clemente State Beach, California. *Journal of American Association of Petroleum Geologists Bulletin*, **86**(9), 1543–1560.

CHEEL, R. J. & LECKIE, D. A. 1993. Hummocky cross-stratification. *In*: WRIGHT, V. P. (ed.) *Sedimentology Review*, **1**, Blackwell, Oxford, 103–122.

CINGOLANI, C. A. & VARELA, R. 1999. The San Rafael Block, Mendoza (Argentina): Rb-Sr isotopic age of basement rocks. *II South American Symposium on Isotope Geology*, Córdoba, Actas, 23–26.

CINGOLANI, C. A. & VARELA, R. 2008. The Rb-Sr low metamorphic age of the Río Seco de los Castaños Formation, San Rafael Block, Argentina. *VI South American Symposium on Isotope Geology*, Bariloche, Argentina (CD-Rom).

CINGOLANI, C. A., BASEI, M. A. S., LLAMBÍAS, E. J., VARELA, R., CHEMALE, F. JR., SIGA, O. JR. & ABRE, P. 2003*a*. The Rodeo Bordalesa Tonalite, San Rafael Block (Argentina): Geochemical and isotopic age constraints. *10° Congreso Geológico Chileno*, Concepción (CD Rom version).

CINGOLANI, C., MANASSERO, M. & ABRE, P. 2003*b*. Composition, provenance and tectonic setting of Ordovician siliciclastic rocks in the San Rafael Block: Southern extension of the Precordillera crustal fragment, Argentina. *Journal of South American Earth Sciences* (Special Issue on the Pacific Gondwana Margin), **16**(1), 91–106.

CINGOLANI, C. A., LLAMBÍAS, E. J., BASEI, M. A. S., VARELA, R., CHEMALE, F. JR., SIGA, O. JR. & ABRE, P. 2005. Grenvillian and Famatinian-age igneous events in the San Rafael Block, Mendoza Province, Argentina: Geochemical and isotopic constraints. *Gondwana 12 Conference*, Abstracts, 102.

COLLINSON, J. D. 1968. Deltaic sedimentation units in the Upper Carboniferous of northern England. *Sedimentology*, **10**, 233–254.

COLLINSON, J. D. 1978. Vertical sequence and sand body shape in alluvial sequences. *In*: MIALL, A. D. (ed.) *Fluvial Sedimentology*. Memoir of the Canadian Society of Petroleum Geologists, Calgary, 577–586.

COLLINSON, J. D. & THOMPSON, D. B. 1989. *Sedimentary Structures*. Unwin Hyman, London.

CRIADO ROQUE, P. & IBÁÑEZ, G. 1979. *Provincia geológica Sanrafaelino-Pampeana*. *In*: TURNER, J. C. (ed.) *Segundo Simposio de Geología Regional Argentina*. Academia Nacional de Ciencias, Córdoba, **I**, 837–869.

CUERDA, A. J. & CINGOLANI, C. A. 1998. El Ordovícico de la región del Cerro Bola en el Bloque de San Rafael, Mendoza: sus faunas graptolíticas. *Ameghiniana*, **35**(4), 427–448.

DEPAOLO, D. J. 1981. Neodymium isotopes in the Colorado Front Range and crust-mantle evolution in the Proterozoic. *Nature*, **291**, 193–196.

DEPAOLO, D. J., LINN, A. M. & SCHUBERT, G. 1991. The Continental Crustal Age Distribution, Methods of determining mantle separation ages from Sm-Nd isotopic data and application to the Southwestern United States. *Journal of Geophysical Research*, **96**, 2071–2088.

DESSANTI, R. 1956. *Descripción geológica de la Hoja 27c, Cerro Diamante (Provincia de Mendoza)*. Dirección Nacional de Geología y Minería, Buenos Aires, Boletín **85**.

DICKINSON, W. R. & SUCZEK, C. A. 1979. Plate tectonics and sandstone composition. *American Association of Petroleum Geologists, Bulletin*, **63**, 2164–2182.

DICKINSON, W. R., BEARD, S. *ET AL.* 1983. Provenance of North American Phanerozoic sandstones in relation to tectonic setting. *Geological Society of America Bulletin*, **64**, 233–235.

DI PERSIA, J. 1972. Breve nota sobre la edad de la denominada Serie de la Horqueta- Zona Sierra Pintada. Departamento de San Rafael, Provincia de Mendoza. *4ª Jornadas Geológicas Argentinas*, **3**, 29–41.

DOTT, R. H. 1964. Wacke, graywacke and matrix-what approach to immature sandstone classification. *Journal of Sedimentary Petrology*, **34**, 625–632.

EDWARDS, D., MOREL, E., POIRÉ, D. G. & CINGOLANI, C. A. 2001. Land plants in the Villavicencio Formation, Mendoza Province, Argentina. *Review of Paleobotany and Palynology*, **116**, 1–18.

FEDO, C. M., NESBITT, H. W. & YOUNG, G. M. 1995. Unraveling the effects of potassium metasomatism in sedimentary rocks and paleosols, with implications for paleoweathering conditions and provenance. *Geology*, **23**(10), 921–924.

FENG, R. & KERRICH, R. 1990. Geochemistry of fine-grained clastic sediments in the Archean Abitibi greenstone belt, Canada: Implications for provenance and tectonic setting. *Geochimica et Cosmochimica Acta*, **54**, 1061–1081.

FINNEY, S., GLEASON, J., GEHRELS, G., PERALTA, S. & ACEÑOLAZA, G. 2003. Early Gondwanan connection for the Argentine Precordillera terrane. *Earth and Planetary Sciences Letters*, **205**, 349–359.

GLEASON, J. D., FINNEY, S. C., PERALTA, S. H., GEHRELS, G. E. & MARSAGLIA, K. M. 2007. Zircon and whole-rock Nd-Pb isotopic provenance of Middle and Upper Ordovician siliciclastic rocks, Argentine Precordillera. *Sedimentology*, **54**(1), 107–136.

GONZÁLEZ BONORINO, G. 1986. Determinación de la profundidad de agua en que se formaron ondulitas simétricas por corrientes oscilatorias. *Primera Reunión Argentina de Sedimentología, Actas*, 221–224.

GONZÁLEZ BONORINO, G. & MIDDLETON, G. N. 1976. A Devonian submarine fan in western Argentina. *Journal of Sedimentary Petrology*, **46**(1), 56–69.

GONZÁLEZ DÍAZ, E. F. 1972. *Descripción geológica de la Hoja 27d San Rafael, Mendoza*. Servicio Minero-Geológico, Buenos Aires, Boletín **132**.

GONZÁLEZ DÍAZ, E. F. 1981. Nuevos argumentos a favor del desdoblamiento de la denominada Serie de La Horqueta del Bloque de San Rafael, Provincia de Mendoza. *8° Congreso Geológico Argentino*, **3**, 241–256.

KAY, S. M., ORRELL, S. & ABBRUZZI, J. M. 1996. Zircon and whole rock Nd-Pb isotopic evidence for a Grenville age and a Laurentian origin for the basement of the Precordillera in Argentina. *Journal of Geology*, **104**, 637–648.

KISCH, H. J. 1980. Incipient metamorphism of Cambro-Silurian clastic rocks from the Jamtland Super group, Central Scandinavian Caledonides, western Sweden: Illite crystallinity and 'vitrinite' reflectance. *Journal of the Geological Society, London*, **137**, 271–288.

KISCH, H. J. 1991. Illite crystallinity: Recommendations on sample preparation, X ray diffraction settings, and interlaboratory samples. *Journal of Metamorphic Geology*, **9**, 665–670.

KOMAR, P. 1974. Oscillatory ripple marks and evaluation of ancient conditions and environments. *Journal of Sedimentary Petrology*, **44**, 169–180.

KRUMM, S. 1994. Winfit 1.0. A computer program for X-ray diffraction line profile analysis. In XIII Conference of Clay Mineralogy and Petrology. *Acta Universitatis Carolinae Geologica*, **38**, 253–261.

MCDANIEL, D. K., HEMMING, S. R., MCLENNAN, S. M. & HANSON, G. N. 1994. Petrographic, geochemical and isotopic constraints on the provenance of the Early Proterozoic Chelmsford Formation, Sudbury basin, Ontario. *Journal of Sedimentary Research*, **64**, 632–642.

MCLENNAN, S. M. 1989. Rare earth elements in sedimentary rocks: Influence of provenance and sedimentary processes. *Mineralogical Society of America Reviews in Mineralogy*, **21**, 169–200.

MCLENNAN, S. M. 1993. Weathering and global denudation. *Journal of Geology*, **101**, 295–303.

MCLENNAN, S. M. & TAYLOR, S. R. 1991. Sedimentary rocks and crustal evolution: Tectonic setting and secular trends. *Journal of Geology*, **99**(1), 1–21.

MCLENNAN, S. M., TAYLOR, S. R., MCCULLOCH, M. T. & MAYNARD, J. B. 1990. Geochemical and Nd-Sr isotopic composition of deep-sea turbidites: Crustal evolution and plate tectonic associations. *Geochimica et Cosmochimica Acta*, **54**, 2015–2050.

MCLENNAN, S. M., TAYLOR, S. R. & HEMMING, S. R. 2006. Composition, differentiation, and evolution of continental crust: Constraints from sedimentary rocks and heat flow. *In*: BROWN, M. & RUSHMER, T. (eds) *Evolution and Differentiation of the Continental Crust*. Cambridge University Press, 92–134.

MANASSERO, M., CINGOLANI, C., ABRE, P. & URIZ, N. 2005. Facies sedimentarias de la Formación Río Seco de los Castaños (Silúrico-Devónico) del Bloque de San Rafael, Mendoza. *XVI Congreso Geológico Argentino*, **1**, 9–16.

MARTINO, R. & CURRAN, A. 1990. Sedimentology, ichnology and paleoenvironments of the Upper Cretaceous Wenonah and Mt Laures Formations, New Jersey. *Journal of Sedimentary Petrology*, **60**(1), 125–144.

MELVIN, J. 1986. Upper carboniferous fine-grained turbiditic sandstones from southwest England: A model for growth in an ancient delta-fed subsea fan. *Journal of Sedimentary Petrology*, **56**(1), 19–34.

MILLER, R. & HELLER, P. 1994. Depositional framework and controls on mixed carbonate-siliciclastic gravity flows: Pennsylvanian-Permian shelf to basin transect, south-western Great Basin, USA. *Sedimentology*, **41**, 1–20.

MOORE, D. M. & REYNOLDS, R. C. 1989. *X-Ray Diffraction and the Identification Analysis of Clay Minerals*. Oxford University Press, New York.

MOREL, E., CINGOLANI, C. A., GANUZA, D. G. & URIZ, N. J. 2006. El registro de Lycophytas primitivas en la Formación Río Seco de los Castaños, Bloque de San Rafael, Mendoza. *9° Congreso Argentino de Paleontología y Bioestratigrafía*, Córdoba. Abstract.

MOULDER, T. & ALEXANDER, J. 2001. The physical character of subaqueous sedimentary density flows and their deposits. *Sedimentology*, **48**(2), 269–299.

NAIPAUER, M., CINGOLANI, C. A., VALENCIO, S., CHEMALE, F. JR. & VUJOVICH, G. I. 2005. Estudios isotópicos en carbonatos marinos del Terreno Precordillera-Cuyania: Plataforma común en el Neoproterozoico-Paleozoico inferior?. *Latin American Journal of Sedimentology and Basin Analysis*, **12**(2), 89–108.

NANCE, W. B. & TAYLOR, S. R. 1976. Rare earth element patterns and crustal evolution I: Australian post-Archean sedimentary rocks. *Geochimica et Cosmochimica Acta*, **40**, 1539–1551.

NESBITT, H. W. & YOUNG, G. M. 1982. Early Proterozoic climates and plate motions inferred from major element chemistry of lutites. *Nature*, **199**, 715–717.

NESBITT, H. W. & YOUNG, G. M. 1989. Formation and diagenesis of weathering profiles. *Journal of Geology*, **97**, 129–147.

NESBITT, H. W., YOUNG, G. M., MCLENNAN, S. M. & KEAYS, R. R. 1996. Effects of chemical weathering and sorting on the petrogenesis of siliciclastic sediments, with implications for provenance studies. *Journal of Geology*, **104**, 525–542.

NUÑEZ, E. 1976. *Descripción geológica de la Hoja Nihuil. Informe Inédito*. Servicio Geológico Nacional, Buenos Aires.

PERALTA, S. 2003. An introduction to the geology of the Precordillera, Western Argentina. *In*: PERALTA, S., ALBANESI, G. & ORTEGA, G. (eds) *Ordovician and Silurian of the Precordillera, San Juan Province, Argentina*. INSUGEO, Tucumán, Argentina, **10**, 7–22.

PERALTA, S. 2005. Formación Los Sombreros: un evento diastrófico extensional del Devónico (Inferior-Medio?) en la Precordillera Argentina. *16° Congreso Geológico Argentino*, **4**, 322–326.

PERALTA, S. H., LEÓN, L. I. & CARTER, C. H. 1995. Estratigrafía de las sedimentitas del Eopaleozoico-Terciario de Pachaco, Precordillera Central sanjuanina. *Revista Ciencias, Facultad Ciencias Exactas, Físicas y Naturales, Universidad Nacional de San Juan*, **4**(6), 41–55.

POIRÉ, D. G. & MOREL, E. 1996. Procesos sedimentarios vinculados a la depositación de los niveles con plantas en secuencias siluro-devónicas de la Precordillera, Argentina. *VI Reunión Argentina de Sedimentología, Actas*, 205–210.

POIRÉ, D. G., CINGOLANI, C. & MOREL, E. 2002. Características sedimentológicas de la Formación Río Seco de los Castaños en el perfil de Agua del Blanco: Pre-Carbonífero del Bloque de San Rafael, Mendoza. *XV Congreso Geológico Argentino*, **I**, 129–133.

POIRÉ, D. G., EDWARDS, D., MOREL, E., BASSETT, M. G. & CINGOLANI, C. A. 2005. Depositional environments of Devonian land plants from Argentine Precordillera, South-West Gondwana. *Gondwana 12 Conference*, Mendoza. Abstract.

RAMOS, V. A. 2004. Cuyania, an exotic block to Gondwana: Review of a historical success and the present problems. *Gondwana Research*, **7**, 1009–1026.

RAMOS, V., JORDAN, T. E., ALLMENDINGER, R. W., MPODOZIS, C., KAY, S. M., CORTÉS, J. M. & PALMA, M. A. 1986. Paleozoic terranes of the central Argentine-Chilean Andes. *Tectonics*, **5**, 855–880.

RAMOS, V. A., DALLMEYER, R. & VUJOVICH, G. 1998. Time constraints on the Early Paleozoic docking of the Precordillera, central Argentina. *In*: PANKHURST, R. J. & RAPELA, C. W. (eds) *The Proto-Andean Margin of Gondwana*. Geological Society, London, Special Publications, **142**, 143–158.

RAPELA, C. W., PANKHURST, R. J., CASQUET, C., BALDO, E., SAAVEDRA, J. & GALINDO, C. 1998. Early evolution of the Proto-Andean margin of South America. *Geology*, **26**(8), 707–710.

READING, H. G. 1996. *Sedimentary Environments: Processes, Facies and Stratigraphy*. Blackwell Science, Oxford.

RUBINSTEIN, C. 1997. Primer registro de palinomorfos silúricos en la Formación La Horqueta, Bloque de San Rafael, Provincia de Mendoza, Argentina. *Ameghiniana*, **34**(2), 163–167.

SATO, A. M., TICKYJ, H., LLAMBÍAS, E. J., BASEI, M. A. S. & GONZÁLEZ, P. D. 2004. Las Matras Block, Central Argentina (37° S-67° W): The southernmost Cuyania terrane and its relationship with the Famatinian Orogeny. *Gondwana Research*, **7**(4), 1077–1087.

STOW, D. A. & PIPER, D. J. 1984. *Fine-grained Sediments: Deep Water Processes and Facies*. Geological Society, London.

TAYLOR, S. R. & MCLENNAN, S. M. 1985. *The Continental Crust: Its Composition and Evolution*. Blackwell, Oxford.

THOMAS, W. A. & ASTINI, R. A. 2003. Ordovician accretion of the Argentine Precordillera terrane to Gondwana: A review. *Journal of South American Earth Sciences*, **16**, 67–79.

TORSVIK, T. H. & COCKS, L. R. M. 2004. Earth geography from 400–250 Ma: A palaeomagnetic faunal and facies review. *Journal of the Geological Society (London)*, **161**, 555–572.

VUJOVICH, G. I., PORCHER, C., CHERNICOFF, C. J., FERNANDES, L. A. D. & PÉREZ, D. J. 2005. Extremo norte del basamento del terreno Cuyania: nuevos aportes multidisciplinarios para su identificación. *Asociación Geológica Argentina, Serie D, Publicación Especial*, **8**, 15–41.

WALKER, R. G., DUKE, W. L. & LECKIE, D. A. 1983. Hummocky cross-stratification: Significance of its variable bedding sequences, discussion. *Bulletin of the Geological Society of America*, **94**, 1245–1249.

WARR, L. N. & RICE, A. H. N. 1994. Interlaboratory standarization and calibration of clay mineral crystallinity and crystallite size data. *Journal of Metamorphic Geology*, **12**, 141–152.

ZUFFA, G. 1984. Optical analysis of arenites, influence of methodology on compositional results. *In*: *Provenance of Arenites*. Nato Series, Reidel Publishing Company, Dordrecht, 165–188.

The Devonian System of China, with a discussion on sea-level change in South China

XUEPING P. MA[1]*, WEIHUA LIAO[2] & DEMING WANG[1]

[1]*Department of Geology, Peking University, Beijing 100871, China*

[2]*Nanjing Institute of Geology and Palaeontology, Chinese Academy of Sciences, Nanjing, Jiangsu 210008, China*

**Corresponding author (e-mail: maxp@pku.edu.cn)*

Abstract: Based on tectonic, lithological and biotic features, 11 regions may be recognized in the Devonian Period of China. The Junggar and Hinggan regions are characterized by thick sequences of clastic rocks associated with volcanic rocks; carbonate deposits were only local, sometimes consisting of isolated reefs. The Tarim region was characterized by intertidal sandstones on the platform and deeper water deposits in its marginal areas. The North China region was mostly barren of Devonian deposits except in some marginal areas, and the Qilian–Qaidam region was a mountainous region mostly with Middle and Upper Devonian continental sediments. Qinling Region was closely related with the South China Region in terms of faunal affinity, probably being a marginal area of the South China Plate. Western Yunnan and the major part of Xizang (including northern Xizang and the northern slope of the Himalayas) featured continuous Silurian–Devonian deposition, generally with carbonates in the Lower Devonian, and different lithologies in different regions for the Middle and Late Devonian. Qinling and Hoh Xil-Bayan Har regions were closely related with the South China Region, yielding common fossils such as brachiopods *Stringocephalus* and *Yunnanella* (= *Nayunnella*) faunas. The Devonian Period of South China comprised deposits of two large transgressive–regressive cycles: Lochkovian to Eifelian and late Eifelian to about the end of the Devonian Period.

In China, Devonian-related studies go back to the middle of the nineteenth century when de Koninck (1846) and Davidson (1853) first described some Devonian brachiopod fossils from South China. During the first half of the twentieth century, significant biostratigraphic works on the Devonian System were made by Grabau (1931*a*, *b*), Tien (1938*a*, *b*), Sun (1935, 1945, 1958), Yoh (1937, 1938, 1956), Wang (1942, 1948) and Sze (1952). Tien (1938*b*) reviewed previous Devonian studies in various provinces and established a basic Devonian framework of China. In summarizing Devonian studies of that time, Wang & Yu (1962) first established Chinese regional stages. Major advances were made through large-scale geological surveying during the 1960s in various provinces. The results were presented at the 1974 'Symposium on the Devonian System of South China' held in Liuzhou of Guangxi, which was a turning point for Devonian research in China. At that time the Devonian stratigraphic sequence was fully and firmly established (IGMR 1978) for South China as well as other parts of China (Yang *et al.* 1981). In the following years, palaeontological atlases and stratigraphic charts of various provinces were also published, including major fossil groups and regional stratigraphic sections of the Devonian Period. In addition, a number of significant studies focusing on various parts of China were published around the 1980s, including South China (e.g. Wang *et al.* 1974, 1979, 1982; Hou 1978; Liao *et al.* 1978, 1979; Bai *et al.* 1982; Wang & Rong 1986; Liao & Ruan 1988; Yu 1988; Zhong *et al.* 1992), Longmenshan (e.g. Hou *et al.* 1988), Tarim (e.g. Liao *et al.* 1992, 2001), northern Xinjiang (e.g. Xiao *et al.* 1992), Western Qinling (e.g. XIGMR & NIGP 1987), western Yunnan (e.g. Yu & Liao 1978; Tan *et al.* 1982; Wang 1994), Xizang (e.g. Liao 1984; Rao & Yu 1985), western Sichuan and eastern Xizang (e.g. RGSTS & NIGP 1982). These achievements were mostly summarized in a monograph (Hou & Wang 1988).

During the last 30 years, major progress has been made regarding international correlation, which has been related to studies of pelagic fossils, including conodonts (e.g. Wang 1989; Ji & Ziegler 1993; Bai *et al.* 1994), dacryoconarids (e.g. Ruan & Mu 1989), entomozoans (e.g. Groos-Uffenorde & Wang 1989), ammonoids (Ruan 1981) as well as refined benthic invertebrate, vertebrate and plant fossil groups (e.g. Hou & Xian 1975; Pan *et al.* 1987; Liao & Birenheide 1984, 1985, 1989; Birenheide & Liao 1985; Hao & Gensel 1998, 2001; Ma 1995; Ma & Day 2000, 2007; Ma *et al.*

From: KÖNIGSHOF, P. (ed.) *Devonian Change: Case Studies in Palaeogeography and Palaeoecology*. The Geological Society, London, Special Publications, **314**, 241–262.
DOI: 10.1144/SP314.13 0305-8719/09/$15.00

2006; Wang *et al.* 2002). All these workers and many others have enhanced pelagic and neritic correlation and marine–non-marine correlation to a certian extent. Recent summarizing works include Hou (2000), Hou *et al.* (2000), Wang (2000), Cai (2000), Liao & Ruan (2003) and Wang *et al.* (2005). In the meantime there have been some advances in relation to sequence stratigraphy and palaeogeography as well as some refined stratigraphic studies, e.g. Wu *et al.* (1997), and Gong *et al.* (2005).

This paper gives a general outline of the Devonian System of China based on previous studies of various workers, and discusses sea-level changes in South China.

Outcrops, palaeoplates and stratigraphic regions

Devonian strata are widely exposed in South China, NW China and NE China except for the North China Platform which has no Devonian deposits as it became land from the Late Ordovician and did not receive deposits until the Carboniferous (Fig. 1).

Based on tectonic, lithological and biotic features, 11 regions may be recognized in the Devonian Period of China (Liao & Ruan 2003; Fig. 1), in addition to Altai Region: (1) Junggar Region, (2) Hinggan Region, (3) Tarim Region (including South Tianshan), (4) Qilian-Qaidam Region, (5) North China Region, (6) Qinling Region, (7) Hoh Xil-Bayan Har Region, (8) Xizang (Tibet) Region, (9) South China Region, (10) Western Yunnan Region and (11) Himalaya Region. These regions represent 11 variously sized palaeoplates (Fig. 1b): (1) the eastern corner of the Siberia Plate, (2) northern part of the Kazakhstan Plate, (3) southern part of the Songhua River-Bureja Plate, (4) Tarim Plate, (5) Qilian-Qaidam Plate (He *et al.* (1994) considered it to belong to the North China Plate), (6) North China Plate, (7) Hoh Xil-Bayan Har Plate, (8) Xizang (Tibet) Plate, (9) South China Plate, (10) western corner of the Sibumasu (= Shan Thai-Malaya) Terrain and (11) northern margin of the India Plate. It should be pointed out that Wang *et al.* (1990, p. 39) indicated that the Tarim, Qaidam and North China plates were united and became a vast northern continent during the Devonian Period.

Palaeogeography

At the beginning of the Early Devonian, most parts of China were above sea level except in some marginal and interarc areas of the above palaeoplates. However, the North China Plate (Fig. 2) was a vast denuded area during the Devonian Period, except for minor marine deposits in its southern margin – the northern Qinling Mountain and along the Chifeng–Aohan Banner–Jilin belt in its northern margin.

In the southern corner of the Siberian Plate, both sides of the present Altai Mountains were covered by shallow sea with frequent acidic volcanic eruptions, while the central part of the Altai Mountains was uplifted in the Devonian Period.

Northern Xinjiang (excluding the Altai area) belonged to the eastern part of the Kazakhstan Plate. The eastern and western Junggar were probably island arc and trench belts respectively. In northern Tianshan transgression developed from east to west along the Eren Habirga Mountain–Karlik Mountain Fault in the Early Devonian, and led to a narrow shelf up to the Late Devonian.

The central part of the Tarim Plate was characterized by a vast littoral plain under an arid climate. The northern margin of the Tarim Plate (South Tianshan) and the southern margin of the Tarim Plate (Tiekelik Mountain) were shallow shelves in the Early to early Late Devonian. However, in the Tiekelik Mountain and the eastern part of the South Tianshan a regression occurred resulting in a littoral–delta environment in the late Late Devonian, whereas a shallow shelf environment remained in the other parts of South Tianshan as before.

The northern Qilian Mountain and Hexi Corridor area (i.e. northern part of Qilian-Qaidam Plate) were an intramontane zone, lacking Lower Devonian deposits, but with molasse formation in the Middle Devonian. It changed into a braided stream–lake district and extended to the southern Qilian Mountain in the Late Devonian. The Qaidam Basin area in the middle part of Qilian-Qaidam Plate was an upwarped land. The areas around the basin became piedmont belts where molasse and volcanic rocks accumulated. The Burhan Budai Mountain and Qimantag area in the southern part of the Qilian-Qaidam Plate were characterized by a shallow shelf environment.

The South China Plate comprised three subregions: Cathayan, Longmenshan and Guangxi-Hunan (including eastern Yunnan and southern Guizhou). The Cathayan area in the east was a vast upwarped district in the Early Devonian. Three intramontane zones developed respectively along the valleys of the Lower Yangtze River, Qiantang River and Minjiang River in the Middle Devonian, where molasse accumulated. The intramontane zones became shallower and changed into shoals or estuaries in the Late Devonian.

The Guangxi-Hunan Subregion was subsiding again and transgression occurred from south to north at the beginning of the Devonian Period. The sea covered only the southern part of this subregion in the late Early Devonian. However, the

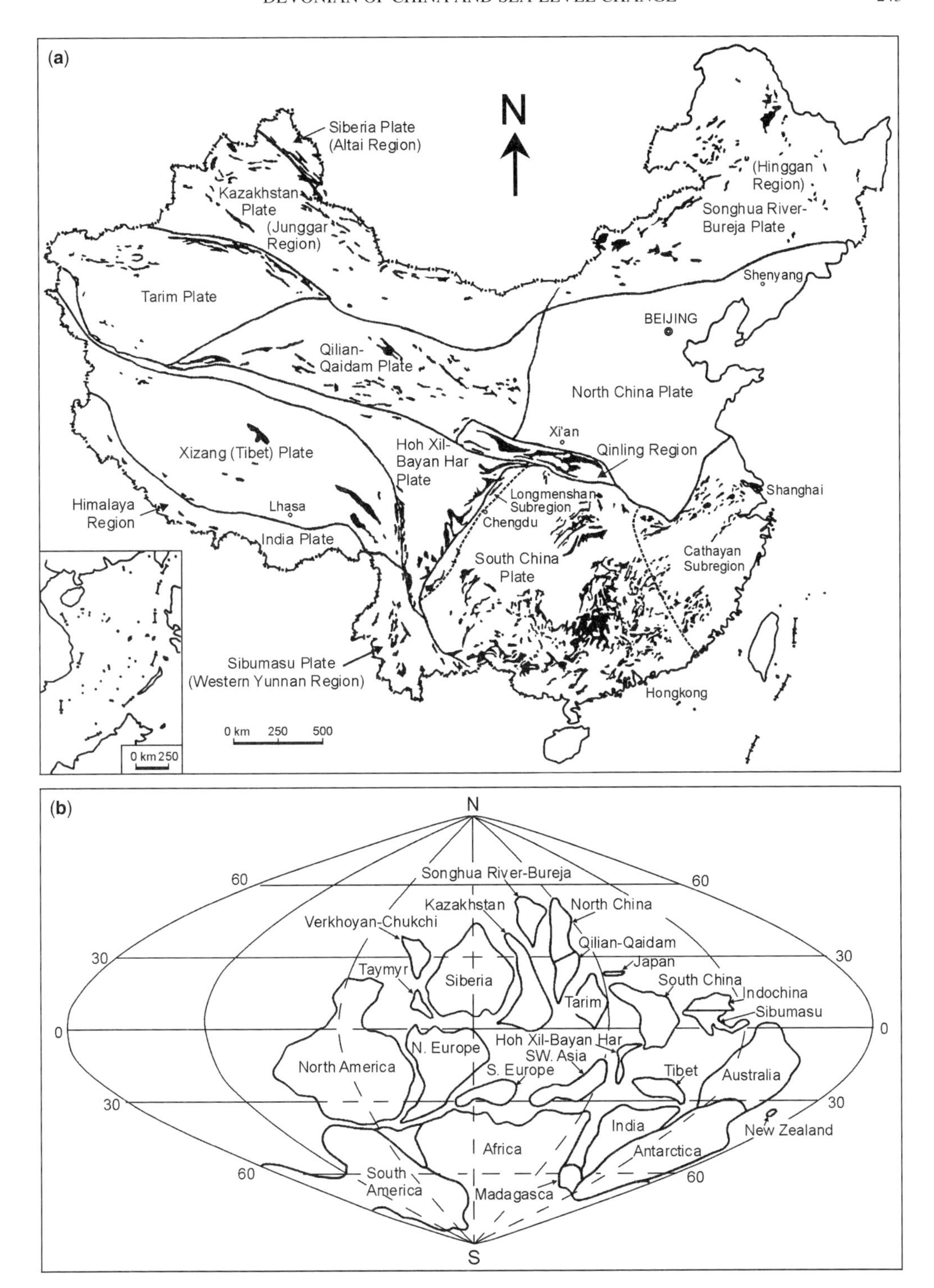

Fig. 1. (**a**) Devonian outcrops, palaeoplates and stratigraphic regions of China. only tratigraphic region names that are different from palaeoplate names are given. Modified from Liao & Ruan (2003, figs. 1, 3). (**b**) Devonian world reconstruction showing spatial position of various China-associated terrains/plates (from Liao & Ruan 2003; modified from Heckel & Witzke 1979).

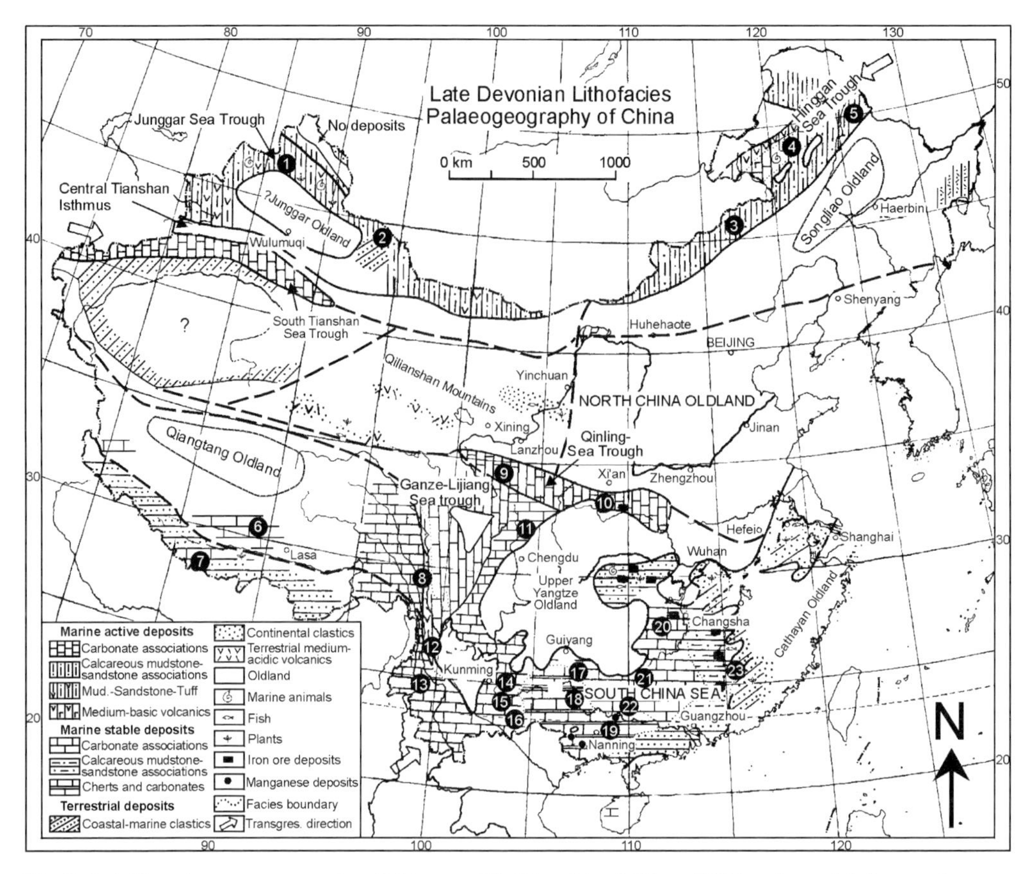

Fig. 2. Late Devonian palaeogeography of China, modified from Hou & Wang (1985) and new data from various sources. Palaeoplate boundary (thick dashed line) based on Liao & Ruan (2003, fig. 3). Selected stratigraphic sections (black circles with numbers): 1, Xar Burd Mountain, western Junggar; 2, eastern Junggar; 3, Dong Ujimquin Qi (= Dong Wuqi); 4, Wunur; 5, Handaqi; 6, Xainza; 7, Yalai, Nyalam County; 8, Mangkang; 9, Dangduogou, Diebu County; 10, Xunyang; 11, Longmenshan; 12, Lijiang; 13, Baoshan; 14, Cuifengshan-Xujiachong, Qujing; 15, Panxi; 16, Wenshan; 17, Dushan; 18, Luofu; 19, Liujing; 20, Qiziqiao-Xikuangshan, central Hunan; 21, Tangjiawan, Guilin; 22, Dale, Xiangzhou County; 23, Sanmentan, Yudu County.

transgression reached the northern part in the Middle and early Late Devonian. In the meantime, some large rift faults also developed and deeper troughs formed as a result. These rift faults cut the epicontinental sea into isolated platforms, forming a network of rift valleys (deeper water) and carbonate platforms (shallow water) (see Ma & Bai 2002, fig. 1). This situation continued in the southern part of the Guangxi-Hunan area in the late Late Devonian, but the northern part was undergoing a regression from north towards south, and platforms gradually changed into shoals and deltas.

Northeastern China belonged to the southern part of the Songhua River-Bureja Plate. Most of northeastern China was an upwarped land. Devonian sea was restricted to the Da (= great) Hinggan Mountain-Hegen Mountain belt (entire Devonian Period), the Mishan-Baoqing belt (entire Devonian), the Yichun-Yuquan belt (Early–Middle Devonian) and Inner Mongolia Prairie belt (Early and Late Devonian).

The Hoh Xil-Bayan Har Plate had been deformed due to long-term ambient pressure from adjacent plates during the post-Devonian. Information about the Devonian palaeogeography in the central part of the plate is lacking. But these were rift valleys at the margin of the plate during the early Early Devonian and the northern part of the valley became a restricted lagoon-bay in the late Early Devonian. The northern and middle parts became an open platform environment and the southern part had a deltaic environment during the Middle and Late Devonian.

The Xizang (Tibet) Plate comprises most of the Xizang Autonomous Region of China, and the area north of the Yarlung Zangbo River was occupied by a wide shallow platform. The belts along the Nujiang and Jinsha River were two denuded lands in the Early Devonian, but changed into rift valleys in the Middle and Late Devonian.

The Himalayas, the northern margin of the India Plate, was occupied by a deeper outer shelf in the Early Devonian, and then became relatively shallow, changing into a shoal and delta environment.

The northern corner of the Sibumasu Plate in western Yunnan was separated by three large rift faults into four upwarped belts in the Early Devonian. In deeper troughs, pelitic and siliceous materials were accumulated with predominantly pelagic faunas. In the Middle and Late Devonian the troughs became shallower and changed into a littoral to shelf environment.

Biostratigraphy

Abundant fossil groups have been studied, especially from South China, and the Devonian framework has been firmly established (e.g. Liao *et al.* 1978; Hou *et al.* 1988; Hou & Wang 1988; Ruan & Mu 1989; Wang 1989; Zhong *et al.* 1992; Bai *et al.* 1994). They are grouped into zones or assemblage zones for correlation of different stratigraphic sections (Fig. 3). Conodonts, dacryoconarid tentaculites, entomozoan ostracods and ammonoids are major pelagic fossils that serve for correlation, whereas graptolites are present only in some local areas, e.g. the Qinzhou area in southern Guangxi.

South China

South China is the most important area for the study of the Devonian System in China. The Chinese regional stages were all established based on stratigraphic studies from this region.

The Longmenshan section represents one of the best-studied sections in China (Hou *et al.* 1988). It may be divided into three lithological intervals. The lower interval is a clastic interval (Fig. 4a), i.e. the Pingyipu Group (Lochkovian–Pragian) including the Guixi to Guanshanpo formations, hosting some fish, ostracods and '*Lingula*' brachiopods and minor plant fragments. The middle interval is an alternating clastic and carbonate interval of the Emsian through the Early Givetian (Bailiuping through Jinbaoshi formations). The upper interval is a carbonate interval of the Middle Givetian through the end of the Famennian. There are three reef-building intervals: late Emsian (Ertaizi Formation), Early and Middle Givetian (upper Jinbaoshi and lower Guanwushan formations), and Late Frasnian (Xiaolingpo Formation).

Qinling Region (excluding the northern Qinling area) was closely related to the South China Region in terms of faunal affinity. The Devonian System in the Qinling region was well documented by XIGMR & NIGP (1987). For general lithological and biological aspects, see Figure 5.

The Cathayan (also known as Southeast China) Subregion is mainly characterized by clastic deposits (with abundant fish and plant fossils) of various ages from the Middle Devonian (Givetian) to Late Devonian, e.g. the Yudu section (Fig. 5).

Devonian strata are widely distributed in the Guangxi-Hunan Subregion. It may be divided into four types based on lithologies and faunal–floral compositions: Qujing, Xiangzhou, transitional and Nandan types (Hou & Wang 1988). See their representative sections (Qujing, Dushan, Liujing and Luofu sections) in Figure 4 for lithologies and fossil compositions.

Qujing type

The Xishancun, Xitun, Guijiatun and Longhuashan formations (in ascending order) in Qujing District of eastern Yunnan represent the type section of the Early Devonian non-marine sediments in South China. Unconformably overlying the Lower Devonian is the Middle Devonian Chuandong Formation of Eifelian age (Cai *et al.* 1994) below and Haikou Formation (= Xichong Formation) above (e.g. Cai *et al.* 1994). The basal dolostone of the Haikou Formation (Fig. 4a) was also named the Shangshuanghe Formation (BGMRYP 1990).

Previously the Miandiancun Formation (shale) (= upper part of the Yulongsi Formation) was considered to be the lowermost Devonian (e.g. Hou & Wang 1988). It is now widely accepted that it belongs to the Silurian (Fang *et al.* 1994; Cai 2000).

The Xishancun Formation (also known as Xiaxishancun Formation) is characterized by greyish-black sandstones and shales (weathering colour is greyish yellow). Fish are widespread in this Lochkovian formation, and include the agnathans (*Yunnanogaleaspis*, *Polybranchiaspis*, *Dongfangaspis*, *Laxaspis*, *Diandongaspis*, *Nochelaspis*) and the antiarchs (*Yunnanolepis*, *Heteroyunnanolepis*, *Chuchinolepis*, *Minicrania*). Plants in this formation include *Zosterophyllum* and some algae, in addition to some gastropods and ostracods (Fig. 4a).

The Xitun Formation is characterized by greyish blue shales and marls, with relatively common bivalves, ostracods, brachiopods ('*Lingula*'), in addition to plants and fish, e.g. *Yunnanolepis*. Fang *et al.* (1994) considered that the upper part of the Xishancun Formation and the Xitun Formation were equivalent to the Lianhuashan Formation in the Liujing section (Fig. 4b) and assigned an age range from Late Lochkovian to

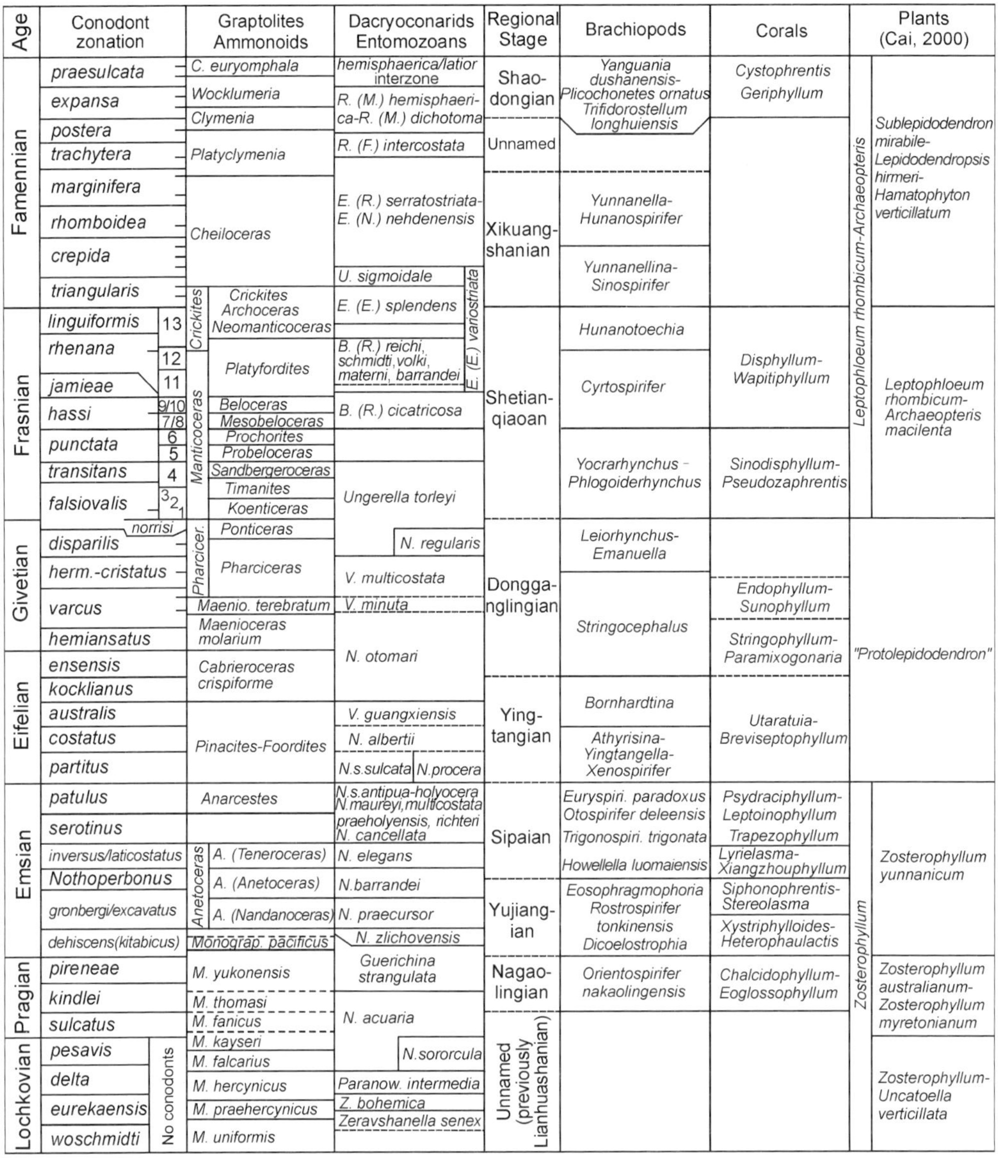

Age	Conodont zonation		Graptolites Ammonoids	Dacryoconarids Entomozoans	Regional Stage	Brachiopods	Corals		Plants (Cai, 2000)
Famennian	praesulcata		C. euryomphala	hemisphaerica/latior interzone	Shaodongian	Yanguania dushanensis-Plicochonetes ornatus Trifidorostellum longhuiensis	Cystophrentis Geriphyllum	Leptophloeum rhombicum-Archaeopteris	Sublepidodendron mirabile-Lepidodendropsis hirmeri-Hamatophyton verticillatum
	expansa		Wocklumeria	R. (M.) hemisphaerica-R. (M.) dichotoma					
	postera		Clymenia						
	trachytera		Platyclymenia	R. (F.) intercostata	Unnamed				
	marginifera		Cheiloceras	E. (R.) serratostriata-E. (N.) nehdenensis	Xikuangshanian	Yunnanella-Hunanospirifer			
	rhomboidea								
	crepida								
	triangularis		Crickites: Crickites, Archoceras	U. sigmoidale; E. (E.) splendens		Yunnanellina-Sinospirifer			
Frasnian	linguiformis	13	Neomanticoceras	E. (E.) variostriata	Shetianqiaoan	Hunanotoechia	Disphyllum-Wapitiphyllum		Leptophloeum rhombicum-Archaeopteris macilenta
	rhenana	12	Manticoceras: Platyfordites	B. (R.) reichi, schmidti, volki, materni, barrandei					
	jamieae	11				Cyrtospirifer			
	hassi	9/10, 7/8	Beloceras, Mesobeloceras	B. (R.) cicatricosa					
	punctata	6, 5	Prochorites, Probeloceras						
	transitans	4	Sandbergeroceras	Ungerella torleyi		Yocrarhynchus-Phlogoiderhynchus	Sinodisphyllum-Pseudozaphrentis		
	falsiovalis	3, 2, 1	Timanites, Koenticeras						
Givetian	norrisi								
	disparilis		Pharciceras: Ponticeras	N. regularis	Dongganglingian	Leiorhynchus-Emanuella			
	herm.-cristatus		Pharciceras	V. multicostata			Endophyllum-Sunophyllum		
	varcus		Maenio. terebratum	V. minuta		Stringocephalus			
	hemiansatus		Maenioceras molarium	N. otomari			Stringophyllum-Paramixogonaria		"Protolepidodendron"
Eifelian	ensensis		Cabrieroceras crispiforme						
	kocklianus								
	australis		Pinacites-Foordites	V. guangxiensis	Yingtangian	Bornhardtina	Utaratuia-Breviseptophyllum		
	costatus			N. albertii		Athyrisina-Yingtangella-Xenospirifer			
	partitus			N.s.sulcata; N.procera					
Emsian	patulus		Anarcestes	N.s.antipua-holyocera N.maureyi, multicostata	Sipaian	Euryspiri. paradoxus	Psydraciphyllum-Leptoinophyllum	Zosterophyllum	
	serotinus			praeholyensis, richteri N. cancellata		Otospirifer deleensis			
	inversus/laticostatus		Anetoceras: A. (Teneroceras)	N. elegans		Trigonospiri. trigonata	Trapezophyllum		Zosterophyllum yunnanicum
	Nothoperbonus		A. (Anetoceras)	N.barrandei		Howellella luomaiensis	Lyrielasma-Xiangzhouphyllum		
	gronbergi/excavatus		A. (Nandanoceras)	N. praecursor	Yujiangian	Eosophragmophoria Rostrospirifer tonkinensis	Siphonophrentis-Stereolasma		
	dehiscens(kitabicus)		Monograp. pacificus	N. zlichovensis		Dicoelostrophia	Xystriphylloides-Heterophaulactis		
Pragian	pireneae		M. yukonensis	Guerichina strangulata	Nagaolingian	Orientospirifer nakaolingensis	Chalcidophyllum-Eoglossophyllum		Zosterophyllum australianum-Zosterophyllum myretonianum
	kindlei		M. thomasi						
	sulcatus		M. fanicus	N. acuaria	Unnamed (previously Lianhuashanian)				
Lochkovian	pesavis	No conodonts	M. kayseri, M. falcarius	N.sororcula					
	delta		M. hercynicus	Paranow. intermedia					Zosterophyllum-Uncatoella verticillata
	eurekaensis		M. praehercynicus	Z. bohemica					
	woschmidti		M. uniformis	Zeravshanella senex					

Fig. 3. Pelagic fossil zones and major faunal and floral assemblages or associations in South China against the standard conodont zonation. Most conodont zones have also been established based on occurrences of index conodonts in South China. The Chinese regional stages are also shown for comparison. Modified from Liao & Ruan (2003, tables 3–5), Cai (2000, table 6-1) and new data. Frasnian Montagne Noir zonation (1–13) is based on Klapper & Becker (1999).

Early Pragian based on analysis of spore, fish, bivalve, ostracod and plant biota.

The Guijiatun Formation consists largely of brownish red (weathering colour) sandstones, siltstones and mudstones; fossils are rare, including minor mollusc, fish and plant fossils.

The Longhuashan Formation (also known as Xujiachong Formation) is composed of greyish yellow (weathering colour) sandstones and shales, with abundant plant and spore fossils. There are diverse views regarding the age of this formation, including Pragian, or Emsian or Pragian to Emsian age (see Wang *et al.* (2002) for a summary of controversies). Comprehensive analyses of spore assemblages (Gao 1981), stratigraphic sequence, floras and faunas (fish and bivalves) and biostratigraphic

(a)

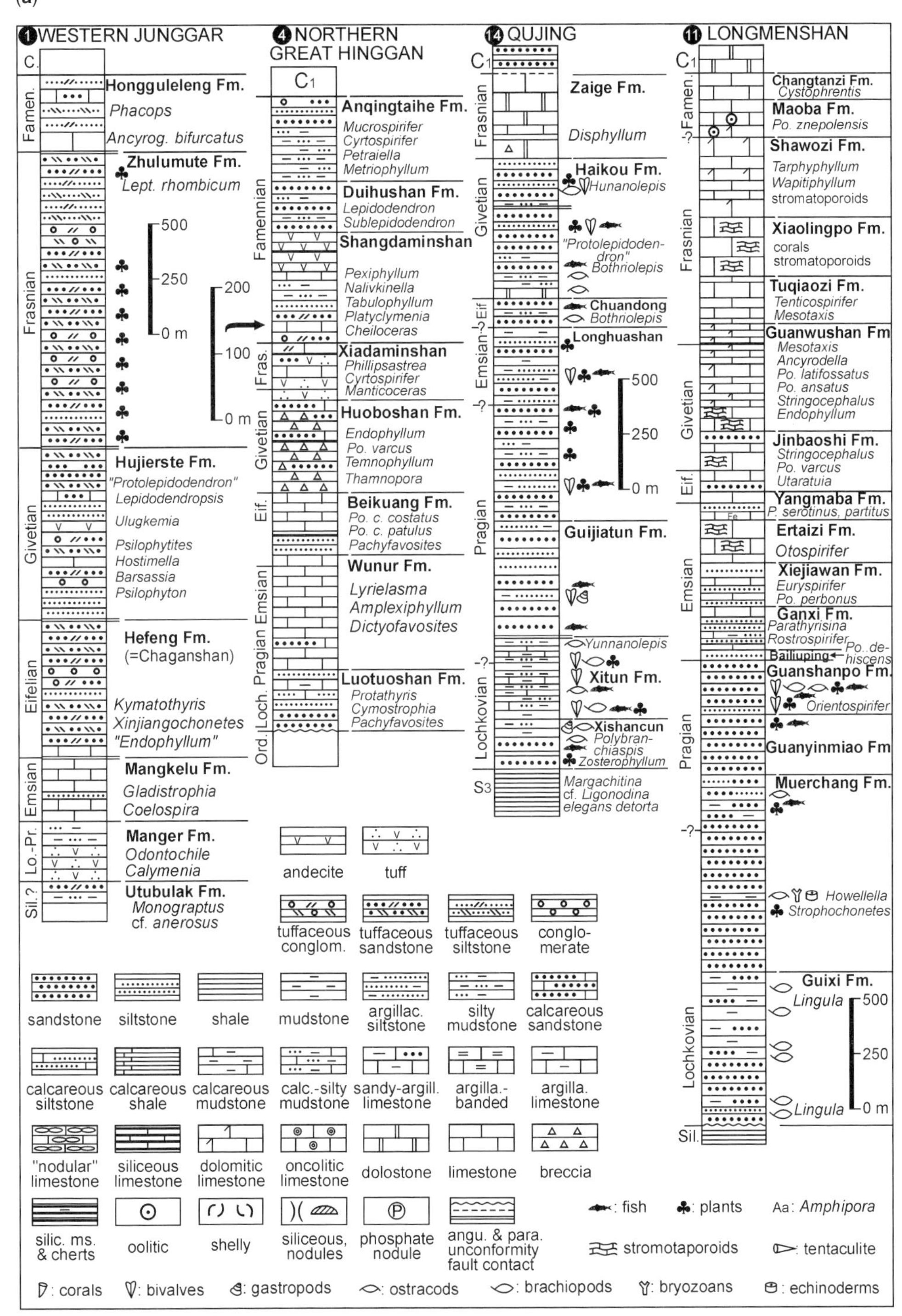

Fig. 4. (**a**) and (**b**) Representative Devonian sections from northwestern, northeastern and South China. Data from various sources, including Liao *et al.* (1979), Hou & Wang (1988), Hou *et al.* (1988), Kuang *et al.* (1989), BGMRNMAR (1991), Xiao *et al.* (1992), Zhong *et al.* (1992), Ji (1994), Hou *et al.* (2000) and Jiang *et al.* (2000).

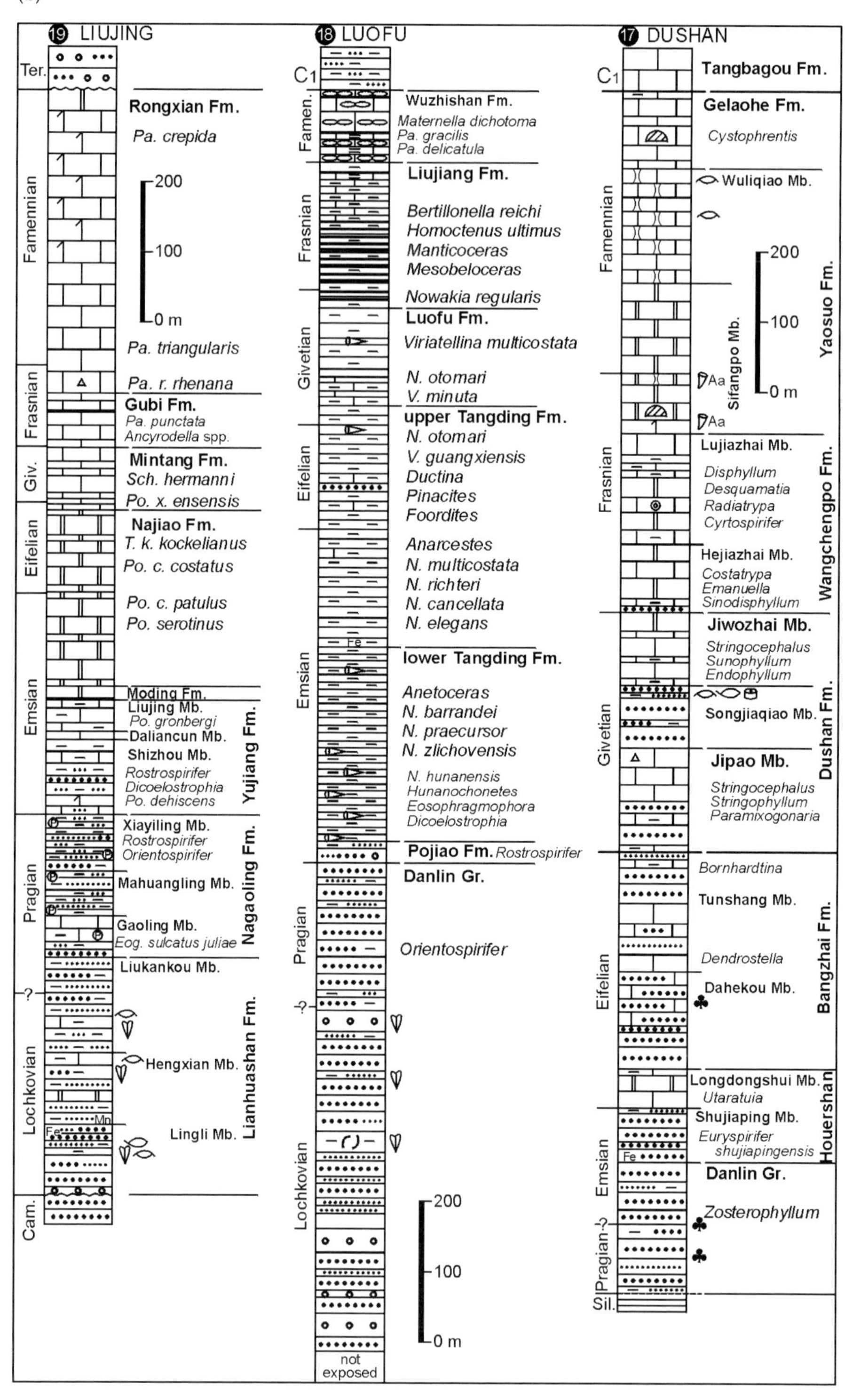

Fig. 4. (*Continued*)

Age	9 Western Qinling	10 Eastern Qinling	15 Panxi (Yunnan)	16 Wenshan (Yunnan)	22 Dale (Guangxi)	21 Tangjiawan (Guangxi)	20 central Hunan	23 Yudu (southern Jiangxi)
Famennian	**Yiwa Fm. (lower)** Limestone, *Cystophrentis* cyrtospiriferids >195 m **Doushishan Fm.** Limestone, minor cherts, *Yunnanella*, *Yunnanellina*, *Polygnathus semicostatus* 660 m	**Tieshan Fm.** Upper part, 304 m: platy limestone and argillaceous limestone, minor calcareous sandstone, shale, and limestone breccia, bearing cyrtospiriferids, *Yunnanella*		?	(eroded upwards) **Rongxian Fm.** 80 m Thick bedded limestone. *Pa. rhomboidea*, *Yunnanellina*	**Etoucun Fm.** 63 m Limestone, foraminifers, stromatoporoids, and corals **Dongcun Fm.** 486 m Dolomitic limestone, with foraminifer *Septatournayella rauserae* near top	**Menggong'ao Fm.** Limestone, shale 100m **Shaodong Fm.** 100m Fine clastics, *S. rauserae* **Oujiachong Fm.** 116m Clastics, fish, plants **Xikuangshan Fm.** Limestone, shale 300m **Changlongjie Fm.** shale, marls 55m	**Zhangdong Fm.** 400m Mudstone, sandstone, siltstone. *Leptophloeum rhombicum*, *Sublepidodendron mirabile* **Sanmentan Fm.** ~50m calcareous siltstone and mudstone, minor sandstone&limestone, hematite. *Yunnanella*&cyrtospiriferids
Frasnian	**Cakuohe Fm.** 547 m Limestone, shale and siltstone. *Cyrtospirifer Palmatolepis gigas*, *P. proversa*, *P. poolei* **Pulai Fm.** 267 m Clastics, minor limestone, *Icriodus symmetricus*, *Nowakia otomari*, *Grypophyllum*	Lower Part, 246 m: Medium bedded to massive, minor thin-bedded limestones, minor carbonaceous limestone and shale, and limestone breccia, *Disphyllum*	**Zaijieshan Fm.** 144m Grey thick bedded dolostone and limestone. Corals *Pseudozaphrentis*; stromatoporoids. **Yidade Fm.** 345 m Argillaceous limestone, limestone; shale at base, brachiopods *Emanuella*, *Cyrtospirifer*, *Desquamatia*, *Spinatrypina*, *Phlogoiderhynchus*	**Rongxian Fm.** 520 m Medium to thick bedded oolitic limestone, minor dolostone and limestone, with argillaceous limestone in the lower part *Amphipora*	**'Gubi' Fm.** 98 m Thick and thin bedded limestone. *Pa. rhenana*, *P. punctata*, Ancyrodella; corals *Pseudozaphrentis*	**Guilin Fm.** 512 m Lagoonal dolomitic limestone, dolostone. Stromatoporoids, corals, brachiopods cyrtospiriferids, *Emanuella*	**Laojiangchong Fm.** Marly limestone, 50 m brachiopods, corals **Qilijiang Fm.** 140 m Reefal limestone **Longkouchong Fm.** Clastics, with 250m corals and brachiopods	**Zhongpeng Fm.** 330m Sandstone, minor shale and siltstone, mainly plants (*Sphenopteridium*, *Leptophloeum rhombicum*, *Cyclostigma kiltorkensis*, *Sublepidodendron*), minor cyrtospiriferid brachiopods and fish *Bothriolepis*
Givetian	**Xiawuna Fm.** 323 m Sandstone-siltstone-shale-limestone cycles *Stringocephalus Geranocephalus Dendrostella*	**Yanglinggou Fm.** Sandstone, slates, limestone, dolostone. 595 m *Stringocephalus*, atrypids and corals **Dafenggou Fm.** Sandstone and limestone *Stringocephalus*, corals 188 m	**Qujing Fm.** 956 m Limestone. Corals, brachs *Stringocephalus*. **Haikou Fm.** 206 m Mudstone, dolomitic limestone, minor sandstone	**Qujing Fm.** 1078 m Limestone, minor dolostone. Brachiopods *Stringocephalus*; Corals *Temnophyllum*, *Grypophyllum*, *Dendrostella*	**Baqi Fm.** 22m Platy limestone and chert intercalations. *Mesotaxis falsiovalis*, *Po. rhenanus* **Donggangling Fm.** 268 m. Limestone and shale. *Stringocephalus*, *Emanuella*; *Sunophyllum*	**Tangjiawan Fm.** 227m Dolostone and limestone; Stromatoporoids, corals *Stringophyllum*, *Sunophyllum*; brachiopods *Stringocephalus*	**Qiziqiao Fm.** 362 m Reefal limestone, corals and stromatoporoids **Yijiawan Fm.** 25 m Shale, marls, with corals *Endophyllum*, *Stringophyllum*; *Stringocephalus* **Tiaomajian Fm.** 50m Clastics, fish, plants	**Yunshan Fm.** 250 m Sandstone, conglomerate, minor siltstone and shale. Plants *Barrandeina* cf. *duslinna*, *Bucheria*, *Lepidodendropsis*, minor fish *Bothriolepis* cf. *kwangtungensis*
Eifelian	**Lure Fm.** 239 m Limestones, minor shale and siltstone. *Uncinulus*-*Indospirifer*-*Productella*	**Shijiagou Fm.** 154 m Argillaceous limestone, reef limestone, dolostone, silty slate, minor siltstone and sandstone. *Athyrisina*, *Sociophyllum*, *Utaratuia*	**Nanpanjiang Fm.** Limestone 75 m Corals *Grypophyllum*; brachiopods *Bornhardtina* *Acrothyris*, *Eoreticularia* **Cuifengshan Gr.** 660 m. Sandstone; base is not exposed	**Gumu Fm.** 420 m Limestone. Corals *Favosites*, *Tryplasma*	**Yingtang Fm** Changcun Mb. 180m shale, mudstone Guche Mb. 173m marly limestone and shale. *I. c. corniger* Gupa Mb. 90m shale, marls	**Xindu Fm.** Sandstone (Covered)		
Emsian	**Dangduo Fm.** 180 m Clastics and limestone. Abundant corals and brachiopods **Gala Fm.** ~600 m Predominently dolomite. Corals *Lyrielasma*, *Siphonophrentis*	**Gongguan Fm.** Medium to thick bedded dolostones. 550 m thick Ostracods, gastropods, and stromatoporoids		**Zhichang Fm.** 176 m Massive dolostone. *Zdimir*, *Megastrophia*; *Nowakia praecursor* **Pojiao Fm.** 174 m Mudstone and shale. *Calceola*, *Dicoelostrophia*, *Rostrospirifer tonkinensis*	**'Sipai' Fm.** 928 m. Marly limestone, shale, minor dolostone *Euryspirifer* **'Yujiang' Fm.** 209 m. Shale, sandstone, minor siltstone. *Rostrospirifer*			
Pragian	**Shangputonggou Fm** Slates, dolomitic siltstone, minor limestone, minor dolostone at top, corals and brachiopods 337 m	**Xichahe Fm.** 189 m Conglomerate, sandstone, slate. Rare bivalves		**Posongchong Fm.** Shale, siltstone, 241m fine sandstone. Plants *Zosterophyllum*	**Dayaoshan Gr.** (= Lianhuashan Fm.) Sandstone and siltstone, minor plant, fish, and bivalve fossils. 1158 m			
Lochkovian	**Xiaputonggou Fm.** Slates, minor limestone, bearing abundant corals, brachiopods etc., with conodont *Icriodus w. woschmidti* at base 175 m		?					

Fig. 5. Devonian stratigraphy of selected section areas from South China. Data are from various sources, including Liao *et al.* (1979), XIGMR & NIGP (1987), Hou & Wang (1988), Zhong *et al.* (1992), Bai *et al.* (1994) and Hou *et al.* (2000).

method indicate that the Longhuashan Formation is of late Pragian–early Emsian age (Wang *et al.* 2002; Wang 2007). Seven plants have been described in detail from this formation: *Hsüa*, *Bracteophyton*, *Huia*, *Hedeia*, *Guangnania*, *Zosterophyllum* and *Drepanophycus*. Among them, *Drepanophycus* is widespread in the horizons. The mid–lower plant assemblage of the Longhuashan Formation is comparable with the upper *Baragwanathia* flora of Australia and the Pragian Posongchong flora of Wenshan District of southeastern Yunnan, with such identical taxa as *Zosterophyllum*, *Huia*, *Guangnania* and *Hedeia*. In the uppermost part of the Longhuashan Formation, there are fish such as agnathans (*Sanchaspis*, *Eugaleaspis*, *Pterogonaspis* and *Gantarostrataspis*), antiarchs (*Mizia* and *Yunnanolepis*), petalichthyids (*Pampetalichthys* and arthrodires (*Szelepis*), associated with some bivalves (*Cimitaria*, *Modiomorpha*, *Antactinodion*).

The Chuandong Formation is composed of sandstones and silty mudstones, with conglomerate at the base. Fish are mostly present in the upper part, including *Eurycaraspis incilis*, *Kenichthys campbelli*, *Heimenia* sp., Wudinolepididae, *Microbrachius chuandongensis*, *Hunanolepis* sp., *Xichonolepis qujingensis* and *Bothriolepis tungseni* (Zhu & Wang 1996; Wang & Zhang 1999; Liu 2002). It is generally considered to be Eifelian in age (Hou & Wang 1988). However, there are also other opinions regarding its age, such as late Emsian (Zhu & Wang 1996) and Givetian (e.g. Liu 2002).

The Haikou Formation contains plants such as *Minarodendron*, *Dimeripteris*, *Panxia*, *Psilophyton* and *Tsaia*, and spores of the *Archaeozonotriletes variabilis–Cymbosporites magnificus* (VM) assemblage zone (Tian & Zhu 2005), and it is generally considered to be Givetian in age (Hou & Wang 1988).

The Upper Devonian Zaige Formation has a varied thickness from tens to hundreds of metres. It is composed of thick-bedded crystalline dolostones, minor limestones and chert nodules, also with some brecciated carbonate layers, yielding corals (*Disphyllum* and *Hunanophrentis*; Frasnian) in the lower part, and ostracods (*Leperditia*) and corals (*Syringopora* and *Michelinia*) in the upper part (Hou & Wang 1988).

Xiangzhou type

The Xiangzhou type represents deposits of shallow marine facies, hosting abundant benthic fossils. Most sections selected from this subregion may be assigned to the present type, including Panxi, Wenshan, Dale (Lower and Middle Devonian) and Qiziqiao-Xikuangshan sections (Fig. 5).

The Dale section is one of the best representative sections (Bai *et al.* 1982, 1994). In this section, the Dayaoshan Group unconformably overlies the Cambrian grey slate with grey sandstone interbeds. The 'Yujiang' Formation may be divided into the Xiaoshan Member (Mbr; also known as Xiaoshan Formation (Fm) by some workers) below and the Tonggeng Mbr (or Fm) above. The Xiaoshan Mbr, about 40–60 m thick, is composed of grey medium- to thick-bedded quartzose sandstone, bearing minor brachiopods, bivalves, plant and fish fragments, with *Orientospirifer wangi* and *Dicoelostrophia punctata* in the uppermost part. The Tonggeng Mbr, about 168–200 m, is made up of shale, siltstone, limestone and minor dolostone, and yields a diverse fauna including brachiopods (*Orientospirifer wangi*, *Dicoelostrophia punctata*, *Rostrospirifer tonkinensis*, *Acrospirifer*, *Athyris*, *Elymospirifer*, *Howellella*, *Nadiastrophia*, *Parathyrisina*, *Cymostrophia*), coral (*Calceola*), trilobites (*Gravicalymene maloungkaensis*, *Basidechenella liujingensis*), tentaculite (*Nowakia zlichovensis*), ostracods and gastropods and bivalves.

The 'Sipai' Formation may be divided into three members or formations, in ascending order: Luomai (or Ertang), Luetang (or Guanqiao), and Dale. The Luomai Mbr, 106–120 m thick, is composed of shale/mudstone interbedded with limestone, yielding the brachiopods *Howellella luomaiensis*, *Acrospirifer*, the corals *Cystolyrielasma*, *Pseudomicroplasma*, *Xiangzhouphyllum*, *Lyrielasma*, the ammonoid *Erbenoceras ellipticum*, the conodonts *Po. perbonus* associated with *Po. dehiscens*, the tentaculite *Nowakia* gr. *praecursor*, and the ostracod *Leperditia elliptica*. The Luetang Mbr, 335 m thick, consists of grey shale intercalated with medium-bedded light grey dolostone, yielding minor fossils including the brachiopods *Trigonospirifer trigonata*, the trilobite *Shipaia hexaspina*, and the very large ostracod *Paramoelleritia xiangzhouensis*. The Dale Mbr (or Fm), 487 m thick, is characterized by marlstones, argillaceous and marly limestones, and minor shales. It can be divided into three submembers (or members) in ascending order: Shipeng, Liuhui and Dingshanling. The Shipeng Submember, about 179 m thick, yields brachiopods (*Trigonospirifer trigonata* and *Reticulariopsis indifferens*), very large ostracods (*Paramoelleritia xiangzhouensis*), cephalopods (*Nothoceras sianghsiangense*) and conodonts (*Po. inversus*). The Liuhui Submember, about 207 m thick, yields diverse brachiopods (including *Otospirifer daleensis*, *Euryspirifer paradoxus*, *Reticulariopsis indifferens*, *Athyrisina*, *Nadiastrophia*, *Xenospirifer*), diverse corals (including *Trapezophyllum cystosus*, *Embolophyllum*, *Pseudomicroplasma*, *Thamnophylloides*, *Lyrielasma acanthophylloides*, *Cystohexagonaria*, *Thamnophyllum*, *Leptoinophyllum*, '*Phillipsastrea*', *Calceola*, the trilobite *Shipaia hexaspina*, the cephalopod *Nothoceras sianghsiangense*, conodonts *Po. serotinus* and *Po. inversus*, and the tentaculite *Nowakia holyensis*. The Dingshanling Submember, about 92 m thick, yields the corals *Zelolasma*, *Tryplasma*, *Cystohexagonaria*, *Leptoinophyllum*, *Psydracophyllum*, *Nardophyllum*, *Trapezophyllum*, and the brachiopods *Euryspirifer kwangsiensis*, *Indospirifer*, *Lazutkinia*, *Reticulariopsis indifferens*, *Athyrisina*, *Nadiastrophia*.

The Yingtang and Donggangling formations, about 440 m and 270 m thick respectively, are characterized by abundant benthic brachiopods and corals, and minor conodonts such as *Eognathodus bipennatus montensis*, *Icriodus c. corniger* of the *Po. costatus costatus* Zone and *I. lindensis* of the Lower *Po. varcus* Zone. Some benthic fossils may also serve for regional correlation, such as the brachiopods *Xenospirifer fongi*, *Yingtangella sulcatilis*, '*Acrospirifer*' *supraspeciosus*, *Stringocephalus burtini*, and the corals *Utaratuia*, *Dendrostella* and *Sunophyllum*.

The interval from the Baqi Formation upwards represents a different environmental setting that should be assigned to the Transitional type (see below). The Baqi Formation is characterized by greyish black platy limestone and interbedded cherts, with abundant pelagic conodonts (from *Po. rhenanus* to *Mesotaxis falsiovalis* Zones) and tentaculites, apparently representing substantial relative deepening (probably resulting from intense rifting progress).

Transitional type

This type represents deposits in a geographic setting transitional between shallow water (benthic) facies and deeper water (pelagic) facies, including ramp and platform marginal facies. Deposits of this type are important for stratigraphic correlation because of the presence of both benthic and pelagic fossils. Liujing section (Fig. 4b: Moding Formation and upwards) is one of the best representatives. There are a number of studies focusing on its biostratigraphy (e.g. Kuang *et al.* 1989; Bai *et al.* 1994).

The Moding Formation is composed of bedded cherts and dolostones, yielding the tentaculites *Nowakia barrandei*, *N. elegans*, *Viriatellina hercynica*, conodonts *Po. perbonus*, and the ammonoids *Erbenoceras*, *Teicherticeras*, *Convoluticeras*. The overlying Najiao Formation yields conodonts of the Lower and Middle Devonian (Fig. 4b). The Eifelian–Givetian, Givetian–Frasnian and Frasnian–Famennian boundaries have been well defined by Bai *et al.* (1994), Jiang *et al.* (2000) and Ji (1994), respectively, in the light of conodont distribution.

Nandan type

This type refers to deposits in a deeper water (pelagic) facies with abundant pelagic fossils such as tentaculites, ammonoids, pelagic ostracods and conodonts. Lithologically this type may be divided into two subtypes: one predominantly shaly (e.g. Luofu section) and the other predominantly carbonate (e.g. Daliantang section in southeastern Yunnan and Nayi section in western Guangxi). The type section of this facies group is shown in Figure 4b (Luofu section). The Danlin Group (greyish white quartzose sandstone) and the Pojiao Formation are composed of fine to coarse clastic rocks, with minor marine fossils and without fish and plant fossils, representing nearshore deposits. The interval from the Tangding Formation through the Wuzhishan Formation represents the typical Nandan type deposits, which suggests that sea level rose dramatically in the early Emsian. Lithologically the lower part of the Tangding Formation is similar to the upper part (previously called the Nabiao or Tangxiang Formation), except that it is characterized by thin-bedded mudstone. The Emsian–Eifelian boundary is recognized based on the occurrence of the ammonoids *Anarcestes* below and the ammonoids *Pinacites-Foordites* and the tentaculite *Nowakia procera* above. The Eifelian–Givetian boundary should lie somewhere below the top of the Tangding Formation as indicated by the presence of *Viriatellina minuta*, a tentaculite that indicates placement within the Middle *varcus* Zone, at the base of the overlying Luofu Formation (Liao & Ruan 2003, p. 262). The Givetian–Frasnian boundary is indicated by the occurrence of the *Nowakia regularis* Zone, which is close to the base of the Frasnian. The Frasnian–Famennian boundary is probably coincident with the Wuzhishan and Liujiang Formation boundary based on the distribution of conodonts, entomozoan ostracods and tentaculites.

Junggar and Hinggan regions

Junggar region

From the Devonian Period onward (He *et al.* 1994, p. 53), the Junggar region was part of the Kazakhstan Plate. There are two areas without Devonian deposits, the present Junggar basin area and central Tianshan, which are known as the Junggar Oldland and Central Tianshan Isthmus respectively (Wang 1985; Fig. 2).

Around the Junggar Oldland, the Devonian System is characterized by geosynclinal volcanic and clastic associations, with great lithofacies and thickness variations. The thickness varies from 1500 m to 12 000 m. Due to subsequent tectonics, continuous sections are rare. A complete Devonian sequence usually consists of two or more isolated, commonly fault-contacted sections. Sometimes a section, thought to be upright, has proved to be overturned. For example, in the Bulongguoer section, the Frasnian terrestrial plant-bearing deposits and volcanic rocks were previously considered to be late Famennian through early Carboniferous due to the reverse dipping of the sequence (Xiao *et al.* 1992).

Two reference sections of the region have been established, the section in the Xar Burd Mountain area in western Junggar and the section in the Kaokesaiergai Mountain area (Figs 4, 6). The Devonian System in western Junggar described below, in descending order (Xiao *et al.* 1992).

Upper Devonian. The **Hongguleleng Fm** is about 200–700 m thick and is characterized by abundant coral and brachiopod fossils as well as a diverse echinoderm fauna (Hou *et al.* 1993). It consists of two members (Xiao *et al.* 1992). The upper member (*c.* 346 m) is mainly a variegated lithological interval, including tuffaceous and siliceous siltstones, minor argillaceous limestones and sandstones, bearing marine fossils (corals, such as *Amplexocarinia*, brachiopods, bryozoans, crinoids, the trilobite *Phacops*). Its top is continuous with the Lower Carboniferous with similar lithology (shale with marly limestone intercalations), yielding conodonts and corals of the Uppermost Famennian (*praesulcata* Zone) including *Protognathodus collisoni*, *P. meischneri*, *Metriophyllum* and *Cyathocarina*. The lower member (89 m) is mainly composed of limestones and fine clastic rocks (shale and tuffaceous siltstone), bearing abundant fossils, including the corals *Nalivkinella*, *Amplexus*, *Hebukephyllum*, *Tabulophyllum*, *Neaxon*, *Amplexocarinia* (Liao & Cai 1987); the brachiopods *Palaeospirifer sinicus*, *Centrorhynchus turanica*, *Mesoplica semiplicata*; the trilobite *Phacops accipitrinus mobilis*; conodonts, crinoids, and bryozoans. Conodont data suggest that the limestone interval of this formation has a range from the Famennian *Pa. crepida* through *Pa. marginifera* Zones (Zhao & Wang 1990) or from the late Frasnian Upper *rhenana* Zone through early Famennian Middle *crepida* Zones (Xia 1997). Minor ammonoids such as *Cymaclymenia* cf. *striata* have been found laterally in other areas.

The **Zhulumute Fm** is about 540–1330 m. It is composed of fine to coarse tuffaceous clastic rocks (siltstones to fine conglomerates) and minor volcanic rocks. There are abundant plants (*Leptophloeum rhombicum*, *Sublepidodendron*, *Lepidodendropsis*, *Callixylon*, *Astralocaulis*; Cai 2000). This formation is probably of Frasnian age.

Middle Devonian. The **Hujierste Fm** is about 273–765 m thick. It is composed of (tuffaceous) siltstone and sandstone, minor conglomerates, volcanic rocks and limestone lenses, bearing '*Protolepidodendron*', *Lepidodendropsis*, *Colpodexylon*?, *Drepanophycus*, *Psilophyton* and *Barsassia* (Cai 2000); the conchostracan *Ulugkemia* is present in the upper part (Xiao *et al.* 1992; Cai 2000). The lower part may belong to the Eifelian and the upper part to the Givetian according to Cai (2000, p. 116, table 6-5). Recent study shows that its upper part may belong to the early Frasnian (Xu 1999).

The **Hefeng Fm** (= Chaganshan Fm) is over 625 m. It is composed of tuffaceous shale, siltstone sandstone, minor limestone, with abundant corals and brachiopods. Corals include *Favosites*, '*Endophyllum*', *Prismatophyllum*, *Cystiphylloides*, *Tryganolites*, *Squameofavosites*; brachiopods include *Xinjiangochonetes*, *Acrospirifer*, *Schuchertella*, *Xinjiangospirifer* and *Kymatothyris* (Xu 1991). It is probably Eifelian in age according to Xu (1991).

Lower Devonian. The **Mangkelu Fm** is 288 m thick. It is mainly composed of fossiliferous sandy limestones, minor siltstone and sandstone, with abundant corals, brachiopods and bryozoans. Corals include *Syringaxon*, *Gurievskiella*, *Orthopaterophyllum*, *Schlotheimophyllum*, *Squameofavosites*, *Pachyfavosites*, *Emmonsia*, *Favosites*, *Heliolites*, *Placocoenites*, *Alveolites*; brachiopods include '*Paraspirifer*' *gigantea*, *Leptaenopyxis bouei*, *Gladostrophia knodoi*, *Coelospira*, *Kozillowskiella*. It is probably Emsian in age.

The **Manger Fm** is 267 m thick. It is composed of marine silty mudstones and tuffs, intercalated with limestone lenses and tuffaceous siltstones, bearing the trilobites *Odontochile sinensis* and *Calymenia*, the coral *Syringaxon* and the brachiopods *Aulacella* and *Resserella*? The lower part of the formation also yields plant fragments. This formation probably ranges from the Lochkovian(?) to Pragian.

The underlying Utubulak Formation is made up of tuffaceous siltstones, sandstones and tuffs, and minor limestones. It yields abundant corals, brachiopods, graptolites and plants. This formation was considered to be transitional in age between the Late Silurian and Early Devonian (Xiao *et al.* 1992). It is now regarded entirely as the Late Silurian by most workers (e.g. Liao & Cai 1987; Hou *et al.* 2000).

In summary, in the Junggar region, the Lower Devonian deposition was continuous with the Late Silurian. Marine transgression reached a highstand during deposition of Mangkelu Formation, during which time the sea invaded eastwards into western Inner Mongolia, resulting in the deposition of the Zhusileng Formation, yielding ammonoids *Anarcestes*. During the Middle Devonian (Givetian), terrestrial deposits with plant fossils occurred in the Hujierste Formation in western Junggar. During

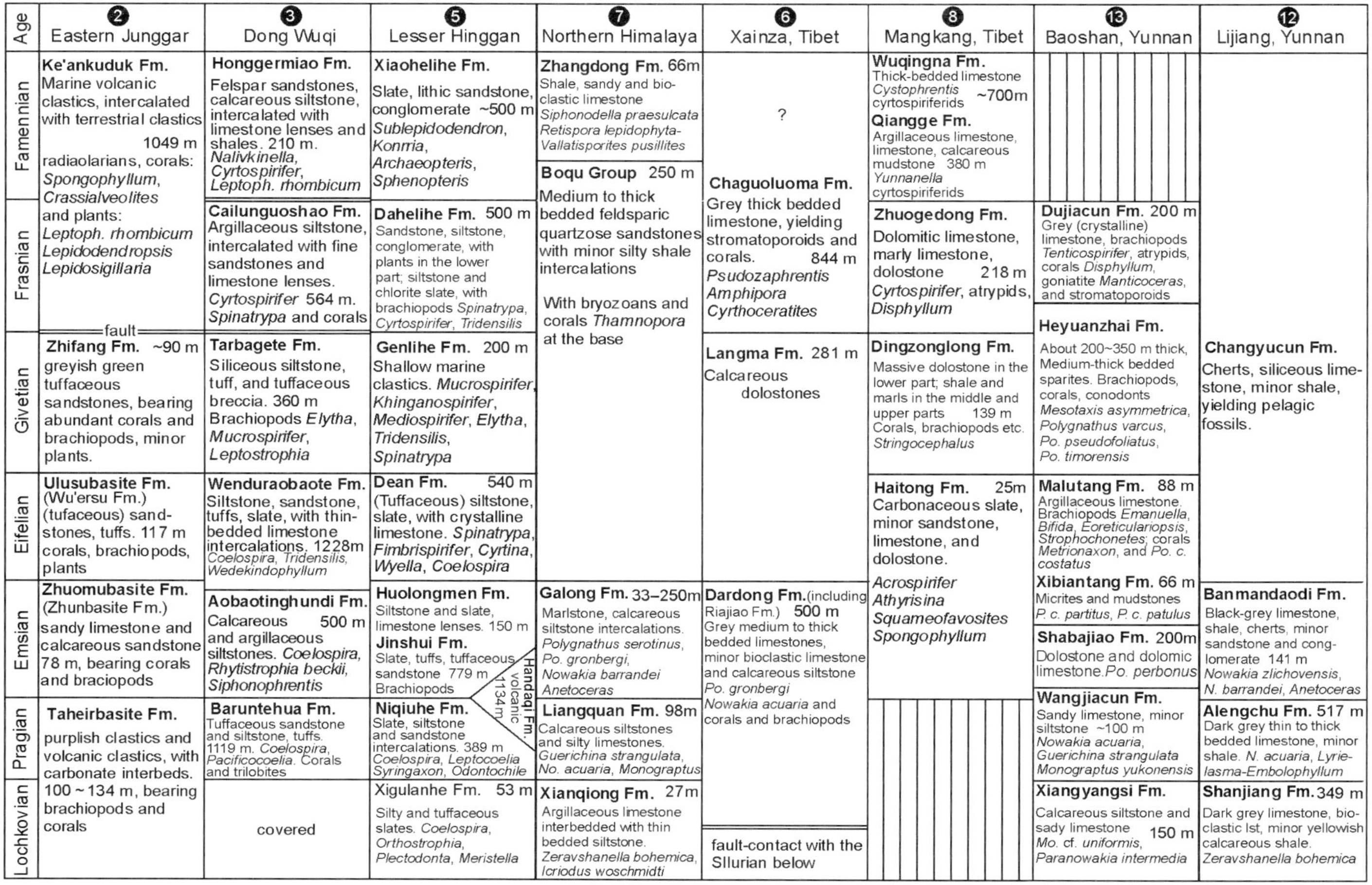

Age	❷ Eastern Junggar	❸ Dong Wuqi	❺ Lesser Hinggan	❼ Northern Himalaya	❻ Xainza, Tibet	❽ Mangkang, Tibet	⓭ Baoshan, Yunnan	⓬ Lijiang, Yunnan
Famennian	**Ke'ankuduk Fm.** Marine volcanic clastics, intercalated with terrestrial clastics 1049 m radiaolarians, corals: *Spongophyllum*, *Crassialveolites* and plants: *Leptoph. rhombicum* *Lepidodendropsis* *Lepidosigillaria*	**Honggermiao Fm.** Felspar sandstones, calcareous siltstone, intercalated with limestone lenses and shales. 210 m. *Nalivkinella*, *Cyrtospirifer*, *Leptoph. rhombicum*	**Xiaohelihe Fm.** Slate, lithic sandstone, conglomerate ~500 m *Sublepidodendron*, *Konrria*, *Archaeopteris*, *Sphenopteris*	**Zhangdong Fm.** 66m Shale, sandy and bio-clastic limestone *Siphonodella praesulcata* *Retispora lepidophyta*-*Vallatisporites pusillites* **Boqu Group** 250 m Medium to thick bedded feldsparic quartzose sandstones with minor silty shale intercalations With bryozoans and corals *Thamnopora* at the base	?	**Wuqingna Fm.** Thick-bedded limestone *Cystophrentis* cyrtospiriferids ~700m **Qiangge Fm.** Argillaceous limestone, limestone, calcareous mudstone 380 m *Yunnanella* cyrtospiriferids		
Frasnian		**Cailunguoshao Fm.** Argillaceous siltstone, intercalated with fine sandstones and limestone lenses. *Cyrtospirifer* 564 m. *Spinatrypa* and corals	**Dahelihe Fm.** 500 m Sandstone, siltstone, conglomerate, with plants in the lower part; siltstone and chlorite slate, with brachiopods *Spinatrypa*, *Cyrtospirifer*, *Tridensilis*		**Chaguoluoma Fm.** Grey thick bedded limestone, yielding stromatoporoids and corals. 844 m *Psudozaphrentis* *Amphipora* *Cyrthoceratites*	**Zhuogedong Fm.** Dolomitic limestone, marly limestone, dolostone 218 m *Cyrtospirifer*, atrypids, *Disphyllum*	**Dujiacun Fm.** 200 m Grey (crystalline) limestone, brachiopods *Tenticospirifer*, atrypids, corals *Disphyllum*, goniatite *Manticoceras*, and stromatoporoids	
Givetian	fault **Zhifang Fm.** ~90 m greyish green tuffaceous sandstones, bearing abundant corals and brachiopods, minor plants.	**Tarbagete Fm.** Siliceous siltstone, tuff, and tuffaceous breccia. 360 m Brachiopods *Elytha*, *Mucrospirifer*, *Leptostrophia*	**Genlihe Fm.** 200 m Shallow marine clastics. *Mucrospirifer*, *Khinganospirifer*, *Mediospirifer*, *Elytha*, *Tridensilis*, *Spinatrypa*		**Langma Fm.** 281 m Calcareous dolostones	**Dingzonglong Fm.** Massive dolostone in the lower part; shale and marls in the middle and upper parts 139 m Corals, brachiopods etc. *Stringocephalus*	**Heyuanzhai Fm.** About 200~350 m thick, Medium-thick bedded sparites. Brachiopods, corals, conodonts *Mesotaxis asymmetrica*, *Polygnathus varcus*, *Po. pseudofoliatus*, *Po. timorensis*	**Changyucun Fm.** Cherts, siliceous lime-stone, minor shale, yielding pelagic fossils.
Eifelian	**Ulusubasite Fm.** (Wu'ersu Fm.) (tufaceous) sand-stones, tuffs. 117 m corals, brachiopods, plants	**Wenduraobaote Fm.** Siltstone, sandstone, tuffs, slate, with thin-bedded limestone intercalations. 1228m *Coelospira*, *Tridensilis*, *Wedekindophyllum*	**Dean Fm.** 540 m (Tuffaceous) siltstone, slate, with crystalline limestone. *Spinatrypa*, *Fimbrispirifer*, *Cyrtina*, *Wyella*, *Coelospira*			**Haitong Fm.** 25m Carbonaceous slate, minor sandstone, limestone, and dolostone. *Acrospirifer* *Athyrisina* *Squameofavosites* *Spongophyllum*	**Malutang Fm.** 88 m Argillaceous limestone. Brachiopods *Emanuella*, *Bifida*, *Eoreticulariopsis*, *Strophochonetes*; corals *Metrionaxon*, and *Po. c. costatus*	
Emsian	**Zhuomubasite Fm.** (Zhunbasite Fm.) sandy limestone and calcareous sandstone 78 m, bearing corals and braciopods	**Aobaotinghundi Fm.** Calcareous 500 m and argillaceous siltstones. *Coelospira*, *Rhytistrophia beckii*, *Siphonophrentis*	**Huolongmen Fm.** Siltstone and slate, limestone lenses. 150 m **Jinshui Fm.** Slate, tuffs, tuffaceous sandstone 779 m Brachiopods **Handaqi Fm.** volcanic 1134 m	**Galong Fm.** 33–250m Marlstone, calcareous siltstone intercalations. *Polygnathus serotinus*, *Po. gronbergi*, *Nowakia barrandei* *Anetoceras*	**Dardong Fm.**(including Riajiao Fm.) 500 m Grey medium to thick bedded limestones, minor bioclastic limestone and calcareous siltstone *Po. gronbergi* *Nowakia acuaria* and corals and brachiopods		**Xibiantang Fm.** 66 m Micrites and mudstones *P. c. partitus*, *P. c. patulus* **Shabajiao Fm.** 200m Dolostone and dolomic limestone.*Po. perbonus*	**Banmandaodi Fm.** Black-grey limestone, shale, cherts, minor sandstone and cong-lomerate 141 m *Nowakia zlichovensis*, *N. barrandei*, *Anetoceras*
Pragian	**Taheirbasite Fm.** purplish clastics and volcanic clastics, with carbonate interbeds. 100 ~ 134 m, bearing brachiopods and corals	**Baruntehua Fm.** Tuffaceous sandstone and siltstone, tuffs. 1119 m. *Coelospira*, *Pacificocoelia*. Corals and trilobites	**Niqiuhe Fm.** Slate, siltstone and sandstone intercalations. 389 m *Coelospira*, *Leptocoelia* *Syringaxon*, *Odontochile*	**Liangquan Fm.** 98m Calcareous siltstones and silty limestones. *Guerichina strangulata*, *No. acuaria*, *Monograptus*			**Wangjiacun Fm.** Sandy limestone, minor siltstone ~100 m *Nowakia acuaria*, *Guerichina strangulata* *Monograptus yukonensis*	**Alengchu Fm.** 517 m Dark grey thin to thick bedded limestone, minor shale. *N. acuaria*, *Lyrie-lasma*-*Embolophyllum*
Lochkovian		covered	Xigulanhe Fm. 53 m Silty and tuffaceous slates. *Coelospira*, *Orthostrophia*, *Plectodonta*, *Meristella*	**Xianqiong Fm.** 27m Argillaceous limestone interbedded with thin bedded siltstone. *Zeravshanella bohemica*, *Icriodus woschmidti*	fault-contact with the Silurian below		**Xiangyangsi Fm.** Calcareous siltstone and sady limestone 150 m *Mo. cf. uniformis*, *Paranowakia intermedia*	**Shanjiang Fm.** 349 m Dark grey limestone, bio-clastic lst, minor yellowish calcareous shale. *Zeravshanella bohemica*

Fig. 6. Devonian stratigraphy of selected section areas from eastern Junggar, Hinggan, Xizang (Tibet) and western Yunnan. Data are from various sources, including Hou & Wang (1988), BGMRNMAR (1991), Xiao *et al.* (1992), BGMRHP (1993), BGMRXAR (1993) and Hou *et al.* (2000).

the Late Devonian, the area of marine deposition shrank, resulting in the huge deposition of the Zhulumute Formation. During the Famennian, local transgression occurred in the western Junggar.

Hinggan region

For a general overview of the Devonian System in this region, three representative sections are shown in Figures 4 and 6. The Beikuang Formation (about 113 m) consists of radiolarian cherts and argillites, siltstones and minor limestones yielding corals, brachiopods and conodonts (*Icriodus c. corninger*, *Po. linguiformis*, *Po. c. costatus*, *Po. c. patulus*), which suggest an Eifelian age.

New geological data show that there are two additional formations overlying the Famennian Shangdaminshan Formation. The Duihushan Fm represents deposition in a littoral facies, yielding the plants *Lepidodendron*, *Sublepidodendron*. The Anqingtaihe Fm (= Hongshuiquan Fm), about 250 m thick, is composed of marine clastic deposits, yielding the brachiopods *Mucrospirifer*, *Cyrtospirifer*, *Tylothyris* and corals *Petraiella*, *Metriophyllum*.

It should be pointed out that the chronological correlation of most formations is based on benthic faunas so that specific correlation is left open for further study, e.g. the Luotuoshan and Wunur formations.

Marine transgression started in the Early Devonian from the NE and reached a highstand during the Emsian, leading to the deposition of the Wunur Formation (reefs). Marine regression began in the Givetian, with deposition of sandstones and breccias of the Huoboshan Formation, hosting plant fragments in addition to marine fossils. During the Late Devonian (Frasnian), a minor transgression probably occurred. The Late Devonian (Famennian) sea-level change was quite similar to that in South China (see below).

Xizang (Tibet) and western Yunnan

Xizang

During the Devonian Period the Xizang Autonomous Region was very complicated in terms of tectonics, sedimentology and palaeobiogeography (Liu *et al.* 1992). It comprised at least three regions on the basis of the Devonian stratigraphic development: (1) Xizang (Tibet) Region (including northern and middle parts of Tibet) (= Qiangtang region of others); (2) Himalaya Region (southern part of Tibet); and (3) Hoh Xil-Bayan Har Region.

Tibet Region. This region, lying north of the Yarlung Zangbo River, comprises most of the Xizang Autonomous Region. A complete section of the Devonian System is well developed in Xainza County (Fig. 2) and is rich in corals, brachiopods, conodonts and dacryoconarids (Yu & Liao 1982; BGMRXAR 1993). The Devonian sequence mainly consists of medium- to thick-bedded limestones and dark grey, thin, sandy limestones (Fig. 6). The Lower Devonian Dardong Formation (including the Riajiao Fm) is over 500 m thick, and yields conodonts (*Polygnathus dehiscens*, *Po. gronbergi*), dacryoconarids (*Sogdina paracuaria*, *Nowakia acuaria*, *N. praecursor*, *N.* aff. *cancellata*, *Guerichina strangulata*), corals *Gurieskiella, Embolophyllum, Martinophyllum*) and diverse brachiopods, including *Leptaena*, *Atrypa*, *Eoglossinotoechia*, *Howellella* and *Linguopugnoides*. The Chaguoluoma Formation yields stromatoporoids and the coral *Pseudozaphrentis*, suggesting a Late Devonian age. However, the cephalopod *Cyrthoceratites* and the conodonts *Icriodus, Polygnathus* and *Bispathodus* indicate a Middle Devonian age.

At Lazhulong, Ngari Prefecture, northwestern Xizang, the Lower and Middle Devonian Yaxier Group is represented by white-grey quartzose sandstone and siliceous limestone intercalated with marbles, about 100–330 m thick, containing the corals *Temnophyllum* and *Heliolites*. The Upper Devonian Lazhulong Formation is composed of grey limestones (alternating with sandstone in the lower part), yielding the corals *Phillipsastraea*, *Peneckiella*, *Coenites*, and some stromatoporoids in the lower part, and the brachiopods *Tenticospirifer*, *Composita*, *Mesoplica*, *Ovatia*, *Whidbornella*, *Cupulorostrum* and *Ptychomaletoechia* and some solitary corals in the upper part. They indicate Frasnian and Famennian ages respectively.

A more complete section of Middle and Upper Devonian rocks is exposed in Qamdo Prefecture in eastern Xizang. The Emsian–Eifelian Haitong Formation is composed of carbonaceous slates, sandstones, argillaceous limestones and dolostones (Fig. 6), about 25 m thick, yielding the brachiopods *Athyrisina*, *Indospirifer*, *Acrospirifer* and corals *Squameofavosites, Spongophyllum*(?). The Givetian Dingzonglong Formation, about 55–150 m thick, is characterized by limestones (sometimes reefs), marls and dolostones and contains brachiopods (*Stringocephalus*, *Emanuella*), and atrypids, corals (*Grypophyllum*, *Temnophyllum*, *Thamnopora*) and some stromatoporoids, gastropods, bivalves and charophytes. The Frasnian Zhuogedong Formation, about 177–218 m thick, consists of marls, limestones (reefs) and dolostones and is rich in brachiopods such as *Hypothyridina*, *Cyrtospirifer*, *Tenticospirifer*, *Spinatrypa*, '*Atrypa*', *Leiorhynchus* and corals such as *Sinodisphyllum*, *Disphyllum*?, *Coenites* and *Alveolitella*, as well as gastropods, bivalves and stromatoporoids. Famennian Qiangge and Wuqingna formations are composed of dark grey limestones intercalated with argillaceous limestones and marls and contain brachiopods, including cyrtospiriferids, *Productella*, *Ptychomaletoechia*, *Ambocoelia* and *Yunnanella* Grabau, 1931*a*. Biogeographically it is apparent that this section is closely related to the South China Region.

Volcanic and carbonate deposits were developed at the eastern boundary of the Xizang Autonomous Region, i.e. along the valley of the Jinsha River.

Himalaya Region. Devonian deposits can be traced here and there along the northern slope of the Himalayas, such as in Zhongba, Nyalam, Tingri and Dinggye of southern Xizang. The Chinese territories are all situated in the Tethyan Himalaya Morphotectonic Zone. The most complete section of the Devonian System is seen near Yalai village of Nyalam county (Fig. 6). The Lower Devonian includes grey siltstones, yellow-grey argillaceous limestones and dark grey, thin-bedded limestones which rest conformably on the Upper Silurian and yield conodonts (*Icriodus* cf. *beckmanni*, *Spathognathodus optimus*, *Polygnathus perbonus, Neoprioniodus latidentatus*, *Ozarkodina remscheidensis remscheidensis*, *Polygnathus serotinus*, *Polygnathus gronbergi*), tentaculites (*Zeravshanella bohemica*, *Nowakia acuaria*, *N. barrandei*, *Guerichina strangulata*, *Metastyliolina, Viratellina, Styliolina*), graptolites *Neomonograptus atopus*, *Monograptus thomasi, Monograptus* cf. *yukonensis*, *Monograptus yaliensis*) and the ammonoid *Anetoceras*. The Middle and Upper Devonian Boqu Group consists of light grey unfossiliferous quartzose sandstones. The latest Devonian Zhangdong Formation (approximately Yaligou Mbr and Goulongri Mbr of the former Yalai Fm) is composed of mostly black shales, containing the coral *Metriophyllum* and the bivalve *Paracyclas*, which pass gradually upwards into a series of medium- to thick-bedded limestones containing the Early Carboniferous ammonoids *Gattendorfia* and *Acutimitoceras* and brachiopods *Tylothyris*, *Pseudosyrinx*, and *Girtyella*.

Hoh Xil-Bayan Har Region. Devonian strata in the region occur as scattered outcrops and are poorly studied. They are commonly slightly metamorphic and poor in fossils. The region is divided into four districts: Rear Longmen Mountain District, Shaluli District, Kunlun Mountains District and Bayan Har District.

Devonian strata are relatively well studied in Zhongza, Batang County, western Sichuan, in the southern part of the Shaluli District, and may be divided into the Gerong, Qiongcuo, Cangna and Talipo formations in ascending order. The Gerong Formation, 132 m thick, is conformable upon the Upper Silurian and consists of medium- to thick-bedded dolostone, containing a few antiarch fish of Early Devonian age and the brachiopod '*Lingula*'. The Lower to Middle Devonian Qiongcuo Formation, with a thickness of 206 m, is composed of medium- to thick-bedded carbonaceous limestone and chert. Abundant fossils include the corals *Parastriatopora*, *Striatopora*, *Favosites*, *Pachyfavosites*, *Alveolites*, *Thamnopora*, *Neocolumnaria* and *Grypophyllum*, and the stromatoporoids *Amphipora* and *Paramphipora*. The Middle Devonian Cangna Formation, about 230 m thick, consists of medium- to thick-bedded limestone and bears rich stromatoporoids *Amphipora* and *Paramphipora*. The Upper Devonian Talipo Formation, 188 m thick, is conformable under the Lower Carboniferous and is mainly composed of dolomitic limestone and subordinate algal limestone.

In Rear Longmen Mountain District, Lower Devonian strata comprise siliciclastic rocks and subordinate dolostone, with occasional intercalated gypsum.

Middle and Upper Devonian strata are mainly composed of terrigenous clastic rocks in the southern part of the region, and of limestone and dolostone in central and northern parts.

Devonian volcanic rocks occur near the borders of the region; for example, intercalated basalt and andesite are present in Lower Devonian strata near Kangding at the eastern margin, basic and intermediate-basic volcanic rocks of both Early and Late Devonian epochs are present along the Jinsha River fault at the western margin, and intermediate and subordinate basic volcanic rocks occur in the Middle Devonian strata in the northern Kunlun Mountains and southern Bayan Har at the northern margin.

Western Yunnan

Western Yunnan Region can be subdivided into Lijiang Subregion and Baoshan Subregion. In the Lijiang Subregion, the Devonian System is mainly composed of deeper water deposits (Fig. 6). The Lower Devonian of Baoshan Subregion consists of argillaceous limestones, marls and siltstones, yielding graptolites and tentaculites. Middle and Upper Devonian strata of Baoshan Subregion are mainly of carbonate deposits, containing benthic corals and brachiopods.

A complete Devonian section is well developed at Malutang and Heyuanzhai, Shidian County, Baoshan Prefecture, western Yunnan. The Devonian sequence is described below in descending order (Hou & Liao in Hou & Wang 1988).

Upper Devonian (Frasnian). The **Dujiacun Formation** is about 200 m thick, and is represented by light grey limestones, containing the brachiopods

Tenticospirifer, '*Atrypa*' and *Pugnax*, the coral *Disphyllum*, the goniatite *Manticoceras* and the stromatoporoid *Stromatoporella*.

The Lower Carboniferous rests unconformably on the Upper Devonian, which means that either Famennian sediments were eroded away or no Famennian sediments were deposited.

Middle–Upper Devonian. The **Heyuanzhai Formation** is about 200–350 m thick, and consists of grey medium- to thick-bedded sparites and minor lime breccias, shelly limestone, and argillaceous limestone. It is rich in brachiopods, such as *Aulacella*, *Uncinulus*, *Devonaria*, *Kayserella*, *Kerpina* and *Indospirifer*, and the corals *Siphonophrentis*, *Macgeea* and *Neoacinophyllum* in the lower part, *Isorthis* (*Tyersella*), *Gypidula* and *Pyramidalia* in the middle part, and *Devonoproductus*, *Pugnoides* and *Spinatrypa* in the upper part. Wang (1994) described the coral fauna from the upper part of the Heyuanzhai Formation, which is characterized by the absence of some typical Old World Realm taxa such as *Disphyllum*, *Cyathophyllum*, *Temnophyllum* and *Endophyllum*. The conodonts *Polygnathus timorensis*, *P. linguiformis*, *P. varcus* and *P. xylus xylus* have been found in the lower part and *Mesotaxis asymmetrica* is present in the upper part.

Middle Devonian. The **Malutang Formation** is about 88 m thick, and is composed of yellow marls, calcareous mudstones and biosparites, containing the brachiopods *Emanuella*, *Aulacella*, *Reticulariopsis*, *Mesodouvillina*, *Strophochonetes*, *Athyris* and *Bifida*, the corals *Metrionaxon*, *Barrandeophyllum* and *Syringaxon*, and the conodont *Polygnathus c. costatus*, which indicates a late Eifelian age.

Lower–Middle Devonian. The **Xibiantang Formation** is about 70–100 m thick, and consists mainly of medium- to thick-bedded micritic limestones intercalated with marls or mudstones, yielding the brachiopod *Strophochonetes* and the conodonts *Polygnathus costatus costatus*, *Polygnathus costatus patulus* and *Polygnathus costatus partitus*.

Lower Devonian. The **Shabajiao Formation** is characterized by dolostones and dolomitic limestones, bearing *Polygnathus perbonus*. It is widely distributed in this subregion.

The **Wangjiacun Formation** is lithologically similar to the underlying Xiangyangsi Formation and is composed of sandy limestones intercalated with siltstones, about 100 m thick. It contains the graptolite *Monograptus yukonensis*, the dacryoconarids *Nowakia acuaria* and *Guerichina strangulata*, and the trilobite *Reedops*, suggesting a Pragian age.

The **Xiangyangsi Formation** is about 0–180 m thick, and is composed of sandy limestones intercalated with siltstones, yielding the conodont *Icriodus woschmidti woschmidti* and the dacryoconarids *Paranowakia intermedia* and *Zeravshanella bohemica*.

Sea-level change in the Devonian Period of South China

Numerous studies deal with Devonian sea-level changes and palaeogeography of China, especially South China, e.g. Zhong *et al.* (1992), Liu *et al.* (1996), Wu *et al.* (1997), Tsien & Fong (1997) and Wang (2001). The following discussion of sea-level changes is exclusively based on data from South China, including Guangxi, Yunnan, Guizhou and Hunan provinces. Generally speaking, the Devonian Period of South China comprises two large transgressive–regressive cycles (= depophases of Johnson *et al.* 1985; Fig. 7), (a) Lochkovian to Eifelian and (b) late Eifelian to about the end of the Devonian Period, a view expressed long ago (Hou & Wang 1985). The first cycle reached its highstand in the mid-Emsian as witnessed by the widespread occurrence of the ammonoid *Erbenoceras* fauna in both benthic and pelagic facies settings. The second cycle reached its highstand probably in the early Late Frasnian during which time the Liujiang Formation (mainly bedded cherts) was most widely distributed and the ammonoid *Manticoceras* fauna was also brought northwards into the central Hunan area, which was characterized by a shallow carbonate platform with some intra-platform deeper areas.

Sea-level change in South China

On a regional scale, marine transgression started right at the beginning of the Devonian Period and invaded the South China Plate from south or SW to north or NE, in response to the basin-scale, intense basement faulting at depth (e.g. Wu Yi in Zhong *et al.* 1992). Therefore, continuous deposition of the Silurian and Devonian periods took place in some portions of southern Guangxi, e.g. in the Qinzhou area. In most other areas of South China, Devonian deposits unconformably overlie various pre-Devonian strata.

During the Early Devonian, marine transgression mainly reached central and western parts of Guangxi, including the Liujing and Dale areas. At that time sea-level changes easily produced frequent lithological changes under a largely nearshore, shallow water environment. These changes are represented by a number of lithological units

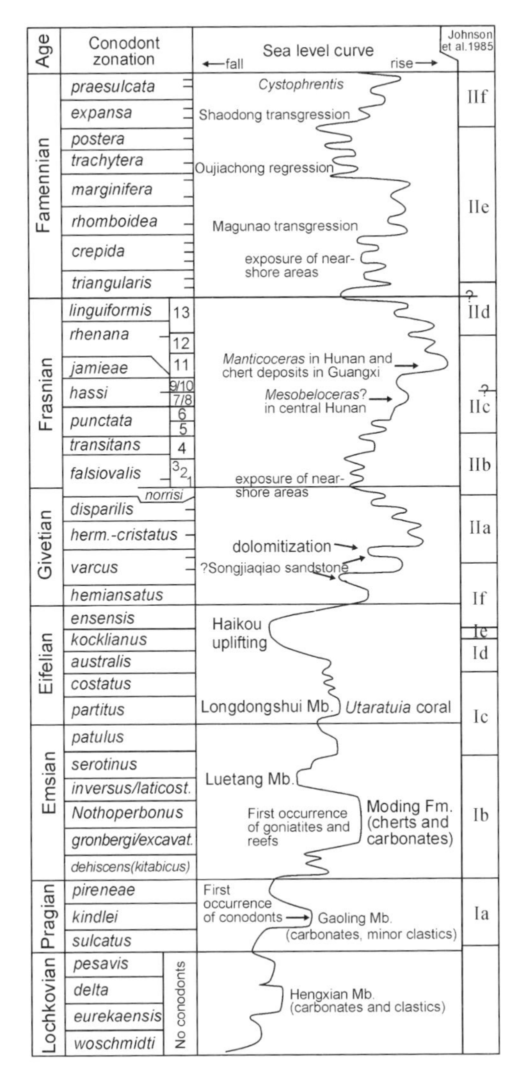

Fig. 7. Devonian sea-level change in the platform facies of South China.

(formations/members, e.g. in the Liujing section; Fig. 4b). There were several transgressive events as witnessed by lithological and faunal signatures. Conodonts first appear in the Gaoling Member limestone (Fig. 4b). The first ammonoids seem to occur in the Moding Formation in the *Nowakia barrandei* Zone, approximately equivalent to the late *Po. gronbergi* to *nothoperbonus* conodont Zones, during which time the *Erbenoceras* ammonoid fauna became widely distributed in most parts of Guangxi, not only in pelagic facies deposits, e.g. in the lower part of the Tangding Formation in the Luofu section (Fig. 4b; Zhong *et al.* 1992), but also in shallow water benthic facies deposits, e.g. in the Luomai Member of the Dale section (Ruan 1981, p. 11). By the very late Early Devonian and early Middle Devonian, the sea gradually covered the NE part of Guangxi (including Guilin) and southern Hunan Province.

According to Wu Yi (in Zhong *et al.* 1992, pp. 286–288), a brief regression occurred by the late Eifelian (Haikou Movement of Hou 1978) and mid-Givetian respectively, whereas the rest of the Givetian and Frasnian was characterized by strong transgression and minor regression. The Late Givetian was another time interval of transgression, resulting in the deposition of the Baqi Formation. However, during the Middle and Late Devonian transition period, it appears that water depth changed in different ways in different areas. In some areas it became deeper, for example the start of the deposition of the Liujiang Formation in some areas, characterized by bedded cherts. In others, this time interval is characterized by regressive deposits or is even represented by an unconformity; in central Hunan, the early Frasnian is mainly composed of sandstones (Fig. 5). This sandstone layer in the Middle–Upper Devonian transition seems widespread in platform facies, e.g. in the Dushan section of Guizhou Province (Liao *et al.* 1979). Tectonic controls might have played an important role as stressed by Chen *et al.* (2001*a*, *b*).

In the later Frasnian, there were two important deepening events, the *hassi* Zone transgression and the Lower *rhenana* Zone transgression, respectively corresponding to the Lower Rhinestreet Event and the *semichatovae* transgression recognized worldwide. During these transgressions the ammonoid *Mesobeloceras*(?) fauna and *Manticoceras* fauna were respectively brought into central Hunan Province (Ma *et al.* 2004).

During the Frasnian–Famennian transition, sea level dropped substantially in platform facies such as in the Leimingdong section of central Hunan (characterized by a thick suite of sandstones) and the Dushan section of Guizhou and Longmenshan section of Sichuan (both characterized by a thick suite of dolostones), whereas in deeper water settings the picture seems complicated: both slight shallowing and deepening trends have been reported (Gong & Li 2001; Wang & Ziegler 2002).

During the Famennian, although the sea level generally dropped somewhat compared with the Frasnian, the basic palaeogeographic framework was little changed. Famennian transgression–regression cycles are well manifested in central Hunan, e.g. in the Xikuangshan section (Figs 5, 7), with three major cycles: (1) Changlongjie Formation–Tuzitang Member–Nitangli Bed; (2) Magunao Member–Oujiachong Formation; and (3) Shaodong Formation–Menggong'ao Formation. Deposits of the first cycle are mostly absent in nearshore areas in central Hunan, such as in the Leimingdong and Qiziqiao sections. Although the Oujiachong Formation is lithologically quite different from the underlying Xikuangshan Formation, it is generally a coarsening-upward sequence representing a regressive phase, with minor marls and lingulid brachiopods in the lower part and varied clastic rocks yielding plants and fish upwards (Huang 1978). It probably represents a prolonged lowstand period. The Chinese Shaodongian (approximately *expansa* through *praesulcata* conodont Zones) represents a renewed gradual transgression, with clastic rocks and minor carbonates in the Shaodong Formation and mainly carbonates in the Menggong'ao Formation (Fig. 5). A widely distributed sandstone–shale layer (with plant fragments and spores), about 2–10 m thick, in the platformal facies ended the Devonian Period in central Hunan, e.g. in the Malanbian and Zhouwangpu sections (Tan 1987). A black shale layer (Changshun Shale), about 10 cm to a few tens of centimetres thick, developed in the deeper water setting at the end of the *praesulcata* conodont Zone, e.g. the Huangmao section of Guangxi (yielding bivalves and spores) and the Muhua section of Guizhou (Bai *et al.* 1994).

Comparison with the Euramerica transgression–regression cycles

The transgression–regression cycles of Johnson *et al.* (1985) are shown on the right of Figure 7 for comparison. The start and end of the IId cycle were placed at the beginning of MN Zones 9–10 and the Famennian, respectively, according to Day & Copper (1998, p. 157). There are a number of similar transgression–regression cycles approximately at the same time intervals, e.g. Ia, Ib, Ic, If, IIb, IIc, IIe and IIf, although some minor discrepancies in timing and magnitude exist in the above-mentioned cycles between South China and Euramerica. For example, the transgression in the early Emsian as represented by the Moding Formation in South China is much more distinct than that in Euramerica.

There are also a number of distinct discrepancies. The two depophases in South China are different from those of Euramerica in timing and magnitude. The first depophase in South China began right at the beginning of the Devonian Period and ended in the late Eifelian due to the supposed Haikou uplifting. A mid-Givetian regression was also widespread in South China, approximately in the *varcus* conodont Zone. This regression resulted in sandstone deposition in nearshore areas, such as the Songjiaqiao Formation in the Dushan section, and possible dolomitization of the previous deposits in some platform and platform-margin areas in Guangxi. Whether subaerial erosion existed with this regression in South China needs further study. A significantly different transgression–regression pattern exists in the Frasnian–Famennian transition period. In South China an interpreted distinct sea-level fall occurred at the Frasnian–Famennian boundary in the platform facies, presumably due to local tectonic activities. In contrast, only a minor regression occurred in South China in comparison with that of Euramerica at the Devonian–Carboniferous boundary.

We thank C. Brett (University of Cincinnati) and M. Zhu (Institute of Vertebrate Paleontology and Paleoanthropology, Beijing) for carefully reviewing the manuscript, improving the English text, and giving constructive suggestions. This work was supported by the National Natural Science Foundation of China (Grants 40872007 and 40830211) and the Major Basic Research Projects of the Ministry of Science and Technology, China (2006CB806400). It is a contribution to IGCP Project No. 499.

References

BAI, S. L., JIN, S. Y. & NING, Z. S. 1982. *Devonian Biostratigraphy of Guangxi and Adjacent Area.* Peking University Press, Beijing [in Chinese].

BAI, S. L., BAI, Z. Q., MA, X. P., WANG, D. R. & SUN, Y. L. 1994. *Devonian Events and Biostratigraphy of South China.* Peking University Press, Beijing.

BGMRHP (BUREAU OF GEOLOGY AND MINERAL RESOURCES OF HEILONGJIANG PROVINCE) 1993. Part 1, Stratigraphy, Chapter 5 Devonian System. *In*: *Regional Geology of Heilongjiang Province.* Geological Publishing House, Beijing [in Chinese].

BGMRNMAR (BUREAU OF GEOLOGY AND MINERAL RESOURCES OF NEI MONGOL AUTONOMOUS REGION) 1991. Part 1, Stratigraphy, Chapter 6 Devonian. *In*: *Regional Geology of Nei Mongol (Inner Mongolia) Autonomous Region.* Geological Publishing House, Beijing, 128–155 [in Chinese].

BGMRXAR (BUREAU OF GEOLOGY AND MINERAL RESOURCES OF XIZANG AUTONOMOUS REGION) 1993. Volume 1, Stratigraphy, Chapter 6 Devonian. *In*: *Regional Geology of Xizang (Tibet) Autonomous Region.* Geological Publishing House, Beijing, 56–71 [in Chinese].

BGMRYP (BUREAU OF GEOLOGY AND MINERAL RESOURCES OF YUNNAN PROVINCE) 1990. Volume 1, Stratigraphy, Chapter 6 Devonian. *In*: *Regional Geology of Yunnan Province.* Geological Publishing House, Beijing, 105–135 [in Chinese].

BIRENHEIDE, R. & LIAO, W. H. 1985. Rugose Korallen aus dem Givetium von Dushan, Provinz Guizhou, S-China. 3: Einzelkorallen und einige Koloniebildner. *Senckenbergiana lethaea*, **65**(3/5), 217–267.

CAI, C. Y. 2000. Non-marine Devonian. *In*: NIGP (Nanjing Institute of Geology, Palaeontology, Chinese Academy of Sciences) (ed.) *Stratigraphical Studies in China*

(1979–1999). University of Science and Technology of China Press, Hefei, 95–127 [in Chinese].

CAI, C. Y., FANG, Z. J. *ET AL.* 1994. New advance in the study of biostratigraphy of Lower and Middle Devonian marine-continental transitional strata in east Yunnan. *Science in China*, **24B**(6), 634–639 [in Chinese].

CHEN, D. Z., TUCKER, M. E., JIANG, M. S. & ZHU, J. Q. 2001*a*. Long-distance correlation between tectonic-controlled, isolated carbonate platforms by cyclostratigraphy and sequence stratigraphy in the Devonian of South China. *Sedimentology*, **48**, 57–78.

CHEN, D. Z., TUCKER, M. E., ZHU, J. Q. & JIANG, M. S. 2001*b*. Carbonate sedimentation in a starved pull-apart basin, Middle to Late Devonian, southern Guilin, South China. *Basin Research*, **13**, 141–167.

DAVIDSON, T. 1853. On some fossil brachiopods, of the Devonian age, from China. *Quarterly Journal of the Geological Society, London*, **9**, 353–359.

DAY, J. & COPPER, P. 1998. Revision of latest Givetian-Frasnian Atrypida (Brachiopoda) from central North America. *Acta Palaeontologica Polonica*, **43**(2), 155–204.

DE KONINCK, L. G. 1846. Notice sur deux especes de Brachiopodes du terrain Paleozoique de la Chine. *Bulletin de l'Academie Royale des Sciences Lettres et Beaux Arts, de Belgique*, **13**(2), 415–426.

FANG, Z. J., CAI, C. Y. *ET AL.* 1994. New advance in the study of the Silurian–Devonian boundary in Qujing, east Yunnan. *Journal of Stratigraphy*, **18**(2), 81–90 [in Chinese with English abstract].

GAO, L. D. 1981. Devonian spore assemblages of China. *Review of Palaeobotany and Palynology*, **34**, 11–23.

GONG, Y. M. & LI, B. H. 2001. Devonian Frasnian/Famennian transitional event deposits and sea level changes. *Earth Science – Journal of China University of Geosciences*, **26**(3), 251–257 [in Chinese with English abstract].

GONG, Y. M., XU, R., TANG, Z. D. & LI, B. H. 2005. The Upper Devonian orbital cyclostratigraphy and numerical dating conodont zones from Guangxi, South China. *Sciences in China Ser. D Earth Sciences*, **48**(1), 32–41.

GRABAU, A. W. 1931*a*. Devonian Brachiopoda of China, I: Devonian Brachiopoda from Yunnan and other districts in South China. *Palaeontologia Sinica, Series B*, **3**(3), 1–538.

GRABAU, A. W. 1931*b*. Problems in Chinese stratigraphy IV-e, The Devonian of China. *The Science Quarterly of the National University of Peking*, **2**(2), 91–162.

GROOS-UFFENORDE, H. & WANG, S. Q. 1989. The entomozoacean succession of South China and Germany (Ostracoda, Devonian). *Courier Forschungsinstitut Senckenberg*, **110**, 61–79.

HAO, S. G. & GENSEL, P. G. 1998. Some new plant finds from the Posongchong Formation of Yunnan, and consideration of a phytogeographic similarity between South China and Australia during the Early Devonian. *Science in China*, **41**, 1–13.

HAO, S. G. & GENSEL, P. G. 2001. The Posongchong Floral Assemblages of Southeast Yunnan, China-Diversity and Disparity in Early Devonian Plant Assemblages. In: GENSEL, P. G. & EDWARDS, D. (eds) *Plants Invade the Land: Evolutionary and Environmental Considerations*. Columbia University Press, 103–119.

HE, G. Q., LI, M. S., LIU, D. Q., TANG, Y. L. & ZHOU, R. H. 1994. *Paleozoic Crustal Evolution and Mineralization in Xinjiang of China*. Xinjiang People's Publishing House, Hong Kong [in Chinese with English summary].

HECKEL, P. H. & WITZKE, B. J. 1979. Devonian world paleogeography determined from distribution of carbonates and related lithic paleoclimatic indicators. *In*: HOUSE, M. R., SCRUTTON, C. T. & BASSETT, M. G. (eds) *The Devonian System*. Special Papers in Palaeontology, **23**, The Palaeontological Association, London, 99–124.

HOU, H. F. 1978. Devonian strata in South China. *In*: Institute of Geology, Mineral Resources, Chinese Academy of Geological Sciences (ed.) *Symposium on the Devonian strata of South China*. Geological Publishing House, Beijing, 214–230 [in Chinese].

HOU, H. F. 2000. Devonian stage boundaries in Guangxi and Hunan, South China. *Courier Forschungsinstitut Senckenberg*, **225**, 285–298.

HOU, H. F. & WANG, S. T. 1985. Devonian palaeogeography of China. *Acta Palaeontologica Sinica*, **24**(2), 186–197 [in Chinese with English summary].

HOU, H. F. & WANG, S. T. 1988. *Stratigraphy of China, No. 7: The Devonian System of China*. Geological Publishing House, Beijing [in Chinese].

HOU, H. F. & XIAN, S. Y. 1975. Early and Middle Devonian brachiopods from Guangxi and Guizhou. *Professional Papers of Stratigraphy and Palaeontology*, **1**, 1–85 [in Chinese].

HOU, H. F., WAN, Z. Q., XIAN, S. Y., FAN, Y. N., TANG, D. Z. & WANG, S. T. (eds) 1988. *Devonian Stratigraphy, Paleontology and Sedimentary Facies of Longmenshan, Sichuan*. Geological Publishing House, Beijing [in Chinese with English summary].

HOU, H. F., LANE, N. G., WATERS, J. A. & MAPLES, C. G. 1993. Discovery of a new Famennian echinoderm fauna from the Hongguleleng Formation of Xinjiang, with redefinition of the formation. *In*: YANG, Z. (ed.) *Stratigraphy and Paleontology of China*, volume 2. Geological Publishing House, Beijing, 1–18.

HOU, H. F., CAO, X. D., WANG, S. T., XIAN, S. Y. & WANG, J. X. 2000. *Chinese Stratigraphic Encyclopedia–Devonian*. Geological Publishing House, Beijing [in Chinese].

HUANG, D. X. 1978. On the division of the "Xuefengshan Sandstone" and the "Shaodong Member". *In*: Institute of Geology, Mineral Resources, Chinese Academy of Geological Sciences (ed.) *Symposium on the Devonian Strata of South China*. Geological Publishing House, Beijing, 90–97 [in Chinese].

IGMR (INSTITUTE OF GEOLOGY AND MINERAL RESOURCES, CHINESE ACADEMY OF GEOLOGICAL SCIENCES) (ed.) 1978. *Symposium on the Devonian System of South China*. Geological Publishing House, Beijing [in Chinese].

JI, Q. 1994. On the Frasnian–Famennian extinction event in South China as viewed in the light of conodont study. *Professional Papers of Stratigraphy and Palaeontology*, **24**, 79–107 [in Chinese with English abstract].

JI, Q. & ZIEGLER, W. 1993. The Lali section – An excellent reference section for Upper Devonian in South

China. *Courier Forschungsinstitut Senckenberg*, **157**, 1–183.

JIANG, D. Y., DING, G. & BAI, S. L. 2000. Conodont biostratigraphy across the Givetian–Frasnian boundary (Devonian) of Liujing, Guangxi. *Journal of Stratigraphy*, **24**(3), 195–200 [in Chinese with English abstract].

JOHNSON, J. G., KLAPPER, G. & SANDBERG, C. A. 1985. Devonian eustatic fluctuations in Euramerica. *Geological Society of America Bulletin*, **96**, 567–587.

KLAPPER, G. & BECKER, R. T. 1999. Comparison of Frasnian (Upper Devonian) conodont zonations. *Bolletino della società Paleontologia Italiana*, **37**, 339–348.

KUANG, G. D., ZHAO, M. T. & TAO, Y. B. 1989. *The Standard Devonian Section of China: Liujing Section of Guangxi*. China University of Geosciences Press, Wuhan [in Chinese with English summary].

LIAO, W. H. 1984. Stratigraphy of Xizang (Tibetan) Plateau, Chapter VIII – Devonian. *In*: Team of the Comprehensive Scientific Expedition to the Qinhai-Xizang Plateau (ed.) *The Series of the Comprehensive Scientific Expedition to the Qinhai-Xizang Plateau*. Science Press, Beijing, 307–310 [in Chinese].

LIAO, W. H. & BIRENHEIDE, R. 1984. Rugose Korallen aus dem Givetium von Dushan, Provinz Guizhou, S-China. 1: "Cystimorpha". *Senckenbergiana lethaea*, **65**(1/3), 1–25.

LIAO, W. H. & BIRENHEIDE, R. 1985. Rugose Korallen aus dem Givetium von Dushan, Provinz Guizhou, S-China. 2: Kolonien der Columnariina. *Senckenbergiana lethaea*, **65**(4/6), 265–295.

LIAO, W. H. & BIRENHEIDE, R. 1989. Rugose corals from the Frasnian of Tushan, Province of Guizhou, South China. *Courier Forschungsinstitut Senckenberg*, **110**, 81–103.

LIAO, W. H. & CAI, T. C. 1987. Sequence of Devonian rugose coral assemblages from northern Xinjiang. *Acta Palaeontologica Sinica*, **26**(6), 689–707 [in Chinese with English summary].

LIAO, W. H. & RUAN, Y. P. 1988. Devonian of East Asia. *Canadian Society of Petroleum Geologists Memoir*, **14–1**, 597–606.

LIAO, W. H. & RUAN, Y. P. 2003. Devonian Biostratigraphy of China. *In*: ZHANG, W. T., CHEN, P. J. & PALMER, A. R. (eds) *Biostratigraphy of China*. Science Press, Beijing, 237–279.

LIAO, W. H., XU, H. K., WANG, C. Y., RUAN, Y. P., CAI, C. Y., MU, D. C. & LU, L. C. 1978. Subdivision and correlation of the Devonian stratigraphy of Southwest China. *In*: Institute of Geology, Mineral Resources, Chinese Academy of Geological Sciences (ed.) *Symposium on the Devonian Strata of South China*. Geological Publishing House, Beijing, 193–213 [in Chinese].

LIAO, W. H., XU, H. K., WANG, C. Y., CAI, C. Y., RUAN, Y. P., MU, D. C. & LU, L.C. 1979. On some basic Devonian sections in Southwestern China. *In*: Nanjing Institute of Geology, Palaeontology (ed.) *Carbonate Biostratigraphy in Southwestern Regions*. Science Press, Beijing, 221–249 [in Chinese].

LIAO, W. H., XIA, F. S., ZHU, H. C., ZHANG, J. & ZHAN, S. G. 1992. Devonian of Tarim and adjacent areas. *In*: ZHOU, Z. Y. & CHEN, P. J. (eds) *Biostratigraphy and Geological Evolution of Tarim*. Science Press, Beijing, 173–201 [in Chinese].

LIAO, W. H., ZHU, H. C., XIA, F. S., LUO, H. & CHEN, Z. Q. 2001. Devonian. *In*: ZHOU, Z. Y., ZHAO, Z. X., HU, Z. X., CHEN, P. J., ZHANG, S. B. & YONG, T. S. (eds) *Stratigraphy of the Tarim Basin*. Science Press, Beijing, 103–118, 345–346 [in Chinese with English abstract].

LIU, W. J., CHEN, Y. R., ZHENG, R. C., WANG, H. F. & LI, X. H. 1996. Devonian sequence stratigraphy and relative sea level changes in Longmenshan area, Sichuan. *Journal of China University of Geosciences*, **7**(1), 80–86.

LIU, X., FU, S. R., YAO, P. Y., LIU, G. F. & WANG, N. W. 1992. *The Stratigraphy, Paleobiogeography and Sedimentary-Tectonic Development of Qinghai-Xizang (Tibet) Plateau in Light of Terrane Analysis*. Geological Publishing House, Beijing [in Chinese with English summary].

LIU, Y. H. 2002. The age and correlation of some Devonian fish-bearing beds of east Yunnan, China. *Vertebrata Palasiatica*, **40**(1), 52–69 [in Chinese with English summary].

MA, X. P. 1995. The type species of the brachiopod *Yunnanellina* from the Devonian of South China. *Palaeontology*, **38**, 385–405.

MA, X. P. & BAI, S. L. 2002. Biological, depositional, microspherule and geochemical records of the Frasnian/Famennian boundary beds, South China. *Palaeogeography, Palaeoclimatology, Palaeoecology*, **181**(1/3), 325–346.

MA, X. P. & DAY, J. 2000. Revision of *Tenticospirifer* Tien, 1938a, and similar spiriferid brachiopod genera from the Late Devonian (Frasnian) of Eurasia, North America, and Australia. *Journal of Paleontology*, **74**, 444–463.

MA, X. P. & DAY, J. 2007. Morphology and revision of Late Devonian (Early Famennian) *Cyrtospirifer* (Brachiopoda) and related genera from South China and North America. *Journal of Paleontology*, **81**, 286–311.

MA, X. P., SUN, Y. L., BAI, Z. Q. & WANG, S. Q. 2004. Advances in the study of the Upper Devonian Frasnian of the Shetianqiao Section, central Hunan, south China. *Journal of Stratigraphy*, **28**(4), 369–374 [in Chinese with English abstract].

MA, X. P., BECKER, R.T., LI, H. & SUN, Y. Y. 2006. Early and Middle Frasnian brachiopod faunas and turnover on the South China shelf. *Acta Palaeontologica Polonica*, **51**(4), 789–812.

PAN, J., HUO, F. C. ET AL. 1987. *Continental Devonian System of Ningxia and its Biotas*. Geological Publishing House, Beijing [in Chinese with English summary].

RAO, J. G. & YU, H. J. 1985. Devonian of southern Xizang (Tibet). *Contributions to the Geology of the Qinghai Xizang (Tibet) Plateau*, **16**, 51–69 [in Chinese with English abstract].

RGSTS (REGIONAL GEOLOGICAL SURVEYING TEAM OF SICHUAN) & NIGP (NANJING INSTITUTE OF GEOLOGY AND PALAEONTOLOGY, ACADEMIA SINICA) (eds) 1982. *Stratigraphy and Palaeontology of Western Sichuan and Eastern Xizang, Vol. I*. People's Publishing House of Sichuan, Chengdu [in Chinese].

RUAN, Y. P. 1981. Devonian and earliest Carboniferous ammonoids from Guangxi and Guizhou. *Memoirs of Nanjing Institute of Geology and Palaeontology*,

Academia Sinica, **15**, 1–151 [in Chinese with English summary].

RUAN, Y. P. & MU, D. C. 1989. Devonian tentaculoids from Guangxi. *Memoirs of Nanjing Institute of Geology and Palaeontology, Academia Sinica*, **26**, 1–234 [in Chinese with English summary].

SUN, Y. C. 1935. On the occurrence of the *Manticoceras* fauna in central Hunan. *Bulletin of the Geological Society of China*, **14**(1), 249–254.

SUN, Y. C. 1945. Devonian subdivision of east Yunnan. *Science Record*, **1**(3–4), 486–491.

SUN, Y. C. 1958. The Upper Devonian coral faunas of Hunan. *Palaeontologia Sinica, New series B*, **8**, 1–28.

SZE, H. C. 1952. Upper Devonian fossil plants from China. *Palaeontologia Sinica, New Series A*, **4**, 1–30 [in Chinese].

TAN, X. C., DONG, Z. Z. & QIN, D. H. 1982. The Lower Devonian strata in Baoshan area of western Yunnan, with reference of the Silurian-Devonian boundary. *Journal of Stratigraphy*, **6**(3), 199–208 [in Chinese].

TAN, Z. X. 1987. Stratigraphy. *In*: Regional Geological Surveying Party, Bureau of Geology and Mineral Resources of Hunan Province (ed.) *The Late Devonian and Early Carboniferous Strata and Palaeobiocoenosis of Hunan*. Geological Publishing House, Beijing, 2–65, 169–179 [in Chinese with English summary].

TIAN, J. J. & ZHU, H. C. 2005. Devonian miospore biostratigraphy of the Longhuashan and Xichong formations in Zhanyi of Yunnan. *Journal of Stratigraphy*, **29**(4), 311–316, 322 [in Chinese with English abstract].

TIEN, C. C. 1938*a*. Devonian Brachiopoda of Hunan. *Palaeontologia Sinica, new series B*, **4**, 1–192.

TIEN, C. C. 1938*b*. The Devonian of China. *Geological Review*, **3**(4), 355–404 [in Chinese].

TSIEN, H. H. & FONG, C. C. K. 1997. Sea-level fluctuations in South China. *In*: HOUSE, M. R. & ZIEGLER, W. (eds) *On sea-level fluctuations in the Devonian. Courier Forschungsinstitut Senckenberg*, **199**, 103–115.

WANG, C. Y. 1989. Devonian conodonts of Guangxi. *Memoirs of Nanjing Institute of Geology and Palaeontology, Academia Sinica*, **25**, 1–212 [in Chinese with English summary].

WANG, C. Y. 2000. The Devonian. *In*: NIGP (Nanjing Institute of Geology, Palaeontology, Chinese Academy of Sciences) (ed.) 2000. *Stratigraphical Studies in China (1979–1999)*, University of Science and Technology of China Press, Hefei, 73–94 [in Chinese].

WANG, C. Y. & ZIEGLER, W. 2002. The Frasnian–Famennian conodont mass extinction and recovery in South China. *Senckenbergiana lethaea*, **82**, 463–493.

WANG, D. M. 2007. Two species of *Zosterophyllum* from South China and application of a biostratigraphic method to an Early Devonian flora. *Acta Geologica Sinica (English edition)*, **81**(4), 525–538.

WANG, D. M., HAO, S. G. & LIU, Z. F. 2002. Plant researches on the Lower Devonian Xujiachong Formation of Qujing District, Eastern Yunnan. *Acta Geologica Sinica (English edition)*, **76**(4), 393–407.

WANG, H. C. 1942. The stratigraphical position of the Devonian fish-bearing series of east Yunnan with a special discussion on the Tiaomachien Formation of central Hunan. *Bulletin of the Geological Society of China*, **22**(3-4), 217–225.

WANG, H. C. 1948. The Middle Devonian rugose corals of East Yunnan. *Contributions of Geological Institute, University of Peking*, **33**, 1–45.

WANG, H. Z. (ed.) 1985. *Atlas of the Palaeogeography of China*. Cartographic Publishing House, Beijing [in Chinese with English summary].

WANG, H. Z., YANG, S. P., ZHU, H., ZHANG, L. H. & LI, Z. 1990. Palaeozoic biogeography of China and adjacent regions and world reconstruction of the palaeocontinents. *In*: WANG, H. Z., YANG, S. N., LIU, B. P. *ET AL.* (eds) *Tectonopalaeogeography and Palaeobiogeography of China and Adjacent Regions*. China University of Geosciences Press, Wuhan, 35–86 [in Chinese with English summary].

WANG, J. Q. & ZHANG, G. R. 1999. New material of Microbrachius from Lower Devonian of Qujing, Yunnan, China. *Vertebrata Palasiatica*, **37**(3), 200–211 [in Chinese with English summary].

WANG, J. X., XIAN, S. Y. & HOU, H. F. 2005. Devonian. *In*: WANG, X. F. & CHEN, X. H. (eds) *Stratigraphic Division and Correlation of each Geologic Period in China*. Geological Publishing House, Beijing, 195–233 [in Chinese].

WANG, X. L. 1994. The rugose coral fauna from the upper part of the Heyuanzhai Formation in western Yunnan, China. *Journal of Faculty of Science, Hokkaido University, series IV*, **23**(3), 343–553.

WANG, Y. & RONG, J. Y. 1986. Yukiangian (Early Emsian, Devonian) brachiopods of the Nanning-Liujing district, central Guangxi, South China. *Palaeontologia Sinica, N. ser. B*, **22**, 1–282 [in Chinese with English summary].

WANG, Y. & YU, C. M. 1962. *The Devonian of China*. Science Press, Beijing [in Chinese].

WANG, Y., YU, C. M. & WU, Q. 1974. Advances in the Devonian biostratigraphy of South China. *Memoirs of Nanjing Institute of Geology and Palaeontology*, **6**, 1–71 [in Chinese].

WANG, Y., YU, C. M., XU, H. K., LIAO, W. H. & CAI, C. Y. 1979. Biostratigraphy of South China. *Acta Stratigrapica Sinica*, **3**(2), 81–89 [in Chinese].

WANG, Y., YU, C. M. *ET AL.* 1982. Correlation chart of the Devonian in China with explanatory text. *In*: Nanjing Institute of Geology, Palaeontology, Academia Sinica (ed.) *Stratigraphic Correlation in China with Explanatory Text*. Science Press, Beijing, 90–108 [in Chinese].

WANG, YUE 2001. On outcrop sequence stratigraphy and sea level changes of Devonian in Dushan, south Guizhou. *Guizhou Geology*, **18**(3), 154–162 [in Chinese with English abstract].

WU, Y., GONG, Y. M. & DU, Y. S. 1997. *Devonian Sequence Stratigraphy and Sea Level Change of South China*. China University of Geosciences Press, Wuhan [in Chinese with English summary].

XIA, F. S. 1997. Marine microfauna (bryozoans, conodonts and microvertebrate remains) from the Frasnian-Famennian interval, northwestern Junggar Basin of Xinjiang in China. *Beriträge zu Palaeontologie (Wien)*, **22**, 1–207.

XIAO, S. L., HOU, H. F. *ET AL.* 1992. *The Researches of Devonian System in North Xinjiang*. Xinjiang

Science Technology & Hygiene Publishing House (K), Urumqi [in Chinese].

XIGMR (Xi'an Institute of Geology and Mineral Resources) & NIGP (Nanjing Institute of Geology and Palaeontology, Academia Sinica). 1987. *Late Silurian–Devonian Strata and Fossils from Luqu-Tewo Area of West Qinling Mountains, China, Vol. 1*. Nanjing University Press, Nanjing [in Chinese with English abstract].

Xu, H. K. 1991. Early and Middle Devonian boundary strata of Hoboksar, west Junggar and their brachiopods. *Acta Palaeontologica Sinica*, **30**(3), 307–333 [in Chinese with English summary].

Xu, H. K. 1999. Discovery of brachiopods from the Hujiersite Formation in west Junggar and its significance. *In*: Nanjing Institute of Geology, Palaeontology, Academia Sinica (ed.) *Palaeozoic Fossils of Northern Xinjiang*. Nanjing University Press, Nanjing, 305–313 [in Chinese with English summary].

Yang, S. P., Pan, K. & Hou, H. F. 1981. The Devonian System in China. *Geological Magazine*, **118**(2), 113–224.

Yoh, S. S. 1937. Die Korallenfauna des Mitteldevon aus der Provinz Kwangsi, Südchina. *Palaeontographica*, **87A**, 45–76.

Yoh, S. S. 1938. Beiträge zur Kenntnis des marinen oberen Unterdevons und unter Mitteldevons Südchina. *Bulletin of the Geological Society of China*, **18**(1), 67–73.

Yoh, S. S. 1956. Subdivision, zonation and correlation of the Devonian formations in Longmenshan area, N. W. Szechuan. *Acta Geologica Sinica*, **36**(4), 443–476 [in Chinese].

Yu, C. M. (ed.) 1988. *Devonian–Carboniferous Boundary in Nanbiancun, Guilin, China – Aspects and Records*. Science Press, Beijing.

Yu, C. M. & Liao, W. H. 1978. Lower Devonian rugose corals from Alengchu of Lijiang, NW Yunnan. *Acta Palaeontologica Sinica*, **17**(3), 245–266 [in Chinese with English abstract].

Yu, C. M. & Liao, W. H. 1982. Discovery of Early Devonian Tetracorals from Xainza, N Xizang (Tibet). *Acta Palaeontologica Sinica*, **21**(1), 96–107 [in Chinese with English abstract].

Zhao, Z. X. & Wang, C. Y. 1990. Age of the Hongguleleng Formation in the Junggar basin of Xinjiang. *Journal of Stratigraphy*, **14**(2), 145–146 [in Chinese].

Zhong, K., Wu, Y., Yin, B. A., Liang, Y. L., Yao, Z. G. & Peng, J. L. 1992. *Stratigraphy of Guangxi, China, Part 1: Devonian of Guangxi*. The Press of the China University of Geosciences, Wuhan [in Chinese with English abstract].

Zhu, M. & Wang, J. Q. 1996. On the Lower-Middle Devonian boundary in Qujing, Yunnan. *Journal of Stratigraphy*, **20**(1), 58–63 [in Chinese with English abstract].

Late Devonian echinoderms from the Hongguleleng Formation of northwestern China

G. D. WEBSTER[1]* & J. A. WATERS[2]

[1]*School of Earth and Environmental Sciences, Washington State University, PO Box 642812, Pullman, WA 99164-28112, USA*

[2]*Appalachian State University, Department of Geology, 195 Rankin Science, 572 Rivers Street, Boone, NC 28608, USA*

**Corresponding author (e-mail: webster@wsu.edu)*

Abstract: The Late Devonian, Famennian, Hongguleleng Formation of northwestern China has yielded one of the most diverse echinoderm faunas known from China. New collections and re-evaluation of earlier collections results in recognition of new taxa, increases the known diversity, provides new morphological information on some of the previously reported taxa, and provides new information on the affinity of the Late Devonian echinoderms of the Hongguleleng Formation.

New crinoid taxa introduced are *Gnarycrinus lanei* n. gen., n. sp., *Anamesocrinus tieni* n. sp., *Histocrinus*? *chenae* n. sp., *Eumhacrinus tribrachiatus* n. gen., n. sp., *Sostronocrinus aberratus* n. sp. and *Labrocrinus granulatus* n. gen., n. sp. One new species of blastoid, *Hadroblastus liaoi*, n. sp. is described.

Previous reports listed 46 echinoderms from the Hongguleleng Formation, including 13 blastoids and 33 crinoids. An additional 11 crinoid species bring the total crinoids to 44 species assigned to 32 genera. Similarly, the one additional species of blastoid brings the totals to 14 species assigned to 12 genera. The new echinoderms described herein include the first report of a dimerocrinitid, amphoracrinid, allagacrinid, glossocrinid, histocrinid, cercidiocrinid, dactylocrinid and neoschismatid from the Devonian of northwestern China. Additional morphologic information is provided for three of the previously described taxa and a revision of species assigned to *Grabauicrinus* is proposed.

The crinoids and blastoids suggest closer affinity with Mississippian faunas than with Devonian faunas, and with North American faunas than European faunas. Collectively the blastoid and crinoid faunas from the Hongguleleng indicate that rediversification happened rapidly after extinction in contrast to current suggestions of a long interval of lowered origination following these extinction events.

Echinoderms from several localities in the Late Devonian, Famennian, Hongguleleng Formation in the Xinjiang-Uyghur Autonomous Region of northwestern China (Fig. 1) were first reported by Hou *et al.* (1994) and Lane *et al.* (1995). Systematic descriptions were made of the Hongguleleng echinoderms by Lane *et al.* (1997) and Waters *et al.* (2003). The Hongguleleng Formation has yielded the most diverse Palaeozoic echinoderm fauna from China. Taxa recognized within it differ from the Middle Devonian fauna from western Yunnan described by Chen & Yao (1993). Comparison of these faunas shows that the Yunnan fauna is dominated by camerate crinoids, with a few blastoids and cladid crinoids and apparently lacks flexible crinoids. In contrast, the Hongguleleng fauna has a diversity of blastoids as well as camerate and flexible crinoids, but is dominated by cladid crinoids with a few disparid crinoids.

The Hongguleleng localities were recollected in August 2005 by a Devonian working group that included Xiuqin Chen, John Talent, Ruth Mawson, Brian and Beverley Green, Thomas Suttner, Jiri and Barbora Fryda, John Pickett, Johnny Waters and Gary Webster. Although all localities, except Ganare, mentioned in this report were collected, efforts were concentrated on the section west of the Bonlongour Reservoir because it was recognized from the earlier reports of Zhao (1986), Hou *et al.* (1988), Lu & Wicander (1988), Xu *et al.* (1990), Zhao & Wang (1990), Xia (1996) and the echinoderm reports mentioned above, as a section rich in both micro- and megafossils making it an important reference section for northwestern China. The new echinoderm fossils discovered increase the known diversity, provide new morphological information on some of the previously reported taxa, and provide new information on the affinity of the Late Devonian echinoderms of the

From: KÖNIGSHOF, P. (ed.) *Devonian Change: Case Studies in Palaeogeography and Palaeoecology*.
The Geological Society, London, Special Publications, **314**, 263–287.
DOI: 10.1144/SP314.14 0305-8719/09/$15.00

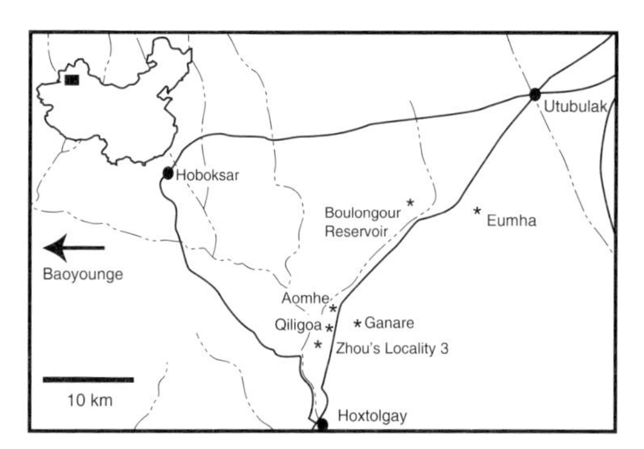

Fig. 1. Map showing Late Devonian echinoderm localities in the northern portion of Xinjiang-Uyghur Autonomous Region. East Eumha is approximately 20 m east of Eumha. Baoyonghe (also given as Ba Yang Ghou) is approximaltely 80 km west of Hoxtolgay.

Hongguleleng Formation. This report is one part of several planned by the members of the working group to document the fauna of the Hongguleleng Formation.

Tectonic and stratigraphic setting

Lane *et al.* (1997) and Waters *et al.* (2003, 2008) reviewed the tectonic, stratigraphic and biostratigraphic setting of the Hongguleleng Formation. The Devonian working group that recollected localities in the Hongguleleng in 2005 did so in part to refine the biostratigraphy of the formation. Although the new synthesis is still forthcoming, Pickett (2007) referenced a personal communication from Mawson on 15 March 2007 confirming the Famennian age (middle to upper *crepida* Biozone, early Famennian) of limestones at Aomuhu that are correlated with the marly beds at the Boulongour Reservoir section.

As discussed by Windley *et al.* (2007), Central Asia has a long, complex geological history, and tectonic models continue to evolve as fieldwork and geochemical and chronological constraints provide additional information. For the purposes of this discussion, the key points from this ongoing work are as follows:

(1) Current models strongly suggest similarity with accretionary orogens of the modern circum-Pacific and specifically point to an Indonesian archipelago-type model (Windley *et al.* 2007). Recent geochemical work on Palaeozoic ophiolites from Xinjiang reported by Wang *et al.* (2003) support the more generalized interpretations from Windley *et al.* (2007).

(2) The Hongguleleng Formation was part of an accretionary complex that likely formed in a complex, tectonically active island arc environment similar to modern Indonesia. Palinspastic reconstructions of the Central Asian Orogenic Belt (Windley *et al.* 2007) place the deposition of the Hongguleleng Formation at approximately 25 to 30 degrees north, a more tropical setting than the previous palaeogeographic reconstructioins reported in Waters *et al.* (2003). The diverse marine fauna found in the Hongguleleng Formation, which may exceed 100 genera when completely described, support a more tropical setting.

(3) Windley *et al.* (2007) refine the concept of the Turkestan Ocean of Seugör *et al.* (1993), by partitioning out a small wedge-shaped ocean called the Junggar–Balkhash Ocean. Between Early Devonian and Late Carboniferous time, the Junggar–Balkhash Ocean was closed by opposite-dipping subduction zones with numerous island arcs (Windley *et al.* 2007). The Hongguleleng Formation contains siliciclastic and carbonate sediments deposited in conjunction with one of these island arcs.

Faunal analysis

Blastoids

As discussed in Waters *et al.* (2003) the Hongguleleng Formation probably is the second most diverse blastoid fauna in the world and contains the oldest occurrence of the Codasteridae, the Astrocrinidae, the Granatocrinidae, and the Orbitemitidae. The new blastoid described herein is the oldest known member of the Neoschmatidae. All these blastoid families are important components of Carboniferous echinoderm communities, particularly in North America and Europe, and suggest that Asia was the nexus of blastoid diversification after the Late Devonian mass extinction events.

Crinoids

A total of 249 crinoid specimens from the Hongguleleng Formation were described in the combined papers of Lane *et al.* (1997) and Waters *et al.* (2003), as reported in the latter. An additional 86 specimens are reported herein. Most of the specimens came from the marly beds in the Boulongour Reservoir and Eumha localities. Table 1 is a species list that also gives the localities from which the specimens were collected and the number of specimens from each. Table 2 is a comparison of the number of genera and species of the four major crinoid groups reported from the combined papers of Lane *et al.* (1997) and Waters *et al.* (2003) with the total specimens including the new specimens reported herein. Both of these tables show that the cladid genera (16) and species (23) are essentially equal in taxonomic diversity to

Table 1. *List of crinoid species from the Hongguleleng Formation of western China*

Species	Locality–no. specimens
Crinoidea	
Camerata	
Diplobathrida	
Dimerocrinitidae	
Gnarycrinus lanei	BRM-3
Monobathrida	
Periechocrinidae	
Athabascacrinus orientale	BRM-1; E-2
Amphoracrindae	
Amphoracrinus sp.	BR-1; BRG-1
Actinocrinitidae	
Abactinocrinus devonicus	E-1; EE-1
Actinocrinites zhaoae	Z-1
Actinocrinid? indeterminate	BRG-1-1
Hexacrinidae	
Hexacrinites pinnulata	BRB-1; BRM-15
Parahexacrinidae	
Agathocrinus junggarensis	BRB-2
Platycrinitidae	
Chinacrinus xinjiangensis	BRM-99; E-2
Chinacrinus nodosus	Z-1; BRM-1; EE-2
Chinacrinus species A	BRM-1
Chinacrinus species B	BRM-1
Disparada	
Calceocrinidae	
Deltacrinus asiaticus	BRM-2; EE-1
Anamesocrinidae	
Anamesocrinus tieni	BRM-7
Cladida	
Euspirocrinidae	
Genus indet.	BRM-1
Botryocrinidae	
Parisocrinus nodosus	E-1
Parisocrinus conicus	E-1
Glossocrinoida	
Glossocrinid indeterminate	BRM-1
Scytalocrinidae	
Julieticrinus romeo	E-5; EE-1
Histocrinus sp.	BRM-1
Histocrinus? chenae	BRM-3
Scytalocrinid indeterminate	EE-1
Bridgerocrinidae	
Bridgerocrinus discus	E-2
Bridgerocrinus delicatulus	BRM-1
Cercidocrinidae	
Eumhacrinus tribrachiatus	EE-1
Aphelecrinidae	
Cosmetocrinus parvus	BRM-3
Sostronocrinidae	
Sostronocrinus quadribrachiatus	BRM-14; E-3
Sostronocrinus minutus	BRM-6; EE-1
Amadeusicrinus subpentagonalis	E-2; BRM-2
Decadocrinidae	
Grabauicrinus xinjiangensis	E-11; BRM-3; EE-1
Grabauicrinus constricutus	BRM-2
Grabauicrinus elongates	BRM-1

(*Continued*)

Table 1. *Continued*

Species	Locality–no. specimens
Grabauicrinus rugosus	BRM-1
'Decadocrinus' usitatus	BRM-6
Graphiocrinidae	
Holcocrinus asiaticus	BRM-3
Cladid indeterminate 1	E-1
Cladid indeterminate 2	BRM-1
Basal Plate indeterminate	EE-1
Flexibilia	
Taxocrinida	
Taxocrinidae	
Eutaxocrinus chinaensis	BRB-36; E-47; EE-1; Q-1
Eutaxocrinus boulongourensis	BRB-4
Eutaxocrinus basellus	BRM-2
Taxocrinus anomalus	BRB-1
Synerocrinidae	
Euonychocrinus websteri	BRM-3; E-10
Sagenocrinida	
Sagenocriidae	
Forbesiocrinus inexpectans	BRM-1; E-1; EE-1
Dactylocrinidae	
Labrocrinus granulatus	BRM-1

Abbreviations: BR, Boulongour Reservoir; BRB, Blastoid Hill; BRG, Ganare; BRM, marly beds; E, Eumha; EE, East Eumha; Q, Qiligoa; Z, Zhou's locality.

the combined numbers (14, 21) of camerates (9, 12), disparids (2, 2) and flexibles (5, 7), but are known from the smallest number of specimens not including the disparids. This reflects the evolutionary radiation of the cladids and beginning of the decline of the camerates, which began in the Late Devonian, although both had major radiations in the Mississippian (Kammer & Ausich 2006).

Table 2. *Comparison of total numbers of crinoids reported by Waters* et al. *(2003) and combination of Waters* et al. *(2003) and 84 new specimens of this work for the Hongguleleng Formation of northwestern China*

	Waters *et al.* (2003)	This work
Camerates	93 specimens	135 specimens
Genera	6	9
Species	9	12
Disparids	2 specimens	10 specimens
Genera	1	2
Species	1	2
Cladids	47 specimens	79 specimens
Genera	11	16
Species	17	22
Flexibles	93 specimens	109 specimens
Genera	4	5
Species	6	7
Total	249 specimens	333 specimens
Genera	22	32
Species	33	44

It should also be noted in Table 1 that two species, *Chinacrinus xinjiangensis* (101 specimens) and *Eutaxocrinus chinaensis* (85 specimens), account for 186 (56%) of the 335 specimens. If the next three most abundant species (*Sostronocrinus quadribrachiatus*, 17 specimens; *Grabauicrinus xinjiangensis*, 15 specimens; *Euonychocrinus websteri*, 13 specimens) are added to the two most abundant species they total 231 specimens or 69% of the 335 specimens. That is, these five species of the 45 species account for 11% of the crinoid diversity and account for 69% of the specimens recovered. This may provide an estimate of the abundance of these species in the fauna provided it does not reflect selective preservation. It should also be noted that these five species are all moderately robust forms that were probably in the higher feeding tier. However, no specimens are known from the Hongguleleng Formation with more than a few of the proximal columnals attached.

Review of the crinoids described by Lane *et al.* (1997) and Waters *et al.* (2003) and identification of recently collected specimens resulted in the recognition of seven endemic genera: *Gnarycrinus* n. gen., *Chinacrinus* Lane *et al.* 1997, *Julieticrinus* Waters *et al.* 2003, *Eumhacrinus* n. gen., *Amadeusicrinus* Waters *et al.* 2003, *Grabauicrinus* Waters *et al.* 2003 and *Labrocrinus* n. gen. All endemic genera are assigned to families that contain genera that have recognized Devonian or Early Mississippian ranges. Among the non-endemic genera the Hongguleleng fauna contains the first record of several taxa from China. These are *Anamesocrinus* Goldring 1923, *Histocrinus* Kirk 1940, an indeterminate glossocrinid, a cercidocrinid (*Eumhacrinus*) and dactylocrinid (*Labrocrinus*). In addition, the Hongguleleng fauna contains the second report of a dimerocrinitid (*Gnarycrinus*) and *Amphoracrinus* from China. The first report of a dimerocrinitid and *Amphoracrinus* was from western Yunnan (Chen & Yao 1993).

New specimens provided significant morphological information resulting in emended descriptions for four species: *Chinacrinus nodosus* Lane *et al.* 1997, *Deltacrinus asiaticus* Lane *et al.* 1997, *Amadeusicrinus subpentagonalis* (Lane *et al.* 1997) and *Grabauicrinus xingiangensis* (Lane *et al.* 1997). Identification of some of the new specimens resulted in reassignment of earlier identifications. For example, the crown of *Gnarycrinus* retained the infrabasals, which were not preserved in the partial crown Waters *et al.* (2003) identified as *Actinocrinites* sp. In addition, new fragmentary specimens include several taxa that are identified to subclass or lower taxonomic rank. These specimens increase the diversity of the fauna. Hopefully future investigations of the Hongguleleng Formation will yield more complete specimens of these taxa allowing species identification.

As noted above, the Hongguleleng Formation was assigned a Frasnian–Famennian age by Xia (1996) on the basis of conodonts and bryozoans. Famennian crinoids are poorly known worldwide as noted by Waters *et al.* (2001, 2003). Thus the Hongguleleng crinoids are important not only for their occurrence, but for understanding the evolutionary lineages between the Middle Devonian and Mississippian faunas. If the age of the Hongguleleng Formation were based only on the crinoids it would probably have been interpreted as Early Mississippian because more of the genera are related to Mississippian faunas than they are to Devonian faunas, and more are related to North American faunas than to European faunas.

Table 3 is a compilation made from the index of Webster (2003) of the Hongguleleng crinoid genera showing the general relationship of each genus to European or North American faunas and the age of those faunas. The seven Hongguleleng endemic genera are listed for completeness, but without the age or relationship to European or North America faunas. Of the nine Hongguleleng genera related to Devonian faunas, three are related to European faunas, five are related to North American faunas, and one is not known in Europe or North America. Of these, *Anamesocrinus* and *Deltacrinus* were known only from the Devonian of North America and *Agathocrinus* was known only from the Devonian of Russia. All others have species reported from the Mississippian and North America, although most species are reported from Europe. Two (*Amphoracrinus* and *Actinocrinites*) of the 12 genera related to Mississippian faunas have approximately the same number of species reported from Europe as from North America. The other nine genera are related to North American Misssissippian faunas. *Athabascacrinus, Amphoracrinus, Abactinocrinus, Histocrinus, Holcocrinus, Cosmetocrinus* and *Forbesiocrinus* were previously known only from the Mississippian, and *Abactinocrinus* and *Histocrinus* were known only from North America. Thus the ranges of these taxa are extended downward into the Famennian from the Mississippian. Two genera (*Bridgerocrinus* and *Taxocrinus*) range from the Devonian into the Mississippian, but most species are known from the Mississippian of North America. *Euonychocrinus* was previously known only from Pennsylvanian strata of North America and its range is also extended downward into the Famennian.

The only other Devonian crinoid fauna described from China is from western Yunnan and is reported as of Middle Devonian age (Chen & Yao 1993). However, more recent conodont data (Wang *et al.* 2007) indicate that a small part of the fauna is of Famennian age. Devonian crinoids from Yunnan are dominated by camerates (six species) with one disparid and three cladids. There are no flexible crinoids as noted by Chen & Yao (1993). The Yunnan fauna lived on the Baoshan block along the north side of Gondwana at approximately 30° south latitude (Metcalf 1996; Wopfner 1996) whereas the Hongguleleng fauna lived on the Kazakhstan block at approximately 40° north (Waters *et al.* 2003). No genera are common to both faunas, except *Hexacrinites.*

Two Famennian faunas from Europe were recently revised. The British fauna, originally described by Whidborne (1896, 1898), was reported by Lane *et al.* (2001*a*) to be dominated by cladids (12 species) with eight camerates and two flexibles. Three genera (*Actinocrinites, Bridgerocrinus* and *Sostroncrinus*) are common to the Hongguleleng and British faunas. The German fauna, described by Schmidt (1930) and revised by Lane *et al.* (2001*b*), is dominated by camerates (ten species) with five cladids and two flexibles. *Taxocrinus* and *Sostronocrinus* are the only two genera in common with the Hongguleleng and German faunas. The absence of disparids is to be noted in the British and German faunas.

Famennian crinoids of North America are known mostly from New York and Colorado. The New York specimens came from a number of localities in the

Table 3. *Comparison of generic affinities of the Famennian Hongguleleng Formation crinoids with North American and European faunas and the Devonian or Mississippian (Miss.) age of those faunas*

Camerata	Endemics	European	N. America	Devonian	Miss.
Gnarycrinus	×				
Athabascacrinus			×		×
Amphoracrinus		×			×
Abactinocrinus			×		×
Actinocrinites		×	×		×
Hexacrinites		×		×	
Agathocrinus				×	
Chinacrinus	×				
Disparada					
Deltacrinus			×		×
Anamesocrinus			×	×	
Cladida					
Parisocrinus		×	×		×
Glossocrinid indet.			×	×	
Bridgerocrinus			×		×
Julieticrinus	×				
Histocrinus			×		×
Eumhacrinus	×				
Cosmetocrinus			×		×
Sostronocrinus		×		×	
Amadeusicrinus	×				
Grabauicrinus	×				
'*Decadocrinus*'			×	×	
Holcocrinus			×		×
Flexibilia					
Eutaxocrinus		×		×	
Taxocrinus			×		×
Euonychocrinus			×		
Forbesiocrinus			×		×
Labrocrinus	×				

Note: The Devonian or Mississippian age of the genus is not the stratigraphic range of the genus. It is the age that most of the species of that genus occur within.

Chemung Formation and are dominated by cladids as compiled by Bassler & Moodey (1943). There is no single locality that has yielded a diverse fauna as such. Genera in common with the Hongguleleng fauna are *Anamesocrinus* and *Eutaxocrinus*. The Colorado fauna is from two beds in the Broken Rib Member of the Dyer Formation and contains crinoids and stelleroids (Webster *et al.* 1999). The crinoids are dominated by cladids (12 species) with three flexibles and one camerate. Only *Taxocrinus* is common with the Hongguleleng fauna.

In summary, the Hongguleleng Formation contains the most diverse Famennian blastoid and crinoid faunas recognized worldwide. The blastoid fauna contains the oldest representatives of five families, all important components of Carboniferous echinoderm communities. The crinoid fauna is dominated by cladids, contains representatives of each of the four crinoid subclasses common in the middle Palaeozoic, contains the oldest representatives of several cladid genera, and shows greatest affinity with North American Mississippian faunas. Collectively the blastoid and crinoid faunas from the Hongguleleng indicate that rediversification happened rapidly after extinction, even if the faunas are not geographically widespread, in contrast to current suggestions of a long interval of lowered origination following these extinction events.

Systematics

The classification follows Simms & Sevastopulo (1993). Crinoid measurement terminology follows Webster & Jell (1999). Anal terminology follows Webster & Maples (2006). Radial facet terms follow Webster & Maples (2008). Specimens are reposited at the Nanjing Institute of Geology and Paleontology (NIGP), Nanjing, China.

Class BLASTOIDEA Say 1825
Order FISSICULATA Jaekel 1918

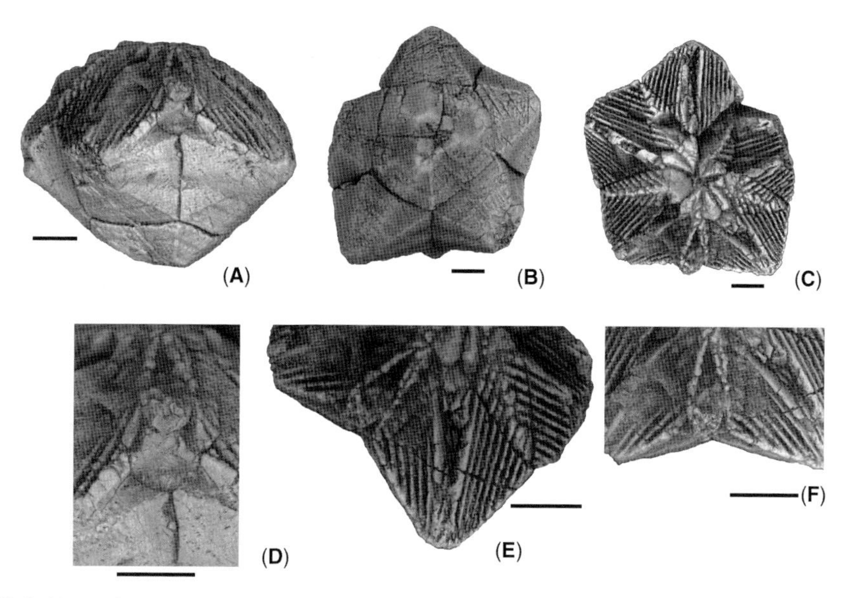

Fig. 2. *Hadroblastus liaoi*, sp. nov. (A–F) Holotype NIGP 148832. (**A**) Lateral view of C-interray showing anal deltoids. (**B**) Basal view, A-ray up, azygous basal in AB interray. (**C**) Oral view, A-ray up, showing pentagonal outline. Ambulacra mostly missing, deltoid crests broken in some rays. (**D**) Close-up of anal deltoids. (**E, F**) Detail of oral view. Eight hydrospire fields developed across full width and length of deltoid body visible in non-anal interrays. Nine hydrospires per field. Anal hydrospires reduced in number to five per field.

Family NEOSCHISMATIDAE Wanner 1940 Nomen correctum Fay (1964) pro NEOSCHISMIDAE
Genus *HADROBLASTUS* Fay 1962
Hadroblastus liaoi sp. nov.
Text Figure 2A–F

Derivation of name. Named for Liao Zhouting in recognition for his invaluable support in conducting fieldwork in China.

Types. Holotype NIGP 148832.

Type locality and horizon. Baoyonghe locality, Hongguleleng Formation. Collected by Z. T. Liao in 2005.

Diagnosis. A species of *Hadroblastus* with biconical theca, pentalobate outline, protuberant interradial areas, weak growth lines with distinct ridges radiating to the apices of the basals, broadly exposed hydrospires with 9 folds per field reduced to 5 in the anal interray.

Description. Theca biconical in lateral view with broad conical pelvis and convex vault. Outline pentagonal to slightly pentalobate in plan view. Interradial areas largely missing, but protuberant where preserved. Greatest width equatorial, at aboral tips of ambulacra. Length 10.3 mm, width 13 mm, vault 6 mm, pelvis 4.3 mm, pelvic angle 110°. Basals 3 in normal position, forming lower third of conical pelvis. Basals pentagonal in plan view. Stem attachment weathered, triangular. Azygous basal rhombic with straight to slightly inflated lateral and distal edges. Basal-radial sector flat normal to and parallel to basal-radial axis. Growth lines weak as preserved. Zygous basal pentagonal with straight to slightly inflated edges. Basal-radial sectors flat normal to and parallel to basal-radial axis crossing over steep ridge that forms triple junction at interradial and radial-basal suture. Growth lines weak as preserved. Azygous basal length 3.6 mm, width 4.5 mm. Length azygous basal basal-radial growth axis 3.2 mm; azygous basal basal-radial growth front 2.4 mm. Zygous basal length 3.7 mm, width 5.6 mm. Length zygous basal basal-radial growth axis 2.4 mm; zygous basal basal-radial growth front 4.2 mm. Radials 5, forming lower half of vault and upper two-thirds of pelvis. Radial pentagonal in plan view, triangular in lateral view. Radial-radial sutures straight. Radial-basal sutures slightly concave in aboral axial view. Radial-basal and radial-deltoid sectors flat to slightly inflated with strong ridges that run from aboral tip of ambulacrum to junction of radial-radial and radial-basal sutures. Growth lines weak as preserved. Radial width at basals 4.0 mm; radial width at deltoid 7.7 mm; radial-basal 5.4 mm; radial-radial 4.6 mm; radial-deltoid 4.6 mm. Deltoids 4, form margins of oral opening in conjunction with epideltoid. Deltoids modified triangular in plan view from above oral axis, form flattened cap to vault in lateral view. Deltoid crests convex in lateral view, broadly exposed, protuberant in lateral view. Deltoid body broad compared to crest and lip, bear exposed hydrospires across entire width of deltoid radial sector. Deltoid length 5.3 mm, greatest width 6.7 mm, greatest adoral width 2.3 mm. Hypodeltoid quadrate with hemispherical suture with radials, straight lateral sides, and short, concave suture with anal opening. Epideltoid surfacial expression same as other deltoids. Cryptodeltoids, if present, not exposed at surface. Specimen not sectioned. Hypodeltoid length 1.55 mm, width 1.86 mm. Epideltoid length 1.8 mm, width 2.4 mm. Ambulacra not preserved. Sinus where ambulacra were located very narrow, occupying approximately half vault. Eight hydrospire fields developed across full width and length of deltoid body, visible in non-anal interrays. Nine hydrospires per field. Anal hydrospires reduced to 5 per field.

Remarks. Hadroblastus is a well-known fissiculate blastoid that previously has been described from the Mississippian of North America, Ireland, and possibly Scotland. This is the first occurrence of the genus from the Devonian and from Asia. In fact, this is the first Devonian occurrence of the Family Neoschismatidae, a family best known from Permian genera in the Tethyan realm.

The following characters can be used to distinguish the species of *Hadroblastus* from one another (Breimer & Macurda 1972), and from *H. liaoi*.

H. blairi: Theca biconical; outline pentagonal – decagonal: slightly protuberant interradial areas; strong growth lines; ridges radiating to apices of basals; ambulacra narrow and lanceolate.
H. convexus: Theca biconical; outline rounded pentagonal; slightly protuberant interradial areas; fine growth lines; slight indentation of interbasal sutures; ambulacra broad, lanceolate.
H. kentuckyensis: Theca biconical; outline pentagonal; growth lines low, broad: ambulacra narrow sublanceolate.
H. whitei: Theca biconical; outline rounded pentagonal to pentagonal; very slight concavity in interambulacral areas; fine growth lines: ambulacra lanceolate.
H. (?) *benniei*: Theca biconical; outline pentagonal; strong growth lines; ambulacra sublanceolate strong rim separates radial deltoid and radial radial sectors.
H. (?) n. sp. from Ireland: Theca biconical; outline pentagonal slightly concave interradial areas; strong growth ridges separating all basal and radial sectors.

Class CRINOIDEA J. S. Miller 1821
Subclass CAMERATA Wachsmuth & Springer 1885
Order DIPLOBATHRIDA Moore & Laudon 1943
Suborder EUDIPLOBATHRINA Ubaghs 1953
Superfamily DIMEROCRINITOIDEA Zittel 1879
Family DIMEROCRINITIDAE Zittel 1879
Genus *GNARYCRINUS* gen. nov.

Derivation of name. Gnary is an anagram derived from N. Gary for N. Gary Lane, in recognition of his extensive work on Palaeozoic crinoids.

Type species. Gnarycrinus lanei new species, by monotypy.

Diagnosis. Distinguished by combination of the medium bowl-shaped cup, lack of ray ridges or other thecal plate ornamentation, no intraray plate between the proximal secundibrachials, arms are free above the second secundibrachial, and a pentagonal axial canal.

Description. See description of *Gnarycrinus lanei.*

Remarks. Gnarycrinus n. gen. is most closely related to *Dimerocrinites* Phillips 1839. It differs by lacking the ray ridge ornament, lacking the nodose or inflated thecal plates, has no plates between the proximal secundibrachials, has a medium bowl-shaped cup instead of a conical cup, and a pentagonal instead of a quinquelobate axial canal. It is judged to have evolved from one of the Devonian species of *Dimerocrinites* by loss of the ornamentation, lowering of the cup to the bowl shape, loss of the intraray plate between the proximal secundibrachials, and the arms becoming free above the second rather than the third secundibrachial. All other genera of dimerocrinitids have more than 10 arms and other characters differing from those of *Gnarycrinus.* As currently known *Gnarycrinus* is endemic to the Kazakhstan block of western China.

Gnarycrinus lanei sp. nov.
Text Figure 3K, L

Actinocrinites sp. Waters *et al.* 2003, p. 938, figs. 9.5, 12.9 (incorrectly labelled as *A. zhaoa* in figure explanations).

Derivation of name. Named for N. Gary Lane in recognition of his work on the Chinese crinoids.

Types. Holotype NIGP 148833; paratype NIGP148834.

Type locality and horizon. Measured section west of Boulongour reservoir, marly beds, Hongguleleng Formation.

Diagnosis. As for the genus.

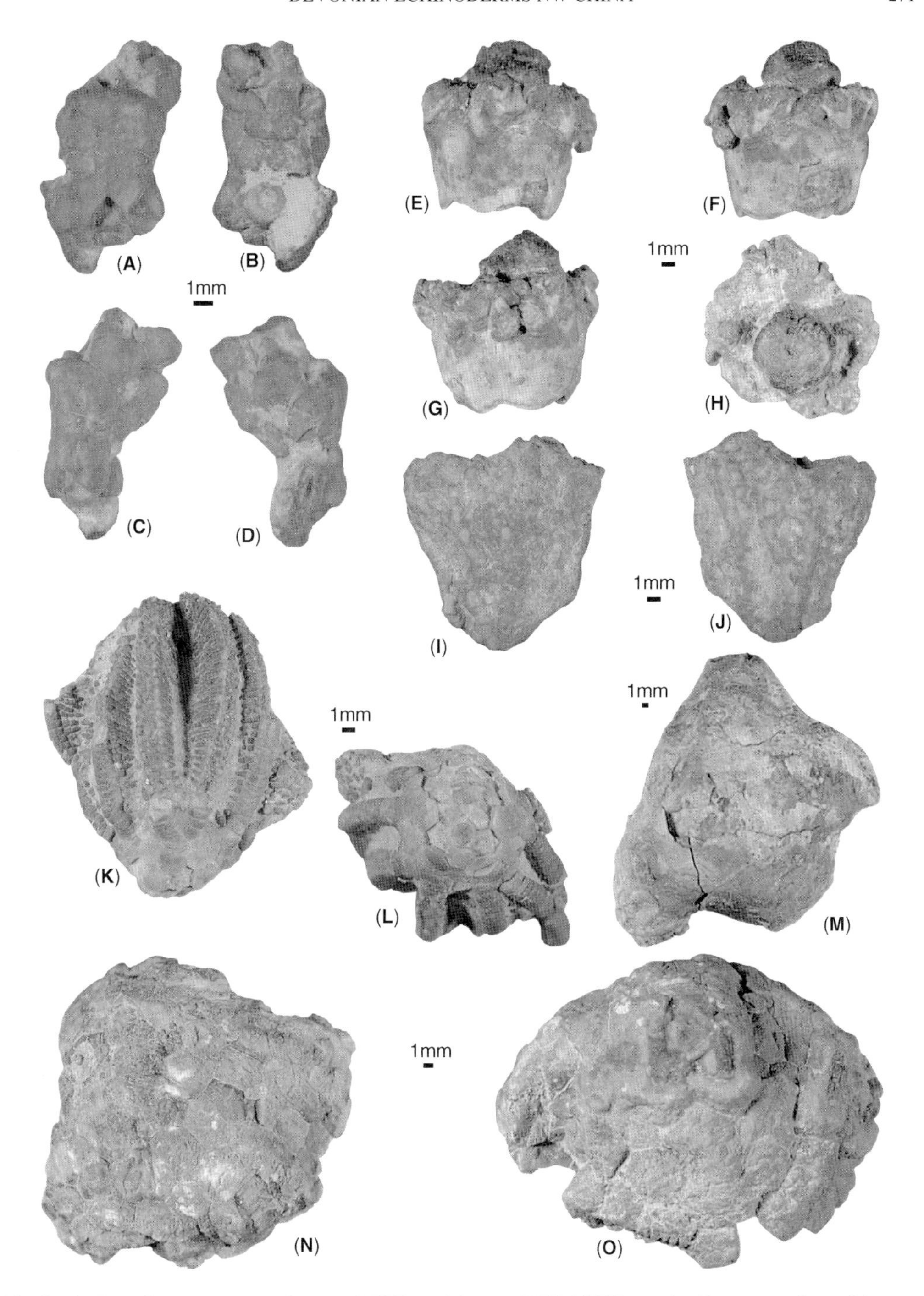

Fig. 3. (**A–D**) *Deltacrinus asiaticus* Lane *et al.* 1997, partial crown NIGP 148839, anterior, D-ray, posterior, and A-ray views. (**E–H**) *Chinacrinus nodosus* Lane *et al.* 1997, partial cup NIGP 148837, A-ray, posterior, E-ray, and oral views. (**I**, **J**) *Sostronocrinus aberratus* sp. nov. Holotype NIGP 148852, posterior and EA-interray views. (**K**, **L**) *Gnarycrinus lanei* gen. et sp. nov., holotype NIGP 148833, C-ray and basal views. (**M–O**), *Amphoracrinus* sp. m, poorly preserved thecae NIGP 148835, oral view. (O, P) Partial theca NIGP 148836, oral and C-ray views.

Description. Crown small, length 24.9 mm, width 20.2 mm (flattened on bedding plane), cylindrical when enclosed. Cup medium bowl shape, shallow basal concavity, sutures flush, no ornament, length 1.4 mm (slightly distorted), width minimum 6.5 mm, maximum 7.3 mm, average 6.9 mm. Infrabasals 5, quadrangular, dart shape, length and width 1 mm, distal tips barely extending beyond stem facet impression. Infrabasal circlet horizontal, diameter 2.1 mm, not visible in lateral view. Basals 5, heptagonal, gently convex longitudinally and transversely, distally truncated to adjoin interray plates; AB-basal length 1.6 mm, width 2.4 mm; CD-basal length 2 mm, width 2.2 mm. Radials 5, pentagonal, wider than long, gently convex transversely and longitudinally; A-radial length 2 mm, width 2.2 mm. Anal series 1:3:3 tegmen plates; primanal largest, length 2.1 mm, width 2 mm. Interray series 1:2:2: or 1:2:3 tegmen plates, incurving sharply at base of secundibrachials. Primibrachials 2, gently convex transversely, slightly convex longitudinally. First primibrachial hexagonal, wider (2.3 mm) than long (1 mm). Axillary second primibrachial pentagonal, branching isotomously distally. Arms 10, 2 per ray, wide, strap-like, flat to slightly convex aboral surface, free above second secundibrachial, incurling at distal tips. Proximal 5 to 11 secundibrachials uniserial, gently cuneate with some wedge-shaped biserial brachials; distal brachials biserial, chisel forms interlocking medially. Arms have no intraray plate between the proximal secundibrachials. Pinnules coarse, roundly convex transversely, straight to concave longitudinally, begin on distal end of second secundibrachial even though it is fixed with interray plates. Some proximal uniserial secundibrachials are hyperpinnulate with one pinnule on each side. Tegmen not exposed. Stem facet round, diameter 1.6 mm; axial canal pentagonal. Stem not preserved.

Remarks. The holotype of *Gnarycrinus lanei* n. sp. is moderately well preserved, the tegmen is not exposed, the arms of the A-ray and anterior part of the E-ray are lost, and the aboral surface of some of the cup plates is partly lost. However, all plate boundaries within the cup are visible, some less obvious than others. The paratype, a partial crown lacking the infrabasal circlet and distal parts of the basals, was identified as *Actinocrinites* sp. by Waters *et al.* (2003). The discovery of the holotype with the clearly exposed infrabasals showed that these specimens are diplobathrids, not monobathrid actinocrinitids, although *Actinocrinites* does occur in the Hongguleleng Formation (Table 1).

The abrupt inward inflection of the interray plates at the base of the arms suggests the tegmen was not highly inflated. It is unknown if an anal tube was present on not. The occurrence of a pinnule on the distal extremity of the second secundibrachials is unusual as the first pinnule is normally above the last fixed brachial.

Suborder MONOBATHRIDA Moore & Laudon 1943
Suborder COMPSOCRININA Ubaghs 1978
Superfamiy PERIECHOCRINOIDEA Bronn 1849
Family AMPHORACRINIDAE Bather 1899
Genus *AMPHORACRINUS* Austin 1848
Amphoracrinus sp.
Text Figure 3M–O

Types. Figured specimen NIGP 148835; mentioned specimen NIGP 148836.

Type locality and horizon. Figured specimen from red limestone beds at top of marly beds, Hongguleleng Formation, south of line of section west of Boulongour Reservoir. GPS 46° 44.597′, E86° 8.195′; second specimen from the lower part of the Hongguleleng Formation, Ganare section.

Description. Partial calyx 27.3 mm long, 29.2 mm diameter; arms grouped, 2 per ray; ray ridges on radials and primibrachials. Basal circlet formed of three plates, slightly upflared, barely visible in lateral view. Radials hexagonal, wider than long. First primibrachial pentagonal, wider than long. Axillary second primibrachial pentagonal or hexagonal, wider than long. First secundibrachials bear arm opening, arms free above. Interray series 1:2:1 tegmen plates. Tegmen strongly inflated, formed of large plates with bulbous centre. Anal opening excentric, surrounded by small plates. Arms, anal series and stem unknown.

Remarks. The figured calyx referred to *Amphoracrinus* sp. is well preserved, but lacks two of the basal plates, C- and D-rays, anal series, and tegmen summit plates thus precluding a specific identification. A second specimen from the Ganare section is larger (length 38.4 mm, incomplete; width 48.4 mm) but has many plates flaked off and is so poorly preserved that most remaining plates cannot be distinguished. This is the first Devonian report of *Amphoracrinus* from China extending the stratigraphic range downward into the Late Devonian from the Early Mississippian.

Suborder GLYPTOCRININA Moore 1952
Superfamily PLATYCRINITOIDEA Austin & Austin 1842
Family PLATYCRINITIDAE Austin & Austin 1842
Genus *CHINACRINUS* Lane *et al.* 1997
Chinacrinus nodosus Lane *et al.* 1997
Text Figure 3E–H

?*Chinacrinus nodosus* Lane *et al.* 1997, p. 19, fig. 6.9.

Types. Figured specimen (crown) NIGP 148837; mentioned specimens basal circlets (2) NIGP 148838.

Type locality and horizon. Measured section west of Boulongour Reservoir, marly beds, Hongguleleng Formation; marly beds, Hongguleleng Formation, Eumha east.

Emended description. Primibrachials 2; secundibrachials 2. Brachials strongly convex transversely, slightly concave to straight longitudinally. First secundibrachials appear to be two fused brachials with partly fused sutures in some rays. Four arms per ray free above first tertibrachials. First pinnules on first primibrachial, one each side. Tegmen formed by central prominently elevated knob and ring of alternating ridges formed of small biserial ambulacrals and troughs formed by large interambulacral plates followed by two smaller plates; ambulacrals and interambulacrals adjoin ring of large plates around the base of the central knob which is formed of small polygonal plates. Anal tube at level of primibrachials projecting laterally. This description supplements the original description of *Chinacrinus nodosus* of Lane *et al.* (1997).

Remarks. Lane *et al.* (1997) questioned the assignment of *Chinacrinus nododus* to the genus. We accept the assignment of the genus but note that *C. nodosus* has only two secundibrachials, although the first secundibrachial appears to be two fused brachials. One new specimen is a partial crown lacking the stem, basal circlet, and distal parts of the arms. It has coarse low nodes on the radials as originally described for *C. nododus.* Two basal circlets, both formed by two equally large plates and one small plate, are from the Eumha east locality and show the faint nodes scattered across the plates.

Subclass DISPARIDA Moore & Laudon 1943
Superfamily CALCEOCRINOIDEA Meek & Worthen 1869
Family CALCEOCRINIDAE Meek & Worthen 1869
Genus *DELTACRINUS* Ulrich 1886

Remarks. We agree with Lane *et al.* (1997) that the arms of *Deltacrinus* Ulrich 1886 are distinctly different from those of *Halysiocrinus* Ulrich 1886 and that both genera should be recognized.

Deltacrinus asiaticus Lane *et al.* 1997
Text Figure 3A–D

Deltacrinus asiaticus Lane *et al.* 1997, p. 26, figs. 8.9, 8.10, 9.7, 9.8.

Types. Figured specimen: Partial crown NIGP 148839.

Type locality and horizon. Marly beds, Hongguleleng Formation, Eumha east.

Emended description. A- and D-ray arms with single axillary primibrachial giving off one small arm towards the E-ray and a larger axillary secundibrachial away from the E-ray. Abanal and first anal plate large. Stem round. To be added to the original description of Lane *et al.* (1997).

Remarks. A new specimen of *Deltacrinus asiaticus* Lane *et al.* 1997 retains the proximal parts of the A- and D-ray arms, abanal, and proximal two columnals.

Superfamily ALLAGECRINOIDEA Carpenter & Etheridge 1881
Family ANAMESOCRINIDAE Goldring 1923
Genus *ANAMESOCRINUS* Goldring 1923

Anamesocrinus Goldring 1923, p. 323. Moore & Laudon 1943, p. 30, fig. 3. Moore & Laudon 1943, p. 147, pl. 52, fig. 1. Ubaghs 1953, p. 746, fig. 140, no. D. Moore 1962, p. 12, fig. 3, no. 5. Moore & Strimple 1978, p. 547.

Emended description. Arm number five or fewer in different rays.

Remarks. When Goldring (1923) described *Anamesocrinus* she recognized that the cup morphology was basically the same plate arrangement as in the Ordovician genus *Ectenocrinus* S. A. Miller 1889, whereas the multiple arms on each radial resembled those in the Early Carboniferous genus *Catillocrinus* Shumard 1865. When describing the arms of the type species *A. lutheri* Goldring 1923, she noted that the arms were distorted and that it was difficult to be certain of the number on each radial, but that there were five on one radial and she suggested that all radials had five arms in the adult stage. She also noted that a smaller form (p. 325, pl. 40, text-figure 6, 7) seemed to have fewer arms. Her plate diagram (fig. 53, p. 324) of *A. lutheri* shows five arms in each ray. This has generally been accepted by later workers as the generic diagnosis; however, the *Treatise on Invertebrate Paleontology* (Moore & Strimple 1978, p. 547) states five arms per ray. As will be noted below, the number of arms is variable in the three specimens of *A. lanei* n. sp. and the C-ray never has more than three arms. Thus the generic description is emended to include forms with fewer than five arms in each ray.

Anamesocrinus tieni sp. nov.
Text Figure 4A–K

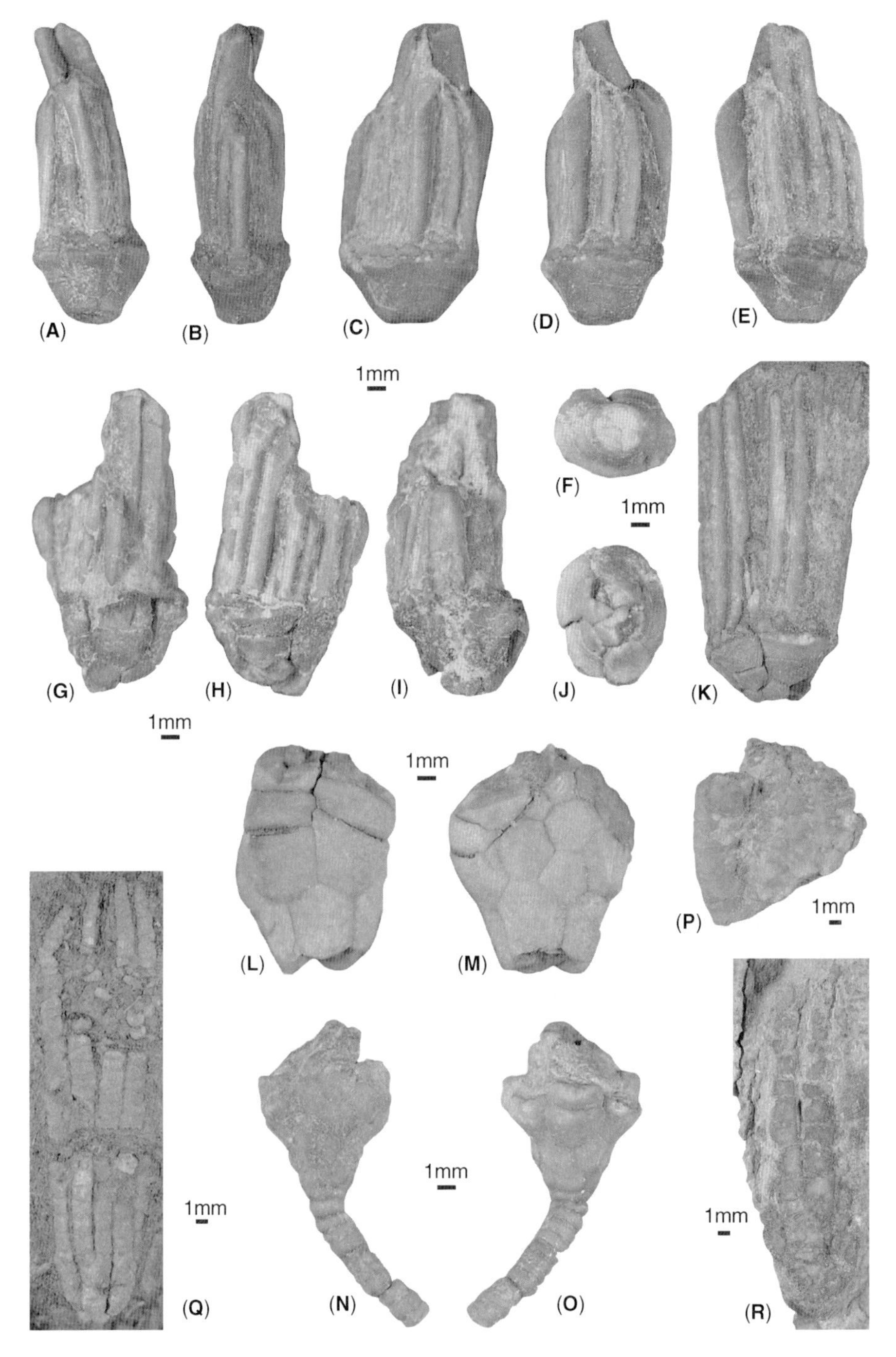

Fig. 4. (**A–K**) *Anamesocrinus tieni* gen. et sp. nov. (A–F) Holotype NIGP 148840, C-ray, B-ray, A-ray, E-ray, D-ray and basal views. (G–I) Paratype 2 NIGP 148842, B-ray, E-ray and C-ray/posterior views. (J, K) Paratype 1 NIGP 148841, basal and BC-interray views. (**L–O**) *Histocrinus*? *chenae* sp. nov. (L, M) Paratype NIGP 148847, DE-interray and posterior views. (N, O) Holotype NIGP 148846 posterior and A-ray views. (**P**) Glossocrinacea indeterminate, tegmen and one arm fragment NIGP 148845, lateral view. (**Q**) *Histocrinus* sp., partial crown NIGP 148848, ray uncertain. (**R**) Scytalocrinid Indeterminate, poorly preserved partial crown NIGP 148849, ray uncertain.

Derivation of name. Named for C. C. Tien in recognition of his early work on Chinese crinoids.

Types. Holotype NIGP 148840, Paratypes 1 and 2, NIGP 148841, 148842, and mentioned specimens NIPG148843, 148844.

Type locality and horizon. Measured section west of Boulongour Reservoir, marly beds, between 61 m and 64 m above the base of the Hongguleleng Formation.

Diagnosis. Distinguished by a shorter cup and only two brachials in each arm.

Description. Crown small, cylindrical. Cup truncated medium bowl, longer in A-ray than in D-ray, flush sutures, no ornament. Basals 5, proximally horizontal, bearing stem facet, distally upturned, visible in lateral view. Basal circlet short. Radials 5, 2 large in A- and D-rays, 3 divided into superradial and subradial in B-, C- and E-rays; gently convex transversely, gently concave longitudinally, widening distally, widest at radial summit; bear multiple arm facets on each radial. Anal on small notch on left side of C-radial. Anal plate large, nearly half length of arms, wide, distal tip rounded. Two brachials in each arm; first brachial short, wider than long, second brachial very elongate, strongly convex transversely, gently convex longitudinally, narrowing to point distally. Arm number variable with growth and differs in different rays. Stem facet round, deeply impressed in basal circlet. Measurements and arm numbers given in Table 4.

Remarks. The holotype is the smallest, but most complete, of the five specimens of *Anamesocrinus tieni* n. sp. It is slightly crushed inwardly along the posterior interradius and the suture between the B-superradial and B-subradial is nearly fused. Paratype 1 lacks two basals and the distal tips of the arms, the cup is distorted, and it retains the anal. Paratype 2, a crown, has the arms slightly splayed distally, lacks two basal plates, and the cup is slightly distorted. The fourth specimen lacks the basal circlet, the radial circlet is distorted, and some radials are broken. In addition, an isolated brachial on a small slab and three brachials on a second slab were easily recognized by the extended length of these plates. Overall the preservation of the specimens is good as the original calcite is well preserved. Specimens are most common in the lower part of the marly beds, but occur throughout the marly beds.

Subclass CLADIDA Moore & Laudon 1943
Order DENDROCRINIDA Bather 1889
Superfamily GLOSSOCRINOIDEA Webster *et al.* 2003
Glossocrinacea indeterminate
Text Figure 4P

Types. Figured specimen NIGP 148845.

Table 4. *Measurements (in mm) and number of arms per ray for* Anamesocrinus tieni *n. sp.*

Specimen	Holotype	Paratype 1	Paratype 2	4th specimen
Crown length	15.5	15.5 (incomplete)	18.4	21 (incomplete)
Crown width	7.4	8.1 (distorted cup)	10.5 (splayed)	6.5 (exposed)
Cup length, A-ray	3.9		3	
Cup length, D-ray	3.1		2.7	
Cup width, maximum	7.3	8.1	6.6	
Cup width, minimum	5.1	7.1	5.8	
Cup width, average	6.2	7.6	6.2	
Basal circlet diameter	3.3	3.3	3.8	
A radial length	2.7	3.3	3	
A radial width	3	3.5	3	
Primibrachial 1, length	0.6	0.9	0.8	
Primibrachial 1, width	1	1.2	1	
Primibrachial 2, length	11	9 (incomplete)	12.9	
Primibrachial 2, width	1.2	0.8	1.5	
Anal length		5.6		
Anal width		2.3		
Stem facet diameter	2.5		2.2	
Arms, A-ray	3	4	4	5
Arms, B-ray	4	3?	5	4
Arms, C-ray	3	2?	3	3
Arms, D-ray	2	2	3	3
Arms, E-ray	4	5	5	5

Type locality and horizon. Measured section west of Boulongour reservoir, marly beds, between 62 m and 64 m above the base of the Hongguleleng Formation.

Remarks. A small fragment of one arm and part of the plicate stellate plated tegmen could belong to any of several genera of the of Glossocrinoidea Webster *et al.* 2003. This is the first report of the glossocrinids in China and adds to the diversity of the Hongguleleng fauna.

Superfamily SCYTALOCRINOIDEA Moore & Laudon 1943
Family SCYTALOCRINIDAE Moore & Laudon 1943
Genus *HISTOCRINUS* Kirk 1940
Histocrinus? *chenae* sp. nov.
Text Figure 4L–O

Derivation of name. Named for Xiuqin Chen who organized the field excursion and collected some the crinoid specimens from the Hongguleleng Formation.

Types. Holotype NIGP 148846; paratype NIGP148847.

Type locality and horizon. Measured section west of Boulongour Reservoir, marly beds, between 62 and 64 m above the base of the Hongguleleng Formation.

Description. Cup medium cone, truncated base, outflaring at base of radials, sutures gently impressed, no ornament, canted downward toward A-ray from posterior. Infrabasals 5, proximally horizontal bearing stem impression, distally up- and outflaring steeply, visible part moderately convex longitudinally and transversely, forms slightly less than one-third of cup wall. Basals 5, hexagonal (EA, AB, BC) or heptagonal (CD, DE), moderately convex longitudinally and transversely, steeply upflared, form middle third of cup wall. Radials 5, pentagonal (C-radial hexagonal), wider than long, moderately outflaring from proximal tip, moderately convex transversely, straight longitudinally. B- and E-radials distal tip above A-radial distal tip; C-radial distal tip above B-radial distal tip; and D-radial distal tip above E-radial distal tip, giving downward cant to cup from posterior toward anterior. Radial facets inplenary, gently declivate, bearing straight transverse ridge two-thirds radial width, moderately deep ligament pit, narrow outer ligament furrow, rounded marginal ridge, and deep wide muscle fields. Ex-interradial notches narrow. Anals 3, menoplax 3 arrangement, all gently convex longitudinally and transversely. Primanal pentagonal. Secundanal hexagonal or heptagonal, distal one-third projecting above radial summit, followed by 2 or 3 tube plates. Tertanal hexagonal or heptagonal, distal half projecting above radial summit, followed by 3 or 4 tube plates. Anal tube plates projecting upwards moderately steeply suggesting presence of anal tube of indeterminate length formed of polygonal plates. First primibrachial wider than long, moderately convex longitudinally and transversely, slightly cuneate. Second primibrachial axillary, bearing entoneural canal. Stem roundly pentagonal, heteromorphic; noditaxis N1. Columnals with moderately convex latus, round lumen. Measurements given in Table 5.

Remarks. The holotype of *Histocrinus*? *chenae* n. sp. is well preserved retaining the proximal columnals, but lacks the arms beyond the first primibrachial. The paratype lacks the infrabasals and the two second primibrachials are cleaved with loss of the distal half. The second primibrachials are much longer than the first primibrachials and appear to be axillary. This suggests a minimum of ten arms if all arms branch uniformly and there is no distal branching. The generic assignment is questioned because the radial facets are inplenary, not peneplenary, and the distal arms are unknown. The cup is distinctive among the Hongguleleng taxa.

Histocrinus sp.
Text Figure 4Q

Types. Figured specimen NIGP 148848.

Type locality and horizon. South side of gully across from the measured section of Lane *et al.* (1997) west of Boulongour Reservoir; approximately 5 m above the base of the Hongguleleng Formation in the bioclastic limestone.

Description. Crown medium cylinder, expanding slightly distally, length 46.7 mm, width12.2 mm at slightly splayed distal arms. Cup medium conical, sutures impressed. Infrabasal circlet very small, diameter 2 mm (estimated), distal tips visible in lateral view. Basals forming major part of cup, poorly preserved. Radials pentagonal, length and width 3 mm, slightly outflaring. Radial facets peneplenary, morphology covered. Interradial notches very narrow. Primibrachials 2, wider than long. Secundibrachials slightly cuneate, wider than long. Arms 10, branching isotomous on second primibrachial. Anals, tegmen and stem unknown.

Remarks. The solution-weathered crown of *Histocrinus* sp. is recrystallized and lacks the infrabasal circlet and most of the basal circlet with only parts of two basals preserved. It is orientated with the E-, A- or B-ray centred, probably the B-ray, because the radial of the ?C-ray on the left has

Table 5. *Measurements (in mm) for* Histocrinus? chenae *n. sp.*

Specimen	Holotype	Paratype
Specimen length	15	11.4 (incomplete)
Specimen width	7	10
Cup length, CD interray	4.9	
Cup length, A ray	4.3	
Cup length, average	4.6	
Cup width, radial summit	6.8	9
Infrabasal circlet diameter	3	4.3 (at base of basals)
Infrabasal length, visible	1.1	
Infrabasal width, visible	2	
AB basal length	1.8	3.7
AB basal width	2	3.2
CD basal length	2.4	4.1
CD basal width	2.7	3.5
A radial length	2.1	
A radial width	3	
First primibrachial length	1	1.3
First primibrachial width	3	4.1
Primanal length	2	3.2
Primanal width	1.8	2.8
Secundanal length	1.7	3.2
Secundanal width	2.1	3
Length proximal stem	9	
Diameter proximal columnal	2	
Diameter distal columnal	1.6	

a slight suggestion of a primanal suture along the proximal left side. The specimen is left in open nomenclature because of incomplete preservation.

Scytalocrinid? indeterminate
Text Figure 4R

Types. Figured specimen NIGP 148849.

Type locality and horizon. Marly beds, Hongguleleng Formation, Eumha east knob.

Description. Crown cylindrical; cup medium cone. Infrabasals and basals upflaring. Radials subvertical, gently convex transversely. Radial facet plenary. Axillary single primibrachials of different lengths, gently convex transversely, straight longitudinally. Branching isotomous, 2 arms per ray. Secundibrachials slightly cuneate, strongly convex transversely. Tegmen and stem unknown.

Remarks. The specimen recognized as Scytalocrinid? indeterminate is recrystallized with botryoidal nodes on all cup and proximal brachials. The ten arms and cup shape suggest relationship with the scytalocrinids. The specimen is left in open nomenclature because the anals are unknown and recrystallization has masked many other features.

Family BRIDGEROCRINIDAE Webster & Lane 2007

Genus *BRIDGEROCRINUS* Laudon & Severson 1953

Bridgerocrinus delicatulus Lane *et al.* 1997

Bridgerocrinus delicatulus Lane *et al.* 1997, p. 28, figs. 8.17, 9.9.

Logocrinus delicatulus (Lane *et al.* 1997) Waters *et al.* 2003, p. 943.

Remarks. Waters *et al.* (2003) transferred *Bridgerocrinus delicatulus* Lane *et al.* 1997 to *Logocrinus* because it has three primibrachials. *Logocrinus* has zyzygial paired secundibrachials, whereas *B. delicatulus* has rectangular to slightly cuneate secundibrachials. The first branching in *Bridgerocrinus* may be on the first, second (most commonly), or third primibrachial and the secundibrachials are slightly cuneate as noted by Webster (1997). Because *B. delicatulus* has the first branching on the third primibrachial in the two rays known and the secundibrachials are slightly cuneate it is considered to be a bridgerocrinid not a logocrinid. If all the rays were known and the branching consistent it is possible that it could represent a new genus. Lacking that information it is left in *Bridgerocrinus*.

Family SOSTRONOCRINIDAE Lane *et al.* 2001*a*

Remarks. Lane *et al.* (2001*a*) removed three genera (*Sostronocrinus* Strimple & McGinnis 1969;

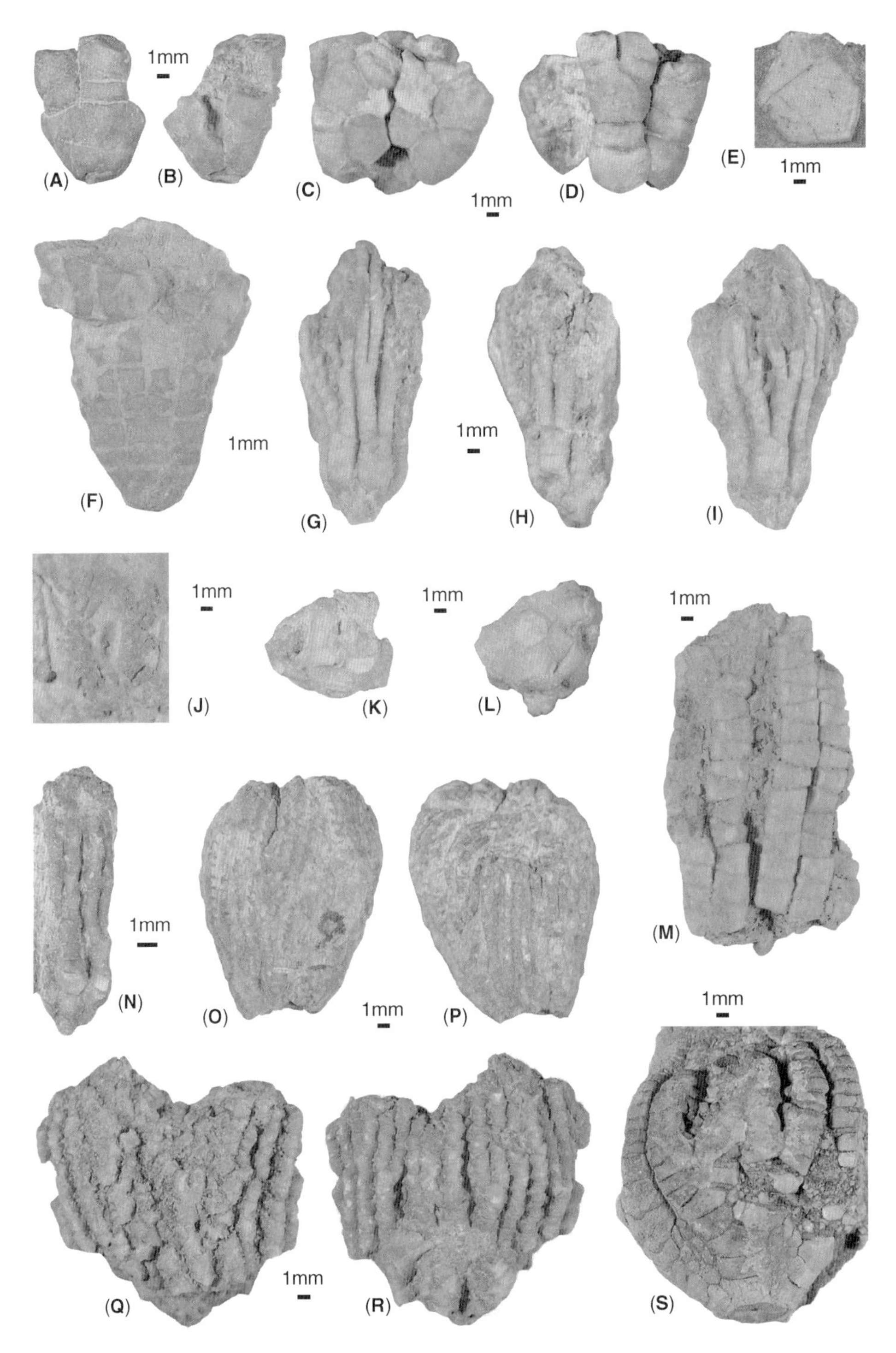

Fig. 5. (**A–D, F**) *Grabauicrinus xinjiangensis* (Lane *et al.* 1997). (A, B) Partial crown NIGP 148859, A-ray and posterior views. (C, D) Partial crown NIGP 148858, posterior and A-ray views. (F) Crown NIGP 148857, A-ray view. (**E**) Basal plate indeterminate NIGP 148862, adoral view. (**G–I**) *Sostronocrinus quadribrachiatus* Waters *et al.* 2003, NIGP 148850, B-ray, posterior and A-ray views of crown. (**J**) Cladid indeterminate 2, poorly preserved fragment of two arms

Tundracrinus Yakovlev 1928; and *Haeretocrinus* Moore and Plummer 1940), each with more than ten arms, from the Scytalocrinidae Moore & Laudon 1943 placing them in the Sostronocrinidae. We agree that these genera do not belong in the Scytalocrindae, but each is probably derived from a different lineage and therefore belong in different families. Reallocation of these three genera is beyond the scope of this study.

Genus *SOSTRONOCRINUS* Strimple & McGinnis 1969

Sostronocrinus quadribrachiatus Waters *et al.* 2003
Text Figure 5G–I

?*Graphiocrinus* species Lane *et al.* 1997, p. 40, figs. 11.4, 16.1, 16.2 only.

Sostronocrinus quadribrachiatus Waters *et al.* 2003, p. 944, figs. 11.1–11.4, 13.6, 13.10, 13.11.

Types. Figured specimens: Crown NIGP 148850, set of arms NIGP 148851.

Type locality and horizon. Measured section west of Boulongour reservoir, marly beds of the Hongguleleng Formation.

Remarks. A number of the specimens referred to ?*Graphiocrinus* species by Lane *et al.* (1997) and Waters *et al.* (2003) have gently cuneate brachials and more than ten arms precluding their assignment to *Graphiocrinus*, which has rectilinear brachials and ten arms. A new set of arms and a crown of *Sostronocrinus quadribrachiatus* Waters *et al.* 2003 show the multiple arms.

Sostronocrinus aberratus sp. nov.
Text Figure 3I, J

Derivation of name. Referring to the aberrant condition of the specimen.

Types. Holotype NIGP 148852, the only specimen.

Type locality and horizon. Marly beds, Hongguleleng Formation, Eumha.

Diagnosis. A *Sostronocrinus* with two primibrachials and distinguished by shallow apical pits, convexly rounded basals and radials, and a medium bowl cup.

Description. Crown small, expanding distally, widest at level of tertibrachials, length 18 mm (incomplete), width 15 mm (flattened on bedding plane). Cup small, length 2.5 mm, with 4.7 mm, medium bowl, base truncated, sutures slightly impressed, shallow apical pits. Infrabasal circlet low, forming basal plane of cup, diameter 1.9 mm. Infrabasals 5, horizontal proximally, distal tips widely up- and outflaring, barely visible in lateral view, barely extending beyond proximal columnal. Basals 5, BC and CD septagonal, all other hexagonal, moderately convex longitudinally and transversely, up- and outflaring; AB-basal length 1.1 mm, width 1 mm. Radials 5, pentagonal, except C-radial hexagonal, wider than long (except E-radial), moderately convex longitudinally, strongly convex transversely, gently outflaring; A-radial length 1.4 mm, width 1.9 mm. Aberrant C-radial axillary for two arms. Aberrant E-radial narrower than long, narrowing distally, terminal with tegmen, armless. Radial facets moderately declivate, ovate outer margin, angustary, facet width/radial width 1.3/1.9 = 0.69. Anals 3, menoplax 3 subcondition. Primanal pentagonal, adjoins C-radial, BC- and CD-basal, secundanal and tertanal. Secundanal quadrate, directly above CD-basal. Tertanal large, distal two-thirds above radial summit. Arms 16, 4 per ray (E-ray armless). Brachials moderately cuneate, sides concave, rounded medial linear ridge, bearing wide pinnule on wide side; first primibrachial wider than long, axillary second primibrachial longer than wide; secundibrachials longer than wide; tertibrachials wider than long. Branching isotomous. Stem impression round, diameter 1 mm.

Remarks. The crown of *Sostronocrinus aberratus* n. sp. is crushed normal to the A-ray-posterior plane of symmetry and lacks the distal parts of the arms and stem. Although aberrant in the C- and E-radials the major morphology of the specimen is well preserved and other specimens of the species could be recognized by these features. It is most similar to the two Famennian species *S.*? *paprothae* and *S.*? *pauli* described by Haude & Thomas (1989) from Germany. However, it differs by the less ornate cup, shallower apical pits, and rounded basals and radials.

Genus *AMADEUSICRINUS* Waters *et al.* 2003

Fig. 5. (*Continued*) NIGP 148861, ray uncertain. (**K, l**) *Eutaxocrinus basellus* Lane *et al.* 1997, NIGP 148864, A-ray and basal views of partial crown. (**m**) Cladid indeterminate 1, set of arms NIGP 148860, ray uncertain. (**N–P**) *Amadeusicrinus subpentagonalis* (Lane *et al.* 1997). (n) Immature crown with 10 arms NIGP 148853, B-ray view. (o, p) Set of arms NIGP 148854, lateral views, rays uncertain. (**Q, R**) *Eumhacrinus tribrachiatus* gen. et sp. nov., paratype NIGP 148856, lateral views of set of arms, rays uncertain. (**S**) *Labrocrinus granulatus* n. gen. et sp., holotype NIGP 148863, C-ray view.

Amadeusicrinus subpentagonalis (Lane *et al.* 1997)
Text Figure 5N–P

?*Pachylocrinus subpentagonalis* Lane *et al.* 1997, p. 32, fig. 11.7–11.10, 11.12, 12.7–12.11.
Amadeusicrinus subpentagonalis (Lane *et al.* 1997); Waters *et al.* 2003, p. 945, figs. 11.7, 11.8, 13.14 (non figs. 11.6, 13.12).

Types. Set of arms NIGP 148853; partial crown NIGP 148854.

Type locality and horizon. Measured section west of Boulongour Reservoir, marly beds of the Hongguleleng Formation.

Emended description. Branching endotomous with tertibrachials. To be added to the description of Waters *et al.* (2003).

Remarks. The holotype cup of *Amadeusicrinus subpentagonalis* (Lane *et al.* 1997) has a very low basal circlet with the distal tips visible in lateral view. The cuneate brachials of other specimens assigned to ?*Pachylocrinus subpentagonalis* Lane *et al.* 1997 have constricted centres with the pinnular facet on the distal edge of the wide side of the brachial. Unfortunately none of these specimens were complete and the distal branching pattern of the arms was uncertain. A new set of arms lacking the cup, primibrachials and proximal secundibrachials shows that the branching is endotomous on secundibrachials 5 to 8 minimally, tertibrachials 9 to 11, and higher quartibrachials. This results in 8 arms per ray and 40 arms total if branching is uniform in all rays.

Three additional specimens were identified as *A. subpentagonalis* by Waters *et al.* (2003) and incorrectly designated lectotypes. The two cups with primibrachials (Waters et al. 2003, figs. 11.7, 11.8, 13.14) from the Eumha locality extended the occurrence of the species. The partial crown (Waters *et al.* 2003, figs. 11.6, 13.12) has a conical cup with the infrabasals upflaring and brachials with the pinnular facets forming large protrusions from the central part of the wide side of the brachial. These two characters preclude inclusion of that specimen in *A. subpentagonalis.*

Family CERCIDOCRINIDAE Moore & Laudon 1943

Emended description. To include genera with more than one primibrachial.

Remarks. The Cercidocrinidae Moore & Laudon 1943 have endotomously branching arms and were considered to have been derived from the Blothrocrinidae Moore & Laudon 1943 by Moore & Strimple (1978). All genera of the Blothrocrinidae have isotomously branching arms. Webster (1997) recognized more than one clade within the Blothrocrinidae based on the different types of brachials and primibrachial branching patterns. He suggested that they were derived from different lineages and noted that the family needed systematic revision. The discovery of *Eumhacrinus* n. gen. with three primibrachials and endotomously branching arms shows relationship with both the Blothrocrinidae and Cercidocrinidae. It is assigned to the Cercidocrinidae because it has endotomously branching arms.

Genus *EUMHACRINUS* gen. nov.

Derivation of name. Named for the locality where the paratype was found.

Type species. Eumhacrinus tribrachiatus n. gen., n. sp. by monotypy.

Diagnosis. Distinguished by the three primibrachials, moderately cuneate brachials, and endotomous branching arms.

Description. See description of *Eumhacrinus tribrachiatus.*

Remarks. Eumhacrinus n. gen. differs from all other genera of the Cercidocrinidae by the first arms branching on the third primibrachial. It extends the range of the family downward into the Famennian and is the first report of a cercidocrinid outside of North America.

Eumhacrinus tribrachiatus sp. nov.
Text Figure 5Q, R

Scytalocrinidae, genus and species indeterminate Lane *et al.* 1997, p. 29, figs. 11.11, 12.1–12.4.

Derivation of name. Referring to the three primibrachials.

Types. Holotype NIGP 148855 (formerly USNM 476164); paratype set of arms NIGP 148856.

Type locality and horizon. Marly beds, Hongguleleng Formation, Eumha east.

Diagnosis. As for the genus.

Description. Crown small, expanding upward, widest at level of tertibrachials or quartibrachials. Dorsal cup small, infrabasals and basals not preserved. Radials approximately equal in length and width, moderately convex transversely, gently convex longitudinally, outflaring. Radial facets plenary or nearly plenary. Anals 3. Arms branching isotomously on third primibrachial and endotomously on eighth secundibrachial and on a higher tertibrachial. Six arms per ray, total of 30 arms if branching uniform in all rays. All brachials strongly convex transversely, zig-zag in appearance, with outer surface elevated just below relatively wide and prominent pinnular facets. Distal end of anal sac narrow and upright. Sac plates thick, short. Larger plates at distal tip. Stem unknown. Measurements of holotype given in Lane *et al.* (1997).

Remarks. Lane *et al.* (1997) recognized that the arms branched on the third primibrachial and again higher (eighth secondibrachial) on the specimen that they referred to the Scytalocrinidae indeterminate, here designated the holotype of *Eumhacrinus tribrachiatus* n. sp. Both of these characters preclude inclusion within the Scytalocrinidae which branch on the first or second primibrachial and have 10 arms; these characters suggest relationship to the Blothrocrinidae. Discovery of a second specimen, a set of arms (28 discernible), shows that the arms branch endotomously, as within the Cercidocrinidae to which the genus is assigned. Branching on the third primibrachial is considered to be a more primitive character than branching on the first primibrachial.

Superfamily DECADOCRINOIDEA Bather 1890
Family DECADOCRINIDAE Bather 1890
Genus *DECADOCRINUS* Wachsmuth & Springer 1880

Remarks. See discussion under *Grabauicrinus* below.

'Decadocrinus' rugosus (Lane *et al.* 1997)

'Decadocrinus' rugosus Lane *et al.* 1997, p. 35, figs. 11.6, 15.2–15.4.

'Decadocrinus' xinjiangensis Lane *et al.* 1997, p. 39, figs. 14.9, 15.13 only.

Grabauicrinus rugosus (Lane *et al.* 1997), Waters *et al.* 2003, p. 927.

Remarks. The cup of *'Decadocrinus' rugosus* has a lower bowl with smaller infrabasals than that of *G. xinjiangensis*, The arms of *'D.' rugosus* are not closely apposed and have a zig-zag appearance formed of moderately cuneate brachials, whereas *G. xinjiangensis* has closely apposed, straight-sided arms made up of rectilinear or slightly cuneate brachials. We consider these differences in the arm structure to be of sufficient importance to merit the assignment of *'D'. rugosus* to a different genus.

Genus *GRABAUICRINUS* Waters *et al.* 2003

Grabauicrinus Waters *et al.* 2003, p. 945.

Remarks. Problems with *Decadocrinus* were discussed by Kammer & Ausich (1993) when they restricted *Decadocrinus* to forms with nine arms (atomous A-ray) and transferred some species previously assigned to the genus they designated *Lanecrinus*. Lane *et al.* (1997) discussed some of the problems of the 29 remaining species assigned to *Decadocrinus* when they named five species assigning them to *'Decadocrinus' sensu lato*. Waters *et al.* (2003) proposed *Grabauicrinus* designating *'D.' xingiangensis* Lane *et al.* 1997 as type species. In addition, they assigned 15 other species previously assigned to *Decadocrinus* or *'Decadocrinus' sensu lato* to *Grabauicrinus*. Waters *et al.* (2003, p. 945) defined *Grabauicrinus* as follows: 'Crown cylindrical or with widely spread arms. Aboral cup bowl-shaped or truncate-cone shaped. Radials with plenary or slightly peneplenary facets, pits may or may not be present at corners of cup plates. Three anal plates in cup. Arms branching on second primibrachials in all rays, resulting in 10 arms, with no higher branching. Brachials quadrate or cuneate, arms straight-sided or zig-zag, closely apposed or widely spreading. Anal sac poorly known. Stem circular.'

Differences in the cup shapes, brachial shapes, and arm morphology included in the species assigned to *Grabauicrinus* by Waters *et al.* (2003) have been used to differentiate genera in several families of the cladids. Thus, we do not accept the inclusion of all of the other species that were transferred to *Grabauicrinus*; however, we do accept *Grabauicrinus* as a valid genus, restricting it to the type species. Our reasons for this are as follows.

Kammer & Ausich (1993) accepted three species as *Decadocrinus sensu stricto* (nine arms): the type species *D. scalaris* (Meek & Worthen 1869), *D. penicilliformis* (Worthen 1882), and *D. tumidulus* (Miller & Gurley 1894). These species share the common morphologic characters of: (1) a bowl-shaped cup with a basal impression; (2) infrabasals not visible in lateral view; (3) three anals in mesoplax 3 arrangement; (4) peneplenary radial facets; (5) 2 primibrachials; (6) weakly to moderately cuneate brachials; (7) 9 arms; (8) a pentagonal, pentalobate or round (*D. tumidulus*) proximal stem; and (10) of Devonian or Mississippian age from North America or England.

Webster (2003), recognizing the revision of *Decadocrinus* by Kammer & Ausich (1993), listed 34 species and two subspecies included under *Decadocrinus*. A literature analysis of those species and subspecies showed that: (1) the cup shape ranges from a low to high bowl or a low to medium cone; (2) infrabasals are visible or not visible in lateral view; (3) brachials are cuneate or rectilinear; (4) one or two primibrachials; (5) 9 or 10 arms; (6) a round or pentagonal proximal stem; and (7) presence or absence of various types of ornamentation.

Decadocrinus multinodosus Goldring 1923 shares the above-listed morphological characters for the genus and was described as having two primibrachials, but one specimen (Goldring 1923 p. 55, fig. 10) has an unbranched A-ray in the proximal three primibrachials present. It is unknown if that arm branched distally. Other specimens of *D. multinodosus* branch on the second primibrachial of the A-ray, thus the species may contain forms with 9 or 10 arms. *Decadocrinus multinodosus* is provisionally left in *Decadocrinus* and may be the most primitive form of the genus as it is from the Middle Devonian, Givetian, of New York. All other species formerly assigned to *Decadocrinus* have morphological characters precluding their inclusion in the genus as recognized by Kammer & Ausich (1993).

Six species ('*D.*' *constrictus* Lane *et al.* 1997; *D. decemnodosus* Goldring 1923; '*D.*' *elongatus* Lane *et al.* 1997; *D. exornatus* Hauser 1999; *D. spinobrachiatus* Goldring 1936; and '*D.*' *usitatus* Lane *et al.* 1997) have10 arms, but otherwise have the morphologic characters of *Decadocrinus*. If *D. multinodosus* is accepted as a bona fide *Decadocrinus* then these six species could be included in the genus and the genus would be considered to contain forms with 9 or 10 arms. If *D. multinodosus* is not accepted as a *Decadocrinus* then a new genus will be needed for it and the other six species listed above.

Two species, *D. fifensis* Wright 1934 and *D. trymensis* Wright 1951 were transferred to *Lanecrinus* by Ausich & Kammer (2006). *Decadocrinus regularis* Strimple 1939 has a bowl-shaped cup with a basal impression, infrabasals not visible in lateral view, 3 anals, plenary radial facets, a single primibrachial, and 10 arms. These are the characters of *Trautscholdicrinus* Yakovlev & Ivanov *in* Yakovlev 1939 and *D. regularis* is herein transferred to *Trautscholdicrinus*.

When Kirk (1940) named *Histocrinus* he noted that the conical cup with the infrabasals visible in lateral view was similar to *Hypselocrinus* Kirk 1940, whereas the arms and tegmen were similar to *Decadocrinus*. He also recognized that most species he assigned to *Histocrinus* have 9 arms (A-ray atomous) with the exception of *Histocrinus juvenis* (Meek & Worthen 1869) with 10. Although a new genus could be erected for the 10-armed forms we provisionally accept the concept of *Histocrinus* with 9 or 10 arms because we recognize that *Decadocrinus multinodosus* has forms with both 9 and 10 arms. The following species, previously assigned to *Decadocrinus*, have the morphological characters of *Histocrinus* and are herein transferred to *Histocrinus*: *H. hughwingi* (Kesling 1964) n. comb., *H. insolens* (Goldring 1923) n. comb., *H. kersadioensis* (Le Menn 1985) n. comb., *H.* lyriope (Hall 1862) n. comb., *H. nereus* (Hall 1862) n. comb., *H. oaktrovensis* (Webby 1961) n. comb., *H. ornatus* (Goldring 1954) n. comb., *H. rugistriatus* (Goldring 1923) n. comb., *H. spinulifer* (Laudon 1936) n. comb., *H. stewartae* (Kier 1952) n. comb., and *H. wrightae* (Goldring 1954) n. comb. The A-ray is unknown on two of these species, *H. isolens* and *H. ornatus*, whereas all others are reported to have 10 arms. The 10-armed forms are considered to be more primitive and, with the exception of *H. spinulifer*, have proximally pentagonal columnals, whereas the 9-armed species have proximally round columnals. The 10-armed species with proximally pentagonal stems are considered the primitive forms and gave rise to the 9-armed forms. Inclusion of the transferred species in *Histocrinus* extends the stratigraphic range downward into the Middle Devonian, Givetian, of New York and Europe from the previously known Mississippian of the Illinois basin and extends the palaeogeographic range to Europe.

Several other species currently assigned have morphological characters precluding their assignment to *Decadocrinus*. Some of these species have similar morphological characters allowing recognition of three groups as listed below. Each of these groups may represent a clade and one or more new genera pending further study. All of the species are currently left in *Decadocrinus sensu lato*. (1) Three species (*D. crassidactylus* Laudon 1936, 10 arms; *D. killawogensis* Goldring 1923, arm number uncertain; and *D. pachydactylus* Laudon 1936, arm number uncertain) have bowl-shaped cups with the infrabasals visible in lateral view, three anals in mesoplax 3 arrangement, two primibrachials, rectilinear brachials, and pentagonal stems. One, *D. pachydactylus*, has plenary radial facets and the other two have peneplenary radial facets. Two genera are probably represented in this group. (2) Three species with 10 arms (*D. aegina* (Hall 1863); *D. brazeauensis* Laudon *et al.* 1952; and *D. gregarious* (Williams 1882)) and two subspecies with an uncertain arm number (*D. multinodosus serratobrachiatus* Goldring 1923 and *D. wrightae silicaensis* Kesling 1971) that have bowl-shaped cups with the infrabasals visible in lateral view, three anals in the mesoplax 3 arrangement, peneplenary radial facets, two primibrachials, and cuneate secundibrachials are excluded from *Decadocrinus* because the infrabasals are visible in lateral view of the cup. (3) Plenary radial facets and visible infrabasals preclude inclusion of *D. baumgardeneri* Laudon & Beane 1937, *D. vintonensis* Thomas 1924, and '*D.*' *rugosus* Lane *et al.* 1997 in *Decadocrinus* or with the two groups characterized above on other morphological characters.

Grabauicrinus xinjiangensis (Lane *et al.* 1997)
Text Figure 5A–D, F

'*Decadocrinus*' *xinjiangensis* Lane *et al.* 1997, p. 39, figs. 14.6, 15.12–17.

Not '*Decadocrinus*' *xinjiangensis* Lane *et al.* 1997, p. 39, figs. 14.9, 15.13. (='*D.*' *rugosus* Lane *et al.* 1997).

Grabauicrinus xingiangensis Waters *et al.* 2003, p. 945, figs. 13.7–9, 13.13.

Types. Figured specimens: crown NIGP 148857, partial crown NIGP 148858, partial crown NIGP 148859.

Type locality and horizon. Marly beds, Hongguleleng Formation, measured section west of Boulongour Reservoir and Eumha east.

Emended description. Brachials rectilinear. Arms apposed, straight-sided, not zig-zag.

Remarks. Lane *et al.* (1997) included a weathered specimen of '*Decadocrinus*' *rugosus* in the specimens upon which they based the description of '*Decadocrinus*' *xinjiangensis*. When Waters *et al.* (2003) defined *Grabauicrinus* they also included the arm characters of that specimen in the description. Most of the specimens that they assigned to the various species of '*Decadocrinus*' are solution weathered and can be difficult to distinguish. All specimens of *G. xinjiangensis*, including three new specimens collected in 2005, have straight-sided arms and rectilinear or only slightly cuneate brachials. The slightly cuneate brachials do not result in a zig-zag arm. Forms with moderately cuneate brachials and zig-zag arms are herein removed from *G. xinjiangensis* and left in '*D.*' *rugosus* n. comb.

Cladid indeterminate 1
Text Figure 5M

?*Graphiocrinus* species Lane *et al.* 1997, p. 40, figs. 11.3, 16.3–16.5 only.
?*Graphiocrinus* species Waters *et al.* 2003, p. 946.

Types. Figured specimen NIGP 148860.

Type locality and horizon. Marly beds of the Hongguleleng Formation, Eumha section.

Remarks. Specimens with slightly cuneate flat brachials and 10 arms were questionably referred to *Graphiocrinus* by Lane *et al.* (1997) and Waters *et al.* (2003). *Graphiocrinus* has flat rectilinear brachials. Therefore the specimens should not be assigned to *Graphiocrinus*. A new set of arms lacking the cup from the Eumha locality is identical to the specimen figured by Lane *et al.* (1997) and indicates that the taxon was common in the Hongguleleng fauna. It may belong to a new genus of the graphiocrinids because some genera, such as *Contocrinus* Knapp 1969, have slightly cuneate brachials that are gently convex transversely. It is referred to the cladids as indeterminate lacking the cup.

Cladid indeterminate 2
Text Figure 5J

Types. Figured specimen NIGP 148861.

Type locality and horizon. Measured section west of Boulongour Reservoir, lower part of marly beds of the Hongguleleng Formation.

Remarks. A poorly preserved set of 10 arms is formed of wedge-shaped biserial brachials that have very tumid transverse ridges and bear elongate pinnules. They are morphologically distinct from all other taxa in the Hongguleleng fauna.

Basal plate indeterminate
Text Figure 5E

Types. Figured specimen NIGP 148862.

Locality. Marly beds, Hongguleleng Formation, Eumha east.

Remarks. A large (length 8.5 mm, width 7.8 mm) hexagonal posterior basal plate with a double row of nodes around the margin is unlike those of other taxa in the Hongguleleng fauna. It is illustrated to show additional diversity in the fauna.

Subclass FLEXIBILIA Zittel 1895
Order TAXOCRINIDA Springer 1913
Superfamily TAXOCRINOIDEA Angelin 1878
Family TAXOCRINIDAE Angelin 1878
Genus *EUTAXOCRINUS* Springer 1906
Eutaxocrinus basellus Lane *et al.* 1997
Text Figure 5K, L

Eutaxocrinus basellus Lane *et al.* 1997, p. 24, figs. 7.9–7.11.

Types. Figured specimen NIGP 148864.

Locality. Marly beds, Hongguleleng Formation, Eumha.

Remarks. Eutaxocrinus basellus Lane *et al.* 1997 was based on a single specimen and distinguished from the other species of the genus recognized in the Hongguleleng Formation on the basis of a narrower base and the infrabasals not entirely covered by the proximal columnal. The new specimen lacks the arms, has four large anals directly above the large posterior basal, and the tips of the infrabasals extend beyond the proximal columnal.

Order SAGENOCRINIDA Springer 1913
Superfamily SAGENOCRINITOIDEA Roemer 1854
Family DACTYLOCRINIDAE Bather 1899
Genus *LABROCRINUS* gen. nov.

Derivation of name. From the Latin *labrum* meaning basin, referring to the shape of the cup.

Type species. Labrocrinus granulatus n. sp. by monotypy.

Diagnosis. A dactylocrinid distinguished by the combination of the posterior basal narrowly separating the C- and D-radials, two primibrachials, two or three secundibrachials, interray plates, but no higher interbrachials and biendotomous distal arm branching.

Description. As for *Labrocrinus granulatus* n. sp.

Remarks. Labronocrinus n. gen. lacks the interbrachial plates present in most dactylocrinid genera, has two or three secundibrachials (most dactylocrinids have three), and there is no other isotomous branching above the secundibrachials (several dactyolocrinids have a third or fourth isotomous branching). *Labrocrinus* probably evolved from *Dactylocrinus* by the loss of the interbrachial plates and reduction in the number of secundibrachials to two in some half rays.

Labrocrinus granulatus sp. nov.
Text Figure 5S

Derivation of name. Referring to the granulate ornament.

Types. Holotype NIGP 148862.

Type locality and horizon. Measured section west of Boulongour Reservoir, upper part of the marly beds of the Hongguleleng Formation.

Diagnosis. As for the genus.

Description. Crown ovoid bowl incurving distally, granular ornament on all exposed cup and interray plates and brachials; length 27.7 mm, incomplete width 26.5 mm. Cup low truncated bowl, length 3.5 mm, diameter 10 mm. Infrabasal circlet horizontal, diameter 2 mm, covered by proximal columnal. Infrabasals 3, 2 equally large, 1 small in C-ray. Basals 5, polygonal shape variable, proximally horizontal, covered by proximal columnal, distally tips upturned, visible in lateral view; AB-basal pentagonal, length and width 2.5 mm; BC-, DE- and EA-basals hexagonal, length 2.7 mm, width 2.5 mm; CD-basal heptagonal, length 3.4 mm, width 2.5 mm. Radials 5, heptagonal, proximal tips covered by proximal columnal, distally widely outflaring; A-radial 3.7 mm long, 6.6 mm wide; C-radial with wider shoulder on posterior side. Anal series unknown, but at least one large anal on shoulders of C- and D-radials. Primibrachials 2, wider than long; first primibrachials length 3.2 mm, width 6.5 mm; axillary second primibrachials length 4.9 mm, width 7.2 mm. Secundibrachials 2 or 3, as large as primibrachials. All brachials strongly convex transversely, straight to gently convex longitudinally, bearing patelloid process. Isotomous branching. Interray plates in contact with radials, series 1:2:3:2. No higher interray or interbrachial plates. Arms abutting, but not fixed; distal branching above secundibrachials biendotomous. Arms incurl distally. Stem facet round, diameter 7.5 mm. Axial canal round, diameter 0.7 mm.

Remarks. The crown of *Labrocrinus granulatus* n. sp. is embedded in a coarse marly grainstone with associated ramose bryozoans, and shows the cup and parts of the arms of the E-, A- and B-rays. The A-ray would have had three secundibrachials (lost) in the right half ray and has two secundibrachials in the left half ray. The B-ray has three secundibrachials in the anterior half ray. The wide shoulder on the posterior side of the C-radial suggests one large anal plate or two smaller plates above the posterior basal, which has a narrow extension between the C- and D-radials.

Our appreciation is extended to the Devonian working group who helped collect the Hongguleleng Formation in 2005. This includes Xiuqin Chen, John Talent, Ruth Mawson, Brian and Beverley Green, Thomas Suttner, Jiri and Barbora Fryda, and John Pickett. This investigation is a contribution of the IGCP 499 project. The reviews of George Sevastopulo and Tom Kammer are gratefully acknowledged.

References

ANGELIN, N. P. 1878. Iconographia Crinoideorum. *In*: *Stratis Sueciae Siluricis fossilium*. Samson and Wallin, Holmiae.

AUSICH, W. I. & KAMMER, T. W. 2006. Stratigraphical and geographical distribution of Mississippian (Lower Carboniferous) Crinoidea from England and Wales. *Proceedings of the Yorkshire Geological Society*, **56**, 91–109.

AUSTIN, T. 1848. Observations on the Cystidea of M. Von Buch, and the Crinoidea generally. *Geological Society of London, Quarterly Journal*, **4**, 291–294.

AUSTIN, T. & AUSTIN, T. 1842. XVIII. – Proposed arrangement of the Echinodermata, particularly as regards the Crinoidea, and a subdivision of the Class Adelostella (Echinidae). *Annals and Magazine of Natural History*, ser. 1, **10**(63), 106–113.

BASSLER, R. S. & MOODEY, M. W. 1943. *Bibliographic and Faunal Index of Paleozoic Pelmatozoan Echinoderms*. Geological Society of America, Special Paper **45**.

BATHER, F. A. 1889. The natural history of the Crinoidea. *Proceedings of the London Amateur Scientific Society*, **1**(1, 2), 32–33.

BATHER, F. A. 1890. British fossil crinoids. II. The classification of the Inadunata. *Annals and Magazine of Natural History*, **5**, 310–334, 373–388, 485–486.

BATHER, F. A. 1899. A phylogenetic classification of the Pelmatozoa. *British Association for the Advancement of Science* (1898), 916–923.

BREIMER, A. & MACURDA, D. B. JR. 1972. The phylogeny of the fissiculate blastoids. *Verhandelingen der Koninklijke Nederlandse Akademie van Webenschappen, Afd. Natuurunde, Eerste Reeks*, **26**(3).

BRONN, H. G. 1848–49. Index palaeontologicus, unter Mitwirking der Herren Prof. H. R. Göppert und H. von Meyer. *Handbuch einer Geschichte der Nature*, **5**, Abt. 1, (1, 2), pt. 3, A. Nomenclator Palaeontologicus, Stuttgart.

CARPENTER, P. H. & ETHERIDGE, R. JR. 1881. Contributions to the study of the British Paleozoic crinoids. – No. 1. On *Allagecrinus*, the representative of the Carboniferous limestone series. *Annals and Magazine of Natural History*, **7**, 281–298.

CHEN, Z-t. & YAO, J-h. 1993. *Palaeozoic Echinoderm Fossils of Western Yunnan, China*. Geological Publishing House, Beijing.

FAY, R. O. 1964. An outline classification of the Blastoidea. *Oklahoma Geology Notes*, **24**(4), 81–90.

GOLDRING, W. 1923. *The Devonian Crinoids of the State of New York*. New York State Museum, Memoir **16**.

GOLDRING, W. 1936. Some Hamilton (Devonian) crinoids from New York. *Journal of Paleontology*, **10**, 14–22.

GOLDRING, W. 1954. Devonian crinoids: New and old, II. *New York State Museum, Circular*, **37**, 1–51.

HALL, J. 1862. *Preliminary notice of some of the species of Crinoidea known in the Upper Helderberg and Hamilton groups of New York*. New York State Cabinet of Natural History 15th Annual Report, 87–125.

HALL, J. 1863. *Preliminary notice, of some species of Crinoidea from the Waverly Sandstone series of Summit Co., Ohio, supposed to be of the age of the Chemung Group of New York*. Preprint of Seventeenth Annual Report of the Regents of the University of the state of New-York, on the Condition of the State Cabinet of Natural History, and the Historical and Antiquarian Collection annexed thereto, State of New York in Senate Document **189**, Albany, Comstock and Cassiday Printers, 50–60.

HAUDE, R. & THOMAS, E. 1989. Ein Oberdevon-/Unterkarbon-Profil im velberter satel (nordliches Rheinisches Schiefergebirge) mit neuen arten von (?) *Sostronocrinus* (Echinodermata). *Bulletin van de Belgische Vereniging voor Geologie*, **98**(3/4), 373–383.

HAUSER, J. 1999. *Die Crinoiden der Frasnes-Stufe (Oberdevon) vom Südrand der Dinant Mulde (Belgische und französische Ardennen)*. Privately published by author.

HOU, H. F., WANG, S.-T., GAO, L.-D. *ET AL*. 1988. *Stratigraphy of China, no. 7: The Devonian System of China*. Geological Publishing House, Beijing.

HOU, H. F., LANE, N. G., WATERS, J. A. & MAPLES, C. G. 1993 [1994]. Discovery of a new Famennian echinoderm fauna from the Hongguleleng Formation of Xinjiang, with redefinition of the formation. *Stratigraphy and Paleontology of China*, **2**, 118.

JAEKEL, O. 1918. Phylogenie und System der Pelmatozoen. *Palaontologische Zeitschrift*, **3**, 1–128.

KAMMER, T. W. & AUSICH, W. I. 1993. Advanced cladid crinoids from the Middle Mississippian of the east-central United States: Intermediate-grade calyces. *Journal of Paleontology*, **67**, 614–639.

KAMMER, T. W. & AUSICH, W. I. 2006. The "Age of Crinoids": A Mississippian biodiversity spike coincident with widespread carbonate ramps. *PALAIOS*, **21**, 238–248.

KESLING, R. V. 1964. *Decadocrinus hughwingi*, a new Middle Devonian crinoid from the Silica Formation in northwestern Ohio. *University of Michigan Contributions from Museum of Paleontology*, **19**, 135–142.

KESLING, R. V. 1971. Arms of *Decadocrinus hughwingi* Kesling. *University of Michigan Contributions from Museum of Paleontology*, **23**, 193–199.

KIER, P. M. 1952. Echinoderms of the Middle Devonian Silica Formation of Ohio. *University of Michigan Contributions from Museum of Paleontology*, **10**, 59–81.

KIRK, E. 1940. Seven new genera of Carboniferous Crinoidea Inadunata. *Journal of the Washington Academy of Science*, **30**, 321–334.

LANE, N. G., MAPLES, C. & WATERS, J. 1995. Paleozoic echinoderms from China. *Maps Digest*, **18**(4), 84–97.

LANE, N. G., WATERS, J. A. & MAPLES, C. G. 1997. *Echinoderm faunas of the Hongguleleng Formation, Late Devonian (Famennian), Xinjiang-Uygur Autonomous Region, China*. Paleontological Society Memoir **47**.

LANE, N. G., MAPLES, C. G. & WATERS, J. A. 2001*a*. Revision of Late Devonian (Famennian) and some Early Carboniferous (Tournaisian) crinoids and blastoids from the type Devonian area of North Devon. *Palaeontology*, **44**(6), 1043–1080.

LANE, N. G., MAPLES, C. G. & WATERS, J. A. 2001*b*. Revision of Strunian crinoids and blastoids from Germany. *Paläontologische Zeitschrift*, **75**(2), 233–252.

LAUDON, L. R. 1936. Notes on the Devonian crinoid fauna of Cedar Valley Formation of Iowa. *Journal of Paleontology*, **10**, 60–66.

LAUDON, L. R. & BEANE, B. H. 1937. The crinoid fauna of the Hampton Formation at LeGrand, Iowa. *University of Iowa Studies*, **17**(6), 227–272.

LAUDON, L. R. & SEVERSON, J. L. 1953. New crinoid fauna, Mississippian, Lodgepole Formation, Montana. *Journal of Paleontology*, **27**, 505–536.

LAUDON, L. R., PARKS, J. M. & SPRENG, A. C. 1952. Mississippian crinoid fauna from the Banff Formation Sunwapta Pass, Alberta. *Journal of Paleontology*, **26**, 544–575.

LE MENN, J. 1985. *Les crinoides du Dévonien inférieur et moyen du massif Armoricain*. Mëmoires de la Société Géologique et Minéralogique de Bretagne, **30**.

LU, L.-C. & WICANDER, R. 1988. Upper Devonian acritarchs and spores from the Hongguleleng Formation, Hefeng District in Xinjiang, China. *Revista Espanola de Micropaleontologia*, **20**(1), 109–148.

MEEK, F. B. & WORTHEN, A. H. 1869. Descriptions of new Crinoidea and Echinoidea from the Carboniferous rocks of the western states, with a note on the genus *Onychaster*. *Proceedings of the Academy of Natural Sciences of Philadelphia*, **21**, 67–83.

METCALF, I. 1996. Pre-Cretaceous evolution of SE Asian terranes. *In*: HALL, R. & BLUNDELL, D. (eds) *Tectonic Evolution of Southeast Asia*. Geological Society, London, Special Publications, **106**, 97–122.

MILLER, J. S. 1821. A natural history of the Crinoidea, or lily-shaped animals; with observations on the genera, *Asteria, Euryale, Comatula and Marsupites*. Bristol, England, Bryan & Co.

MILLER, S. A. 1889. *North American Geology and Paleontology*. Western Methodist Book Concern, Cincinnati.

MILLER, S. A. & GURLEY, W. F. E. 1894. Upper Devonian and Niagara crinoids. *Illinois State Museum, Bulletin*, **4**, 1–37.

MOORE, R. C. 1952. Evolution rates among crinoids. *Journal of Paleontology*, **26**, 338–352.

MOORE, R. C. 1962. *Ray Structures of some Inadunate Crinoids*. University of Kansas, Paleontological Contributions, Echinodermata, Article **5**, 1–47.

MOORE, R. C. & LAUDON, L. R. 1943. *Evolution and Classification of Paleozoic Crinoids*. Geological Society of America, Special Paper **46**.

MOORE, R. C. & PLUMMER, F. B. 1940. *Crinoids from the Upper Carboniferous and Permian Strata in Texas*. University of Texas Publication **3945**.

MOORE, R. C. & STRIMPLE, H. L. 1978. Superfamily & Allagecrinacea Carpenter and Etheridge, 1881. *In*: MOORE, R. C. & TEICHERT, C. (eds) *Treatise on Invertebrate Paleontology*, Part T, *Echinodermata 2, Crinoidea*. Geological Society of America and University of Kansas, 537–548.

PHILLIPS, J. 1839. Encrinites and zoophytes of the Silurian System. *In*: MURCHISON, R. T. (ed.) *The Silurian System*, 670–675.

PICKETT, J. W. 2007. Astraeospongium (Porifera:Calcarea) from the Late Devonian of northwestern China and the late ontogeny of the genus. *Memoirs of the Association of Australasian Palaeontologists*, **34**, 331–342.

ROEMER, C. F. 1852–54. Erste Periode, Kohlen-Gebirge. *In*: BRONN, H. G. (ed.) *Lethaea Geognostica, 1851–1856* (3rd edn.) E. Schweizerbart, Stuttgart, **2**, 210–291.

SAY, T. 1825. On the species of the Linnaean genus Asterias inhabiting the coasts of the United States. *Academy of Natural Sciences, Philadelphia, Journal*, **5**, 141–154.

SCHMIDT, W. E. 1930. Die Echinodermen des deutschen Unterkarbons. *Abhandlungen der Preussichen Geologischen Landesanstalt*, n. s., **122**(1), 1–92.

SENGÖR, A. M. C., NATAL'IN, B. A. & BURTMAN, V. S. 1993. Evolution of the Altaid tectonic collage and Palaeozoic crustal growth in Eurasia. *Nature*, **364**, 299–307.

SHUMARD, B. F. 1865–1866. Catalogue of Palaeozoic fossils, Part 1, Echinodermata. *Transactions of the St. Louis Academy Science*, **2**(2), 334–394 (1865), 395–407 (1866).

SIMMS, A. J. & SEVASTOPULO, G. D. 1993. The origin of articulate crinoids. *Palaeontology*, **36**, 91–109.

SPRINGER, F. 1906. Discovery of the disk of *Onychocrinus* and further remarks on the Crinoidea Flexibilia. *Journal of Geology*, **14**, 467–523.

SPRINGER, F. 1913. Crinoidea. *In*: Zittel, K. A., von, *Text-book of Paleontology* (translated and edited by C. R. EASTMAN; 2nd edn). Macmillan, London, **1**, 173–243.

STRIMPLE, H. L. 1939. A group of Pennsylvanian crinoids from the vicinity of Bartlesville, Oklahoma. *Bulletins of American Paleontology*, **4**(87), 1–26.

STRIMPLE, H. L. & MCGINNIS, M. R. 1969. New crinoid from the Gilmore City Formation, Lower Mississippian of Iowa. *In*: *Fossil Crinoid Studies*. University of Kansas Paleontological Contributions, Paper **42**(5), 21–22.

THOMAS, A. O. 1924. *Echinoderms of the Iowa Devonian*. Iowa Geological Survey, Annual Reports 1919 and 1920, **29**, 385–552.

UBAGHS, G. 1953. Classe des Crinoides. *In*: PIVETEAU, J. (ed.) *Traité de paleontology*. Masson & Cie, Paris, **3**, 658–773.

UBAGHS, G. 1978. Suborder Comosocrinina. *In*: MOORE, R. C. & TEICHERT, C. (eds) *Treatise on Invertebrate Paleontology*, Part T, *Echinodermata* 2, *Crinoidea*. Geological Society of America and University of Kansas, T440–T487.

ULRICH, E. O. 1886. *Remarks upon the names Cheirocrinus and Calceocrinus, with descriptions of three new generic terms and one new species*. Minnesota Geology and Natural History Survey, Annual Report, **14**, 104–113.

WACHSMUTH, C. & SPRINGER, F. 1880–1886. Revision of the Palaeocrinoidea. *Proceedings of the Academy of Natural Sciences of Philadelphia* Pt. I. The families Ichthyocrinidae and Cyathocrinidae (1880), 226–378. Pt. II. Family Sphaeroidocrinidae, with the sub-families Platycrinidae, Rhodocrinidae, and Actinocrinidae (1881), 177–411. Pt. III, Sec. 1. Discussion of the classification and relations of the brachiate crinoids, and conclusion of the generic descriptions (1885), 225–364. Pt. III, Sec. 2. Discussion of the classification and relations of the brachiate crinoids, and conclusion of the generic descriptions (1886), 64–226.

WANG, S., SUN, S., LI, J., HOU, Q., QIN, K., XIAO, W. & HAO, J. 2003. Paleozoic tectonic evolution of the northern Xinjiang, China: Geochemical and geochronological constraints from the ophiolites. *Tectonics*, **22**(2), 1014.

WANG, X. D., SUGIYAMA, T., CAO, C. & LI, Y. 2007. *Peri-Gondwanan Carboniferous to Permian sequences in the Baoshan Block, west Yunnan—faunal, climatic, and geographic changes*. The 16th International Congress on the Carboniferous and Permian, Guidebook Excursion C1, Nanjing.

WANNER, J. 1940. Neue Blastoideen aus dem Perm vonTimor mit einen Beitrag zur Systematik der Blastoideen. *Geological Expedition fo the Lesser Sunda Islands under the leadership of H. A. Brouwer*, **1**, 217–277.

WATERS, J. A., LANE, N. G. MAPLES, C. G. & WEBSTER, G. D. 2001. Late Devonian echinoderms: Controls on post-extinction rebound and biogeographic radiation. *Geological Society of America, Abstracts with Programs*, **33**(6), 248.

WATERS, J. A., MAPLES, C. G., *ET AL*. 2003. A quadrupling of Famennian pelmatozoan diversity: New Late Devonian blastoids and crinoids from northwest China. *Journal of Paleontology*, **77**, 922–948.

WATERS, J. A., MARCUS, S. A. *ET AL*. 2008. An overview of Paleozoic stemmed echinoderms from China. *In*: AUSICH, W. I. & WEBSTER, G. D. (eds) *Echinoderm Paleobiology*. Indiana University Press, Bloomington, 346–367.

WEBBY, B. D. 1961. A Middle Devonian inadunate crinoid from West Somerset, England. *Palaeontology*, **4**, 538–541.

WEBSTER, G. D. 1997. *Lower Carboniferous Echinoderms from Northern Utah and Western Wyoming*. Utah Geological Survey Bulletin **128**, Paleontology Series, **1**, 1–65.

WEBSTER, G. D. 2003. *Bibliography and Index of Paleozoic Crinoids, Coronates, and Hemistreptocrinoids, 1758–1999*. Geological Society of America, Special Paper, **363**.

WEBSTER, G. D. & JELL, P. A. 1999. New Carboniferous crinoids from eastern Australia. *Memoirs of the Queensland Museum*, **43**(1), 237–278.

WEBSTER, G. D. & LANE, N. G. 2007. New Permian crinoids from the Battleship Wash patch reef in southern Nevada. *Journal of Paleontology*, **81**, 951–965.

WEBSTER, G. D. & MAPLES, C. G. 2006. Cladid crinoid (Echinodermata) anal conditions: A terminology problem and proposed solution. *Palaeontology*, **49**(1), 187–212.

WEBSTER, G. D. & MAPLES, C. G. 2008. Cladid crinoid rdial facets, brachials, and arm appendages—A terminology solution for studies of lineage, classification, and paleoenvironment. *In*: AUSICH, W. & WEBSTER, G. D. (eds) *Echinoderm Paleobiology*. Indiana University Press, Bloomington, 197–227.

WEBSTER, G. D., HAFLEY, D. J., BLAKE, D. B. & GLASS, A. 1999. Crinoids and stelleroids (Echinodermata) from the Broken Rib Member, Dyer Formation (Late Devonian, Famennian) of the White River Plateau, Colorado. *Journal of Paleontology*, **73**, 461–486.

WEBSTER, G. D., MAPLES, C. G., MAWSON, R. & DASTANPOUR, M. 2003. *A cladid-dominated Early Mississippian crinoid and conodont fauna from Kerman Province, Iran and revision of the glossocrinids and rhenocrinids*. Journal of Paleontology, Memoir 60, **77**, supplement to 3, 1–35.

WHIDBORNE, G. F. 1896. A preliminary synopsis of the faunas of the Pickwell Down, Baggy, and Pilton Beds. *Proceedings of the Geologists' Association*, **14**, 371–377.

WHIDBORNE, G. F. 1898. *A monograph of the Devonian fauna of the south of England. The fauna of the Marwood and Pilton Beds*. Palaeontographical Society, Monograph 52, **3**, 3, 214–236.

WILLIAMS, H. S. 1882. New crinoids from the rocks of the Chemung period of New York State. *Proceedings of the Academy of Natural Sciences of Philadelphia*, **33**, 17–34.

WINDLEY, B. F., ALEXEIEV, D., XIAO, W. J., KRONER, A. & BADARCH, G. 2007. Tectonic models for accretion of the Central Asian Orogenic Belt. *Journal of the Geological Society, London*, **164**, 31–47.

WOPFNER, H. 1996. Gondwana origin of the Baoshan and Tengchong terranes of west Yunnan. *In*: HALL, R. & BLUNDELL, D. (eds) *Tectonic Evolution of Southeast Asia*. Geological Society, London, Special Publications, **106**, 539–547.

WORTHEN, A. H. 1882. *Descriptions of fifty-four new species of crinoids from the Lower Carboniferous limestones and Coal Measures of Illinois and Iowa*. Illinois State Museum of Natural History, Bulletin, **1**, 3–38.

WRIGHT, J. 1934. New Scottish and Irish fossil crinoids. *Geological Magazine*, **71**, 241–268.

WRIGHT, J. 1950–1960. *The British Carboniferous Crinoidea*. Palaeontographical Society, Monograph, **1**(1), 1–24, 1950; **1**(2), 25–46, 1951*a*; **1**(3), 47–102, 1951*b*; **1**(4), 103–148, 1952*a*; **1**(5), 149–190, 1954*a*; **2**(1), 191–254, 1955*a*; **2**(2), 255–272, 1955*b*; **2**(3), 273–306, 1956*b*; **2**(4), 307–328, 1958; **2**(5), 329–347, 1960 (posthumously, by Wright & Ramsbottom, 1960).

XIA, F. 1996. New knowledge on the age of Hongguleleng Formation in the northwestern margin of Junggar Basin, northern Xinjiang. *Acta Micropalaeontologica Sinica*, **13**(3), 277–285.

XU, H.-K., CAI, C.-Y., LIAO, W.-Y. & LU, L.-C. 1990. Hongguleleng Formation in western Junggar Basin and the boundary between Devonian and Carboniferous. *Journal of Stratigraphy*, **14**(4), 292–301.

YAKOVLEV, N. N. 1928. *Two new genera of crinoids (Poteriocrinidae) of the Upper Paleozoic*. Trudy Akademii Nauk SSSR, Geologicheshogo Muzeya, **3**, 1–8 [in Russian].

YAKOVLEV, N. N. 1939. Class marine lilies-Crinoidea. *In*: GORSKY, I. (ed.) *Atlas of the Leading Forms of the Fossil Faunas of USSR, 5, Middle and Upper Carboniferous*. Central Geological and Prospecting Institute, Leningrad, 64–67 [in Russian].

ZHAO, Z. X. 1986. The conodonts from Hobokasar Formation of Aljiati hill, northeren Xinjiang. *Shiyou Dizhi*, **7**(3), 8107.

ZHAO, Z. X. & WANG, C. Y. 1990. Age of the Honggulelong Formation in the Junggar Basin of Xinjiang. *Journal of Stratigraphy*, **14**, 145–147.

ZITTEL, K. A. von. 1876–80. Echinoderms. *In*: *Handbuch der Palaeontologie*, v. 1, Palaeozoologie. R. Oldenbourg, München, (1879), **1**, 308–560.

ZITTEL, K. A. von. 1895. *Grundzüge der Palaeontologie (Palaeozoologie)*. R. Oldenbourg, München.

Index

Page numbers in *italics* refer to Figures. Page numbers in **bold** refer to Tables.